COMPOSITE MATERIALS:
TESTING AND DESIGN

A symposium
presented at a meeting of
Committee D-30 on High Modulus
Fibers and Their Composites
AMERICAN SOCIETY FOR
TESTING AND MATERIALS
New Orleans, La., 11-13 Feb. 1969

ASTM SPECIAL TECHNICAL PUBLICATION 460

Symposium on Composite Materials: Testing and Design, New Orleans, 1969

List price $31.00

AMERICAN SOCIETY FOR TESTING AND MATERIALS
1916 Race Street, Philadelphia, Pa. 19103

NOTE

The Society is not responsible, as a body,
for the statements and opinions
advanced in this publication.

Printed in York, Pa.
December 1969

Foreword

The papers in the Symposium on Composite Materials: Testing and Design were presented at a meeting held in New Orleans, La., 11-13 February 1969. The symposium was sponsored by The American Society for Testing and Materials through its Committee D-30 on High Modulus Fibers and Their Composites. Steven Yurenka, McDonnell Douglas Corp., presided as symposium chairman.

Related
ASTM Publications

Metal Matrix Composites, STP 438 (1968), $18.50

Fiber-Strengthened Metallic Composites, STP 427 (1967), $12.75

Contents

Fatigue and Creep

Design and Application

Fracture Mechanics

Introduction

S. YURENKA

The purpose of the American Society for Testing and Materials Conference on Composite Materials: Testing and Design was to provide a forum for the discussion of analytical and experimental studies of advanced composites. Emphasis was on the development of a basic understanding of the mechanical and structural behavior of both organic and metal matrix composites. The program brought together for the first time, comprehensive, detailed, and significant work in the areas of mechanical properties testing, as well as examples of design application. The ultimate aim of the society, speakers, and participants was uniformly the same one; to pave the way for quicker and more widespread utilization of the use of composite materials in military and commercial aircraft, missiles, and space vehicles, and naval and other structural component applications.

In recent years advanced composite materials reinforced with high-strength, high-modulus filaments have received increasing attention. This activity has been motivated by the potential for superior performance in structural and other applications afforded by the unique capabilities of such composite materials. The very nature of these advanced composites, namely, the heterogeneity and the high stiffness and strengths, however, has been introducing new problems in both testing and in the development of appropriate design procedures. Thus, innovations in testing techniques and design philosophy have been needed urgently. A short range goal of the sponsoring ASTM Subcommittee on Composites, D-30, has been to derive ideas for new test methods and design and analysis procedures from the technological fallout of this conference.

The technical papers which were presented at the conference and which are included in this volume were divided into four sessions and were in some cases given in simultaneous presentations over a period of three days. The four sessions were entitled "Testing, Fatigue and Creep, Design and Application, and Fracture Mechanics," and each session was opened with a keynote address by an outstanding expert in his field. Although questions and answers were not recorded and published, a lively question and answer period followed each paper.

ASTM conferences and papers fall into a special category of their own which represent a refreshing deviation from, but compromise with, the distractions and commercialism of purely product or service oriented symposia, the restrictions and secrecy of some military oriented reviews, or the dryness and tedium of academic university type seminars. When the emphasis is on testing methods or design procedures and on an increased understanding of the mechanical behavior of promising new materials, a neutral ground is reached where everyone can contribute generously to, and freely benefit from, the knowledge of the whole. The outstanding success of this ASTM conference and the magnitude of the problems that continue unresolved have assured that such a forum will be repeated in the future.

STEVEN YURENKA

McDonnell Douglas Corp.,
Long Beach, Calif. 90801;
symposium chairman.

Testing

Composite Materials' Testing

S. P. PROSEN[1]

Reference:

Prosen, S. P., "Composite Materials' Testing," *Composite Materials: Testing and Design*, ASTM STP 460, American Society for Testing and Materials, 1969, pp. 5–12.

Abstract:

This paper will discuss some of the tests now being used in composites testing. The problems of testing, getting the information sought after, the pitfalls, and some of the controversy surrounding testing will be presented in detail. The added difficulty of testing brought about by the use of high-modulus (advanced) fibers is composites will be discussed thoroughly. For instance, the shear strength test for composite materials has become an extremely controversial subject now that we are dealing with high-modulus fibers. We thought we understood shear testing when we tested glass filament reinforced plastics—at least we all understood one another—now we do not understand shear in this complex advanced composites field.

Key Words:

graphite fibers, carbon fibers, interlaminar shear, graphite composites, composites materials, polymer composites, mechanical properties, whiskers (single crystals), whisker composites, evaluation, tests

Testing can be covered most broadly by the definition "subjection to conditions that show the real character of things" [1].[2] For composites and for our purposes, this can be limited to the mechanical properties of composites; to their strength and stiffness characteristics. When we try to find "the real" character of things, we soon come to understand the importance of testing on the scientific and engineering communities. Testing allows us to progress in our endeavors, since the results of tests precisely tell us where we stand so that we can take the next step. So, it is obvious that for one to do well in his scientific or engineering endeavors he must understand the philosophy of testing, its rules, and some of the pitfalls and dangers that may be encountered along the way.

TESTING OF COMPOSITES

Testing of materials is not simple, even with homogeneous type materials, but composites are anisotropic and a most complex combination of materials. The states of stress within the composite when it is under load in test is complicated by the varied elastic constants of the constituent materials. This is complicated further by the directional strength and stiffness brought about by orientation of the reinforcement in the composite. In some cases the reinforcement itself may have anisotropic properties which further complicates composite testing. If we are involved at all in materials' studies of this complicated material, we must be aware of the extent of the complications and how they might affect our approach to testing. Since we all tend to lean on others and on previous work, we probably design our tests and methods after those used by the metals community; we even may take our techniques directly from metal technology. Even though metals prop-

[1] Manager, Advanced Composites Section, Fiberite Corp., Winona, Minn. 55987.
[2] The italic numbers in brackets refer to the list of references appended to this paper.

erties vary somewhat with the machining direction and even though they may evidence grain structure, they are still essentially single-phase materials in the broad sense. Their general characteristics are different from the advanced composite materials.

The advanced fibrous composites, as said earlier, are extremely anisotropic. They consist of three distinct elements: (1) the fiber, (2) the resin, and (3) the interface between the fiber and the resin. Salkind [2] states that the behavior of an aligned fiber composite is dependent upon, and in many cases can be predicted from, the properties of the two phases—fiber and matrix (resin). The third entity, he states, is the interface, and it plays a most important role in both fabrication and subsequent behavior of the material (composite). It most certainly plays a most important role in testing.

TYPE OF TESTS

Testing in general can be divided into three types: (1) scientific tests, (2) research and development tests, and (3) control tests. Tsai [3] on the other hand divides tests into categories dependent on the information gained as follows: design data generalization, product assurance, manufacturing control, and sub- and full-scale structural performance check. Regardless of the type test, it should be understood that, to gain the maximum from any test, the nature of the answer sought must be established first. Then a test should be devised which will provide this answer and, if possible, give a result which relates to performance.

The nature of the answer sought concept is very important to our understanding of the choice of methods of test of the various disciplines in the composites community. Perhaps an attempt to look at the various reasons for testing would be in order to aid our understanding.

WHY DO WE TEST?

Each of us tests our materials for different reasons. We each have our particular need, and we each look at testing and test methods in a different way. Because we, the scientists, the materials or composites researchers, the developers, the designers, and those of us responsible for control or specifications type of information do not expect or want to get the same information from our testing, we should not expect complete agreement and unanimity in test methods. We can not even expect our philosophical approaches to testing and what we expect to get from our tests to be the same.

SHEAR

A prime example of this difference of opinion on testing is in the area of shear.

The shear strength test for composite materials has become an extremely controversial subject now that we are dealing with high-modulus fibers. We thought we understood shear testing when we tested glass filament reinforced plastics—at least we all understood one another—now we find we do not agree on the definition of shear in this complex advanced composites field

Pogano [4] discusses this at length and states that the term "shear" strength needs clear definition. He adopted the following: the shear strength of a material is defined as the shear stress at failure in an experiment in which a state of pure shear is induced. He goes on to state that this fact has been ignored in many experiments. This, of course, is true, but this fact does not necessarily mean that all such "impure" shear tests are useless or will give useless information. On the contrary, some of these tests have helped in advancing the state of the art of fibrous composites.

Let us look at the various shear tests now being used in composites work. (There are various other tests that we could examine such as tension or compression tests or a whole series of polyaxial tests, but for sheer variety and varied thinking we need look no further than shear.) The shear test is a prime example of how differently the various disciplines look at testing and search the "nature of the answer sought."

SCIENTIFIC TESTS FOR MATERIALS RESEARCH

In materials research work, measurements carefully are sought to delineate fundamental properties and physical constants of materials. For instance in advanced composites work, the researcher may be looking at very fundamental properties of resins, or he may be studying the surface of a graphite fiber. He may very well be interested in the interface of the fiber and the matrix or the coupling agent. In any case, he must lean on standard tests, or he may have to devise completely new techniques and tests.

The researcher who is interested in the interface and its effects on composites performance, of course, will want to measure in some way the effect of his work. He most likely is not interested in a test that will be accepted by designers or development engineers, he is interested in measuring the improvement in adhesion of the fiber and the resin, if indeed this is what he is researching. In this way he measures his performance. He wants to know if the ingredients he puts into and the technology used in attempting to improve interface performance actually improves the performance. He must measure the interfacial shear strength to measure his own performance. Salkind very aptly discussed the various methods of test to do this—the tests ranged from the pullout test [5] to wetting measurements [6]. Other tests [7,8] were described also. (Figure 1 pictorially describes some of the tests.)

Now the materials researcher is interested, in this case, in improving the adhesion at the interface, and he uses not only tests and measurements he devises but also other researchers' tests, since this allows for direct comparisons of his work with others in the same discipline. As stated before, he most likely is not concerned that pure shear is induced even though he is looking for what he calls interfacial or interlaminar shear strength.

CONTROL TESTS (COMPOSITE MATERIALS RESEARCH AND DEVELOPMENT)

What about the man doing development work in the composite materials field? He also is interested in the interface and the interlaminar shear strength among other things, but he must be concerned about how this interface or shear strength affects his composite material. He must measure the effect of the interface by a test conducted on the composite. His test methods are usually more standard because he must be able to compare his material with others and most of all convince the user (structural designer) that he does indeed have a material which will meet the basic requirements of the user. The developer of materials (composites) spends a lot of time in processing and handling the materials. He therefore runs tests on the composite, both unidirectional tests and cross-plied tests. For the most part though he leans on tests such as short-beam or scissors shear [9] or other shear tests [10] (Fig. 2).

Ofttimes he will make up special tests so that he can prove the worth of his material. Here again he is not concerned that the pure state of shear exists since more

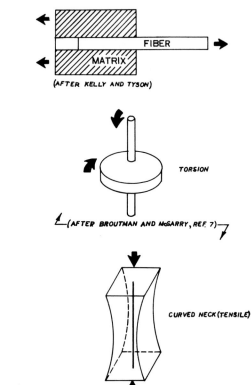

FIG. 1—Is this shear?

often than not he is comparing materials on a day-to-day basis where the stiffness and strength characteristics of the overall composite are the same but where very minor changes have been made at the interface. And since the stiffness and strength characteristics are the same from day to day, the "impure" states of stress are the same for all practical purposes. This reasoning makes his tests valid. His tests are as valid to his discipline as are the tests of the materials researcher cited earlier. He, however, is still not too interested in providing data useful to the structural designer. In fact, he should not be burdened to do so. He should be free to develop his materials with as simple a test as possible.

RESEARCH AND DEVELOPMENT FOR DESIGN INFORMATION

The individual who tests to acquire design data must rely on still different tests. He of course is interested in the interlaminar shear strength since it affects his other properties. He, however, uses tests that will give numbers that the designer can rely on. His job is probably more difficult in this area of testing than the jobs of the two individuals discussed earlier. He must know of and be able to interpret the tests and test results of the other disciplines. He also must be able to take the information from their tests and make it fit his needs. However, he must still satisfy the designer and user. His tests are more sophisticated; he is concerned about the materials behavior as a structural element. The element sees complicated stress and strain fields. To acquire design data, the individual must be able to predict how his materials will stand up to these forces. He therefore needs to do mechanics analysis type work and develop the elastic constants of

Composite Materials

SHORT BEAM

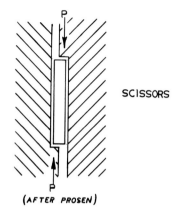

SCISSORS

(AFTER PROSEN)

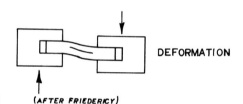

DEFORMATION

FIG. 2—Is this shear?   (AFTER FRIEDERICY)

his material. Work of this nature has been discussed by many [11–13]. Some of the tests he uses are described by Steingiser and Hanna [15] and Waddups [16] and are shown in Fig. 3.

Now we have seen how the various disciplines look at shear. As said before, Pogano adopted a definition of shear strength which the purists like. The man responsible for design data also agrees with this, but if he is concerned at all with interfacial or interlaminar shear as known by the materials people, then he is forced to rationalize—he must attempt to get numbers to grade his materials even though complex states of stress exist in the interlaminar shear tests.

TESTING GUIDES

Obviously the workers in composite materials are interested in advancing these materials to the state where they are accepted in the various use communities. Even though the researcher and the materials developer have concentrated on materials studies and probably have relegated testing to a secondary status, they must not forget the basic rules or precepts of testing—and they should concentrate on following the rules. This of course, is mandatory for the person who generates design information. For this reason it is well to reiterate some of the rules and guides for testing previously discussed by Barnet and Prosen [1].

 ROD TORSION

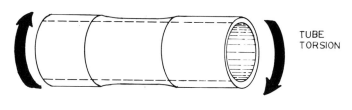 TUBE
TORSION

(AFTER STEINGISER (14))

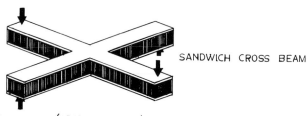 SANDWICH CROSS BEAM

FIG. 3—Is this shear?   (AFTER WADDUPS (15))

RULES FOR TESTING

Concepts of materials usually are idealized and greatly oversimplified. Therefore, test results have inherent limitations. They give only a measure or indication of the material's true properties, as evidenced in the particular specimen selected and when tested under specific circumstances. To gain the maximum from any test, therefore, the nature of the answer sought must be established first. Then a test should be devised which will provide this answer and, if possible, give a result which will relate to performance. Limitations of the test and its necessary precision also should be considered carefully. Lastly, specimen type and number of replicates required should be determined. Due consideration also must be given to the statistical nature of data to be accumulated. If these precepts are followed, one should arrive at a carefully planned and well-defined technique tailored to his needs.

CRITERIA FOR MEASURING PROPERTIES

The technical man must be assured that he is measuring adequately the properties of interest to him. The following criteria should be included:

1. The property can be defined with sufficient exactness.
2. The material under test is of known composition or purity.
3. The attending test conditions are standard or known.
4. The test methods are correct theoretically.
5. The test observations and their reductions are made with thoroughness and care.
6. The order of accuracy of the results is known.

All these criteria should be met in the ideal situation. On the practical level, one should strive to observe as many of them as possible. Obviously, good testing is not just dashed off in one's spare time. It is not a task of a secondary nature but a key activity of science and engineering. Even with ideal homogeneous materials, testing is not simple. It takes much thought and time and careful, diligent professional effort to assure meaningful, accurate results.

ANALYSIS OF TEST DATA

The final part of any test procedure is analysis and reporting of the data. The importance of this operation can not be emphasized too strongly. If analyzed inadequately, the data and conclusions are worse than nothing at all. In fact, one easily can argue that, if the work is not to be presented properly, it should be left undone. At least then, no one can be misled.

Obviously, because of the nature of advanced composites, the composite and specimens cut from it must be described in detail. The test method also must be made fully known. In presenting the data, the variation in these data must be carefully set forth. Lastly, some indication of the significance of the results must be expressed as related to acceptable control values or other reference information. These necessities of reporting, when violated, as they are too frequently, only serve to lessen industry's confidence in laminate materials.

Stepping back and taking a broad look at mechanical testing may lead to some rather hollow phrases about what is desirable in the future. There are, of course, many needed improvements which can be noted quickly. Simplification should be introduced where possible as to specimens, test jigs, and procedures. A great deal of standardization is needed to make methods more specific and precise.

TESTING IS NO SUBSTITUTE FOR THINKING

Tests and test data are not substitutes for thinking. They are only appropriate experimental and engineering aids or tools, as pointed out by Davis et al [14]. The scientist or engineer must understand the purpose of his test. He should try to visualize the results to be expected—of what takes place in the material during the test. He should know his opportunities for error and always be alert to the unusual.

Finally, common sense should be used in the interpretation of data. One should not allow himself to automatically consider test methods as "good" because they are standard. He should be ready to question them severely to be certain that the method he is using is the right one for his program.

Testing will continue to be a part of technology. Engineers and scientists should not permit it to have a hypnotic effect on their work but use it as a stimulus toward more profound and meaningful efforts.

References

[1] Barnet, F. R. and Prosen, S. P., "Validity of Mechanical Tests for Glass Reinforced Plastics," *Materials Protection*, Vol. 3, No. 6 June 1964.

[2] Salkind, M. J., "The Role of Interfaces in Fiber Composites," *Proceedings*, 14th Sagamore Army Materials Conference, Syracuse University Press, Syracuse, N. Y., 1967.

[3] Tsai, S. W., "A Test Method for the Determination of Shear Modulus and Shear Strength," Report AFML-TR-66-372, Air Force Materials Laboratory, Jan. 1967.

[4] Pogano, N. J., "Observations on Shear Test Methods of Composite Materials," Technical Memorandum MAN 67-16, Sept. 1967.

[5] Kelly, A. and Tyson, W., *High Strength Materials*, Wiley, New York, 1965, Chapter 13.

[6] Schwartz, R. T. and Schwartz,

H. S., eds., *Proceedings of Conference on Fundamental Aspects of Fiber Reinforced Plastic Composites,* Dayton, Ohio, 1967.

[7] Broutman, L. J. and McGarry, F. J., *Modern Plastics,* Vol. 40, Sept. 1962, p. 161.

[8] Gotfreund, K., Broutman, L., and Joffee, E., *Proceedings,* 10th Symposium, Society of Aerospace Materials and Process Engineers, San Diego, 1966, pp. 25–E40.

[9] Prosen, S. P. and Simon, R. A., "Carbon Fibre Composites for Hydro and Aerospace," *Filament Winding Conference Plastics Institute,* London, Oct. 1967.

[10] Friedericy, J. A., Davis, J. W., and Rentch, B. W., "A Test Method for Determination of Shear Properties of Reinforced Epoxy Structures," AFML-TR-66-274, Air Force Materials Laboratory, Sept. 1966.

[11] Tsai, S. W., "Mechanics of Composite Materials," AFML-TR-66-149 Part 1, Air Force Materials Laboratory, June 1966.

[12] Adams, D. F., Doner, D. R., and Thomas, R. L., "Mechanical Behavior of Fiber-Reinforced Composite Materials," AFML-TR-67-96, Air Force Materials Laboratory, 1967.

[13] Whitney, J. M. and Riley, M. B., "The Elastic Properties of Fiber Reinforced Composite Materials," *Journal,* American Institute of Aeronautics and Astronautics, Vol. 4, 1966, p. 1537.

[14] Davis, H. E., Troxell, G. E., and Wiskocil, C. T., *The Testing and Inspection of Engineering Materials,* McGraw-Hill, New York, 1955.

[15] Steingiser, S. and Hanna, G., "Proposed Method of Test for Torsion Shear Properties of Reinforced Plastics," ASTM-D-30-V, American Society for Testing and Materials.

[16] Waddups, M. E., "Characterization and Design of Composite Materials," *Composite Materials Workshop,* Technomic, Stanford, Conn., 1968.

Mechanical Properties and Test Techniques for Reinforced Plastic Laminates

S. DASTIN,[1] G. LUBIN,[1]
J. MUNYAK,[1] AND
A. SLOBODZINSKI[2]

Reference:

Dastin, S., Lubin, G., Munyak, J., and Slobodzinski, A., "Mechanical Properties and Test Techniques for Reinforced Plastic Laminates," *Composite Materials: Testing and Design, ASTM STP 460*, American Society for Testing and Materials, 1969, pp. 13–26.

Abstract:

Various test specimen shapes for determining tensile, compressive, and edgewise shear properties of reinforced plastic laminates are compared. Tests were performed and data collected for the various shape specimens from three specific resin content panels at various temperature in both dry and wet condition. Recommendations as to the most reliable specimen shapes are given based upon evaluation of tests performed.

Key Words:

reinforced plastics, laminates, balanced layup, compression brooming, edge crushing, in-plane (edgewise) shear, delaminating, shear modulus, tensile coupon, mechanical properties, evaluation, tests

Presently, the basic mechanical properties of fiber glass reinforced plastic laminates are given in MIL-HDBK-17, and were developed on the first generation resin matrices, reinforcements, finishes, and test methodology.

The material improvements of reinforced plastics has led to their increased usage for highly loaded aircraft parts; thus, a greater need exists for more accurate data on the tensile, compressive, shear strength, and stiffness properties of these composites. To satisfy this need, government and industry agencies have undertaken the revision of MIL-HDBK-17 under the following guidelines:

1. Data to be generated at the B-value statistical level (90 percent probability and 95 percent confidence limits).

2. Test methodology to maximize coupon rather than element specimens.

3. Properties to be measured over a thermal range -65 F to a temperature 50 F higher than the maximum temperature recommended for use of the material being tested.

4. Data to be generated in both the dry and wet conditions including wet specimens tested at elevated temperatures.

5. Generate data with improved test methods, and verify that such "new" tests provide proper failure modes and minimize data scatter.

This presentation describes the initial phases of research on the development of one specific reinforced plastic laminate, that is, Woven Style 7781-550 fiber

[1] Staff, materials engineering, plastics consultant, and project engineer, respectively, Grumman Aerospace Corp., Bethpage, N. Y. 11714. Mr. Munyak is a personal member ASTM.
[2] Materials engineer, Technical Evaluation Center, Picatinny Arsenal, Dover, N. J.

13

glass reinforced F-161 epoxy-novolac high-temperature resin laminated plastic, fabricated by autoclave molding.

PROGRAM PLAN AND TEST SPECIMENS

To provide the maximum program payoff at the minimum elapsed time and minimum costs, only one material, one process, and one laminate thickness was utilized. The material, Trevarno preimpregnated (prepreg) F-161/7781-550, was fabricated into four different ranges of cured resin content (nominally 25, 28, 31, and 34 percent by weight) to fully characterize this composite for various aircraft applications (Table 1).

Table 1—*Prepreg identification.*

Manufacturer's Designation	7781-60 in., F161-505A	7781-60 in., F161-505B	7781-60 in., F161-505C	7781-60 in., F161-505D
Type resin	F161-505A	F161-505B	F161-505C	F161-505D
Fabric style	181/75	181/75	181/75	181/75
Resin content, weight %	36.0	29.1	27.7	24.8
Flow, %	12.2	3.5	2.3	3.1
Volatiles, %	0.57	0.57	0.61	0.81
Gel time	not determined	not determined	not determined	not determined
Batch No	49827	49828	49829	49830
Roll No	01	01	01	01
Roll width	60 in.	60 in.	60 in.	60 in.
Cured resin specific gravity	1.26	1.26	1.26	1.26
Fabric manufacturer	Clark-Schwebel	Clark-Schwebel	Clark-Schwebel	Clark-Schwebel
Fabric designation	7781	7781	7781	7781
Yarn count and No., warp	57/ECDE-75-1/0	57/ECDE-75-1/0	57/ECDE-75-1/0	57/ECDE-75-1/0
Yarn count and No., fill	54/ECDE-75-1/0	54/ECDE-75-1/0	54/ECDE-75-1/0	54/ECDE-75-1/0
Finish	550	550	550	550

The laminated plastic test panels were all processed by autoclave molding, utilizing the balanced layup (nested), as shown in Fig. 1, and eight plies of parallel positioned reinforcement. The test panels were cured at 350 F under 60-psi pressure by the controlled vertical bleed system assembly shown in Fig. 2. Footnote[3] details this fabrication method and indicates the relationship required between the physical properties of the prepreg to the final cured resin content and residual voids.

Test specimens were then machined from the molded laminates; they being tensile, compressive, and shear coupons both parallel and perpendicular to the warp of the reinforcement.

Three types of tensile coupons were prepared: the industry standard ASTM Test for Tensile Properties of Plastics (D 638-68) (Fig. 3), a straight-sided coupon (Fig. 4), and a modified bow tie (long neck) coupon (Fig. 5).

The compression specimen was newly developed based on the original ASTM Test for Compressive Properties of Rigid Plastics (D 695-63T) supported specimen. The modifications included a fixture to prevent end "brooming" and to provide the unsupported section in the middle of the length rather than at the upper ends of the length, Figs. 6 and 7.

The shear specimen for determination of in-plane (edgewise) shear strength and stiffness was a modified picture frame loaded coupon shown in Figs. 8 and 9. The location points for the compressometer needed to determine the shear strain were determined by a series of tests utilizing strain gages to locate the specimen average strain. This specimen and type of loading method is a modification of the procedure described in ASTM Testing of Veneer, Plywood, and Other Glued Veneer (D 805-63).

[3] "Determination of Principal Properties of Fiberglass-Epoxy Laminates for Aircraft," *1st and 2nd Quarterly Reports*, Contract DAAA 21-68-C-0404, Feb. and June 1968, Picatinny Arsenal, Dover, N. J.

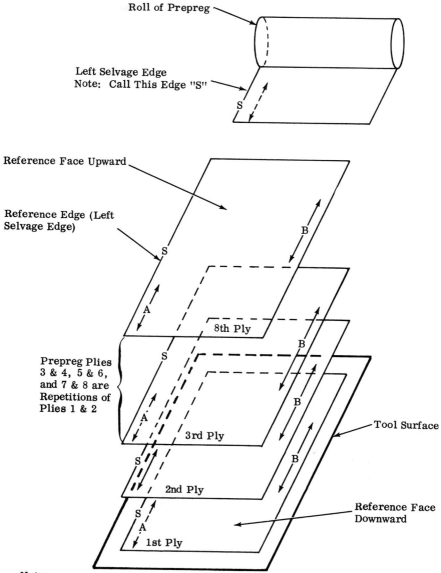

Roll of Prepreg

Left Selvage Edge
Note: Call This Edge "S"

S

Reference Face Upward

Reference Edge (Left
Selvage Edge)

S

B

A

8th Ply

B

Prepreg Plies
3 & 4, 5 & 6,
and 7 & 8 are
Repetitions of
Plies 1 & 2

S

B

A

3rd Ply

Tool Surface

B

S

2nd Ply

B

Reference Face
Downward

S

A

1st Ply

Notes:

A ⤴ – Indicates direction of shaft yarn, facing downward, and parallel to warp.

A ⤴ – Indicates direction of shaft yarn, facing upward, and parallel to warp.

B ⤴ – Indicates direction of warp.

Fig. 1—Prepreg layup sequence (balanced layup).

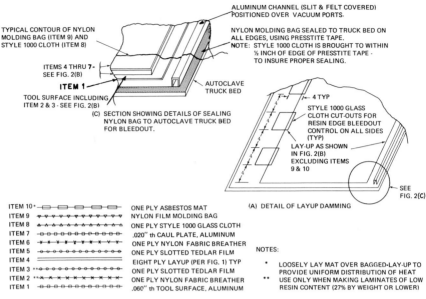

ALUMINUM CHANNEL (SLIT & FELT COVERED)
POSITIONED OVER VACUUM PORTS.

TYPICAL CONTOUR OF NYLON
MOLDING BAG (ITEM 9) AND
STYLE 1000 CLOTH (ITEM 8)

NYLON MOLDING BAG SEALED TO TRUCK BED ON
ALL EDGES, USING PRESSTITE TAPE.
NOTE: STYLE 1000 CLOTH IS BROUGHT TO WITHIN
½ INCH OF EDGE OF PRESSTITE TAPE -
TO INSURE PROPER SEALING.

ITEMS 4 THRU 7-
SEE FIG. 2(B)

ITEM 1

TOOL SURFACE INCLUDING
ITEM 2 & 3 - SEE FIG. 2(B)

AUTOCLAVE
TRUCK BED

(C) SECTION SHOWING DETAILS OF SEALING
NYLON BAG TO AUTOCLAVE TRUCK BED
FOR BLEEDOUT.

4 TYP

STYLE 1000 GLASS
CLOTH CUT-OUTS FOR
RESIN EDGE BLEEDOUT
CONTROL ON ALL SIDES
(TYP)

LAY-UP AS SHOWN
IN FIG. 2(B)
EXCLUDING ITEMS
9 & 10

SEE
FIG. 2(C)

(A) DETAIL OF LAYUP DAMMING

ITEM 10 *	ONE PLY ASBESTOS MAT
ITEM 9	NYLON FILM MOLDING BAG
ITEM 8	ONE PLY STYLE 1000 GLASS CLOTH
ITEM 7	.020" th CAUL PLATE, ALUMINUM
ITEM 6	ONE PLY NYLON FABRIC BREATHER
ITEM 5	ONE PLY SLOTTED TEDLAR FILM
ITEM 4	EIGHT PLY LAYUP (PER FIG. 1) TYP
ITEM 3	ONE PLY SLOTTED TEDLAR FILM
ITEM 2	ONE PLY NYLON FABRIC BREATHER
ITEM 1	.060" th TOOL SURFACE, ALUMINUM

NOTES:

* LOOSELY LAY MAT OVER BAGGED-LAY-UP TO
PROVIDE UNIFORM DISTRIBUTION OF HEAT
** USE ONLY WHEN MAKING LAMINATES OF LOW
RESIN CONTENT (27% BY WEIGHT OR LOWER)

FIG. 2—Bagged layup assembly for vertical bleed.

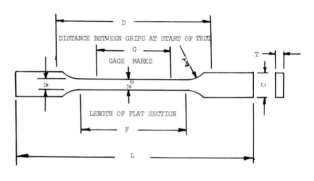

D
DISTANCE BETWEEN GRIPS AT START OF TEST

G
GAGE MARKS

T

LENGTH OF FLAT SECTION

F

L

FIG. 3—ASTM D 638, tension test specimen.

C -	Width overall	0.750
W -	Width of flat section	0.500
F -	Length of flat section	2.25
G -	Gage length	2.00
D -	Distance between grips	4.5
L -	Length overall	8.5
Rad-	Radius of fillet	3

TEST RESULTS AND DISCUSSION

TENSION TESTS

The initial tests of this program were designed to select the type of tensile coupon. Therefore, utilizing three specific cured resin content panels (24, 32, and 36 percent by weight) the three phase tension specimen test was conducted. The results of this segment of testing is given in Table 2. As is noted by the Phase II tests of Table 2, the strength and stiffness levels of the long neck specimens was either equal to or superior to either the ASTM D 638 specimen or the tab-ended (straight-sided) specimen. These results were encouraging in that the failure mode

Table 2—Tensile Strength versus specimen geometry (all specimens 0.080 th, nom.).

Type	Dimensions, in. Length	Gage Width	Radius	E_{tu}, ksi Av	Low	High	E, 10^4 psi Av	Low	High	Laminate S/N	Warp Direction, deg	Resin Content, %	Failure Remarks
PHASE I TESTS													
ASTM D 638....	8½	0.500	3	55.9	...	...	3.40	3.33	3.51	3	0	36.1	100% in radius area
ASTM D 638....	8½	0.500	3	42.6	40.6	43.8	3.75	3.65	3.94	90	3	36.1	100% in radius area
ASTM D 638....	8½	0.500	3	55.9	54.8	56.6	3.45	3.27	3.62	4	0	36.2	100% in radius area
ASTM D 638....	8½	0.500	3	44.5	43.8	45.1	3.43	3.35	3.51	4	90	36.2	100% in radius area
ASTM D 638....	8½	0.500	3	46.3	44.5	48.8	3.46	...	...	6	90	34.8	100% in radius area
Tab ended[c]	10½	0.500	straight length	52.1[a]	51.9	52.3	...	...	...	4	0	36.2	100% in tab area
Tab ended[c]	10½	0.500	straight length	61.0[b]	...	...	...	...	...	4	0	36.2	100% in tab area
Tab ended[c]	10½	0.500	straight length	33.5	32.3	35.5	3.23	3.14	3.38	4	90	36.2	100% in tab area
Tab ended[d]	10½	0.500	straight length	60.8	59.4	62.0	2.73	...	...	4	0	36.2	100% in tab area
PHASE II TESTS													
ASTM D 638....	8½	0.500	3	73.7	70.7	76.7	...	...	...	8	0	24.2	100% in radius area
ASTM D 638....	9¾	0.500	8	72.6	70.1	75.2	...	...	...	8	0	24.2	100% in radius area
ASTM D 638....	11½	0.500	18	73.9	73.0	74.8	...	...	...	8	0	24.2	100% in radius area
Long neck[e,f]	8½	0.500	tapered neck	78.1	74.9	82.1	4.61	4.49	4.78	8	0	24.2	75% in gage area
Long neck[e,f]	11½	0.500	tapered neck	79.7	76.5	83.3	4.58	4.36	4.76	8	0	24.2	100% in gage area
Tab ended[f]	10½	0.500	straight length	64.2	62.8	65.7	4.58	4.19	5.07	8	0	24.2	100% in tab area
Tab ended[f]	9	0.500	straight length	52.5	52.8	57.7	...	...	...	8	0	24.2	100% in tab area
Long neck[e]	8½	0.250	tapered neck	73.5	70.7	76.2	5.10	4.90	5.20	8	0	24.2	100% in gage area
Long neck[e]	11½	0.250	tapered neck	76.9	76.3	77.5	4.40	4.36	4.43	8	0	24.2	100% in gage area
PHASE III TESTS													
Long neck[e,f]	8½	0.500	tapered neck	72.6	71.9	73.0	3.78	3.76	3.80	10	0	32.2	100% in gage area
Long neck[e]	8½	0.500	tapered neck	72.6	71.9	73.0	4.48	4.37	4.55	10	0	32.2	100% in gage area
Long neck[e,f]	11½	0.500	tapered neck	75.3	74.0	77.0	3.60	3.42	3.79	10	0	32.2	100% in gage area
Long neck[e,f]	11½	0.500	tapered neck	75.3	74.0	77.0	4.20	4.17	4.23	10	0	32.2	100% in gage area
Long neck[e,f]	8½	0.500	tapered neck	67.2	59.5	72.2	3.74	3.54	3.99	6	0	34.8	100% in gage area

[a] Partial grip in machine jaws.
[b] Full grip in machine jaws.
[c] Fiber glass tabs.
[d] ABS plastic tabs.
[e] See Figs. 2–7.
[f] Extensometer is placed on edges of specimen. All others are face mounted.

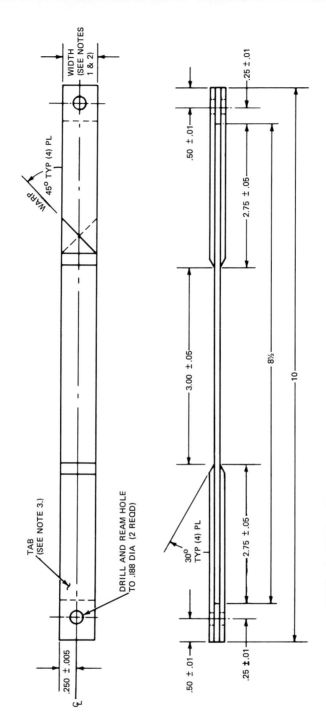

FIG. 4—Straight-sided tension specimen.

NOTES: 1. NOMINAL SPECIMEN WIDTH —: .496-.500
2. EDGES PARALLEL WITHIN −.002: SQUARE TO FACES WITHIN±3°; FREE OF DEFECTS
3. TABS: MAT'L — SCOTCHPLY 1002 (8 CROSS PLIED LAYERS)
ADH. — METLBOND 329
ORIENTATION — 45 - 135° TO MEAN DIRECTION OF SPECIMEN AXIS

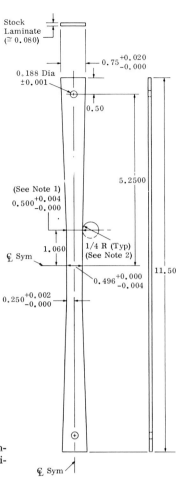

Stock
Laminate
(≈ 0.080)

$0.75^{+0.020}_{-0.000}$

0.188 Dia
± 0.001

0.50

5.2500

(See Note 1)
$0.500^{+0.004}_{-0.000}$

1.060

1/4 R (Typ)
(See Note 2)

℄ Sym

11.50

$0.496^{+0.000}_{-0.004}$

$0.250^{+0.002}_{-0.000}$

℄ Sym

FIG. 5—Standardized, elongated bow tie tension specimen dimensions.

Notes:
1. The width outward from 0.496 shall be increased gradually and equally on each side up to 0.500 so that no abrupt changes in dimensions result.
2. Transition from straight-sided center section to tapered section shall be smoothly joined in the area of the 1/4 rad.

of the long neck specimen was consistantly within the gage area, while the other shaped specimens failed in areas where the stress level was thought to be lower than at the center of the specimen. To evaluate the ASTM D 638 specimen failure mode at the shoulder radius, a series of modified specimens were prepared. The modification was to change the shoulder radius from the standard 3 in. to both 8 in. and 18 in. As shown in Fig. 10, these radius changes did not shift the location of the failure, that is, failures still occurred at the shoulder. To further characterize failure initiation of the various types of tension specimens, high-speed motion pictures were obtained while specimens were being loaded. Figure 11 shows the sequence of both the ASTM D 638 and long neck, bow tie, and Fig. 12 are the frames of the motion picture of the straight-side, tab-ended tension specimen at the failure points.

The motion pictures of the tensile failures were studied in detail at 50, 500, and 5000 frames per second, and they show:

A microexplosion occurs at the time of the break with the fiber glass reinforcement shattering and particles of glass and resin liberated. This explosion, as experienced in the end point of the bow tie and ASTM D 638 specimens, was in the

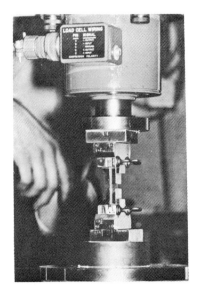

FIG. 6—Compression specimen, typi- FIG. 7—Compression specimen, typical setup prior to mounting com- cal setup after compressometer is
pressometer. located.

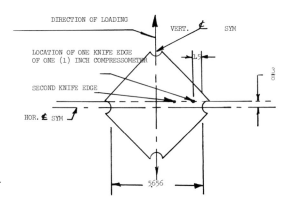

FIG. 8—In-plane shear compressometer location.

plane of the specimen, while the tab-ended, straight-sided specimen explosion was
out of the plane (normal to the thickness).

Immediately prior to failure, formation of a dark spot in the film is noted (see
arrows in Fig. 11) which is thought to be a microdelamination which thus scatters
reflected and transmitted light used during the photographs. It was interesting to
note that this micodelamination was not at the cut edge of the specimen and
that the progression of the delamination was inward and not to the edge of the
specimen.

The failure initiation of the specimens (microdelamination) of both the bow
tie and ASTM D 638 shapes occurred at only one location, that is, the location
of final failure, the central region of the former and the shoulder region of the
latter.

Composite Materials

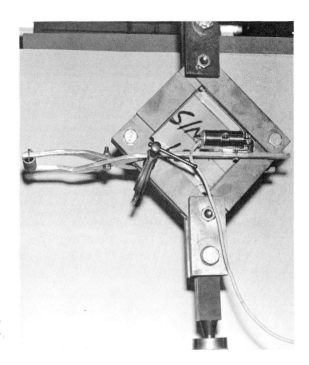

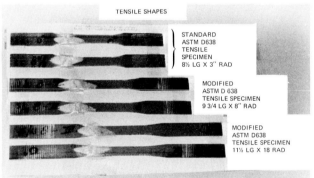

TENSILE SHAPES

STANDARD
ASTM D638
TENSILE
SPECIMEN
8½ LG X 3" RAD

MODIFIED
ASTM D 638
TENSILE SPECIMEN
9 3/4 LG X 8" RAD

MODIFIED
ASTM D638
TENSILE SPECIMEN
11½ LG X 18 RAD

Fig. 10—Effect of shoulder radius variation on location of failure in gage area.

From the results of these tension specimen studies, it was decided to utilize the bow tie shape for determination of tensile properties for the remainder of the program. Table 3 presents a summary of tensile properties obtained to date utilizing the bow tie specimen over a temperature range of −65 to 450 F. Further, the location of failures are given in the table, and, although not centrally located for all test temperatures or for all resin content laminates, it does show that the new specimen configuration, bow tie, is superior to specimens previously utilized. Typical tensile properties are given in Fig. 13.

COMPRESSION TESTS

The results of compression testing in the modified compression fixture are given in Table 4. The failure modes are satisfactory for most conditions in that the fixture inhibited edge crushing, end brooming, and buckling. The ultimate stresses

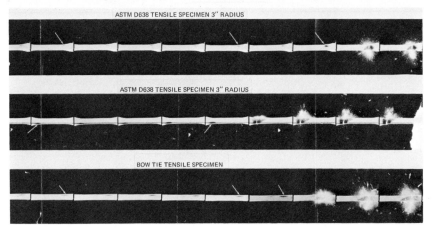

FIG. 11—High-speed photos showing failure propagation using bow tie and ASTM specimens.

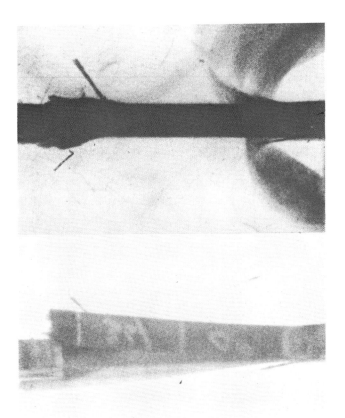

FIG. 12—Typical failure of tension specimen having doubler reinforced tabs.

Table 3—Summary of tensile properties.

Resin Content, %	Test Temperature, deg F	Wet (W) or Dry (D) Specimen	F_M, ksi		Primary E, 10^1 psi		Secondary E, 10^1 psi		Failure Mode Analysis, 1/total, %	
			0 deg	90 deg	0 deg	90 deg	0 deg	90 deg	0 deg	90 deg
25............	450	D	47.50	27.80	3.34	2.71	2.93	2.31	80	100
	75	W	61.40	50.30	4.27	3.96	3.17	2.76	70	80
	−65	D	91.80	74.80	4.50	4.19	3.09	2.75	20	100
	−65	W	80.50	62.20	4.71	4.35	3.28	2.88	20	40
28............	75	W	64.80	53.50	3.85	3.45	2.89	2.42	90	90
	−65	D	93.50	60.90	4.64	4.26	3.66	2.63	70	80
	−65	W	82.00	68.30	4.17	4.00	2.93	2.53	100	80
31............	75	W	62.80	53.50	3.83	3.76	3.07	2.68	100	60
	−65	D	85.30	70.00	4.08	3.81	2.95	2.62	30	30
	−65	W	82.30	67.40	4.34	4.18	3.04	3.10	90	70
34............	450	D	36.90	26.30	2.79	2.33	2.61	2.38	10	10
	75	W	55.50	48.90	3.58	3.30	3.04	2.72	60	90
	−65	D	63.90	68.70	3.80	4.15	2.83	2.65	90	60
	−65	W	73.00	63.90	4.00	3.75	3.01	2.75	90	90

NOTE—Failure mode analysis-legend:
1. Central area.
2. Edge of grips.
3. Inside of grips.
4. Radius area.
5. Delamination.

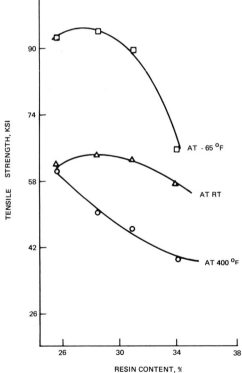

FIG. 13—Typical tensile properties.

Dastin et al on Reinforced Plastic Laminates

obtained were repeatable, and the scatter band was small. Measurement of the stiffness using a linear variable differential transformer (compressometer) attached to the edges of the specimen showed erratic results at various strain levels and in some specimens a stiffness reversal. It appears that with the compression fixture utilized the jig interferes with the specimen strain at higher load levels, and therefore only the initial strain are accurate. Typical compressive properties are given in Fig. 14.

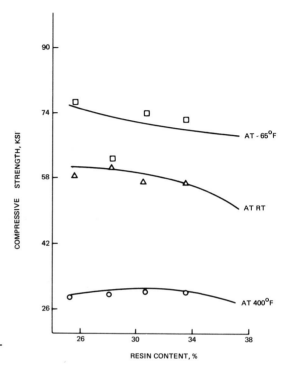

FIG. 14—Typical compressive properties.

The advantages found with the improved compression specimen and fixture were:

1. A reasonable size specimen is utilized, that is, ¾ in. wide by 3 in. long applicable over a thickness range of 0.06 to 0.125 in.

2. Direct measurement of strains utilizing a simple and rapid clip-on compressometer.

3. Strength properties were consistant, and failure modes were generally satisfactory for a compression coupon.

4. Useable over a temperature range of −65 to 450 F; being simple, and inexpensive to conduct.

IN-PLANE (EDGEWISE) SHEAR TESTS

The results of the in-plane shear tests obtained to date are given in Table 5. The picture frame type test was utilized rather than other test methods such as torsion tube, rail shear, or cruciform beam since the latter methods require specialized equipment, are costly to conduct, and are difficult to instrument for elevated temperature tests. Typical shear properties are given in Fig. 15.

The specimen and fixture used to apply the shear loading proved to be successful in that failures occurred without buckling. The strengths were found to be consistent with micromechanical analyses over the temperature range of −65 to 450 F for

Table 4—*Summary of compressive properties.*

Resin Content, %	Test Temperature, deg F	Wet (W) or Dry (D) Specimen	F_{cu}, ksi 0 deg	F_{cu}, ksi 90 deg	E, 10^6 psi 0 deg	E, 10^6 psi 90 deg	1 + 6/ Total, % 0 deg	1 + 6/ Total, % 90 deg	1 + 2 + 3 + 6/ Total, % 0 deg	1 + 2 + 3 + 6/ Total, % 90 deg
25.......	450	D	14.6	11.90	3.80	2.88	40	80	100	100
	75	W	56.30	47.50	4.21	4.17	40	50	100	100
	−65	D	74.70	63.50	4.39	4.17	40	60	100	100
	−65	W	73.80	55.30	4.66	4.33	10	0	100	90
28.......	75	W	58.60	46.40	3.85	4.03	40	60	100	100
	−65	D	71.90	57.70	4.48	3.79	0	80	80	100
	−65	W	58.20	50.10	4.19	3.98	70	60	100	100
31.......	75	W	52.00	48.20	4.10	3.94	90	80	100	100
	−65	D	73.10	58.90	3.73	3.48	40	50	100	100
	−65	W	73.70	64.90	3.98	3.76	20	20	100	100
34.......	450	D	17.50	11.70	2.79	2.34	20	80	80	100
	75	W	55.10	47.20	3.87	3.64	20	10	90	90
	−65	D	76.20	55.00	3.41	3.11	60	60	100	100
	−65	W	68.80	52.90	4.04	3.74	50	40	100	100

NOTE—Failure mode analysis–legend:
1. Central shear.
2. Central compression.
3. End shear.
4. End compression.
5. End brooming with end cracking.
6. Shear inside support plates.
7. Buckling.

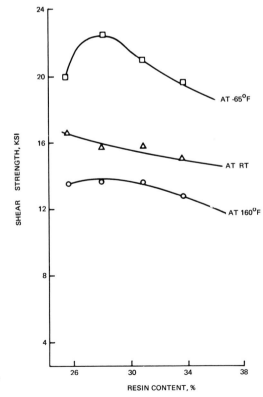

FIG. 15—Typical shear properties.

Table 5—*Summary of shear properties (dry).*

Resin Content, %	Test Temperature, deg F	F_{su}, ksi	G, 10^6 psi
25..............	160	13.50	1.43
	75	16.4	1.89
	−65	19.9	2.02
28..............	160	13.70	1.37
	75	15.80	1.81
	−65	22.9	1.99
31..............	160	13.7	1.24
	75	15.9	1.56
	−65	20.9	1.91
34..............	450	3.1	0.203
	160	12.70	0.957
	75	15.0	1.33
	−65	19.6	1.53

all resin content laminates tested. Strain measurements obtained from the compressometer at initial deformation are satisfactory for determination of shear modulus, but at the higher strain levels of 0/90-deg laminate Style 181 reinforcement the resultant stress-strain diagram is questionable.

CONCLUSIONS AND RECOMMENDATIONS

Based on the program herein outlined the following are the conclusions and recommendations:

1. The bow tie (long neck) tension test specimen enables more uniform load introduction and repeatable strengths and stiffnesses, along with a greater frequency of desired failure mode than either the ASTM D 638 or the straight-sided specimen for bidirectional woven fiber glass reinforced plastics. Therefore, it is recommended that this type specimen and test method be adopted as an industry standard.

2. Compressive strengths obtained using the improved test fixture result in satisfactory failure modes and ultimate strengths. Additional improvements in the fixture are required for obtaining accurate strain measurements; therefore, it is recommended that additional test methods be investigated for developing reliable compression, stress-strain curves for design purposes.

3. In-plane (edgewise) shear strengths of bidirectional reinforced plastic laminates can be determined accurately utilizing the picture frame test. The resultant ultimate stresses are repeatable and consistant with acceptable theory, but the strains beyond the initial load level are questionable. It is recommended that this test be utilized for material characterization rather than design.

Special Problems Associated with Boron-Epoxy Mechanical Test Specimens

R. N. HADCOCK[1] AND
J. B. WHITESIDE[1]

Reference:

Hadcock, R. N. and Whiteside, J. B., "Special Problems Associated with Boron-Epoxy Mechanical Test Specimens." *Composite Materials: Testing and Design, ASTM STP 460,* American Society for Testing and Materials, 1969, pp. 27–36.

Abstract:

Boron-epoxy composite material specimens were developed and tested to provide basic mechanical property data for design of the Grumman R&D FB-111 boron composite wing box extension. Special problems associated with tension, compression, and shear tests are presented in this paper.

Considerable development was required to achieve consistent results even in the apparently simple case of the tension of a unidirectional laminate. Specimen geometry and loading tab configuration have a considerable effect upon the failure stress and the location of the fracture origin, and several designs were tried before consistent failures were produced in the central portions of the specimens.

Problems were encountered also in testing compression coupon specimens. "Brooming" failures occurred when the specimen ends were not encapsulated securely. Considerable scatter evident in early tests indicated that compression specimens failed prematurely due to combined flexure-transverse shear and eccentric loading effects. This was confirmed by theoretical analysis.

The testing of flat panels in shear by the standard picture frame method presented difficulties. An anisotropic finite element analysis yielded stresses which were in good agreement with strain gage readings but indicated a severe stress concentration at the corner radius. Additional shear data were obtained from a modified picture frame method, the rail shear test.

Experience gained in these programs indicates that the testing of high-strength, high-modulus, anisotropic composites with low normal shear properties and negligible ductility lead to special problems in the design and analysis of specimens.

Key Words:

composite materials, boron, testing, test specimens, tension, compression, shear, mechanical properties, evaluation, tests

Tension, compression, and in-plane shear tests were conducted to obtain boron-epoxy mechanical properties data for design of the Grumman R&D FB-111 wing box extension. Some problems, peculiar to the boron-epoxy material, were encountered in both testing and test data evaluation.

TENSION TEST SPECIMEN DEVELOPMENT

It has long been apparent that variations in boron-epoxy composite tension test results are caused primarily by differences in testing methods and test specimen configuration. Honeycomb sandwich beam specimens have produced consistent re-

---

[1] Structural composites group leader and structural development engineer, respectively, Structural Mechanics Section, Grumman Aircraft Engineering Corp., Bethpage, N. Y. 11714.

sults [1],[2] but the cost is high. It was felt that a simple parallel sided coupon specimen, with development, could produce satisfactory results at far less cost.

Initial specimens were made, as shown in Fig. 1. Failures occurred predominantly within the tab region, and there were large variations in test results.

In order to eliminate the effect of laminate thickness on strength and modulus

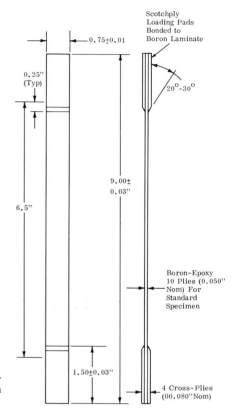

Fig. 1—Original configuration of longitudinal tension test specimen.

of elasticity, all results were evaluated on a load/unit width/layer basis or on the basis of stress calculated using a standard layer thickness of 0.0051 in./layer. An apparent scatter of nearly 14 percent is avoided by eliminating the actual layer thickness variations from results as the epoxy matrix contributes little to strength.

Various modifications to specimen tab geometry and material were investigated. Best results were obtained from specimens with long, fine taper tabs, but failures still tended to occur within the tabs. It was felt that, due to low interlaminar shear stiffness, load was not being fully transferred to interior fibers within the gripping length, thereby resulting in failure of the outer fibers within the tab region. (Analysis by Lenoe et al [2] has shown since that stress concentrations in the outer fibers do exist in the tab region.)

Further experiments [3] indicated that a thinner specimen with lengthened tapered tabs should give improved results. Consequently, six-ply specimens with 1½- and 3½-in. fiber glass tabs were manufactured and tested. The specimens with the longer tabs showed slightly higher average strength. Results were more consistent than those from the ten-ply specimens.

[2] The italic numbers in brackets refer to the list of references appended to this paper.

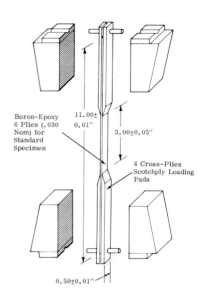

FIG. 2—Specimen configuration for tensile properties of boron-epoxy composite laminates.

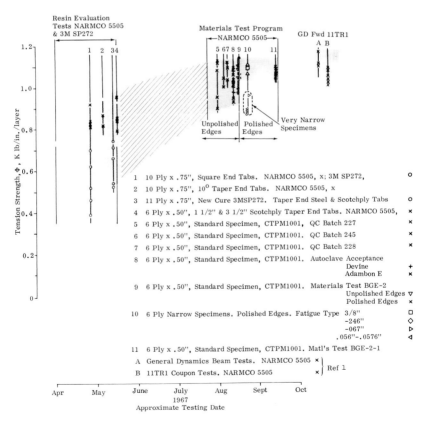

FIG. 3—History of longitudinal tension test results on unidirectional boron-epoxy composite. Grumman room-temperature tests.

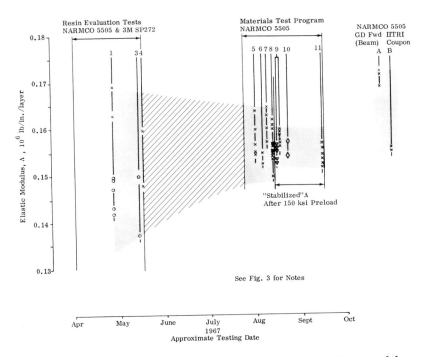

FIG. 4—History of longitudinal tensile modulus results on unidirectional boron-epoxy composite. Grumman room-temperature tests.

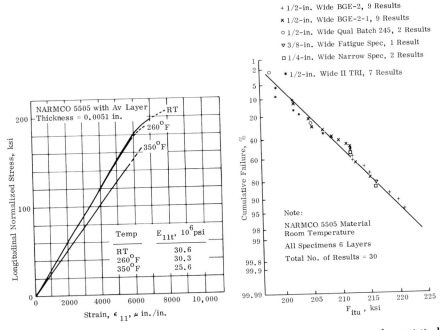

FIG. 5—Typical longitudinal tensile stress-strain curves for unidirectional Narmco 5505 boron-epoxy.

FIG. 6—Tension test result statistical distribution.

Table 1—*Longitudinal tension test results, room temperature.*

	Panel	Specimen No.	Dimensions: T (in.)	Dimensions: W (in.)	Ultimate Strength: P (lb)	Ultimate Strength: F (ksi)	Ultimate Strength: Φ (k lb/in. layer)	Modulus: E (10^6 psi, Final Load Run)	Modulus: Λ (10^6 pil, Final Load Run)	ϵ at Failure (in./in.)	Remarks
GRUMMAN TEST RESULTS (POLISHED EDGES)	BGE-2	6	0.0336	0.495	3230	194.5	1.090	27.6	0.1590	0.0073	
		8	0.0340	0.494	3330	198.0	1.120	27.4	0.1555	0.0075	
		10	0.0334	0.494	3070	186.0	1.055	27.9	0.1555	-	150-Ksi preload
		12	0.0353	0.495	3350	191.0	1.125	27.0	0.1590	0.0074	2nd Run E,
		15	0.0341	0.494	3210	190.0	1.080	27.7	0.1590	0.0072	
		2	0.0340	0.495	3360	200.0	1.130	27.6	0.1550	0.0074	5th Run E
		3	0.0340	0.495	3300	196.3	1.112	27.7	0.1570	0.0073	Various Preloads 8th Run E
		4	0.0340	0.496	3200	190.3	1.080	28.0	0.1585	0.0070	4th Run E
		5	0.0340	0.497	3250	193.5	1.095	27.7	0.1565	0.0071	8th Run E
	BGE-2-1	6	0.0320	0.500	3100	194.0	1.030	29.3	0.1556	-	
		8	0.0321	0.500	3130	195.0	1.042	28.8	0.1540	0.0067	
		9	0.0324	0.500	3230	199.5	1.075	28.2	0.1520	0.0070	
		10	0.0325	0.499	3260	200.5	1.083	28.8	0.1555	0.0070	150-Ksi preload
		11	0.0322	0.495	3260	204.5	1.095	29.5	0.1590	0.0071	2nd Run E,
		12	0.0322	0.500	3240	201.0	1.078	28.4	0.1521	0.0070	Examination of No. 7
		13	0.0320	0.500	3200	200.0	1.062	28.6	0.1518	0.0071	after failure in-
		14	0.0320	0.497	3200	201.0	1.068	28.9	0.1535	0.0069	dicated scored surface at fracture
		15	0.0320	0.497	3150	198.0	1.050	-	-	-	
	Static Test on Fatigue Specmen	1	0.0294	0.375	-	224.0	1.100	32.4	0.1590	0.0075	No preload
	Special Narrow Specimens	1	0.0334	0.246	1625	197.9	1.100	28.3	0.1572	-	No preload
		2	0.0300	0.246	1590	215.5	1.078	30.9	0.1545	-	No preload
	Qualification Batch 245	1	0.0297	0.498	3120	210.5	1.045	31.2	0.1545	-	No preload
		2	2.0299	0.498	3010	202.0	1.010	30.6	0.1525	-	No preload
IITRI TEST RESULTS	5505-UT	1	0.029	0.5 Nom	-	219	1.055	-	-	-	No preload
		3	0.029	0.5 Nom	-	214	1.035	-	-	-	No preload
		5	0.029	0.5 Nom	-	213	1.030	-	-	-	No preload
	(Ref 1,	7	0.029	0.5 Nom	-	226	1.092	-	-	-	No preload
	Table XVII	9	0.029	0.5 Nom	-	223	1.078	-	-	-	No preload
	P. 72)	16	0.029	0.5 Nom	-	210	1.015	32.2	0.1562	0.00648	No preload
		17	0.029	0.5 Nom	-	210	1.015	31.8	0.1540	0.00660	No preload

NOTE: Six unidirectional layers.

Difficulties had been experienced in gripping and aligning the specimen during tests carried out in an environmental chamber. These were overcome by extending the fiber glass end tabs to accommodate pins, seated in V-grooves on the grips, as shown in Fig. 2.

Specimens were tested with both as-machined and polished edges. Higher average strength and less scatter in both strength and modulus were found with polished edges, and it was decided to specify polished edges as standard. Figure 3 shows a history of unidirectional tensile strengths. Results from Ref *1* are included for comparison.

Preloading to about 75 percent of ultimate tension strength has the effect of raising the proportional limit stress and reducing and stabilizing the modulus of elasticity on subsequent loading runs. These effects are discussed in Ref *4*. From mid-August 1967 all specimens were preloaded, resulting in considerably reduced variations in the (stabilized) elastic modulus. A history of unidirectional boron-epoxy tension moduli results is shown in Fig. 4.

Examination of Fig. 3 and 4 shows that slight modification has resulted in considerable improvement of results. The current specimen is still simple, cheaper, and gives more consistent results than the honeycomb sandwich beams. Typical stress-strain curves are shown in Fig. 5, results are given in Table 1, and statistical distribution is shown in Fig. 6. From −67 to 260 F, failures occurred predominantly outside the tab region. However, because at 350 F about half the failures occurred within the tab region, further development of the high-temperature specimen has been initiated.

COMPRESSION TEST SPECIMENS

Coupon Specimen Development

Compression coupon specimens were manufactured, as shown in Fig. 7. Preliminary tests showed that edge restraint was required to prevent "brooming" failures. This was provided by a clamping arrangement in the loading fixture.

Euler buckling loads for this configuration were calculated for frictionless end conditions at 292 ksi and for fully supported end conditions at 623 ksi. These were felt to be sufficient to preclude specimen column failure.

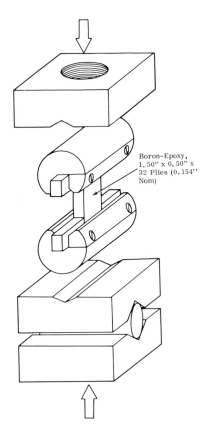

Boron-Epoxy, 1.50" x 0.50" x 32 Plies (0.154" Nom)

FIG. 7—Specimen configuration, compression properties of boron-epoxy laminates.

Theoretical analysis [5] showed that the combined effects of low normal shear stiffness of the specimen and loading eccentricity could account for at least a 37 percent reduction in Euler stress.

The variation of theoretical upper (supported) and lower (frictionless) buckling stresses with load eccentricity are shown in Fig. 8. Examination of the figure shows that all test results lie within these bounds.

Various modifications were then made to the specimen geometry: reducing the specimen length to 1.20 in., increasing the specimen thickness to 50 plies, reducing end flatness tolerance to within ±0.0002 in., and reducing the loading eccentricity to within ±0.001 in. Results were discouraging, showing only minor improvements.

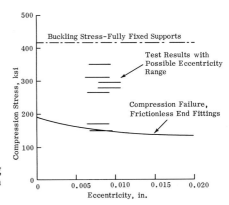

FIG. 8—Theoretical buckling stresses and test results from compression coupons.

ENCAPSULATED COUPON SPECIMENS

Very short (⅛ to ½ in.) specimens then were manufactured from the 50-ply material and were encapsulated in a low-modulus resin. These failed at nominal stresses in excess of 500 ksi. As the effects of restraint provided by the surrounding resin were uncertain, it was considered that these results could not be related to practical structural applications.

In view of the difficulties encountered in the coupon specimen tests and the uncertain stress state in the encapsulated specimens, it was decided to use the sandwich beam compression specimen to provide design allowable data. Excellent results currently are being obtained using this specimen.

SHEAR TEST SPECIMEN DEVELOPMENT

Shear testing of composite materials generally is concerned with two more or less distinct areas of interest: (1) the in-plane shear properties and (2) the interlaminar, or normal, shear properties.

The picture frame specimen, the cruciform sandwich beam, and the torsion tube

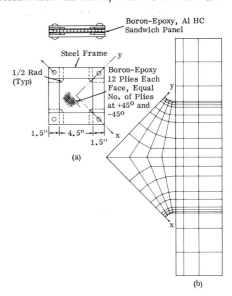

FIG. 9—Picture frame shear panel finite element analysis.

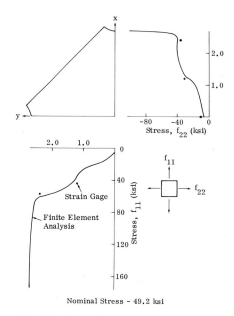

Nominal Stress - 49.2 ksi

FIG. 10—Stress distribution—picture frame shear panel.

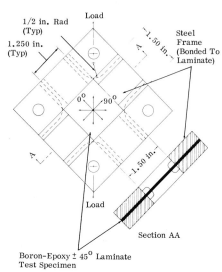

Boron-Epoxy ± 45° Laminate
Test Specimen

FIG. 11—Small picture frame shear test specimen (1½ by 1½ window area).

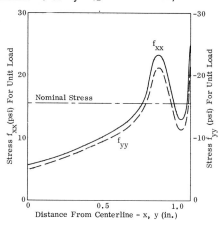

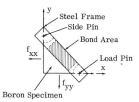

Segment of Small Picture
Frame Specimen (1.5 x 1.5
Window) Selected For
Analysis

FIG. 12—Small picture frame shear specimen, stress distribution.

Composite Materials

are well-known concepts of in-plane shear testing. In addition, the rail shear test method, first proposed by Floeter and Boller [6], shows promise of becoming an accepted test method for composites.

The picture frame configuration initially used for in-plane shear tests on boron-epoxy laminates has a 4.5-in. "window" within bonded steel members. Premature failures occurred in tests on 12-ply laminates. Increasing the thickness of the laminate sufficiently to preclude buckling would have increased the shear failure loads beyond both the testing machine capacity and the ultimate strength of the frame bond. Hence, to achieve the required stability, a sandwich construction was employed, as shown in Fig. 9a. Still, low strengths were obtained from such specimens.

An anisotropic finite element analysis of this specimen was run using the gridwork shown in Fig. 9b. This analysis, Fig. 10, agrees reasonably well with strain gage readings but indicates severe stress concentrations at the corner radii.

A second configuration, Fig. 11, employed a single six-ply laminate with 1.5-in. window. Here the corner radius lies beneath the steel frame. Analysis, Fig. 12, showed a somewhat less severe stress concentration than that in the large specimen. A second peak coincides with the corner of the stepped bond face of the steel frame.

These analyses indicate that the proportion of applied load carried by the side pins was 12.1 percent with the 4.5-in. configuration and 20.6 percent with the 1.5-in. configuration.

Because the stress distribution in these tests is evidently nonuniform, these tests have been used only for purposes of materials qualification and selection.

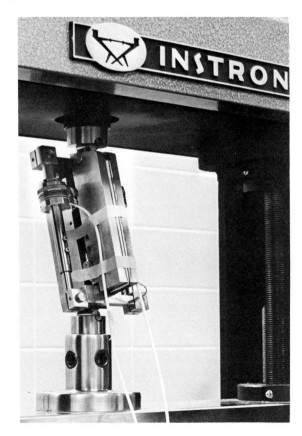

Fig. 13—Rail shear apparatus mounted in testing machine.

Some rail shear tests have been carried out [7] with encouraging results. A capacitance transducer was used to measure deformations, as shown in Fig. 13. An extensive experimental investigation [8] is currently under way to determine the stress distribution in this specimen.

The short-beam shear tests and the notched coupon test are well-known interlaminar shear test methods. The stress and strain distribution in the notched coupon specimen has not been determined accurately and is dependent on notch geometry and means of fabrication, that is, molded or machined notches. The short-beam shear specimen presents less problems in fabrication and test. But an elastic anisotropic finite element analysis of the short-beam shear specimen [5] has shown that simple analyses may lead to significant errors in stress and modulus determination.

References

[1] Structural Airframe Applications of Advanced Composite Materials," IITRI, TEI, General Dynamics Fort Worth Division, Contract No. AF 33(615)5257, Fourth Quarterly Progress Report, June 1967.

[2] Lenoe, E. M., "Testing and Design of Advanced Composite Materials," paper presented at American Society of Civil Engineers' meeting, Pittsburgh, Pa., 30 Sept.–4 Oct. 1968.

[3] Donohue, P., "Grumman Participation in the Round Robin Investigation of a Proposed Method for Determining Tensile Strength of Advanced Composites," Report GSTR 007, Grumman Aircraft Engineering Corporation, Bethpage, N. Y., May 1967.

[4] Hadcock, R. N., "Advanced Composite Wing Structures, Boron-Epoxy Composite Material Preliminary Design Data," Advanced Development Report FSR-AD2-01-68.09, Grumman Aircraft Engineering Corporation, Bethpage, N. Y., Feb. 1968.

[5] Hadcock, R. N. et al, "Evaluation of Test Data on Unidirectional Boron-Epoxy Material," FSR-AD2-01-68.07, Grumman Aircraft Engineering Corporation, Bethpage, N. Y., Jan. 1968.

[6] Floeter, L. H. and Boller, K. H., "Use of Experimental Rails to Evaluate Edgewise Shear Properties of Glass-Reinforced Plastic Laminates," U.S. Forest Products Laboratory, Madison, Wis., Task Force for Test Methods, Industry Advisory Group, MIL-HDBK-17, 28 April 1967.

[7] Rutherford, J., Bossler, F., and Swain, W., "Shear Properties of Boron-Epoxy Composites," Contract No. 9-69929, June 1968, Singer-General Precision, Inc. sponsored by Grumman Aircraft Engineering Corporation.

[8] Rutherford, J. et al, "Design Allowables—Boron-Epoxy Composites," Phase 1 Report, Contract No. 9-81329, Nov. 1968, Singer-General Precision, Inc. sponsored by Grumman Aircraft Engineering Corporation.

Characterization of Anisotropic Composite Materials

J. C. HALPIN,[1]
N. J. PAGANO,[1]
J. M. WHITNEY,[1] AND
E. M. WU[2]

Reference:

Halpin, J. C., Pagano, N. J., Whitney, J. M., and Wu, E. M., "Characterization of Anisotropic Composite Materials," *Composite Materials: Testing and Design,* ASTM STP 460, American Society for Testing and Materials, 1969, pp. 37–47.

Abstract:

Characterization procedures for anisotropic composite materials are examined critically with reference to obtaining a uniform procedure for both stiffness and ultimate property characterization. An experimental procedure employing thin-walled anisotropic cylinders for both uniaxial and multiaxial characterization is proposed.

Key Words:

characterization, mechanical properties, composite materials, anisotropic cylinders, failure. evaluation, tests

Characterization procedures for anisotropic composite materials must provide not only experimental data for moduli or stiffness properties, but also must characterize the complete mechanical response of the material up to and including the ultimate failure of the system. The analytical and experimental results presented here clearly demonstrate that many classical notions associated with mechanical characterization experiments, extended from the technology of isotropic solids, must be revised in connection with the testing of anisotropic solids. The failure of certain current accepted test procedures, as shown in this paper, necessitates new and unique testing procedures for both unidirectional and multidirectional composites. Using basic principles of anisotropic mechanics, difficulties encountered in finding proper characterization procedures are discussed.

STIFFNESS MEASUREMENTS

The constitutive equations with respect to the material symmetry axes of a unidirectional composite sheet are given by

$$\epsilon_1 = S_{11}\sigma_1 + S_{12}\sigma_2 \dots \dots \dots (1)$$

$$\epsilon_2 = S_{12}\sigma_1 + S_{22}\sigma_2 \dots \dots \dots (2)$$

$$\epsilon_6 = S_{66}\sigma_6 \dots \dots \dots (3)$$

for a state of plane stress in the plane of the sheet. In these equations the subscripts 1 and 2 refer to normal stress and strain, 6 indicates shear components (ϵ_6 is the *engineering* shear strain), and

$$S_{11} = \frac{1}{E_{11}}, S_{22} = \frac{1}{E_{22}}, S_{12} = -\frac{\nu_{12}}{E_{11}}, S_{66} = \frac{1}{G_{12}} \dots \dots \dots (4)$$

---

[1] Air Force Materials Laboratory, Wright-Patterson Air Force Base, Ohio. 45433.
[2] Materials Research Laboratories, Washington University, St. Louis, Mo. Personal member ASTM.

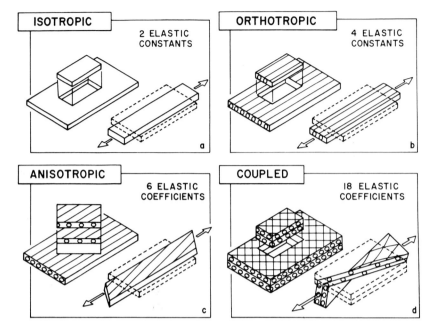

FIG. 1—Various types of materials showing number of elastic constants associated with them.

For isotropic materials, only two of these material coefficients are independent. These cases are illustrated in Figs. 1a and b.

For unidirectional composite materials in which the fibers are oriented at an arbitrary angle to the edges of the sheet (Fig. 1c), the constitutive equations take the form

$$\epsilon_1' = S_{11}'\sigma_1' + S_{12}'\sigma_2' + S_{16}'\sigma_6' \dots\dots\dots\dots\dots\dots\dots\dots (5)$$

$$\epsilon_2' = S_{12}'\sigma_1' + S_{22}'\sigma_2' + S_{26}'\sigma_6' \dots\dots\dots\dots\dots\dots\dots\dots (6)$$

$$\epsilon_6' = S_{16}'\sigma_1' + S_{26}'\sigma_2' + S_{66}'\sigma_6' \dots\dots\dots\dots\dots\dots\dots\dots (7)$$

where primed quantities are measured with respect to axes parallel to the edges of the sheet and

$$S_{16}' = \frac{\eta_{16}'}{E_{11}'}, \, S_{26}' = \frac{\eta_{26}'}{E_{22}'} \dots\dots\dots\dots\dots\dots\dots\dots (8)$$

By use of the standard transformation equations [1],[3] S_{ij}' can be expressed in terms of S_{ij} and fiber orientation. As a consequence of the shear coupling compliances S_{16}' and S_{26}', the response of an "off-angle" composite (that is, a specimen in which the filaments are neither parallel or perpendicular to the direction of the applied force) under uniaxial tension appears as shown in Fig. 1c.

For laminated materials, the elastic properties of the individual layers are expressed in terms of the reduced stiffffness matrix Q_{ij} and from classical laminated plate theory [1], the laminate constitutive relations are expressed as

$$\begin{bmatrix} N_1 \\ N_2 \\ N_6 \\ M_1 \\ M_2 \\ M_6 \end{bmatrix} = \begin{bmatrix} A_{11} & A_{12} & A_{16} & B_{11} & B_{12} & B_{16} \\ A_{12} & A_{22} & A_{26} & B_{12} & B_{22} & B_{26} \\ A_{16} & A_{26} & A_{66} & B_{16} & B_{26} & B_{66} \\ B_{11} & B_{12} & B_{16} & D_{11} & D_{12} & D_{16} \\ B_{12} & B_{22} & B_{26} & D_{12} & D_{22} & D_{26} \\ B_{16} & B_{26} & B_{66} & D_{16} & D_{26} & D_{66} \end{bmatrix} \begin{bmatrix} \epsilon_1 \\ \epsilon_2 \\ \epsilon_6 \\ \kappa_1 \\ \kappa_2 \\ \kappa_6 \end{bmatrix} \dots\dots\dots\dots (9)$$

[3] The italic numbers in brackets refer to the list of references appended to this paper.

where

$$(A_{ij}, B_{ij}, D_{ij}) = \int_{-h/2}^{h/2} Q_{ij}(1, z, z^2) \, dz \dots\dots\dots\dots\dots\dots (10)$$

$$(N_i, M_i) = \int_{-h/2}^{h/2} \sigma_i(1, z) \, dz, \ i, j = 1, 2, 6 \dots\dots\dots\dots (11)$$

Examination of Eqs 9 reveals a coupling phenomenon between bending, twisting, and extension, for example, a two-layer angle-ply composite under uniaxial tension twists, as shown in Fig. 1d. In the event that Q_{ij} is an even function of z (symmetric layup), B_{ij} vanishes, and the constitutive equations become uncoupled.

The principal problems encountered in experimental characterization arise from the coupling coezcients S_{16}, S_{26}, Q_{16}, Q_{26}, D_{16}, D_{26}, and B_{ij} in the constitutive equations. The lack of appreciation of the physical consequences of these coupling terms has resulted in the current deficiencies in characterization and analysis. The concept of coupling phenomena is peculiar to anisotropic and layered systems, and it is for this reason that the technology developed for isotropic solids must be modified for anisotropic bodies.

FRACTURE CHARACTERIZATION

The characterization of the fracture properties of anisotropic solids involves the development of a strength criterion as well as fracture mechanics or toughness and fatigue characterization. Knowledge of the state of stress, which is preferably uniform, particularly when material response is nonlinear, in strength determinations is absolutely essential if we are to have a large base of experimental evidence upon which strength and fatigue theories may be deduced. The development of a failure criterion for fiber reinforced materials involves a great amount of off-angle testing of unidirectional composites in order to establish transformation properties of "strength." Furthermore, while it is recognized generally that strength properties in tension and compression are functions of fiber orientation, it is not appreciated

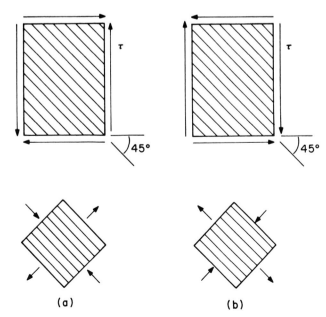

Fig. 2—Illustration of positive and negative srear.

        (a)   (b)

Table 1—*Dependence of the strength of an orthotropic solid on the sign of the shear stress.*

θ	15 deg	45 deg	60 deg
$+\tau_{xy}$ ......3820 psi		5320 psi	2690 psi
$-\tau_{xy}$...... 1930 psi		436 psi	1150 psi

generally that the shear strength of an anisotropic material is strongly dependent on the direction of the shear stress [2]. For example, consider the unidirectional 45-deg composite shown in Fig. 2 under states of pure shear of opposite sense. Owing to the directional properties of the material, the nature of the fracture, as well as the ultimate value of τ, in these two cases will be vastly different. For the state of stress in Fig. 1a, the fracture surface would tend to align parallel to the filaments, while in Fig. 1b, the filaments themselves would be fractured. This feature of anisotropic material response recently has been demonstrated experimentally by Halpin and Wu [3] (Table 1).

The prediction of the lifetime and failure modes for damaged anistropic bodies involves an experimental fracture mechanics program. The goal of a program of this nature is to specify the applied stresses (or loads) and geometry (crack length, specimen dimensions) to simulate in the laboratory the conditions which will cause crack extension and failure of a structure in service. A typical experiment in this area, similar to that performed by Wu [4], studies crack extension under (a) tension and shear, (b) pure shear, and (c) compression and shear (Fig. 3).

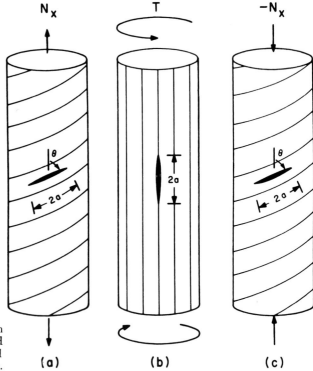

FIG. 3—Crack extension under (a) tension and shear, (b) pure shear, and (c) compression and shear.

ANALYSIS OF CHARACTERIZATION

TENSION TEST OF OFF-ANGLE SPECIMEN

One of the most elementary concepts in elasticity theory is that of a uniform state of stress. Producing such a state of stress in the characterization experiment is absolutely essential to determine fundamental material properties; however, this is no trivial task. For example, the common tension test of off-angle specimens can yield strength and stiffness data which are greatly erroneous [5,6]. Suppose that the tension test of an off-angle composite is used as the basis for determining E'_{11}, the composite modulus of elasticity parallel to the long edges of the specimen. If the effects of end constraint are not taken into account, this modulus will be recorded erroneously as E^*_{11}, where [5]

$$E^*_{11} = \frac{1}{S'_{11}} \left(\frac{1}{1 - \eta} \right) \quad\quad\quad\quad (12)$$

with

$$E'_{11} = \frac{1}{S'_{11}} \quad\quad\quad\quad (13)$$

and

$$\eta = \frac{6{S'_{16}}^2}{S'_{11} \left(6S'_{66} + S'_{11} \dfrac{l^2}{h^2} \right)} \quad\quad\quad\quad (14)$$

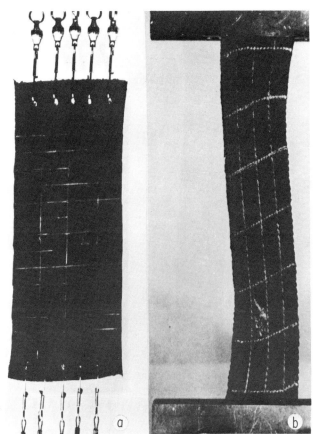

FIG. 4—Off-angle specimen under (a) pure tension and (b) clamped tensile load.

The term involving η represents a conservative estimate of the error in this experiment. The error results because conventional clamping devices induce severe perturbations (bending and shear) in the stress field [5,7]. These large effects are a direct consequence of the shear coupling compliance, S'_{16}. The response of an off-angle composite under uniform tension is shown in Fig. 4a, while the corresponding clamped specimen is depicted in 4b. Uniform states of stress and strain are only possible when $S'_{16} = 0$. This condition, of course, is satisfied for 0 or 90-deg fiber orientation so that these specimens can be utilized to evaluate the compliance coefficients S_{11}, S_{12}, and S_{22} and the analogous engineering constants and strength properties. The shear compliance S_{66} must be determined from an additional experiment.

TORSION OF A UNIDIRECTIONAL SPECIMEN

An acceptable experiment for defining the shear modulus G_{12} and interlaminar shear stress (at least to the point of material nonlinearity) is the pure torsion of a solid circular cylindrical rod with the fibers oriented parallel to the longitudinal axis. The analysis of this specimen is identical to that of an isotropic cylinder under torsion. This test configuration has been employed by Thomas et al [13] among others for the determination of unidirectional composite shear properties. The maximum stress and shear modulus can be computed from the classical strength of materials relations

$$\tau_{\max} = \frac{2T}{\pi R^3} \dots\dots\dots\dots\dots\dots\dots\dots\dots (15)$$

$$G_{12} = \frac{2T}{\pi R^3 \epsilon_6} \dots\dots\dots\dots\dots\dots\dots\dots (16)$$

where ϵ_6 is the maximum shear strain on the surface of the rod.

FLEXURE TEST

Next to tension testing the most common experiment employed to define the response of composite materials is the flexure test. Although this test is generally invalid for strength determination owing to material nonlinearity and stress concentrations in the region of an applied force, there are instances where flexural experiments can yield useful information if the data are interpreted properly.

In 1949, Hoff [8] presented flexural formulas for laminates, and recently Pagano [9] has modified these expressions to account for shear deflection. Although these formulas are based upon elementary strength of materials considerations, they appear quite adequate to describe the special cases of deflection of undirectional and bidirectional (0/90 deg) beams. While the neglect of shear deflection in long isotropic beams is a valid assumption, shear deflection in composite beams can be very significant due to the low ratio of shear modulus to longitudinal modulus for most composites.

Despite the correction for shear deflection, however, the flexural analysis is currently incomplete since it cannot describe the complex response of a laminated beam of arbitrary layer orientations and stacking sequence.

We now consider the deflection of a unidirectional (off-angle) beam induced by pure bending moments. Two cases are of interest to us: (1) the fibers are parallel to the front face of the member as in Fig. 5a, and (2) the fibers are parallel to the top surface as shown in Fig. 5b.

For Case 1, Figs. 5a and 6a, it can be shown [10] that the center line vertical displacement, v, is independent of S_{16} and corresponds to that given by the ele-

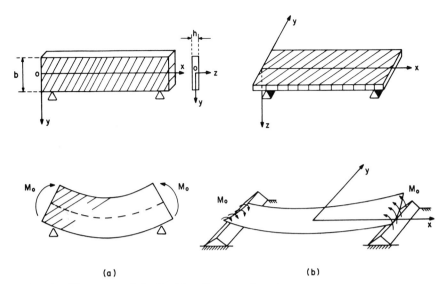

FIG. 5— Illustration of idealized bending experiments.

mentary Euler-Bernoulli beam theory. If the ends, $x = o$, l, are pinned, the deflection surface is given by

$$v = \frac{M S_{11}}{2I} (lx - x^2) \dots\dots\dots\dots\dots\dots\dots(17)$$

where it is understood that the compliance S_{11} is the primed quantity given in Eq 5. It should be noted however that the Euler-Bernoulli hypothesis itself is not satisfied since the horizontal displacement u is a nonlinear function of y (u also depends upon S_{16}).

Consider now the orientation of Case 2, Figs. 5b and 6b, and let the vertical central-plane displacement be represented by w. The three dimensional elasticity solution for the maximum deflection w_0 is given as

$$w_0 = \frac{3 M_0 b^2}{2h^3} (S_{11}C^2 + S_{16}C + S_{12}) \dots\dots\dots\dots\dots\dots(18)$$

when C is the length-to-width ratio and b is the beam width. Equation 18 is based upon certain assumed boundary conditions on w. Although these boundary conditions are plausible, it is quite likely that the boundary conditions in a physical experiment are impossible to define owing to the warping exhibited by the beam, as shown in Fig. 6b. The term involving S_{16} involves a significant deviation from classical beam theory; it is because of the existence of S_{16} that the beam twists and lifts off the support, as shown in Figs. 5b and 6b. If one attempts to suppress the lift-off from the supports by employing double knife edge supports, twisting moments will be induced in the member. As the value of C increases, the influence of S_{12} diminishes rapidly, while the effect of S_{16} dissipates only for very large values of C. In fact the apparent flexural modulus of a unidirectional 10-deg off-angle specimen is higher than that of a 0-deg specimen for the material parameters considered in Ref 12.

Although we have only discussed pure bending here, we can expect that beams loaded by concentrated forces would display, at least qualitatively, similar response characteristics. Unfortunately, the exact elasticity solutions for the configurations

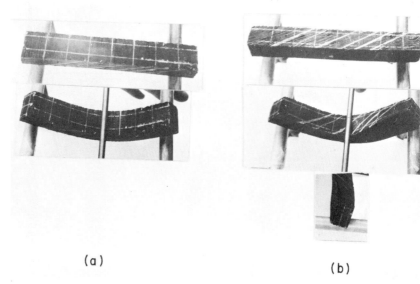

(a)   (b)

Fɪɢ. 6—Experimental bending observations.

of Cases 1 and 2 under concentrated forces do not exist. This fact, together with the other objections noted above, lead us to reject these modes of testing for precise material characterization, in particular, Case 2. However, an approximate elasticity solution for Case 1 under a central load may be derived from the cantilever beam solution presented by Hashin [11]. Hence, this configuration may merit some attention as a means of *stiffness* characterization if specimen stability does not prove to be critical.

Use of Cylindrical Tubes

In the preceding discussion we have demonstrated some of the practical complications resulting from the unique response characteristics of anisotropic and layered media. If we are required to measure shear modulus or produce a well-defined combined state of stress, or even a state of uniaxial stress, we are led to the consideration of yet another test configuration, namely, the thin-walled hollow anisotropic cylinder. This specimen, that is, a helical wound composite tube serves as the basis for a general characterization procedure (Fig. 3). Again the primary concern is the existence of a uniform state of stress under practical testing conditions. While the elasticity analysis shows that a uniform state of stress cannot exist in such a specimen unless S'_{16} vanishes [14], the stress field is nearly uniform for thin tubes. An acceptable characterization procedure for uniaxial tension or combined loading can be achieved in the laboratory by allowing for the unconstrained rotation and longitudinal motion of one end of the tube [14]. The rotation of the helical wound tube under tension or compression is a natural consequence of the shear coupling compliance S'_{16}. Conversely the state of pure shear induced by torsion results in either an increase or decrease in the height of the cylindrical tube. If these motions are prohibited, extraneous stresses are induced in the specimen. Photographs of deformed helical wound tubes under uniform tension and internal pressure are shown in Figs. 7a and b, respectively.

The next step is the analysis of layered shells under combined loading conditions. Using the well known Donnell approximations employed by Dong et al [15] in their analysis of thin laminated anisotropic cylindrical shells, a relatively straight-

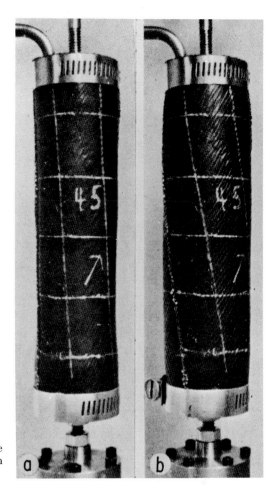

FIG. 7—Helical wound tube under (a) uniform tension and (b) internal pressure.

forward solution can be obtained [16] under combined tensile (or compressive), shear, and internal pressure. In this study by Whitney and Halpin, all six components of the in-plane stiffness matrix are defined in terms of experimentally observable quantities for arbitrary layer orientations.

$$\epsilon_x^0 = (A_{11}^* + A_{16}^* K_1)N_x + (A_{12}^* + A_{16}^* K_2)Rp + A_{16}^* K_3 T \dots \dots (19)$$

$$\epsilon_y^0 = (A_{12}^* + A_{26}^* K_1)N_x + (A_{22}^* + A_{26}^* K_2)Rp + A_{26}^* K_3 T \dots \dots (20)$$

$$\epsilon_{xy}^0 = (A_{16}^* + A_{66}^* K_1)N_x + (A_{26}^* + A_{66}^* K_2)Rp + A_{66}^* K_3 T \dots \dots (21)$$

where

$$K_1 = \frac{B_{16}^*}{R}, \ K_2 = \frac{B_{26}^*}{R}, \ K_3 = \frac{1}{2\pi R^2} \dots \dots \dots \dots (20)$$

and

$$\left. \begin{array}{l} A^* = A^{-1} \\ B^* = -A^{-1}B \end{array} \right\} \dots \dots \dots \dots \dots \dots (21)$$

For symmetric layups $(B_{ij} = 0)$

$$K_1 = K_2 = 0 \dots \dots \dots \dots \dots \dots (22)$$

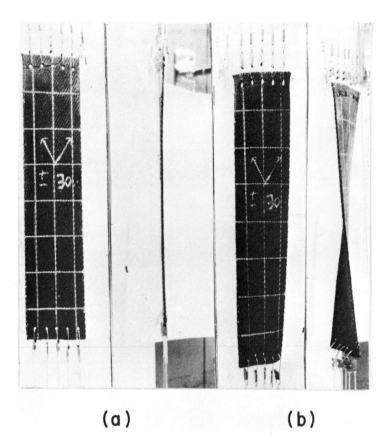

Fig. 8—Observation of coupling between bending and stretching.

The quantities R, p, T, and N_x are the radius, internal pressure, applied torque, and longitudinal normal stress resultant, respectively. For a symmetric composite, Eqs 19 to 22 reveal that all components of A_{ij}^* can be determined by applying combinations of axial force, internal pressure, and torsion to the specimen. As a consequence, all of the equivalent engineering constants can be evaluated in this series of experiments. It should be noted however, that certain difficulties arise in defining the engineering constants of laminated materials, as pointed out by Whitney [17]. These constants can be expressed always in terms of the in-plane compliance matrix, but not in terms of the in-plane stiffness matrix alone unless the laminate is balanced and A_{ij} is orthotropic. This fact often leads to confusion in the reporting of experimental data. For example, owing to the variation of stresses through the thickness, the shear modulus of a ±45-deg symmetric angle-ply composite is not equal to the shear modulus of a 45-deg unidirectional layer, even though the latter shear modulus is uniform through the thickness.

EFFECT OF NONSYMMETRY

The response of a laminate which is not symmetrical through the thickness (a two-layer angle ply) under uniaxial tension is illustrated in Fig. 1d and shown experimentally in Fig. 8. Despite the fact that the fiber orientations are symmetric with respect to the load direction, coupling between twisting and extension exists owing to B_{16}, Eq 9. Obviously, the constraints exerted by conventional clamping devices on such a specimen will induce a highly complex state of stress in the specimen. Similar phenomena have been observed in flexure experiments for *symmetric* angle-ply composites [12]. For in this case, despite the fact that $B_{ij} = 0$, the flexural moduli D_{16} and D_{26} do not vanish, which can lead to serious errors in the interpretation of flexural experiments.

CONCLUSIONS

For stiffness, creep, and strength characterization of fiber reinforced systems possessing arbitrary fiber orientations and lamination sequences, the experimental procedures outlined here employing thin-walled cylindrical test specimens [14,16] constitute the most unified procedure available. Most of the other experiments discussed suffer from a lack of a well-defined stress-strain field and are influenced widely by conventional clamping systems. For the special case of characterization of limited quantities of unidirectional materials, we recommend coupon testing of 0 and 90-deg composites combined with a solid rod torsion experiment for shear properties.

References

[1] *Composite Materials Workshop,* Tsai, Halpin, and Pagano, eds., Technomic Publishing, Stanford, Conn., 1968.

[2] Gol'denblat, I. I., and Kapnov, V. A., "Strength of Glass Reinforced Plastics in Complex Stress States," *Mech. Polimerov,* Vol. 1, No. 70, 1965.

[3] Halpin, J. C. and Wu, E. M., *Journal of Composite Materials,* 1969.

[4] Wu, E. M., in Ref. *1* and work reported by Corten, H. T. in *Fundamental Aspects of Fiber Reinforced Plastic Composites,* Schwartz and Schwartz, eds., Interscience, New York, 1968.

[5] Pagano, N. J. and Halpin, J. C., "Influence of End Constraint in the Testing of Anisotropic Bodies," *Journal of Composite Materials,* Vol. 2, No. 18, 1968.

[6] Kicher, T., Case Institute, unpublished results.

[7] Wu, E. M. and Thomas, R. L., "Off-Axis Test of a Composite," *Journal of Composite Materials,* Vol. 2, No. 523, 1968.

[8] Hoff, N., *Engineering Laminates,* Dietz, A. G. H., ed., Wiley, New York, 1949.

[9] Pagano, N. J., "Analysis of the Flexture Test of Bidirectional Composites," *Journal of Composite Materials,* Vol. 1, No. 336, 1967.

[10] Lekhnitskii, S. G., *Anisotropic Plates,* Gordon and Breach, New York, 1968.

[11] Hashin, Z., "Plane Anisotropic Beams," *Journal of Applied Mechanics,* Vol. 34, No. 257, 1967.

[12] Whitney, J. M., in preparation.

[13] Thomas, R. L., Doner, D., and Adams, D., "Mechanical Behavior of Fiber-Reinforced Composite Materials," AFML-TR-67-96, Air Force Materials Laboratory, May 1967.

[14] Pagano, N. J., Halpin, J. C., and Whitney, J. M., "Tension Buckling of Anisotropic Cylinders," *Journal of Composite Materials,* Vol. 2, No. 154, 1968.

[15] Dong, S. B., Pister, K. S., and Taylor, R. L., "On the Theory of Laminated Anisotropic Shells and Plates," *Journal of Aeronautical Sciences,* No. 969. 1962.

[16] Whitney, J. M. and Halpin, J. C., "Analysis of Laminated Anisotropic Tubes Under Combined Loading," *Journal of Composite Materials,* Vol. 2, No. 360, 1968.

[17] Whitney, J. M., "Engineering Constants of Laminated Composite Materials," *Journal of Composite Materials,* Vol. 2, No. 261, 1968.

Test Methods for High-Modulus Carbon Yarn and Composites

J. T. HOGGATT[1]

Reference:

Hoggatt, J. T., "Test Methods for High-Modulus Carbon Yarn and Composites," *Composite Materials: Testing and Design, ASTM STP 460,* American Society for Testing and Materials, 1969, pp. 48–61.

Abstract:

Several test methods used for the evaluation of carbon filaments and composites are presented. Included are methods for determining tensile, compressive, interlaminar shear, and bearing strength properties of flat and tubular carbon composites.

The tensile and compressive properties of composite materials having varying fiber volume functions are reported. The use of both tubular and flat specimens made it possible to determine the effect of narrow (flat) specimens versus wide (cylindrical) specimens on the strength properties of the composite.

Interlaminar shear strength (ILS) studies of carbon composites, using flat short-beam specimens, indicated that the strength and the mode of failure obtained are very sensitive to the selected span-to-depth ratio (SDR). Results also show that the apparent ILS increases approximately 25 percent by changing the SDR from eight to four.

The bearing strength of a multi-directional composite was found to be 38,000 psi, which is 90 percent of the tensile strength.

In addition to methods for composite properties evaluation, material control test methods are described. These include a yarn tension test, resin content, and void content determination.

Key Words:

carbon composites, fiber composites, composite materials, carbon yarn, mechanical properties, evaluation, tests

The introduction of advanced composites has necessitated the development of new techniques for material evaluation and quality control. Available test methods for glass reinforced composites are generally inadequate for these purposes, since they fail to distribute efficiently the load to the high-modulus reinforcements, thereby yielding results not representative of the actual material capabilities. This paper summarizes the test methods currently used by The Boeing Company[2] for evaluation of advanced composites. The methods are generally refinements and modifications of ASTM techniques, but when no applicable ASTM test methods existed, as in the case of resin content determination, new methods were developed. The test methods were developed for the primary purpose of obtaining data for preliminary design, but most of the test specimens are simple and sufficiently low in cost so that they can be used for quality control purposes.

YARN TENSION TEST METHOD

One of the current quality control procedures for determining carbon yarn tensile strength is to test a limited number of filaments, usually less than ten, selected at random from the yarn bundle containing 1440 filaments. Because such a small

---

[1] Research specialist, Missile Materials and Processes Organization, The Boeing Company, Seattle, Wash. 98124.

[2] Missile and Information Systems Division.

specimen is not considered representative of the population, a strand test was developed, based on the single-end glass yarn tension test developed and reported by Pirzadeh and Kennedy [1].[3] In this test the entire impregnated carbon yarn strand is tested in tension and the average filament strength computed. The test is sensitive to yarn imperfections caused by manufacturing and handling techniques, and therefore can be used to evaluate the effects of equipment and processing procedures on yarn quality.

GRIPS

The grips and test specimen holder shown in Fig. 1 consist of three parts: (1) *mounting tabs* with scribed reference lines for yarn centering and for defining gage length (the holes in the tabs are countersunk from the back to provide a knife edge at the front face to minimize the amount of eccentric loading on the specimen during test); (2) a *grooved plate* for keeping the tabs at a fixed distance while the specimen is bonded to the tabs; and (3) *pinch clamps* for holding the tabs to the plate during specimen mounting.

ADAPTERS

Two adapters for connecting the two ends of the test fixture to the heads of a universal test machine are required (Fig. 1). These adapters each have a pin with a V-groove for seating and positioning the knife edge of the mounting tabs.

TEST SPECIMEN PREPARATION

The most commonly used method for impregnating the yarn prior to test is by hand. This method is used on incoming yarn or in situations where the quality of the undisturbed yarn is to be determined. A second method is to impregnate the yarn during an actual winding operation to assess the effects of processing on yarn strength.

Hand Impregnation

With minimum handling, the material is draped loosely across a wooden or metallic frame and the ends secured with masking tape, taking care not to add or subtract twists from the yarn. The rack is placed in an air circulating oven at 150 F and oriented with the fibers vertical. The catalyzed resin system, preheated to lower its viscosity, is liberally applied to the upper end of the strand and allowed to run down the length of the fibers. Approximately 0.5 g of the mixed resin per foot of strand is adequate. Slight tension is then applied to the impregnated specimens to straighten them. The specimens are cured according to the resin manufacturer's recommended cure cycle, then cut from the rack and stored in a stoppered glass tube until ready for mounting.

Machine Impregnation

The yarn coming from the resin impregnating pot of the filament winding system is wound onto a flat frame approximately 15 in. wide by 15 in. long by 1½ in. thick at a spacing of about one thread per inch. The edges of the frame are beveled to prevent damage to the yarn. The impregnated yarn then is cured on the frame at the designated cure cycle for the resin system. Specimens containing beads or deformation are discarded.

SPECIMEN MOUNTING

The cured impregnated strand is mounted on the assembled mounting fixture shown in Fig. 2 using an 8-in. gage length. The specimen is bonded to the

---

[3] The italic numbers in brackets refer to the list of references appended to this paper.

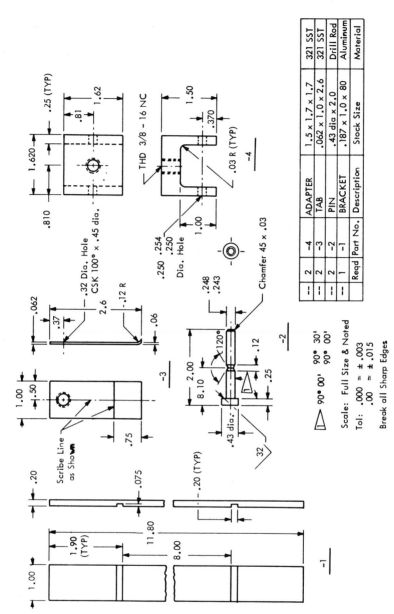

Reqd	Part No.	Description	Stock Size	Material
--	-4	ADAPTER	1.5 × 1.7 × 1.7	321 SST
--	-3	TAB	.062 × 1.0 × 2.6	321 SST
--	-2	PIN	.43 dia × 2.0	Drill Rod
1	-1	BRACKET	.187 × 1.0 × 80	Aluminum

Scale: Full Size & Noted

Tol: .000 = ±.003
.00 = ±.015

Break all Sharp Edges

FIG. 1—Tension test fixture for yarn.

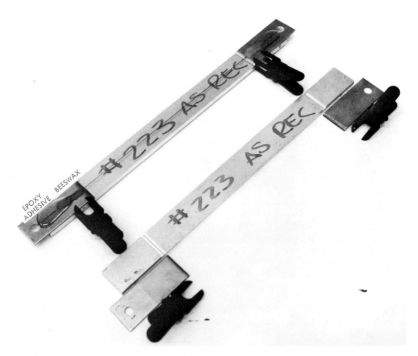

Fig. 2—Yarn tension test specimen holder.

tabs between the hole and the cross scribe with a room-temperature curing liquid epoxy adhesive, such as Epoxy 220 by Hughes Associates, taking care that the specimen is aligned with the axial scribe line. After the adhesive is cured, molten beeswax is applied over the remaining length of the strand which is in contact with the tabs to reduce the stress concentrations at the resin-fiber interface.

TESTING

The test specimen assembly, with the backing plate still in place, is mounted and aligned in the test fixture. V-grooves in the loading pins and the knife edge on the tabs provide uniform loading for the specimen. To prevent inadvertent stressing and damaging of the specimen, the backing plate is kept in place until just prior to loading. The specimen is loaded to failure using a crosshead speed of 0.05 in./min. Specimens that break at the edges of the bonding adhesive are discarded and retests made. The ultimate load in pounds is recorded, and, from this value and the calculated cross-sectional area, the tensile strenth in psi is determined.

RESULTS

Tensile strength data on 240 individual fllament tests[4] were compared with the yarn strength data obtained from the same material rolls to determine the statistical variation of the yarn strength and to establish a correlation between individual filament and yarn bundle strengths.

A frequency plot made from the filament data showed a normal distribution [2] with average filament strength of 280,000 psi.

A theoretical bundle (yarn) strength was calculated from the filament strength

---

[4] Thornel 50 filament data supplied by Union Carbide Corp.

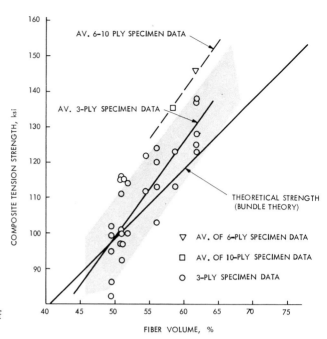

FIG. 3—Tensile strength of unidirectional composites.

distribution using the Ekvall method [3] which was based on the work of Daniels [4] and Pierce [5].

$$\frac{S_a}{\Delta_s} = \frac{\sum\limits_{i=a+1}^{n} N_i}{N_a}$$

where:

N_a = number of filaments at strength level a,
S_a = average strength of the group of filaments at strength level a,
Δ_s = incremental difference between strength levels, and

$\sum\limits_{i=a+1}^{n} N_i$ = number of filaments grouped in strength levels above a.

This theory, which assumes that the individual filaments in the yarn are either encased in a matrix or evenly tensioned to provide a uniform stress distribution between filaments, predicted the average theoretical bundle strength to be 225,000 psi based on an average single filament strength of 280,000 psi. The filament bundle strength (225 ksi) can be converted to composite tensile strength using the equation:

$$F_{TU} = S_B V_F R_B$$

where:

F_{TU} = expected ultimate composite tensile strength,
S_B = filament bundle strength (225 ksi),
V_F = fiber volume fraction, and
R_B = fraction of the fibers intact at S_B (obtained from filament strength distri-
       bution curve).

With a fiber volume of 50 percent, the ultimate tensile strength of the composite should be approximately 98,500 psi, as was borne out by the test results shown in Fig. 3.

When this method is applied to the strength distribution of the impregnated yarn tested as a single strand (not as a composite), the predicted yarn bundle breaking load is 17 lb, which translates to a composite filament stress of 231,000 psi. This compares well to 225,000 psi from the individual filament data [6].

COMPOSITE TENSION TEST METHOD

Flat specimens for determining the tensile properties of carbon composites are shown in Fig. 4. The Type 1 specimen shown is used to determine the longitudinal (0 deg) tensile strength of unidirectional composites, while the longitudinal and transverse properties of multidirectional composites and the transverse properties of unidirectional composites are determined using the Type 2 specimen. Both types of specimens have end reinforcing tabs, for gripping purposes and effective load transfer, that are made from three plies of 181 Volan A glass fabric impregnated with Epon 828/Versamid 125 (50 parts per hundred resin). The same resin is used to bond the tabs to the composite.

The specimens are tested in self-aligning, self-tightening wedge grips with serrated faces which completely enclose the end tabs. The load is applied at a crosshead speed of 0.05 ± 0.005 in./min, and strain readings are taken with an ASTM Class B-2 extensometer or equivalent accurate to 0.02×10^{-3} in./in. strain. A transverse SR-4 type strain gage or equivalent is employed if determination of Poisson's ratio is desired.

Figures 3 and 5 compare the actual results and predicted properties of unidirectional composites. The data shows good correlation of strength and modulus at various fiber volumes. The effect of fiber volume is linear for the Epon 828 as expected. In Fig. 5 the curve for $K = 1.0$ is the expected modulus of the composite where K is the misalignment factor for the Tsai micromechanics stiffness equation [7]. The equation for longitudinal stiffness is:

$$E_L = K[E_m + (E_f - E_m)V_f]$$

where:
- K = misalignment factor,
- E_L = composite tension modulus,
- E_f = filament modulus,
- E_m = matrix modulus, and
- V_f = fiber volume fraction.

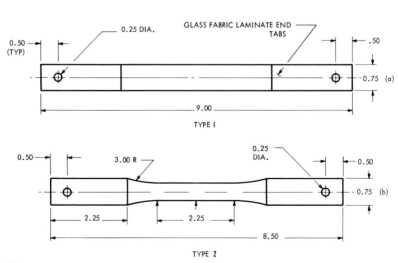

FIG. 4—Tension test specimen.

For straight filaments $K = 1.0$. However, the carbon yarns are twisted resulting in an effective K of 0.92.

A three-ply laminate was principally used with the Type 1 specimen to conserve material and to provide data realistic of thin gage unidirectional laminates. In the majority of the structural applications involving composites, the unidirectional (0 deg) plies are distributed between the off-angle plies resulting in a laminate composed of thin discrete layers. The tensile properties of these thin unidirectional layers are determined using the thin gage specimen.

The effect of laminate thickness on tensile strength is shown in Fig. 3. Specimens in the five to ten ply thickness range fail at statistically equivalent stress levels, while the thinner three-ply specimens fail according to the bundle theory and at a slightly lower stress level. In the latter specimens broken filaments in the yarn bundles cannot effectively distribute their load to adjacent filaments, following the theory of Scop and Argon [8,9]. The thinner specimens also gave a sightly lower modulus translation (Fig. 5).

The thickness limitation on the Type 1 specimen is about 0.080 in. With thicker specimens, the tabs shear off at the tab specimen interface due to the low shear strength of the carbon composite. Limited tests with tapered end tabs bonded to a scarfed specimen show that specimens 0.100 in. thick can be satisfactorily tested. The upper thickness limit has not been explored.

The Type 2 specimen, for multidirectional composites, has been tested satisfactorily up to thicknesses of 0.110 in. However, it should be noted that the ultimate loads were about the same as those of the 0.080-in.-thick unidirectional specimens. The tapered end tab concept is also applicable to this specimen, but again the maximum thickness limitation has not been determined.

Figure 6 compares tensile data on a multidirectional composite obtained using a tubular specimen and the Type 2 specimen shown in Fig. 4. The data from

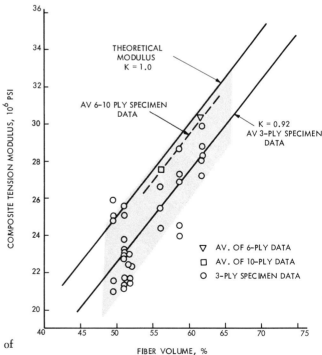

Fig. 5—Tension modulus of unidirectional composites.

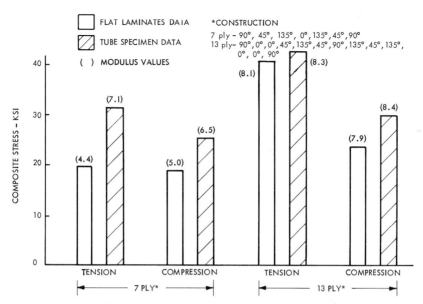

FIG. 6—Tension and compression properties—multidirectional composites.

the Type 2 specimens are significantly lower, indicating that representative tensile results on multidirectional composites cannot be obtained using the cheaper flat specimens. The tubular tension specimen used to obtain these results was 2.25 in. in diameter and 8 in. long with a 4-in. test section. The ends are reinforced with circumferentially wound glass yarn to provide a gripping surface. The end reinforcement uses the same matrix system as the tube and is wound and cured integrally with the tube.

COMPRESSION TEST METHOD

It was necesssary to select two different specimen configurations, Fig. 7, to determine composite compression properties, since any one design could not satisfy the accuracy desired for both strength and modulus. The straight sided specimen, Fig. 7b, provides an accurate modulus measurement, but end crushing—even with restraining clamps—prevents ultimate strength measurements. Therefore, a necked-down specimen, with a shortened gage length for stability, Fig. 7a, is used for ultimate strength tests. This specimen is not suitable for modulus measurement because a 1-in. gage length extensometer is not accurate enough on the carbon material.

Both specimen configurations are tested in the fixture shown in Fig. 7. Side supports prevent buckling, while adjustable clamps on each end of the specimen prevent crushing or brooming. A recommended end clamp, which accommodates a variety of specimen sizes, is shown in Fig. 7. The edges must be beveled to prevent damage to the specimen.

To determine the ultimate compression strength, the necked-down specimen (Fig. 7a) is loaded to failure at crosshead speed of 0.05 in./min, and the maximum load is recorded.

For modulus determinations an ASTM Class B-2 extensometer is mounted on the straight-sided specimen which is then loaded to approximately 50 percent of ultimate at a crosshead speed of 0.05 in./min. If desired, the specimen for

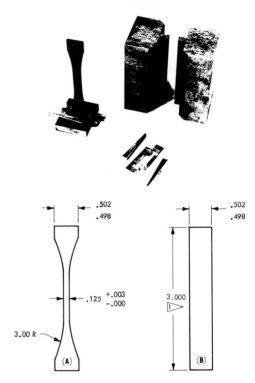

.502
.498

.502
.498

.125 $^{+.003}_{-.000}$

3.000

3.00 R

(A)

(B)

Fig. 7—Compression test specimen and fixture.

ENDS ARE TO BE FLAT AND PARALLEL WITHIN ±.00025 AND PERPENDICULAR TO THE LONGITUDINAL AXIS OF THE SPECIMEN WITHIN .001".

the ultimate compressive strength test can be machined from straight-sided specimen after determination of the elastic modulus.

RESULTS

Figures 8 and 9, respectively, show the compressive strength and modulus of unidirectional carbon/epoxy composites at varying fiber volumes. Both the strength and modulus of the composite increase with increasing fiber fraction, as expected. However, the values are considerably lower than the theoretical compressive values probably due to the twisted carbon yarns. All failures with the necked-down specimen failed in the center gage section and proved quite satisfactory for undirectional laminates. For comparison, the compressive properties of unidirectional and multidirectional carbon/epoxy composite material was determined using sandwich beam and tubular type specimens, respectively. The sandwich beam gave slightly higher properites, while the tubular specimens gave significantly higher strength and modulus than the corresponding flat specimens, as shown in Figs. 6, 8, and 9. The greatest difference was in the multidirectional tubular specimen.

SHORT-BEAM INTERLAMINAR SHEAR TEST

The interlaminar shear strength of carbon/epoxy composites is determined using a modification of the ASTM Test for Apparent Horizontal Shear Strength of Reinforced Plastic by Short-Beam Method (D 2344-67).

The ASTM method was satisfactory except that it was necessary to increase

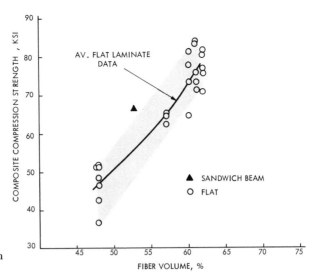

FIG. 8—Compressive strength of unidirectional composites.

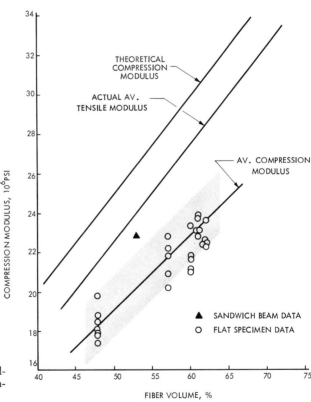

FIG. 9—Compressive modulus of unidirectional composites.

the recommended span-to-depth ratio (SDR) of four and five to eight in order to achieve horizontal shear failures in the carbon composites. The effect of SDR ratio on the apparent interlaminar shear strength of the composite is illustrated in Fig. 10. The lower the SDR, the higher the apparent interlaminar shear strength that can be obtained. Therefore, it is important that an SDR be selected that

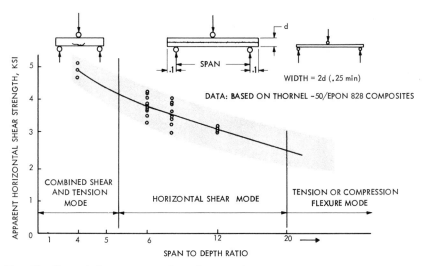

FIG. 10—General characteristics of interlaminar shear test.

provides true interlaminar shear failure, otherwise the results can be misleading. The interlaminar shear strength values obtained on carbon composites[5] using an SDR of eight have been substantiated in subsequent joint and lap bond tests where the primary mode of failure was interlaminar shear in the composite. Unfortunately, having to vary the SDR depending upon the composite material does not lend itself to test method standardization and makes data comparisons very difficult. Also, the short-beam interlaminar shear test is reliable only for unidirectional composites. The flexure stresses in multidirectional laminates generally do not follow the simple beam theory since the normal stresses do not vary linearly across the beam depth [10]. Depending upon the laminate construction, an SDR may not be found which will force a horizontal shear failure. In this case the resulting test value is not a true representation of the composite interlaminar shear strength.

The test specimen is shown in Fig. 10. The test span of the short-beam shear specimen is calculated from the measured specimen thickness using a chosen SDR. The overall specimen length is determined by adding 0.2 in. to the test span. The width of the specimen is 0.25 in. or twice the thickness of the laminate, whichever is greater.

A single point loading device with ¼-in.-diameter loading points is used for testing the short-beam shear specimens. A crosshead speed of 0.05 ± 0.005 in./min is used.

The interlaminar shear strength is calculated in pounds per square inch as follows:

$$\tau = \frac{3}{4}\frac{P}{A}$$

where:
τ = interlaminar shear strength in psi,
P = load in lb,
A = cross-sectional area of specimen in in.², $(b \times d)$,
b = width in in., and
d = thickness in in.

[5] Epon 828/Thornel 50 composite.

Composite Materials

BEARING TEST

The bearing specimen configuration shown in Fig. 11 is per Ref *11*, Method 6-73786. One hole is slightly larger than the other to concentrate the failure at one end; however, a constant distance to hole diameter ratio of two is maintained. The specimen is loaded at 0.05-in./min head travel.

Bearing tests have been conducted on 7 and 13-ply multidirectional carbon composites to determine the feasibility of using mechanical fasteners on composites. The results of these tests are shown in Fig. 11. The bearing strengths of the two laminates were approximately equal, and, surprisingly, the longitudinal and transverse strength were equal in both bases despite the fact that the laminates have an unbalanced construction. This indicates that the ±45-deg layers have a dominating role in determining the bearing strength of the composite. A 2 percent deformation in the hole diameter generally is used as the criteria for bearing failure. Bearing strengths as high as 38,000 psi were obtained on the seven-ply laminate which has a tensile strength of 20,000 psi (Fig. 6).

RESIN CONTENT DETERMINATION

Current popular practices in industry for conducting resin content of carbon composites are: (1) chemical digestion, (2) thermogravimetric analysis, and (3)

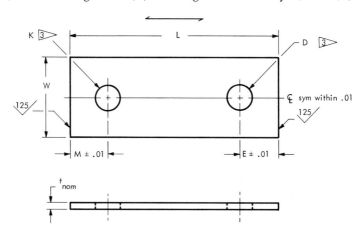

No. of PLIES [4]	t_{nom} (in.)	E (in.)	D [2] (NOM) (in.)	W (in.)	L (in.)	"K" [1] (in)	"M" (in.)	Av. Bearing Yield Strength Ksi	
								Long. Dir.	Trans. Dir.
7	.060	.500	.250	1.25	4.00	.290	.580	37.3	37.8
13	.110	1.000	.500	2.50	5.00	.567	1.000	32.4	35.5

[1] HOLE TOLERANCE: (+. 010, -.000)

[2] HOLE TOLERANCE: (+ .004, -.000)

[3] HOLES MUST HAVE SMOOTH FINISH EQUIVALENT TO REAMING.
REMOVE BURRS BUT DO NOT CHAMFER. HOLES MUST HAVE SHARP EDGES.

[4] CONFIGURATION:
7-ply 0°, ±45°, 90°, ±45°, 0°

13-ply 9°, 0, 0, 45, 135, 45, 90, 135, 45, 135, 0, 0, 90°

Fig. 11—Bearing specimen and test results.

air burnout. After evaluating each method, the latter technique was chosen because of simplicity, low cost, and, most of all, accuracy.

The air burnout method used consists of placing a composite specimen (10 g minimum weight) for 8 to 10 h in a muffle furnace set at 780 ± 10 F and then measuring the weight loss. Care must be taken to maintain the temperatures within the stated limits since the carbon oxidizes rapidly above 800 F. Below the lower limit incomplete burnout is achieved. It is not advisable to extend the burnout time at 780 F beyond 10 h since limited oxidation of the carbon can occur. Tests have shown up to 2 percent weight loss in the carbon yarn after 24 h exposure at 780 F, compared to <0.5 percent at 8 h. The advantage of this method is that the equipment requirements are simple, and reasonably good accuracy (± 0.5 percent) is obtained. The long time required before the results are available (8 h) can be a disadvantage. Also, complete resin burnout cannot be obtained below 800 F with some of the more heat resistant matrix systems. In such situations another burnout technique is utilized. This method, which requires a Burrell furnace (Model T-2-9) or equivalent equipment having 1000 C temperature capabilities and provisions for an argon purge, is as follows:

A 1.5-g composite specimen is placed in a combustion boat which has been preconditioned at 1000 C for 30 min and stored in a desiccator. The boat containing the specimen is weighed and placed in the combustion chamber of the Burrell furnace in an argon atmosphere for 30 min ± 1 min at 1000 C $\pm$ 20 C. After this period, the boat and specimen are immediately removed, allowed to cool in a desiccator, and then reweighed to obtain the weight loss of the specimen. All weighings are made to the nearest 0.1 mg.

The procedure is then repeated using a 1.5-g specimen of pure resin casting, comparable to the resin system contained in the composite. (This step need only be performed once for each resin system formulation.)

The resin content by weight is determined as follows:

$$\text{percent resin content} = \frac{W_c}{W_r} \times 100$$

where:

W_c = percent weight loss of composite specimen $\left(\dfrac{\text{weight before} - \text{weight after}}{\text{weight before}} \times 100 \right)$

W_R = percent weight loss of resin specimen $\left(\dfrac{\text{weight before} - \text{weight after}}{\text{weight before}} \times 100 \right)$

A minimum of three determinations are recommended. Tests to date have shown the method to be accurate to within 1 percent.

SPECIFIC GRAVITY/VOID CONTENT DETERMINATION

Specific gravity of the carbon composites is determined using the ASTM Test for Specific Gravity and Density of Plastics (D 792-66). The specific gravity of the yarn is determined by the same method execept for the following modifications:

A loosely wound skein of unimpregnated heat-cleaned yarn is preconditioned for 2 h in a desiccator (using Anhydrone) before weighing in air and soaked for 4 h in distilled water at 100 F prior to weighing in water. The specific gravity is calculated per the ASTM method using the measured weight.

Void content is calculated from the differences in the actual and theoretical specific gravity of the composite using the measured specific gravity of the resin and fiber. Microscopic examination of composite cross sections has proven good for obtaining an estimation of the void content and general laminate quality.

CONCLUSIONS

1. The yarn strand tension test produces strengths representative of the filaments in the yarn, and the data can be used to estimate the tensile strength of a composite laminate. Data from individual filament tests, the yarn strand test, and the thin composite tension test can be correlated by the bundle theory of failure.

2. The test methods reported with the exception of ILS, provide good data for preliminary design and should be satisfactory for general quality control.

ACKNOWLEDGMENT

Portions of the data reported herein were obtained under Air Force Contract AF33615-67-C-1641, "Development of Carbon Composite Structural Elements," sponsored by the Advanced Composites Branch of the Air Force Materials Laboratory, Wright-Patterson Air Force Base, Ohio, with F. J. Fechek as project manager.

The author gratefully acknowledges the efforts of Matthew House, D. A. Anderson, and J. E. Bell, in the development of the methods and data reported.

References

[1] Pirzadeh, N. and Kennedy, P. B., "Single-End Glass Yarn Tension Test," *Standards for Filament-Wound Reinforced Plastics, ASTM STP 327,* American Society for Testing and Materials, 1963, pp. 165–177.

[2] Hoggatt, J. T. and Bell, J. E., "Development of Carbon Composite Structural Elements for Missile Interstage Application," Technical Management Report No. 5, D2-125559-6, The Boeing Company, Seattle, Wash., Sept. 1968.

[3] Ekvall, J. C., "Structural Behavior of Monofilament Composites," AIAA Conference, Palm Springs, Calif., 5–7 April, 1965, pp. 250–263.

[4] Daniels, H. E., "The Statistical Theory of the Strength of Bundles of Threads," *Proceedings,* Royal Society, London, 1945, pp. 183, 405.

[5] Pierce, F. T., "Tensile Tests for Cotton Yarns and 'the Weakest Link'—Theorems on the Strength of Long and of Composite Specimens," *Journal Textile Institute,* 1926, pp. 17, 355.

[6] Hoggatt, J. T. et al, "Development of Processing Techniques for Carbon Composites in Missile Interstage Application," Technical Report AFML-TR-68-155, The Boeing Company, June 1968.

[7] Tsai, S. W., "Structural Behavior of Composite Materials," NASA CR-71, National Aeronautics and Space Administration, July 1964.

[8] Scop, P. M. and Argon, A. S., "Statistical Theory of Strength of Laminated Composites," *Journal of Composite Materials,* Vol. I, No. 1, Jan. 1967, p. 92.

[9] Scop, P. M. and Argon, A. S., "Statistical Theory of the Tensile Strength of Laminates," *10th National Symposium and Exhibit,* Society of Aerospace Materials and Process Engineers, San Diego, Calif., 9–11 Nov. 1966.

[10] Hoggatt, J. T. and Bell, J. E., "Development of Carbon Composite Structural Elements for Missile Interstage Application," Technical Management Report No. 4, D2-125559-5, The Boeing Company, Seattle, Wash., June 1968.

[11] Nygren, R., "Specimen Configurations—Materials Properties Tests," Document D-17056, The Boeing Company, Seattle, Wash., Sept. 1965.

The Effect of Geometry on the Mode of Failure of Composites in Short-Beam Shear Test

S. A. SATTAR[1] AND
D. H. KELLOGG[1]

Reference:

Sattar, S. A. and Kellogg, D. H., "The Effect of Geometry on the Mode of Failure of Composites in the Short-Beam Shear Test," *Composite Materials: Testing and Design, ASTM STP 460*, American Society for Testing and Materials, 1969, pp. 62–71.

Abstract:

The interlaminar shear strength of composite materials is often determined by the short-beam shear test method. This type of test is satisfactory only so long as a pure shear failure occurs. Nonshear or mixed-mode failures, hence invalid data, can result if the interlaminar shear strength is too high with respect to the ultimate flexure strength, or if span-to-depth ratio or width-to-depth ratio is improperly chosen.

Since the flexural stress is dependent on span length, various span-to-depth ratios of interest were investigated using beam theory to determine the interaction of shear and flexure stresses present in the short-beam specimen. These interactions are shown as a function of the direct stress in the short beam.

Examination of the shear-flexure interaction was made by employing a failure criterion based on distortion energy. Failure was found limited to either shear mode at the center of the beam or flexural mode at the outer fiber. The combination of shear and flexure at intermediate locations is not of sufficient magnitude to cause failure.

Since the probability of nonshear failure increases as interlaminar shear strength is improved without accompanying increases in flexural strength a means of choosing proper length-to-depth ratios knowing the approximate flexural-to-shear strength ratio is presented. The possibility of a flexural failure is minimized by incorporating a flexural stress to flexural strength ratio which will not be exceeded at the chosen length-to-depth ratio.

The effect of the width-to-depth ratio on the shear stresses in the specimen is developed by a theory of elasticity solution. The formula currently being used to calculate the shear stresses is based on beam theory and assumes that the shear stresses are constant along the width of the specimen. The exact solution for an orthotropic beam shows a significant effect of the width-to-depth ratio on the magnitude and distribution of shear stresses in the short beam. The interlaminar shear stress is highest at the edges of the specimen and lowest at the middle.

Calibration curves for graphite-polymer, boron-polymer, and boron-aluminum have been developed to correct the shear stresses as given by the conventional beam theory formula for various values of width-to-depth ratios. Experimental data has been obtained on a graphite-polymer system for fiber orientations of both zero and ±45 deg. Conventional beam theory calculations show the interlaminar shear strength of the ±45-deg cross ply to be reduced significantly over that of the unidirectional specimens. Application of the theory of elasticity correction factors to these data indicate that the actual interlaminar shear strength for both orientations are approximately equal.

Key Words:

composite materials, shear tests, laminates, shear strength, mechanical properties, evaluation, tests

---

[1] Assistant project engineer and senior analytical engineer, Pratt and Whitney Aircraft, Division of United Aircraft Corp., East Hartford, Conn. 06108.

NOMENCLATURE

A Beam cross-sectional area
E Modulus of elasticity in tension and compression
G Modulus of elasticity in shear
I Moment of inertia of beam cross section
L Short-beam shear span length
M Bending moment
P Applied load
S Material strength
$[S]$ Elastic tensor
a Beam half depth
b Beam half width
x,y,z Rectangular coordinates
ϵ Tensile strain
γ Shear strain
ν Poisson's ratio
σ Normal stress
τ Shear stress

Composite materials offer tremendous advantage for aerospace application in that they have high tensile strength and modulus coupled with an extremely low density. The weak link in the chain has been and continues to be a weak matrix-reinforcement bond which results in a low interlaminar shear strength (ILSS). This makes interlaminar shear strength one of the most important material properties to be considered when designing components with advanced composites.

The most widely accepted indicator of ILSS in use today is the short-beam shear test because it is a relatively easy and economical test (and has traditionally given accurate results). A great deal of research to improve the ILSS has led to recent significant improvements in composite shear strength. This fact along with the inherent anisotropy of the system leads to two distinct problems when dealing with the short-beam shear strength.

Significant improvements in the ILSS without corresponding improvements in the flexural strength can lead to a flexure failure in a short-beam shear test unless the span/depth ratio is chosen properly. It is important to define proper span/depth ratios such that true ILSS values may be obtained.

A second problem concerns the high degree of anisotropy of the composites. This anisotropy can lead to error when calculating ILSS using simple beam theory if the specimen is too wide. An analytical study has been carried out using the mathematical theory of elasticity to determine the effect of width-to-depth ratio on the stresses in short-beam shear specimens.

DISCUSSION

As a part of an overall effort to develop improved test techniques, an analytical study was initiated to better understand the stresses and failure modes in the short-beam shear tests. The first phase of the study involved the investigation of the effect of the length-to-depth ratio on the failure mode in the short-beam shear tests. The second phase of this overall effort dealt with the analytical investigation of the effect of width-to-depth ratio on the stress distribution in short-beam shear tests.

Span/Depth Ratio

Consider the short-beam shear specimen shown in Fig. 1a. According to elementary theory for narrow beams, the shear stress distribution along the depth (x-direc-

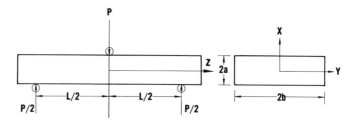

FIG. 1a—Short-beam shear specimen.

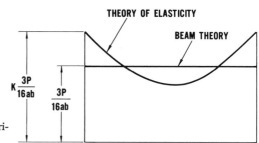

FIG. 1b—Shear stess distribution at $X = 0$.

tion) is parabolic in nature, reaching a maximum value of $3P/4A$ at the center of the specimen $(x = 0)$. Here P is defined as the total load at failure and A as the cross-sectional area. This shear stress is independent of span length. A bending stress distribution of Mx/I also exists in the specimen reaching a maximum of $3/2(P/A)$ times the span/depth ratio at center span under the applied load. For increased span/depth ratios, the tensile and compressive stresses due to bending eventually will become critical, and failure will occur in a flexural mode.

To examine the interaction of flexural and shear stresses in the short-beam specimen, the distortion energy theory for anisotropic materials in plane stress[2] was employed. This relationship can be expressed as

$$\left(\frac{\sigma_z}{S_z}\right)^2 - \left(\frac{S_x}{S_z}\right)\left(\frac{\sigma_z}{S_z}\right)\left(\frac{\sigma_x}{S_x}\right) + \left(\frac{\sigma_x}{S_x}\right)^2 + \left(\frac{\tau_{xz}}{S_{xz}}\right)^2 = C$$

and failure is attained when $C \geq 1$.

For the short-beam shear specimen, the above summation reduces to:

$$\left(\frac{\sigma_z}{S_z}\right)^2 + \left(\frac{\tau_{xz}}{S_{xz}}\right)^2 = C$$

Figures 2 and 3 show a plot of this summation versus depth $(0 \leq x \leq a)$ at the center span location for flexural-to-shear-strength ratios of 10 and 20 and for various values of span/depth ratio. For convenience, the shear strength was assumed at $3/4(P/A)$, thus demanding a failure for each span/depth ratio chosen. It is noted that it is possible for a combination of shear and flexural failure to occur simultaneously at their respective locations for example, (in Fig. 2, summation equals 1 both at $x = 0$ and $x = a$ for span/depth ratio = 5).

Figure 4 is a plot of the minimum value of flexural-to-shear strength ratio $(S_z/\bar{S}_{xz})$ required to ensure shear failure at a given span/depth ratio. This graph illustrates the valid-invalid span/depth ratios for short-beam shear testing if the flexural-to-shear strength ratios are known approximately. The graph also contains a margin restricting the outer fiber flexural stress to 70 percent of the flexural

[2] Tsai, S. W., "Strength Characteristics of Composite Materials," NASA CR-224, National Aeronautics and Space Administration, April 1965.

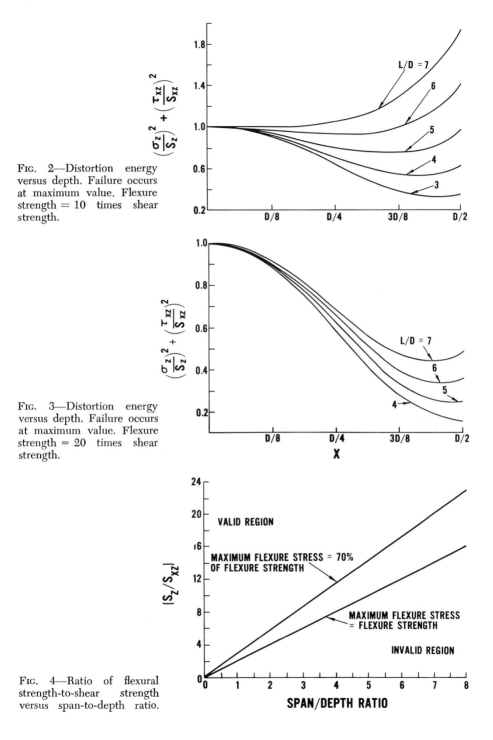

FIG. 2—Distortion energy versus depth. Failure occurs at maximum value. Flexure strength = 10 times shear strength.

FIG. 3—Distortion energy versus depth. Failure occurs at maximum value. Flexure strength = 20 times shear strength.

FIG. 4—Ratio of flexural strength-to-shear strength versus span-to-depth ratio.

strength, thereby giving added assurance that pure shear failure will occur. Using this criterion, a span/depth ratio of 5 would not be used unless the flexural-to-shear strength ratio was at least 14.5. An expected flexural-to-shear strength ratio of 10 would suggest the use of a span/depth ratio of 3.5 instead of 5.

This curve can be useful in determining future span/depth ratios as composite ILSS values increase without corresponding increases in flexural strengths or if inter-laminar shear values are desired for layups where flexural strengths are reduced considerably. In either case, a reduced span/depth ratio can be chosen to yield a valid ILSS test.

Width/Depth Ratio

In elementary beam theory, the only nonzero stresses are σ_z and τ_{xz}. This is a very good approximation so long as the width-to-depth ratio of the beam is small. As soon as the width-to-depth ratio of the beam exceeds two, the elementary beam theory for calculating shear stresses becomes significantly inaccurate for many composites.

The following assumptions are made for the analytical development that follows. It will be shown that with these assumptions we can arrive at a solution that satisfies all of the equations of the theory of elasticity.

Assumptions

1. The composite material can be represented by a homogeneous orthotropic material.
2. The only nonzero stresses are σ_z, τ_{xz}, and τ_{yz}.
3. The σ_z stress distribution is linear through the thickness, that is,

$$\sigma_z = P \frac{(z - L)}{I} x$$

A general orthotropic material can be characterized completely by the following constitutive relationship:

$$(\epsilon) = [S](\sigma)$$

where $[S]$ is given by

$$[S] \begin{bmatrix}
\dfrac{1}{E_x} & -\left(\dfrac{\nu_{xy}}{E_y}\right) & -\left(\dfrac{\nu_{xz}}{E_z}\right) & 0 & 0 & 0 \\[2ex]
-\left(\dfrac{\nu_{yx}}{E_x}\right) & \dfrac{1}{E_y} & -\left(\dfrac{\nu_{yz}}{E_z}\right) & 0 & 0 & 0 \\[2ex]
-\left(\dfrac{\nu_{zx}}{E_x}\right) & -\left(\dfrac{\nu_{zy}}{E_y}\right) & \dfrac{1}{E_z} & 0 & 0 & 0 \\[2ex]
0 & 0 & 0 & \dfrac{1}{G_{xy}} & 0 & 0 \\[2ex]
0 & 0 & 0 & 0 & \dfrac{1}{G_{xz}} & 0 \\[2ex]
0 & 0 & 0 & 0 & 0 & \dfrac{1}{G_{yz}}
\end{bmatrix}$$

In the above relationship the only independent elastic constants are E_x, E_y, E_z, ν_{xy}, ν_{xz}, ν_{yz}, G_{xy}, G_{xz}, and G_{yz}. When $E_x = E_y$, $\nu_{xz} = \nu_{yz}$, and $G_{xz} = G_{yz}$ we obtain the transversely isotropic case (unidirectional composite). It will be seen that the only elastic constants that will affect the final results of this analysis are E_z, ν_{yz}, G_{xz}, and G_{yz}.

In terms of the three nonzero stresses

$$
\left.
\begin{aligned}
\epsilon_x &= -\nu_{xz}\frac{\sigma_z}{E_z}; & \gamma_{xy} &= 0 \\[6pt]
\epsilon_y &= -\nu_{yz}\frac{\sigma_z}{E_z}; & \gamma_{xz} &= \frac{\tau_{xz}}{G_{xz}} \\[6pt]
\epsilon_z &= \frac{\sigma_z}{E_z}; & \gamma_{yz} &= \frac{\tau_{yz}}{G_{yz}}
\end{aligned}
\right\}
\quad\quad\quad\quad (3)
$$

Four of the six compatibility equations are satisfied identically by Eq 3, leaving the following two equations:

$$
2\frac{\partial^2 \epsilon_y}{\partial x \partial z} = \frac{\partial}{\partial y}\left(\frac{\partial \gamma_{yz}}{\partial x} - \frac{\partial \gamma_{xz}}{\partial y}\right) \quad\quad\quad\quad (4a)
$$

$$
2\frac{\partial^2 \epsilon_x}{\partial y \partial z} = \frac{\partial}{\partial x}\left(-\frac{\partial \gamma_{yz}}{\partial x} + \frac{\partial \gamma_{xz}}{\partial y}\right) \quad\quad\quad\quad (4b)
$$

Putting Eq 3 into Eq 4a gives

$$
-2\frac{\nu_{yz}}{E_z}\frac{\partial^2 \sigma_z}{\partial x \partial z} = \frac{\partial}{\partial y}\left\{\frac{\partial}{\partial x}\left(\frac{\tau_{yz}}{G_{yz}}\right) - \frac{\partial}{\partial y}\left(\frac{\tau_{xz}}{G_{xz}}\right)\right\} \quad\quad\quad\quad (5)
$$

Since the beam is assumed to be homogeneous macroscopically, G_{yz} and G_{xz} can be factored from the derivative as constants.

$$
-2\frac{\nu_{yz}}{E_z}\frac{P}{I} = \frac{1}{G_{yz}}\frac{\partial^2}{\partial x \partial y}\tau_{yz} - \frac{1}{G_{xz}}\frac{\partial^2}{\partial y^2}\tau_{xz} \quad\quad\quad\quad (6)
$$

Similarly putting Eq 3 into Eq 4b gives

$$
-\frac{1}{G_{yz}}\frac{\partial^2}{\partial x^2}\tau_{yz} + \frac{1}{G_{xz}}\frac{\partial^2}{\partial x \partial y}\tau_{xz} = 0 \qu\quad\quad\quad\quad (7)
$$

The equations of equilibrium are satisfied by taking a stress function $\phi(x,y)$ such that

$$
\left.
\begin{aligned}
\tau_{xz} &= \frac{\partial \phi}{\partial y} - \frac{Px^2}{2I} + f(y) \\[6pt]
\tau_{yz} &= -\frac{\partial \phi}{\partial x}
\end{aligned}
\right\}
\quad\quad\quad\quad (8)
$$

where ϕ is a function of x and y and f is a function of y alone which will be determined from boundary conditions.

Putting Eq 8 in Eq 6

$$
-2\frac{\nu_{yz}}{E_z}\frac{P}{I} = \frac{\partial}{\partial y}\left\{-\frac{1}{G_{yz}}\frac{\partial^2 \phi}{\partial x^2} - \frac{1}{G_{xz}}\left(\frac{\partial^2 \phi}{\partial y^2} + \frac{df}{dy}\right)\right\}\frac{\partial}{\partial y}\left(\frac{1}{G_{yz}}\frac{\partial^2 \phi}{\partial x^2} + \frac{1}{G_{xz}}\frac{\partial^2 \phi}{\partial y^2}\right)
$$

$$
= 2\frac{\nu_{yz}}{E_z}\frac{P}{I} - \frac{1}{G_{xz}}\frac{d^2 f}{dy^2} \qu\quad\quad\quad\quad (9)
$$

Putting Eq 8 in Eq 7

$$
\frac{\partial}{\partial x}\left(\frac{1}{G_{yz}}\frac{\partial^2 \phi}{\partial x^2} + \frac{1}{G_{xz}}\frac{\partial^2 \phi}{\partial y^2}\right) = 0 \qu\quad\quad\quad\quad (10)
$$

Sattar and Kellogg on Short-Beam Shear Test

From Eq 9 and Eq 10 we get by integrating

$$\frac{1}{G_{yz}}\frac{\partial^2 \phi}{\partial x^2} + \frac{1}{G_{xz}}\frac{\partial^2 \phi}{\partial y^2} = 2\frac{\nu_{yz}}{E_z}\frac{Py}{I} - \frac{1}{G_{xz}}\frac{df}{dy} + C \dots\dots\dots\dots (11)$$

It can be shown that if the x axis is an axis of symmetry $C = 0$. Now since the boundary of the section is free of any stresses

$$\tau_{xz}\frac{ds}{dy} - \tau_{yz}\frac{dx}{ds} = 0$$

where ds is an element of length of the boundary curve containing the section

$$\frac{\partial \phi}{\partial s} = \frac{\partial \phi}{\partial x}\frac{ds}{dx} + \frac{\partial \phi}{\partial y}\frac{dy}{ds} = \left[\frac{Px^2}{2I} - f(y)\right]\frac{dy}{ds}\dots\dots\dots\dots\dots (12)$$

The equation for the boundary of a rectangle is $(x^2 - a^2)(y^2 - b^2) = 0$
If

$$f(y) = \frac{Pa^2}{2I} \qquad \text{then,}\quad \frac{\partial \phi}{\partial s} = 0 \qquad \text{at } x = \pm a, y = \pm b$$

The differential equation, Eq 11, now becomes:

$$\frac{1}{G_{yz}}\frac{\partial^2 \phi}{\partial x^2} + \frac{1}{G_{xz}}\frac{\partial^2 \phi}{\partial y^2} = 2\frac{\nu_{yz}}{E_z}\frac{Py}{I}\dots\dots\dots\dots\dots\dots (13)$$

Whereas for the isotropic case $G_{yz} = G_{xz} = G$ and $G = E/2(1 + \nu)$ and the differential equation becomes:

$$\frac{\partial^2 \phi}{\partial x^2} + \frac{\partial^2 \phi}{\partial y^2} = \frac{\nu}{1 + \nu}\frac{Py}{I}\dots\dots\dots\dots\dots\dots (14)$$

Examination of Eqs 13 and 14 shows that by a proper coordinate transformation the solution for an isotropic beam can be used for calculating stresses in the orthotropic beam if x in the solution of Eq 14, as given in footnote[3], is replaced by

$$x^* = x\sqrt{G_{yz}/G_{xz}} \qquad \text{and } \phi \text{ by} \qquad \phi^* = \phi 2\frac{G_{xz}}{E_z}\nu_{yz}\left(\frac{1 + \nu}{\nu}\right)$$

Then if $\phi(x,y)$ is a solution of Eq 14, then $\phi^*(x^*,y)$ is the corresponding solution of Eq 13 and is given by:

$$\phi^* = -2\frac{G_{xz}\nu_{yz}}{E_z}\frac{P}{I}\frac{8b^3}{\pi^4}\sum_{m=0}^{\infty}\sum_{n=1}^{\infty}\frac{(-1)^{m+n-1}\cos\dfrac{(2m + 1)}{2a}\pi x \sin n\pi y/b}{(2m - 1)\left[\dfrac{(2m + 1)b^2}{4(G_{yz}/G_{xz})a^2} + n^2\right]}\dots (15)$$

Having this stress function, the actual shearing stress τ_{xz} can be calculated using Eq 8, and a correction factor K can be defined as the ratio of τ_{xz} as obtained from the above stress function to τ_{xz} as obtained from the elementary beam theory.

The theory of elasticity solution for an orthotropic beam shows that there is a significant effect of the width-to-depth ratio (b/a) on the maximum shear stress in the short-beam shear tests. The interlaminar shear stress is not uniform across the width of the specimen but is highest at the edges and lowest in the middle as shown in Fig. 1b.

Three advanced composite systems, namely, graphite-polymer, boron-polymer, and Borsic-aluminum were studied to determine the effect of various fiber orientation

[3] Timoshenko, S. and Goodier, J. N., *Theory of Elasticity*, McGraw-Hill, New York, 1951.

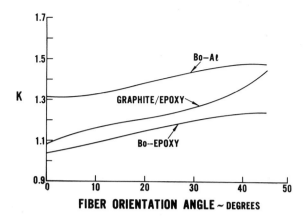

FIG. 5—Correlation factor K versus fiber orientation angle. $b/a = 2.3$.

FIBER ORIENTATION ANGLE ~ DEGREES

angles. The fibers are cross plied in the zy-plane, and the angle is measured from the z-axis (Fig. 1). Where actual measured material properties were not available, theoretical elastic constants were used. Figure 5 shows a plot of the multiplying correction factor K as a function of fiber orientation angle for $b/a = 2.3$ (currently used b/a). It readily is seen that for a given width-to-depth ratio the increase in shear stress at the edge of the specimen is minimum for unidirectional composite (transversely isotropic) and becomes larger as the cross-ply angle increases.

Figures 6, 7, and 8 show the variation of the correction factor K as a function of b/a for three fiber orientation angles. It can be seen that for unidirectional carbon-polymer composite and $b/a = 2.3$ (currently used) the maximum shear stress is 8 percent greater than that calculated by the elementary beam theory. However, for the ±45-deg carbon epoxy composite and the same b/a, the maximum shear stress is 45 percent greater than that calculated by the elementary beam theory.

Table 1 shows the interpretation of some experimental short-beam shear data by the beam theory and the theory of elasticity solution. These tests were made on specimens with the width-to-depth ratio of 2.3. Interpretation by beam theory indicated that ±45-deg composite had significantly lower (23 percent) interlaminar shear strength when compared to the unidirectional composite. However, the theory of elasticity solution shows that there is little difference between the interlaminar shear strength of unidirectional and ±45-deg composites.

Table 2 shows the material constants for advanced composite systems.

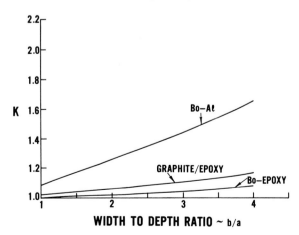

FIG. 6—Correction factor K versus width-to-depth ratio. Fiber orientation angle = 0 deg.

WIDTH TO DEPTH RATIO ~ b/a

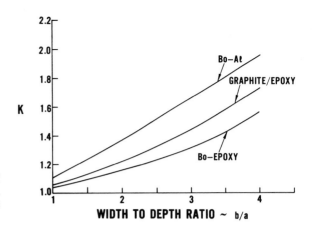

Fig. 7—Correction factor K versus width-to-depth ratio. Fiber orientation angle = 30 deg.

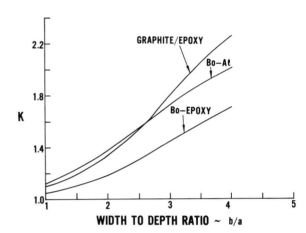

Fig. 8—Correction factor K versus width-to-depth ratio. Fiber orientation angle = 45 deg.

CONCLUSIONS

1. Improper span/depth ratios can lead to flexural rather than shear failures.

2. For the presently used span/depth ratios, the probability of flexural failure will increase as the ILSS of composite materials is improved without corresponding incrases in flexural strength.

3. Proper span/depth ratios can be chosen to minimize the possibility of invalid ILSS values due to nonshear failures.

4. Width-to-depth ratio of the short-beam shear specimen has a significant effect on the shear stresses in the specimen.

5. The shear stress distribution across the width of the beam is nonuniform.

6. The actual maximum shear stress in the specimen is higher than that calculated by the conventional beam theory.

Table 1—Interpretation of experimental short-beam shear data[a]

	Unidirectional Fiber Layup	±45-Deg Fiber Layup
Beam theory.......	3650 psi	2765 psi
Exact theory.......	3950 psi	4020 psi

[a] Represents an average of five specimens each for unidirectional and ±45-deg Thornel 50-epoxy composite (at room temperature).

Table 2—Material constants for advanced composite systems

System	Fiber Orientation, deg	$E_z \times 10^{-6}$	$G_{xz} \times 10^{-6}$	$G_{yz} \times 10^{-6}$	ν_{xz}
Graphite polymer.......	0	19.0	0.8	0.8	0.425
	30	5.6	0.8	4.52	1.68
	45	1.35	0.8	5.81	0.85
Boron polymer..........	0	30.0	0.8	0.8	0.31
	30	7.3	0.8	5.85	2.0
	45	2.0	0.8	7.7	0.88
Borsic-aluminum........	0	35.0	8.1	8.1	0.285
	30	27.0	8.1	10.1	0.366
	45	21.8	8.1	11.1	0.342

Characterization of the Mechanical Properties of a Unidirectional Carbon Fiber Reinforced Epoxy Matrix Composite

E. A. ROTHMAN[1] AND
G. E. MOLTER[2]

Reference:

Rothman, E. A. and Molter, G. E., "Characterization of the Mechanical Properties of a Unidirectional Carbon Fiber Reinforced Epoxy Matrix Composite," *Composite Materials: Testing and Design*, ASTM STP 460, American Society for Testing and Materials, 1969, pp. 72–82.

Abstract:

An exploratory characterization program was conducted on one high-modulus unidirectional composite, carbon fiber reinforced epoxy. Existing test methods were used with minimum modification except in the cases of dynamic moduli and flexural fatigue strength where new methods were developed.

The test methods were generally satisfactory. Two notable exceptions were the short-beam shear test for determining interlaminar shear strength and the flat plate test for static shear modulus.

Key Words:

carbon fibers, epoxy laminates, fiber composites static strength, dynamic moduli, static moduli, fatigue strength, mechanical properties, evaluation, tests

With early attempts to apply high-modulus, filament reinforced composites to hardware, it was necessary to characterize the available systems for design purposes. The methodical development of test methods for materials with longitudinal and transverse strength and elastic properties 50 to 60:1 had not occurred; so, characterization studies and design application are proceeding simultaneously at the same rapid pace. This paper discusses an early effort to characterize certain mechanical properties of unidirectional graphite yarn reinforcement in an epoxy matrix.

Existing test methods were used with minimum modification wherever possible to provide the widest base for data comparison. Several test methods were developed during this program when no known test method existed for the specific property sought. The static properties evaluated were: tensile strength and modulus, flexural strength and modulus interlaminar shear strength, and shear modulus. The dynamic properties evaluated were: flexural modulus, torsional modulus, and flexural fatigue strength.

MATERIAL AND FABRICATION

The reinforcement used in this program was Thornel 50, a continuous two-ply graphite yarn supplied by the Carbon Products Division of Union Carbide Corp. The yarn is formed by twisting together two plies of 720 filaments each at 1.5 turns per inch. The nominal yarn diameter is 0.015 in. The yarn is supplied with

---

[1] Senior design project engineer and senior analytical engineer, respectively, Hamilton Standard Division of United Aircraft Corp., Windsor Locks, Conn. 06096. Mr. Molter is a personal member ASTM.

[2] The italic numbers in brackets refer to the list of references appended to this paper.

a polyvinyl alcohol PVA sizing to enhance resistance to abrasion during handling. No attempt was made to remove the sizing since it was required to facilitate collimation of the yarn into dry tape form. Thornel 50 filaments have a nominal tensile strength of 285,000 psi and a nominal tensile modulus of 50×10^6 psi.

The epoxy resin system used was ERL-2256, supplied by the Union Carbide Corp. This is a difunctional, low-viscosity resin system. The curing agent was an eutectic mixture of aromatic amines supplied under the trade name Sonite-41.

The composite fabrication was performed in an aluminum vacuum injection-type matched die. The die cavity was 8.625 by 5.125 by 0.125 in. All yarn was dry-collimated on an automatic lathe, forming single layer tape sections for charging the mold. The fully loaded and closed die was preheated for 1 h at 180 F, evacuated, and then pressure-filled with resin. The laminate was cured in the die for 2 h at 200 F followed by 3 h at 300 F.

Seven panels were fabricated with a nominal fiber volume of 58.2 percent. Property test specimens were selected at random from the seven panels. All specimen machining was performed with conventional tooling. All edges were left as machined except in the case of the fatigue specimens where the edges were dressed with a fine abrasive cloth.

TEST METHODS AND RESULTS

The test methods and results are described in the following sections. The individual test results from each of 87 tests performed on 47 specimens are not presented individually. Instead, the typical properties obtained from each of the seven panels are tabulated in Table 1 along with the fiber volume fraction and number of specimens tested.

The test methods employed in this program do not in general conform to ASTM test methods or other accepted standards since none were available for this material at the start of the program. Where possible, test methods were patterned after those commonly used and discussed in the literature and proposed ASTM test methods with modifications based on the dimensions of the material and test facilities available.

TENSILE STRENGTH AND MODULUS

A standard tension test specimen for carbon epoxy was not available at the start of this program. A "dog bone" type specimen with bonded fiber glass gripping pads was designed so that at the maximum anticipated tensile strength of the material the average shear stress in the gripping area was below the minimum anticipated interlaminar shear strength of the composite. The specimen configuration and stressing criteria are indicated in Fig. 1. The design ratio of 54 was based on a maximum anticipated tensile strength of 135,000 psi and a minimum anticipated interlaminar shear strength of 2500 psi. The nominal panel thickness of 0.125 in. set the specimen thickness. A test section width of twice the thickness, 0.25 in., was established arbitrarily as providing a reasonable test area. The required gripping length of 2.25 in. was determined based on the design ratio.

Testing was performed on a standard laboratory tensile machine using conventional vise grips. Prior to testing, Budd C12-111 foil strain gages were bonded with Eastman 610 cement on the horizontal and longitudinal center of the test section of all four surfaces of each specimen. During preliminary work both strain gages and a conventional extensometer were used. When the extensometer functioned properly, the simultaneous readings from the strain gages and extensometer were in close agreement. Due to slippage of the extensometer in several tests, however, the strain gage readings were considered the more reliable. The strain gages were zeroed at an initial load of 200 lb, and the strain versus load was recorded in 200-lb load increments up to 1200 lb. The relative readings of the

Table 1—Summary of test results Thornel 50-ERL 2256 epoxy with unidirectional fibers.

Panel Number	Percent Fiber by Volume	Typical Static Properties							Typical Dynamic Properties[a]		
		Tensile		Flexural			Shear		Torsional modulus, psi	Flexural modulus, psi	10^7 Cycle fatigue strength, psi
		Strength, psi	Modulus, psi	Strength, psi	Modulus,[b] psi	Modulus,[c] psi	Strength, psi	Modulus, psi			
T-6	52.1	119.2×10^3(1)	24.1×10^6(1)	106.3×10^3(1)	22.5×10^6(1)	21.5×10^6(2)	4.54×10^3(2)	...	0.698×10^6(2)	16.7×10^6(2)	53.5×10^3(2)
T-7	57.8	108.3(1)	21.2(1)	105.7(1)	24.5(1)	22.8(4)	3.27(4)	...	0.769(4)	19.5(4)	62.0(4)
T-8	59.9	113.0(1)	25.2(1)	108.5(1)	25.0(1)	22.8(4)	3.71(2)	...	0.755(4)	20.0(4)	56.6(4)
T-9	59.2	118.2(1)	25.0(1)	120.7(2)	25.6(2)	23.9(3)	3.66(4)	...	0.806(3)	20.8(3)	49.0(3)
T-10	60.2						3.90(2)		...	...	...
T-11		97.8(2)	24.2(2)	56.2(1)	25.0(1)		3.55(2)		...	...	...
T-12							2.86(2)	0.710×10^6(1)			
Average all panels		109.0(6)	24.0(6)	103.0(6)	24.7(6)	22.8(13)	3.50(21)	0.710(1)	0.762(13)	19.5(13)	55.8(13)

[a] Numbers in parentheses indicate number of specimens tested.
[b] Four-point beam method.
[c] Cantilever beam method.

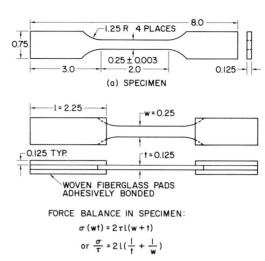

FIG. 1—Carbon epoxy tension specimen.

(a) SPECIMEN

l = 2.25
w = 0.25
0.125 TYP.
t = 0.125
WOVEN FIBERGLASS PADS ADHESIVELY BONDED

FORCE BALANCE IN SPECIMEN:

$$\sigma(wt) = 2\tau l(w + t)$$

$$\text{or } \frac{\sigma}{\tau} = 2l(\frac{1}{t} + \frac{1}{w})$$

WHERE σ = TENSILE STRESS IN TEST SECTION
τ = AVERAGE SHEAR STRESS IN GRIPPING AREA

(b) SPECIMEN WITH BONDED GRIPS

four strain gages were examined under the initial load to detect bending; the specimens were regripped, if required, to maintain the bending strain at less than 5 percent of the average tensile strain.

After the last gage reading was taken at 1200 lb, the load was increased at a uniform crosshead rate of 0.05 in./min until tensile fracture occurred.

The tensile strength was calculated using the original cross-sectional area of the test section and the fracture load, which was the maximum load in all instances. The tensile modulus was determined from the slope of the stress-strain curve using the average of all four strain gage readings.

A total of six tension specimens was tested yielding an average tensile strength of 109,000 psi and an average tensile modulus of 24×10^6 psi. The range in tensile strength was 87,000 to 119,200 psi, and the range in tensile modulus was 21.2×10^6 to 25.8×10^6 psi.

Fiber fractures were not on a single plane normal to the axes of the specimens but occurred randomly and were joined by axial separations indicative of an interlaminar shear mode of fracture. The jagged fracture surfaces in some instances extended into the end radius areas.

STATIC FLEXURAL STRENGTH AND MODULUS—FOUR-POINT METHOD

A four-point beam technique, similar to that of Ref 1, was utilized to determine the flexural strength and modulus. The rectangular beam specimens were 6.0 in. long by 0.75 in. wide by 0.125 in. thick. Figure 2 schematically shows the loading and methods of calculating strength and modulus. Initially, two methods were utilized in determining the elastic modulus. The first method was the common method of generating a load-deflection curve and calculating modulus using the strength of materials relationship between load and deflection. The second method used strain gages to generate a load-strain curve and determining modulus as the ratio of measured surface strain to calculated bending stress. Both methods gave comparable results. The latter method, however, proved more repeatable and was adopted.

A total of six specimens was tested. The average flexural strength was 103,000

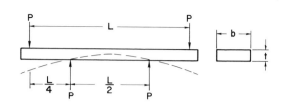

$$\sigma = \frac{6Pa}{bt^2} \text{ AND } E = \frac{6Pa}{\epsilon bt^2}$$

WHERE
 σ = FLEXURAL STRESS, PSI
 E = ELASTIC MODULUS, PSI
 P = REACTION LOAD, LB
 t = SPECIMEN THICKNESS NOMINALLY 0.125 IN.
 b = SPECIMEN WIDTH, NOMINALLY 0.75 IN.
 ϵ = STRAIN
 a = L/4, NOMINALLY 1.25 IN.

FIG. 2—Flexural strength and modulus: four-point load method.

psi with a range of 56,300 to 125,400 psi. The average flexural modulus was 24.7×10^6 psi with a range of 22.5 to 26.2×10^6 psi.

Except for the low value of 56,300 psi, the flexural strength and modulus values agreed favorably with the tensile strength and modulus values. Examining the fractures offered no explanation for the one low-strength fracture. However, five of the six fractures, including the low-strength fracture, occurred at one of the points of support. The location of these fractures probably resulted from the local shear and contact stresses.

INTERLAMINAR SHEAR STRENGTH

A short-beam shear test, similar to that of Ref 1, was used for determining the interlaminar shear strength. A schematic of the loading, pertinent dimensions, and the method of calculating the strength are presented in Fig. 3. A nominal span-to-thickness ratio of 5 was maintained. The contact radii of the two supports were $\frac{1}{16}$ in., and the contact radius of the loading fixture was $\frac{1}{8}$ in. The strength calculation assumes a parabolic distribution of shear stresses in the beam.

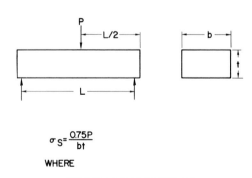

$$\sigma_S = \frac{0.75P}{bt}$$

WHERE

 σ_S = MAXIMUM SHEAR STRESS, PSI
 P = APPLIED LOAD, LB
 b = SPECIMEN WIDTH, NOMINALLY 0.25 IN.
 t = SPECIMEN THICKNESS NOMINALLY 0.125 IN.
 L = SPAN LENGTH, NOMINALLY 0.625 IN.

FIG. 3—Short-beam shear test.

A total of 21 short-beam specimens from seven panels was tested. The average interlaminar shear strength was 3500 psi with a standard deviation of 650 psi; the strength range was 2610 to 4600 psi.

The short-beam shear test for determining interlaminar shear strength has been a subject of controversy. It does not produce pure shear because of the bending of the beam and contact stressing at the points of support and loading. The fractures produced in this program were confused by signs of crushing at the points of loading and support and tensile fractures on the surface. It was difficult to determine which mode of fracture occurred first. The lower-strength fractures, however, were predominantly shear, and the higher-strength specimens showed signs of tensile fractures in addition to shear.

FLAT PLATE SHEAR MODULUS

The four-point loaded flat plate method discussed in the literature [2,3] was used for determining static shear modulus. A schematic of the loading and the method of calculating shear modulus are presented in Fig. 4.

The 4.0 by 4.0 by 0.125-in. specimen was supported by point contact at each of three corners. The remaining free corner was dead weight loaded. An initial load of 1 lb was used as a reference point for the deflection measurements, which was then increased in 1-lb increments to the maximum load. The center deflections of the plate were measured at each respective load level, and, after each level, the load was returned to the reference point to check the deflection zero point. To insure accuracy during the test the plate center deflection was kept at or below one half the plate thickness.

A single specimen was tested yielding a shear modulus of 710,000 psi. The load-deflection curve was linear over the deflection range. This modulus value is in reasonable agreement with the average value of 762,000 psi for the 13 dynamic torsional modulus tests reported on in the following section.

Subsequent to this test, 4.0-in.^{2} plate specimens of aluminum and titanium were tested in an attempt to establish the validity of the test. The shear moduli determined for the aluminum and titanium were 79 and 89 percent, respectively, of the known values for these materials.

[2] The italic numbers in brackets refer to the list of references appended to this paper.

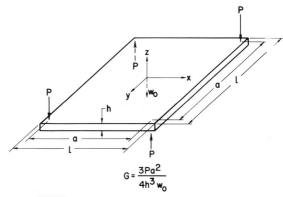

$$G = \frac{3Pa^2}{4h^3 w_0}$$

WHERE

G = SHEAR MODULUS, PSI
P = APPLIED LOAD, LB
a = DISTANCE BETWEEN REACTION POINTS, NOMINALLY 3.75 IN.
L = WIDTH OF PLATE, NOMINALLY EQUAL TO 4.0 IN.
h = AVERAGE SPECIMEN THICKNESS, 0.127 IN.
w_0 = PLATE CENTER DEFLECTION, IN.

FIG. 4—Flat plate shear test.

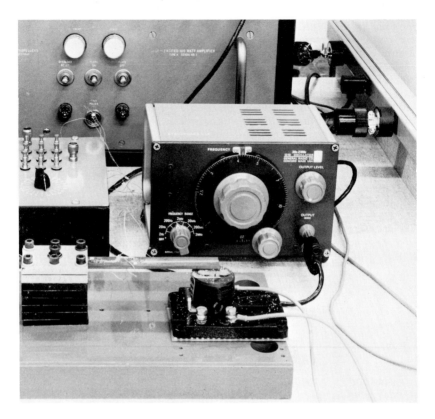

FIG. 5—Dynamic flexural and torsional modulus test setup.

Goble and Campbell et al [4] report that on the basis of total strain energy considerations, the equation of Fig. 4 would be more accurate if corrected by the factor a^2/l^2, where a is the edge distance between loads, and l is the total width of the plate. This correction factor is less than unity and produces an even greater error in the shear modulus values determined for aluminum and titanium.

The theoretical deflected surface of the plate is a hyperbolic paraboloid or saddle [2] derived on the basis of uniform twisting moments acting along the edges. It allows no bending at the edges. The distributed moments are replaced by an equivalent couple produced by point loading at the corners for this test. It is possible, although not investigated in this program, that these point loads are producing local bending that would add to the deflections due to twisting and result in lower apparent shear modulus values.

STATIC AND DYNAMIC FLEXURAL AND DYNAMIC TORSIONAL MODULUS—CANTILEVER BEAM METHOD

A resonant vibrating beam technique was used to determine the dynamic flexural and torsional moduli with a rectangular cross-section beam specimen, 8.0 in. long by 0.75 in. wide by 0.125 in. thick. Although this test technique was developed primarily for determining dynamic moduli, the static flexural modulus was determined on each specimen with the same basic test setup. The results were a check on values determined with the four-point test and provided a comparison between static and dynamic flexural moduli on the same specimen.

This test setup is shown in Fig. 5. The 8-in.-long specimen was clamped firmly as a cantilever beam with 1 in. in the clamp providing an effective beam length of 7 in. A thin metal strip was bonded to the beam tip, on the center line for flexural vibration and off the center line for torsional vibration, to permit electromagnetic excitation. Each specimen was instrumented with two Budd C12-121 strain gages using Eastman 610 cement, one at 0 deg and one at 45 deg to the axis of the specimen, to assist in accurately peaking the flexural and torsional modes of vibration. An electromagnetic coil driven by an oscillator induced a cyclic force in the pole piece and, thus, excited the beam. A frequency scan was made to identify the resonant modes of interest, that is, the first flexural mode for determining flexural modulus and the first torsional mode for determining torsional modulus. Holding the power constant, the oscillator was adjusted to peak the strain gage output, therefore accurately defining the resonant frequencies.

The elastic moduli were determined from the resonant frequencies using the familiar equations for the vibration of prismatical bars [5].

Solving these equations for modulus, the equations used are as follows.

Dynamic torsional modulus:

$$E_d = \frac{3.192 f_f^2 W l^4}{gI}$$

where:
E_d = dynamic flexural modulus, psi;
f_f = flexural resonant frequency, Hz;
W = weight of beam per unit length, lb/in.;
l = effective length of beam, in.;
g = gravitation acceleration, in./s^2; and
I = bending moment of inertia, in.^{4}.

Dynamic torsional modulus:

$$G_d = \frac{16 f_T^2 l^2 \gamma I p}{gk}$$

where:
G_d = dynamic torsional modulus, psi;
f_T = torsional resonant frequency, Hz;
l = effective length of beam, in.;
γ = weight density of beam, lb/in.^{3};
I_p = polar moment of inertia, in.^{4};
g = acceleration of gravity, in./s^2; and
K = rectangular cross section torsional stiffness parameter, in.^{4}, see Ref 6.

The static flexural modulus was determined by dead weight loading the tip of the specimen and using the following strength of materials relationship between load and deflection.

$$Es = \frac{PI^3}{3\delta I}$$

where:
Es = static flexural modulus, psi;
P = applied tip load, lb;
l = effective length of beam, in.;
δ = tip deflection at load P, in.; and
I = bending moment of inertia, in.^{4}.

The flexural and torsional moduli determined on a total of 13 specimens yielded a typical static flexural modulus of 22.8 $\times$ 10^6 psi with a range of 21.0 to 25.4 $\times$ 10^6 psi, a typical dynamic flexural modulus of 19.5 $\times$ 10^6 psi with a range of 16.5

to 22.0×10^6 psi, and a typical dynamic torsional modulus of 0.762×10^6 psi with a range of 0.662 to 0.870×10^6 psi.

The dynamic flexural modulus values were approximately 80 percent of those determined in the static tension tests and four-point beam flexure tests. The static cantilevered beam flexural modulus values were very slightly lower than the static tensile and four-point beam values. Shear deflections in both the static and dynamic tests and clamp area damping in the dynamic tests are possible reasons for the lower-modulus values. Preliminary efforts to include shear in the beam equations for isotoropic materials indicate this would not account for the discrepancies in either the static or dynamic cantilever beam moduli. The idea of treating the beam as a laminated material was sugested but not explored. To check the effect of clamp damping on the dynamic modulus, a specimen was run statically and dynamically as a clamped cantilever beam and also dynamically as a free-free beam, that is, with no clamping. The free-free beam modulus value agreed with the static cantilever beam value, while the dynamic cantilever value was 85 percent of the static value. It appears, therefore, that the clamp should be avoided in dynamic modulus determinations.

FLEXURAL FATIGUE STRENGTH

Attempts at fatigue testing both necked and constant width unidirectional fiber reinforced specimens using resonant cantilever beam techniques proved unsuccessful. It was recognized quickly that variations in the specimen cross section would have to occur in an area unaffected by stressing introduced at the points of loading and retention. A constant section beam vibrating in its first free-free mode of flexural vibration met the desired characteristics.

A free-free beam fatigue testing technique utilizing electromagnetic excitation was developed during this program. The test setup is shown in Fig. 6. The length of panels available and the method of external loading caused some deviations from the ideal constant section specimen.

The fatigue specimens were made from the cantilever beam static and dynamic moduli specimens, and, therefore, the dynamic flexural moduli were known. Steel extension bars were clamped to each end to increase the effective length of the beam, and tip weights were added to increase mass. Steel blocks were bonded to the bottom of the clamps to permit electromagnetic excitation. The specimen assembly was supported on rubber strips at its two nodal points. Paired electromagnetic coils driven by a common oscillator produced a sinusoidal force in the steel blocks, exciting the specimen assembly. The oscillator frequency and nodal supports were adjusted to produce a pure free-free beam resonant mode of vibration for each assembly. Budd C12-121 foil strain gages were bonded with Eastman 610 on the test section. A dynamic calibration of strain versus center deflection was

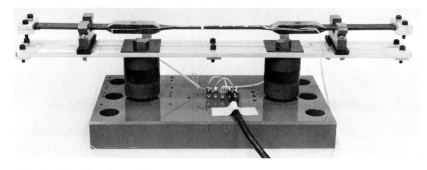

Fig. 6—Free-free beam fatigue test setup.

performed for each specimen. This was converted to stress using the previously determined dynamic flexural moduli. The fatigue tests were then monitored on center deflection.

The inertia loading of the beam extensions, tip weights, clamps, and pole pieces produced a moment distribution equivalent to what would result from point loads at the tips reacted by point loads at the pole pieces. The moment across the test section, therefore, essentially was constant. The bending stressing in the specimen was a minimum at the clamps gradually increasing to a peak value at the center. The essentially constant moment minimized the possibility of shear fracture.

Thirteen specimens were fatigue tested, and the results are presented on the S-N diagram of Fig. 7. The average- and low-boundary fatigue strengths at 10,000,000 cycles were approximately 55,000 and 45,000 psi, respectively. Figure 8 shows an example of a fractured specimen. Fractures were characterized by "sliver-type' cracks initiating the edges of the minimum section and propagating toward the clamp in a manner indicative of a shear mode of separation.

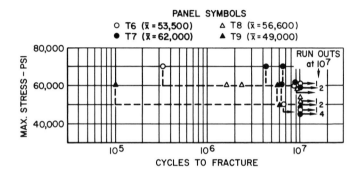

Fig. 7—Thornel/epoxy composite specimens fatigue test data ($R = -1$) tested at room temperature.

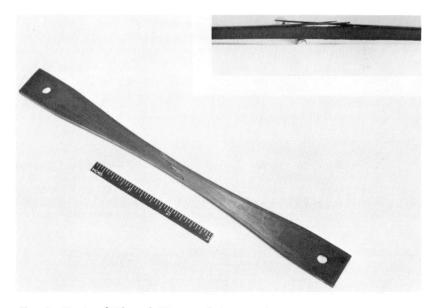

Fig. 8—Fractured Thornel 50-epoxy fatigue specimen.

CONCLUSIONS AND RECOMMENDATIONS

The following conclusions and recommendations are drawn from this initial effort at characterizing certain mechanical properties of a unidirectional carbon fiber reinforced epoxy composite:

1. Tension tests using a dogbone specimen with bonded gripping pads provided a satisfactory method for determining the tensile strength and modulus of this composite. The estimated tensile-strength-to-shear-strength ratio was satisfactory for the carbon epoxy evaluated. For other systems where the ratio is different, the specimen geometry should be adjusted accordingly.

2. The four-point beam flexure test method provided satisfactory flexural modulus values. The flexural strength values are questionable, however, because a large percentage of fractures occurred at the supports where the stressing is complicated by the local contact.

3. The short-beam shear test for interlaminar shear strength did not produce shear fractures consistently. Although it is simple and economical to perform, data generated by this method should be used only in a relative sense and only then with due consideration to the modes of fracture produced.

4. Using the flat plate shear test for determining shear modulus is questionable in view of the inaccurate results for homogeneous-isotropic metals. There is evidence that the simplified load-deflection equation used to calculate shear modulus does not describe adequately the deflected surface of the plate.

5. The electromagnetically excited resonant beam techniques offer promise as methods of determining dynamic flexural and torsional moduli.

6. The free-free beam flexure fatigue testing technique gave consistent results and offers attractive features for testing other composite materials. Refinements in the specimen geometry and specimen fabrication method should be considered to reduce the tendency for sliver-type fractures.

References

[1] "Structural Design Guide for Advanced Composite Applications," (intermediate draft), Advanced Composites Division, Air Force Materials Laboratory, Air Force Systems Command, Wright Patterson Air Force Base, Ohio, Sept. 1967, Section VIII.

[2] Hennessey, J. M., Whitney, J. M., and Riley, M. B., "Experimental Methods for Determining Shear Modulus of Fiber Reinforced Composite Materials," AFML-TR-65-42, Research and Technology Division, Air Force Materials Laboratory, Air Force Systems Command, Wright Patterson Air Force, Base, Ohio, Sept. 1965, pp. 11–13.

[3] Whitney, J. M., "Experimental Determination of Shear Modulus of Laminated Fiber-Reinforced Composites," *Experimental Mechanics*, Vol. 7, No. 10, Oct. 1967, pp. 447–448.

[4] Goble and Campbell, F. et al, "Integrated Research on Carbon Composites," AFML-TR-66-310 Part II, Air Force Materials Laboratory, Air Force Systems Command, Wright Patterson Air Force Base, Ohio, Dec. 1967, pp. 230–234.

[5] Timoshenko, S., *Vibration Problems in Engineering*, 3rd ed., Van Nostrand, Princeton, 1955.

[6] Roark, R. J., *Formulas for Stress, and Strain*, 3rd ed., McGraw-Hill, New York, 1954, p. 174.

A Simplified Method of Determining the Inplane Shear Stress-Strain Response of Unidirectional Composites

P. H. PETIT[1]

Reference:

Petit, P. H., "A Simplified Method of Determining the Inplane Shear Stress-Strain Response of Unidirectional Composites," *Composite Materials: Testing and Design, ASTM STP 460,* American Society for Testing and Materials, 1969, pp. 83–93.

Abstract:

A simple and relatively inexpensive technique is presented for determining the inplane shear stress-strain response of unidirectional composites. The method utilizes the stress-strain results from an uniaxial test on a ±45-deg laminate which is symmetrically laminated about the midplane. Through the use of relations derived from laminated plate theory, expressions are presented which allow the inplane laminae shear stress-strain curve to be incrementally reproduced from the ±45-deg laminate stress-strain curve. The simple coupon and uniaxial sandwich beam specimens, which may be utilized to obtain the data, are described. Comparisons are presented between boron/epoxy and graphite/epoxy shear stress strain curves obtained by this technique and the curves from cross-sandwich beam tests.

Key Words:

composite materials, boron, fiber composites, graphite, mechanical properties, shear tests, shear strength, evaluation, tests

The task of determining the shear stress-strain response of a material always has presented a challenging test problem. These problems are caused primarily by the difficulty in obtaining a pure shear stress state of known magnitude in the test specimen. The requirement of a known and equal biaxial tension-compression stress state, which will result in a pure shear stress state at an angle of 45 deg to the tension-compression axes, is difficult enough to obtain in homogeneous isotropic materials. Because of the necessity of obtaining the shear stress-strain response of the constituent laminae of highly orthotropic fibrous composites, the additional variable of material orthotropy has been added to the test problem. The highly orthotropic nature of the fibrous composites causes additional test problems, particularly, in determining the inplane shear stress-strain response.

Over the past few years, several test methods have been proposed for determining the inplane shear stress-strain response for the advanced composite materials [1,2][2]. Each of these techniques has its relative advantages, disadvantages, and areas of applicability and validity. All of the current shear test methods require expensive and complex test specimens or specialized test fixtures. The technique for determining the inplane shear stress-strain response, which will be presented in this paper, has the advantages of utilizing relatively inexpensive ·straight-sided tension coupon or sandwich beam specimens and conventional test equipment and of being applicable to all the new advanced composite materials.

---

[1] Senior aircraft structures engineer, Lockheed-Georgia Co., Marietta, Ga. 30061.
[2] The italic numbers in brackets refer to the list of references appended to this paper.

TEST TECHNIQUE

In this section, the method of determining the inplane shear stress-strain response of unidirectional composites from ±45-deg angle-ply uniaxial tension specimens will be presented. First, however, a brief review will be given of the accepted methods of determining the material behavior of fibrous composites or orthotropic materials. Then the test technique, which is the subject of this paper, will be presented and the test specimens discussed.

MATERIAL CHARACTERIZATION

In a filamentary composite, the constituent laminae have three mutually perpendicular planes of elastic symmetry, as shown in Fig. 1. A material which has three planes of material property symmetry is called "orthotropic." In addition, the thickness of each ply is very small relative to its other dimensions; therefore, the laminae may be considered to be in a state of plane stress. If the fiber/resin geometry is ignored, then the laminae can be considered to be quasi-homogeneous on the macroscale. Thus, the individual laminae can be characterized as a quasi-homogeneous orthotropic medium subjected to a plane stress state.

Since the unidirectional lamina is orthotropic and in a plane stress state, four elastic constants are required to determine its elastic behavior instead of the two which are required for isotropic materials. Three of these elastic constants may be determined from two unidirectional tests on the material. From a longitudinal test, with longitudinal and transverse strain gages, the initial Young's modulus in the fiber direction, E_{11}, and the major Poissons ratio, γ_{12}, may be determined from the longitudinal stress-strain curve. From a transverse test, the initial Young's modulus transverse to the fiber direction, E_{22}, may be determined from the transverse stress-strain curve. The compression stress-strain response of the laminae normally will be quite different than the tension behavior; therefore, both longitudinal and transverse stress-strain curves must be determined for tension and compression. Thus, with two simple uniaxial tensile or compressive coupon or sandwich beam tests and appropriate strain gage instrumentation, three of the four required elastic constants can be determined. The fourth elastic constant, the inplane shear modulus, G_{12}, and the inplane shear stress-strain curve is more difficult to determine.

To completely characterize the laminae of a fibrous composite, the five stress-strain curves described previously are required. From these curves, the four elastic constants and proportional limits can be determined which will allow a linear or limit strength analysis to be performed on any laminate composed of the characterized laminae. It is also possible to accomplish a nonlinear or ultimate strength analysis on a laminate by utilizing the total stress-strain curves. Such an analysis is presented in Refs 3 and 4. Thus, the current approach to the analysis of fibrous composites in the aircraft industry is predicated on the determination of the five principal

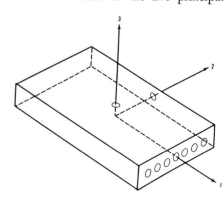

FIG. 1—Lamina principal axes.

lamina stress-strain curves. The determination of the longitudinal and transverse tension and compression curves is accomplished relatively easily with straight-sided coupon specimens or uniaxial sandwich beams. If the shear stress-strain response could be determined in a similar manner, the task of characterizing various fibrous composite materials would be easier and less expensive. In the next section, such a technique will be presented.

SHEAR STRESS RESPONSE

The technique for determining the laminae shear stress-strain response, which will be presented in this paper, is based on laminated plate theory [5,6]. As mentioned above, the individual laminae are characterized as a quasi-homogeneous orthropic medium subjected to a plane stress state. The Kirchoff-Love hypothesis is envoked to ensure that the strain distribution through the laminate thickness is a constant or linear function of the distance from the middle surface and the transverse shear deformation may be neglected. Therefore, if the four elastic constants and the five principal lamina stress-strain curves are known, one should be able to predict the average elastic moduli and stress-strain response of any laminate. This technique has proved quite successful in predicting moduli and strengths of laminates composed of the advanced composite materials when the laminae properties were determined correctly.

Tsai [7] has presented a technique for determining the initial shear modulus of the laminae of unidirectional composites from longitudinal and transverse tests and a unidirectional tests on a +45-deg laminate. However, this technique can not be utilized to obtain the full lamina shear stress-strain curve because the +45-deg uniaxial test is not an acceptable strength test since the principal axes of orthotropy are inclined to the load direction. An explanation of this behavior is given by Pagano and Halpin [8]. A ±45-deg angle-ply laminate tested in uniaxial tension is an acceptable test specimen, and a valid stress-strain curve (σ_x versus ϵ_x) may be obtained. The average laminate stresses, $[\sigma]$, which act on a laminate, are given by the following constitutive relationship [4]:

$$[\bar{\sigma}] = \frac{1}{t}[A][\epsilon] = [\bar{A}][\epsilon] \dots \dots \dots \dots \dots \dots (1)$$

where, the laminate stiffness matrix is given by Ref 6 and

$$A_{ij} = \sum_{k=1}^{m} \bar{Q}_{ij}^{(k)}(h_{k+1} - h_k) \dots \dots \dots \dots \dots (2)$$

where t = total laminate thickness.

Transforms similar to those of Ref 4, which are based on laminated plate theory, may be applied to give a relationship between the Young's modulus for the ±45-deg laminate, E_{xx}, and the shear modulus of the constituent laminae, G_{12}.

$$G_{12} = \frac{2U_1 E_{xx}}{8U_1 - E_{xx}} \dots \dots \dots \dots \dots \dots (3)$$

where

$$U_1 = \frac{1}{8(1 - \nu_{12}\nu_{21})}[E_{11} + E_{22} + 2\nu_{21}E_{11}] \dots \dots \dots (4)$$

and the minor Poisson's ratio, ν_{21}, is given by the reciprocality relations:

$$\nu_{21} = \nu_{12}E_{22}/E_{11} \dots \dots \dots \dots \dots \dots (5)$$

The elastic constants which comprise U_1 are determined from uniaxial tests in the longitudinal and transverse directions. In order to obtain the complete shear

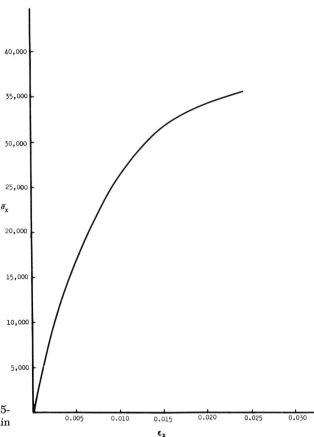

FIG. 2—Boron/epoxy ±45-deg laminate stress-strain curve.

stress-strain curve, the ±45-deg stress-strain curve may be broken down into incre-ments. Over each increment, the curve is assumed to be linear, and the above transforms may be applied over each increment. That is, a value of the laminae shear modulus, G_{12}, is determined over each increment from the tangent modulus, E_{xx}, of the ±45-deg stress-strain curve. The next step is to determine for each increment a relationship between the normal strain, ϵ_x, and the laminae shear strain, γ_{12}, which is referred to the laminte reference axes. The relationship between a shear strain in one set of coordinates and a normal strain in a second coordinate system can be determined for any continuum from elementary elasticity as follows:

$$\gamma_{12} = (1 + \nu_{xy})\epsilon_x \dots\dots\dots\dots\dots\dots\dots\dots\dots\dots(6)$$

With the above relationships and the stress-strain data from the ±45-deg uniaxial tension test, it is possible to reproduce incrementally a shear stress-strain curve for the constituent laminae.

As an illustration of the procedure, an example will be presented. In Fig. 2, the results of a uniaxial sandwich beam tension test on a ±45-deg boron/epoxy (3M SP272 matrix material) laminate are presented. In Table 1, the results of breaking the stress-strain curve down into 13 increments are presented. For this particular boron/epoxy material, the three elastic constants, E_{11}, E_{22}, and ν_{12}, re-

Table 1—*Boron/epoxy ±45-deg laminate stress-strain data.*

Increment No.	σ_x	ϵ_x	ϵ_y	$E_{xx}^{\pm 45\text{-}deg}$	ν_{xy}
1..........	4 000	0.001	0.00083	4.0×10^6	0.83
2..........	7 900	0.002	0.00168	3.9	0.84
3..........	11 200	0.003	0.00252	3.3	0.84
4..........	14 100	0.004	0.00340	2.9	0.85
5..........	16 800	0.005	0.00430	2.7	0.86
6..........	19 200	0.006	0.00522	2.4	0.87
7..........	23 300	0.008	0.00704	2.05	0.88
8..........	26 500	0.010	0.00900	1.6	0.90
9..........	28 900	0.012	0.01092	1.2	0.91
10..........	31 000	0.014	0.01274	1.05	0.91
11..........	32 500	0.016	0.01472	0.75	0.92
12..........	34 000	0.020	0.01820	0.375	0.92
13..........	35 500	0.0241	0.02208	0.36	0.92

quired for this analysis were determined from uniaxial sandwich beam tests. The values for the initial moduli and major Poisson's ratio are as follows:

$$E_{11} = 30.8 \times 10^6$$

$$E_{22} = 3.8 \times 10^6$$

$$\nu_{12} = 0.36$$

When these values are substituted into Eqs 3 and 4, the following is obtained:

$$U_1 = 4.741 \times 10^6$$

$$G_{12} = \frac{(4.741 \times 10^6) E_{xx}^{\pm 45deg}}{(37.928 \times 10^6) - E_{xx}^{\pm 45deg}}$$

By utilizing Eqs 3 and 6 and the values of ϵ_x, ν_{xy}, and $E_{xx}^{\pm 45deg}$ for each increment from Table 1, the values of G_{12} and γ_{12} are determined for each increment and presented in Table 2. The values of τ_{12} are calculated from the tangent shear modulus, G_{12}, and the increment in shear strain. In other words,

$$\Delta \tau_{12} = G_{12} \Delta \gamma_{12}$$

Table 2—*Lamine shear stress-strain data.*

Increment No.	G_{12}	γ_{12}	τ_{12}
1............	1.118×10^6	0.00183	2.046
2............	1.087	0.00367	4 054
3............	0.904	0.00552	6 064
4............	0.785	0.00740	7 139
5............	0.727	0.0093	8 520
6............	0.640	0.0112	9 736
7............	0.542	0.0150	11 795
8............	0.417	0.0190	13 463
9............	0.310	0.0229	14 672
10............	0.270	0.0267	15 698
11............	0.191	0.0307	16 462
12............	0.0947	0.0384	17 191
13............	0.0935	0.0462	17 920

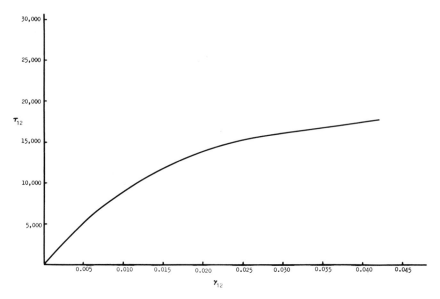

FIG. 3—Boron/epoxy shear stress-strain curve.

From the incremental data tabulated in Table 2, the laminae shear stress-strain curve was plotted and is shown in Fig. 3.

It must be emphasized that the laminates from which the laminae elastic constants, E_{11}, E_{22}, and ν_{12}, were determined and the laminate from which the ±45-deg stress-strain data were obtained all had the same laminae or ply thicknesses. This is required in order that the laminae or ply data are obtained for the same fiber volume content.

APPROXIMATIONS

It must be emphasized that there are certain approximations which are inherent with this procedure. One approximation is caused by the lack of a pure shear stress or strain state in the constituent laminae of the ±45-deg test specimen. The second approximation comes about due to the fact that in the angle ply ±45-deg specimen there is a small region near the boundary of the specimen where the assumptions of the laminated plate theory are not valid. These two approximations will be discussed below.

It can be seen from Table 1 that the laminate Poisson's ratio for this boron/epoxy ±45-deg laminate is approximately 0.85. If the normal strains, ϵ_x and ϵ_y, which are

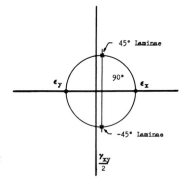

FIG. 4—Mohr's circle of strain.

referred to the laminate reference axes, are plotted on a strain Mohr's circle, the results shown in Fig. 4 are obtained. Since the laminate Poisson's ratio is not exactly 1.0 and hence ϵ_x is not quite equal to the negative of ϵ_y, the strain state at 45-deg to the X-Y axes is not quite pure shear. Small positive or tensile strains exist in addition to the relatively large shear strains in the principal directions of the laminae. This results in tensile stresses existing in the longitudinal and transverse directions of the laminae. This condition is illustrated in Fig. 5. It will be shown in the Results section that these tensile stresses across the shear plane will result in the laminae shear curve, which is obtained from this technique, having a lower ultimate stress than the shear curve determined from a cross-sandwich beam test.

As was pointed out earlier, one basic assumption in laminated plate theory is that the strains are a constant or linear function of the distance, Z, from the laminate middle surface. This is part of the Kirchoff hypothesis, and it ensures strain compatibility between laminae. As a result of this assumption, the laminae of angle-ply laminates in uniaxial tension develop an inplane shear stress, τ_{xy}. In Fig. 6, an illustration is given of the nature of the laminae deformation with and without the Kirchoff hypothesis. This shear stress alternates in sign from the laminae oriented at $+\theta$ to the laminae oriented at $-\theta$; therefore, the shear stress resultant is zero. The inplane shear stress, which is a result of the geometry of the laminate, must go to zero at the edges of the test specimen since there is

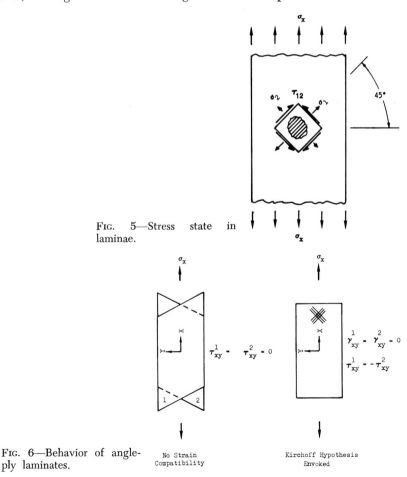

FIG. 5—Stress state in laminae.

FIG. 6—Behavior of angle-ply laminates.

No Strain Compatibility

Kirchoff Hypothesis Envoked

Petit on Unidirectional Composites

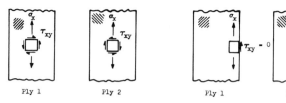

Ply 1   Ply 2   Ply 1   Ply 2

Fig. 7—Lamina shear log effect.

no applied shear stress (Fig. 7). Thus, there is an area near the edges of the test specimen where the basic assumption of the analysis are not valid. The effects of this behavior may be neglected if the width of the test specimen is relatively large compared to the affected edge zone. This behavior has been discussed in Ref 9.

TEST SPECIMENS

In this section, a brief discussion will be made of test specimen configurations. The straight-sided tension coupon specimen and the uniaxial sandwich beam specimen, which could be utilized to obtain the data required for this technique, will be discussed. The cross-sandwich beam specimen, which is also used as a method of determining laminae shear stress-strain response, is discussed briefly since a comparison will be made between the data obtained by this technique and data obtained by the ±45-deg test described previously.

The straight-sided tensile coupon, Fig. 8, can be utilized to obtain the longitudinal and transverse stress-strain data required to determine E_{11}, E_{22}, and ν_{12}. It also serves as a satisfactory specimen for determining the ±45-deg laminate stress-strain curve. For best results, the specimen width, W, should be a minimum of 1.0 in. on the transverse and the ±45-deg specimens. On the ±45-deg specimen, a compromise must be attained between a specimen so thin that the edge effects are significant to a specimen so wide that an even load introduction through the tabs is difficult. The 1-in. wide specimen represents such a compromise. A specimen width of 0.5 in. has been found satistactory on the longitudinal test.

The uniaxial sandwich beam specimen, Fig. 9, is probably a better test specimen than the straight-sided coupon specimen because of the smooth load introduction into the test section of the composite skin. Consistently higher ultimate strengths are obtained with the sandwich beam specimens. The specimen is quite satisfactory for longitudinal, transverse, and the ±45-deg angle-ply test. Beam lengths and heights can vary with the strength and modulus of the composite test skin. Beam widths of 1 in. have proved quite satisfactory. The honeycomb core density should

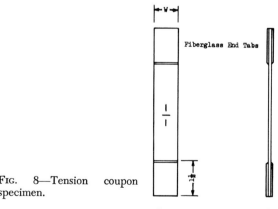

Fiberglass End Tabs

Fig. 8—Tension coupon specimen.

Composite Materials

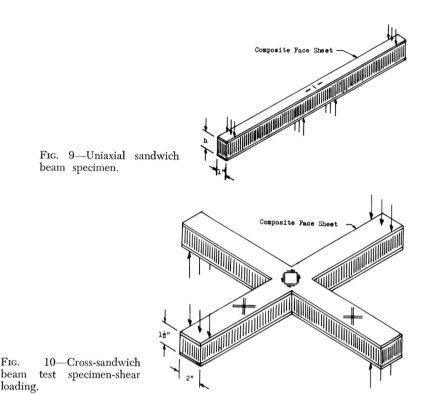

FIG. 9—Uniaxial sandwich beam specimen.

FIG. 10—Cross-sandwich beam test specimen-shear loading.

be as light as possible. A compromise should be met between beam length, beam height, core density, and composite face thickness to minimize beam deflection and preclude core or adhesive failures.

The cross-sandwich beam specimen, Fig. 10, has been used quite extensively to determine the shear stress-strain response of the laminae of fibrous composites. The specimen configuration and the loading for the shear test is shown in Fig. 10. From the loading shown, an equal tension-compression stress state is induced on a ±45-deg laminate symmetrically laminated about the midplane. At an angle of 45 deg to the beam legs, the stress state in the test section is pure shear. Thus, the end result is a pure shear stress state on a 0/90-deg laminate. This is equivalent to a pure shear test on the unidirectional laminae since each laminae is in a state of pure shear.

In the next section, a comparison will be made between the test results obtained from a cross-sandwich beam test and the ±45-deg test outlined in this paper.

RESULTS

In this section, a comparison will be made between the shear stress-strain curves obtained from the ±45-deg test and the curves obtained from cross-sandwich beam tests on boron/epoxy and graphite/epoxy composite materials.

In Fig. 11, the boron/epoxy (3M's SP272) shear stress-strain curve presented in Fig. 3 is shown superimposed with a shear curve obtained from a cross-sandwich beam test. In Fig. 12, a comparison of the shear curves for a graphite/epoxy material (Morganite II fiber and Narmco 5605 resin) is given. Along the initial portions of the curves, excellent agreement was achieved. In the areas considerably

Petit on Unidirectional Composites

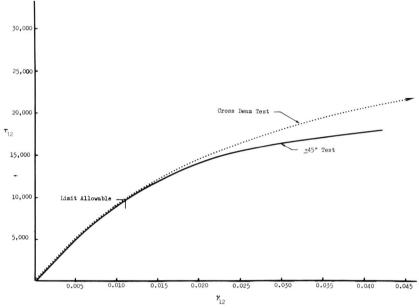

FIG. 11—Comparison of results—boron epoxy.

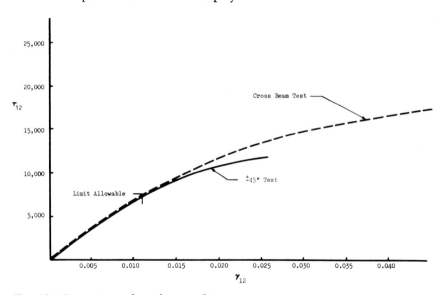

FIG. 12—Comparison of results—graphite epoxy.

beyond the limit allowable, the curves obtained from the ±45-deg test begin to deviate from the crossbeam shear curves.

For increasing values of shear strain, the curves obtained from the ±45-deg test are more compliant than the curves obtained from the crossbeam tests. The apparent loss in stiffness of the ±45-deg specimen for large strains can be attributed partially to a growth of the edge zone where strain compatibility is not achieved. The laminae tensile stresses across the shear plane in the ±45-deg specimen contribute to its failure prior to reaching the stress and strain levels achieved in the cross-sandwich beam test. However, the strain levels achieved in the cross-sandwich beam test are considerably beyond the levels where the assumptions of engineering strain are valid. The boron/epoxy stress-strain curve obtained from the cross-sandwich beam test continues to an ultimate stress of 28,000 psi at a strain level of over 100,000 μin./in.

CONCLUSIONS

A technique has been presented for incrementally reproducing the laminae shear stress-strain curve of fibrous composites from the stress-strain resulsts of a $\pm$45-deg laminate test. It has been shown that there is good agreement between the shear curves obtained from this technique and cross-sandwich beam tests in the area of the stress-strain curve which is most pertinent for design. That is, in the areas of the stress-strain curve where a limit allowable can be set and at strain levels immediately beyond this point, this simplified method of testing will produce very acceptable results.

The principal advantage of this technique is the implicity of test specimens and test technique. It provides an inexpensive method of determining the laminae shear stress-strain response in a manner similar to that which is used normally for determining the longitudinal and transverse properties of fibrous composites. Thus, the complete laminae characterization of a fibrous composite material can be accomplished with simple uniaxial coupon or sandwich beam specimens.

References

[1] Adams, D. F. and Thomas, R. L., "Test Methods for the Determination of Unidirectional Composite Shear Properties," *Twelfth National Symposium*, Society of Aerospace Materials and Process Engineers, Anaheim, Calif., Oct. 1967.

[2] Shockey, P. D. and Waddoups, M. E., "Strength and Modulus Determination of Composite Materials With Sandwich Beam Tests," General Dynamics/Ft. Worth FZM 4691, American Ceramic Society Meeting, Sept. 1966.

[3] Petit, P. H., "Ultimate Strength of Laminated Composites," General Dynamics/Ft. Worth FZM 4977, Air Force Contract Number AF 33(615)-5257, Dec. 1967.

[4] Petit, P. H. and Waddoups, M. E., "A Method of Predicting the Nonlinear Behavior of Laminated Composites," *Journal of Composite Materials*, Vol. III, No. 1, Jan. 1969.

[5] Tsai, S. W., "Structural Behavior of Composite Materials," NASA CR-71, National Aeronautics and Space Administration, July 1964.

[6] Ashton, J. E., Halpin, J. C., and Petit, P. H.. *Primer on Composites: Analysis*, Technomic, Stamford, Conn., April 1969.

[7] Tsai, S. W., "A Test Method for the Determination of Shear Modulus and Shear Strength," Air Force TR AFML-TR-66-372, Air Force Materials Laboratory, Jan. 1967.

[8] Pagano, N. J. and Halpin, J. C., "Influence of End Constraint in the Testing of Anisotropic Bodies," *Journal of Composite Materials*, Vol. II, No. 1, Jan. 1968.

[9] Lenoe, E. M., "Evaluation of Test Techniques for Advanced Composite Materials," AVCO Corp., Quarterly Report No. 5 Air Force Contract F33(615)-C-1719, Sept. 1968.

Angle-Plied Boron/Epoxy Test Methods—A Comparison of Beam-Tension and Axial Tension Coupon Testing

**R. B. LANTZ[1] AND
K. G. BALDRIDGE[1]**

Reference:

Lantz, R. B. and Baldridge, K. G., "Angle-Plied Boron/Epoxy Test Methods—A Comparison of Beam-Tension and Axial Tension Coupon Testing," *Composite Materials: Testing and Design, ASTM STP 460,* American Society for Testing and Materials, 1969, pp. 94–107.

Abstract:

A discussion of necessary considerations for tension test specimen design of laminated angle-plied composite materials is presented. Particular attention is paid to the straight-sided axial coupon and the long-beam flexure test. A large amount of test data, including strain gage measurements, has been compiled to provide a basis of comparisons of tensile strengths and elastic properties of angle-plied boron/epoxy laminates by the two test methods.

Key Words:

composite materials, tension, strength, elastic properties, fiber composites, boron epoxy laminates, laminates, evaluation, tests

The modern advanced composites with large diameter, stiff fiber reinforcements, are accentuating testing difficulties yet commanding more attention due to their tremendous promise as lightweight efficient structure. Difficulties encountered in gripping or supporting such test specimens, compounded by the material's inherent sensitivity to axial alignment, result in inconsistent failure loads and failure modes. Some of the most difficult problems in test specimen design and analysis of test results have been encountered during off-axis loadings and loading of angle-plied laminates. This problem is recognized in the aircraft industry, and several methods of solution are being proposed. Two of the most promising approaches involve the use of the long-beam flexure test and wide-axial coupons, both of which will be examined in this paper.

DATA REQUIREMENTS

Two types of test data commonly are required when developing the properties of composite materials. These are the elastic and strength properties. Knowledge of one will not allow the prediction of the other. Traditionally, both quantities have been obtainable from a single test, depending on the sophistication of the test instrumentation. Early investigations of highly oriented materials such as unidirectional fiber glass reinforced epoxy demonstrated to the researchers that often separate test methods had to be developed, depending upon whether strength or elastic properties were to be obtained. Having developed the required techniques

[1] Senior engineer, Analytical Section, and supervisor, Application and Research Laboratory, respectively, AVCO Aerostructures Division, Nashville, Tenn. 37202. Mr. Baldridge is a personal member ASTM.

to determine accurately the properties of such basic-ply unidirectional composites, the extensions of these techniques to materials formed from combinations of oriented layers of such basic plies pose entirely new and distinct problems.

BALANCED-PLY LAMINATES

The measured properties of such angle-plied laminates are influenced by whether or not the plies form a "balanced-ply arrangement" about the test axis. That is, the laminate must possess axi-symmetry about the test axis to be considered a balanced-ply test arrangement.

Lenoe [1][2] demonstrated through the use of a shear lag analysis that, for a balance-ply arrangement, the interlaminar shear stress developed to maintain compatibility of opposing shear strains of adjacent plies due to the applied tensile stress can approach appreciable fraction of the available interlaminar shear strength, depending upon the orientations of the adjacent "balanced plies." Obviously, such superposition of stresses can influence test results. However, these opposing shear strains are the mechanism by which *uniform* loading is maintained during a test of angle-plied laminates.

Pagano and Halpin [2] and Wu and Thomas [3] describe the *nonuniform* state of stress and strain induced in a nonbalanced-ply material due to the necessary constraints imposed by normal test fixtures. These fixtures are clamping devices which limit the end rotation of the specimen during testing, adding to the complexity of the stress or strain state near the ends. Both references present novel fixtures and techniques to attempt to alleviate the problem. This paper will deal only with balanced-ply laminates with the exception of one data point obtained from a coupon tension test of a 0/±45-deg laminate tested at 45 deg from the 0-deg ply axis.

NONLINEAR CONSIDERATIONS

Angle-plied laminates exhibit varying degrees of nonlinear stress-strain response depending on the constituent properties and angle orientation. As was concluded in Ref 4, the amount of nonlinearity of the composite at a certain stain level can exceed the nonlinear response of all the constituents taken individually and examined at the same strain level. The explanation postulated was that due to close spacing of the filaments, the matrix material in the narrow zone between the filaments would be stressed to high levels causing elasto-plastic yielding. This yielding would be evident as a change in elastic properties of the laminate, that is, the exhibition of nonlinear stress-strain response. It was further speculated [4] that there would be a time-dependent phase during which the matrix material would yield and redistribute the load. Because such yielding would result in the establishment of a residual stress field in the matrix, there would be some tendency for the material to recover partially during a static period following a loading. This nonlinear response (partially time dependent), has been observed during many tests of the newer advanced composite laminates as well as with the more common fiber glass reinforced plastics and support such postulation. The data presented in this paper will exhibit the nonlinear response; however, the tests were conducted as "static" tests, and the material was not allowed to relax.

WIDE AXIAL COUPONS

When testing angle-plied composites, it is recognized that width effects influence the test results. If a long narrow angle-plied tension specimen with the fibers oriented at a significant angle, say ±45 deg, is tested, the fibers in the middle of the test section length will be loaded only through shear of the matrix/fiber

---

[2] The italic numbers in brackets refer to the list of references appended to this paper.

interface bond. The matrix or bond will fail prior to filament failure, and the specimen shape can then be classified as *matrix critical*. If a wide angle-plied tension specimen were wide enough so that certain fibers were continuous from one end to the other along the length and through the test section, the specimen would then become *filament critical*. Somewhere between these two cases, the interface bond length would be sufficent to allow the filament to load to failure, causing this specimen to be also fiament critical. Matrix critical and filament critical conditions are specimen design considerations which directly influence strength data.

Considerations which influence elastic property determination are the presence of no lateral restraint or full lateral restraint, as defined by Dickerson and DiMartino [5] for narrow or wide specimens, respectively. These restraints occur in the center of the specimen due to width, orientation, and constituent properties, the same

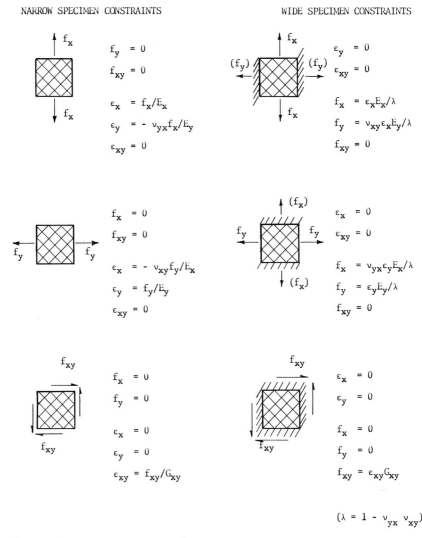

NARROW SPECIMEN CONSTRAINTS

$f_y = 0$
$f_{xy} = 0$

$\varepsilon_x = f_x/E_x$
$\varepsilon_y = - \nu_{yx} f_x/E_y$
$\varepsilon_{xy} = 0$

$f_x = 0$
$f_{xy} = 0$

$\varepsilon_x = - \nu_{xy} f_y/E_x$
$\varepsilon_y = f_y/E_y$
$\varepsilon_{xy} = 0$

$f_x = 0$
$f_y = 0$

$\varepsilon_x = 0$
$\varepsilon_y = 0$

$\varepsilon_{xy} = f_{xy}/G_{xy}$

WIDE SPECIMEN CONSTRAINTS

$\varepsilon_y = 0$
$\varepsilon_{xy} = 0$

$f_x = \varepsilon_x E_x/\lambda$
$f_y = \nu_{xy} \varepsilon_x E_y/\lambda$
$f_{xy} = 0$

$\varepsilon_x = 0$
$\varepsilon_{xy} = 0$

$f_x = \nu_{yx} \varepsilon_y E_x/\lambda$
$f_y = \varepsilon_y E_y/\lambda$
$f_{xy} = 0$

$\varepsilon_x = 0$
$\varepsilon_y = 0$

$f_x = 0$
$f_y = 0$

$f_{xy} = \varepsilon_{xy} G_{xy}$

$(\lambda = 1 - \nu_{yx} \nu_{xy})$

Fig. 1—Coupon specimen internal constraints.

location as is commonly used for strain gage determinations of the elastic properties. While actual use of the relations presented in Ref 5 is limited somewhat to a micromechanics analysis, the knowledge that a certain specimen design possesses a particular restraint will aid in evaluating whether a lower or upper bound has been obtained for the elastic constant being examined. A narrow specimen, no lateral restraint, generally will yield lower-bound elastic constants while a wide specimen, full lateral restraint, generally will yield upper-bound elastic constants. Figure 1 shows the constraints present in the two types specimens as originally presented in Ref 5.

The axial coupons that were tested and are reported here were simply straight-sided flat coupons, 9 in. long. The widths were varied from ¾ to 3 in. in an attempt to bracket the various constraints and conditions inherent in an angle-plied specimen design as just discussed. The ends were reinforced with bonded fiber glass/epoxy laminate tabs.

LONG-BEAM FLEXURE SPECIMENS

The long-beam flexure specimen, as developed for use as an advanced composite tension test specimen [6], is a sandwich beam composed of two face sheets separated by a deep honeycomb core. One face is the test specimen, composed of the composite to be tested, with the proper orientation and thickness. The other face sheet is chosen judiciously to preclude any undue influence of the test results. The long beam, when tested in flexure, strains the test face in tension. The beam test attempts to alleviate the problem of laminate gripping and load introduction present in coupon testing.

Analysis of a beam test specimen currently is accomplished by assuming the face sheet resists the applied bending moment and neglecting any effects of the core except in transmitting shear (the same as a web in an I beam).

The beams dimensions, core, and opposite face sheet materials were derived to allow the composite face sheet material to be the critical test parameter. The design considerations included the opposite face sheet strength; the core shear strength, weight, and cell size; the adhesive strength and curing temperature; the beam span and amount of allowed midspan deflection; and the beam overhang and test section length. The three types laminates being tested required three beam configurations. Figure 2 describes the beams for these three laminates: uniaxial, 0-deg laminates; unaxial, 90-deg laminates; and angle-plied $\pm\alpha$-laminates where $\alpha = 30$, 45, and 60 deg. $0/\pm45$-deg laminates were tested using the $\pm\alpha$-design. $0/90$-deg laminates were tested using the uniaxial 0-deg design.

TEST METHODS

TEST EQUIPMENT

A universal testing machine having full scale load ranges from 1200 to 60,000 lb was used for load application. Crosshead movement, load, and strain pacers were available for controlling testing rates.

The honeycomb beam test fixture, as described in Ref 7, had reaction points set on 21 or 14 in. centers and load points set on 1.5 or 2 in. centers, respectively, depending on the test specimen. Repeatable alignment of the loading nose as related to the reaction points was assured through an upper plate-bushing-load rod arrangement indexed to the base plate. Steel pads were used between the ⅛-in.-radius load reaction points and the beam face sheet to eliminate localized stresses and to allow the face sheet to rotate freely as deflection occurred. Axial tension tests were performed in self aligning grips with fine toothed jaw faces.

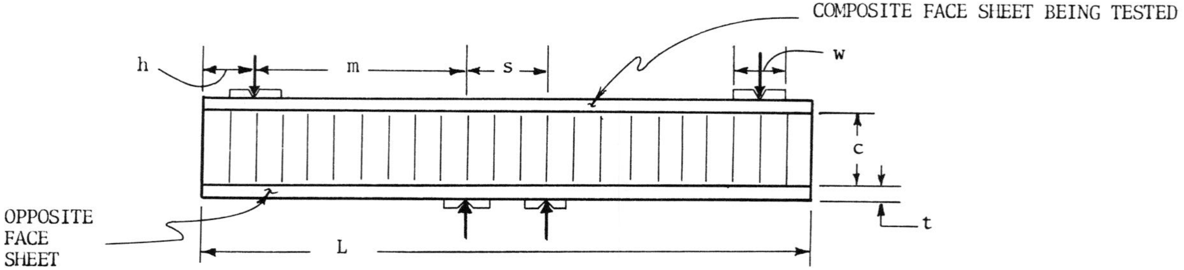

COMPOSITE FACE SHEET CONSTRUCTION	OPPOSITE FACE SHEET		CORE		ADHESIVE	WIDTH	BEAM DIMENSIONS				LOAD PAD WIDTH
	MATERIAL	THICK. t	MATERIAL	THICK. c			TEST SPAN s	MOMENT ARM m	TOTAL LENGTH L	OVER-HANG h	w
		(in.)		(in.)		(in.)	(in.)	(in.)	(in.)	(in.)	(in.)
0°, 0°/90°	7075-T6 Alum.	0.125	1/8-5052-.006-NP (23.4)	1.5	FM1000	1	1.5	9.75	22	0.5	1
90°	FRP/Epoxy	0.080	1/8-5052-.001-NP (4.5)	1.5	FM1000	1	1.5	9.75	22	0.5	1
±α°, 0°/±α°	6AL-4V Titanium	0.040	1/8-5052-.002-NP (8.1)	1.0	FM1000	1	2.0	6.00	16	1.0	1

FIG. 2—Sandwich beam design.

Strain gage equipment used was comprised of the following interconnected apparatus: (1) a digital strain indicator, (2) a printer control unit, (3) a switch and balance unit with provision for 30 channels, and (4) a digital printout recorder. This equipment can be sequenced automatically through each channel on signal or can be sequenced manually through the channels holding on each channel as desired.

INSTRUMENTATION

BLH PFAER-25R-12 rosette strain gages were mounted on the boron face of the honeycomb tension beams and on one face of the axial tension coupon with the rosette axis centered on the test section and aligned to measure the longitudinal, transverse, and 45-deg strains. Axial gages (BLH PFAE-12-12) were mounted at one end of the rosette gage on the longitudinal axis of the honeycomb beam. Axial gages also were mounted to one side and at one end of the rosette gage, on the transverse and longitudinal axis, respectively, of all axial tension coupons with an additional axial gage mounted on the coupon face opposite the axial gage on the longitudinal axis. The axial gages were used to ascertain uniform loading patterns.

TEST PROCEDURE

Rate of load application was controlled by crosshead movement at 0.022 in./min. The number of strain gages and the type of strain gage equipment used required that the load be held at predetermined levels while strain was recorded. The test load levels selected were such that a minimum of ten stress-strain relationships could be determined on each coupon. These test levels obviously varied with fiber orientation.

DATA REDUCTION

The only data handling problem of note appeared during the reduction of the quantities of strain gage readings. Many of the strain gages were three arm rosettes, requiring resolution to principal strain values. The sequence of arm readings during a test was arranged to expedite the null balancing of the wheatstone bridge and was accomplished by grouping the readings according to strain value and not by rosette number. To handle these data, the computer program listed in Ref 8 was used on an IBM 1130. The input to the program was organized so that the raw strain gage data could be keypunched directly from the test lab printout tape. The strain readings could be in any order as long as appropriate keying had been done to allow the computer to sort the strain gage arm readings and assign them to the proper rosette. If a reading was negative, the program calculated the compressive strain indicated. The program handled axial gages as single arm rosettes, biaxial gages as double arm (at 90 deg) rosettes, and assumed a three arm gage to be a rectilinear rosette with arms oriented 0, 45, and 90 deg to the rosette axis.

For any set of strain readings, the corresponding load reading was input. Depending on the strain gage arms available, the program calculated the stress, secant modulus, Poisson's ratio, principal strains, and the angle (ϕ) between the major principal strain and the rosette axis. The gage arm readings also were reduced from the raw test data and displayed. The stress calculations were based on input test specimen parameters and whether the test was an axial or beam test.

Inspection of the output allowed an immediate evaluation of the reliability of the test measurements. Since most of the test laminates were orthotropic and only unidirectional loadings were imposed, the principal axis of the rosette should have coincided with the test axis, provided the rosette was oriented properly. Any large value of the calculated ϕ would make that test suspect. The laminate either was not oriented properly, the specimen machined from the panel properly, or the gage not installed properly. Even though some of the data could still be valid, inspection of the data was expedited by using the ϕ measurement.

From the printout of the data, stress/axial strain and stress/transverse strain curves were constructed. Indications of initial moduli and initial Poisson's ratios were obtained. Strength values were compiled according to normal practice.

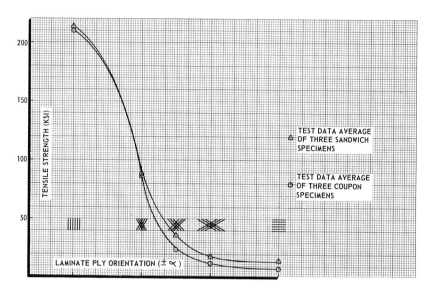

Fig. 3—±α-deg laminates, tensile strength versus laminate ply orientation (8-ply boron/SP 272).

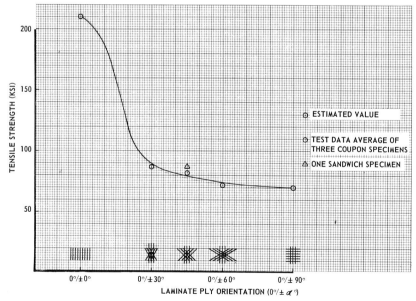

Fig. 4—0/±α-deg laminates, tensile strength versus laminate ply orientation (9-ply boron/SP 272).

EXPERIMENTAL RESULTS AND EVALUATION

Tensile Strength Results

The strength results of all the $\pm\alpha$ and $0/\pm\alpha$-deg laminates (where α equals 0, 30, 45, 60, and 90) are presented in Figs. 3 and 4. The results of the 1½-in.-wide axial coupons are compared to the results of the 1-in.-wide beams. An examination of the failure modes of the specimens indicated no matrix critical failures.

The increase in strength due to the presence of the 0-deg ply in the $0/\pm\alpha$-deg series is interesting although expected. The consistent increase in strength of the

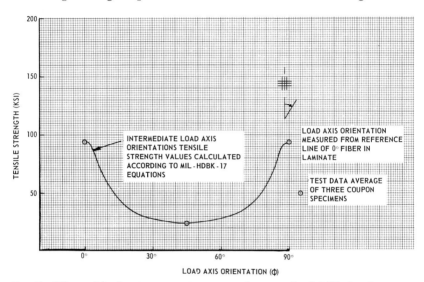

FIG. 5—Effect of load axis orientation on tensile strength of 0/90-deg. laminates (8-ply boron/SP 272).

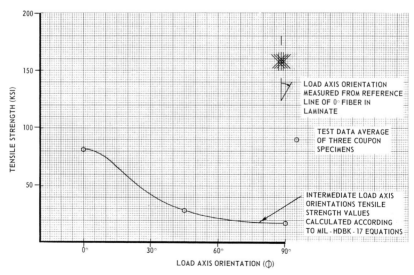

FIG. 6—Effect of load axis orientation on tensile strength of 0/±45-deg laminates (9-ply boron/SP 272).

Lantz and Baldridge on Coupon Testing 101

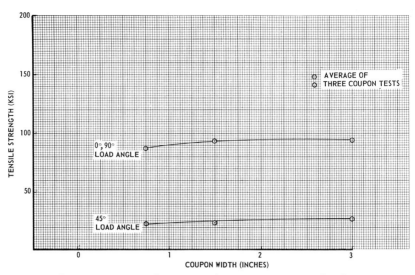

Fɪɢ. 7—Effect of coupon width on tensile strength of 0/90-deg laminates (8-ply boron/SP 272).

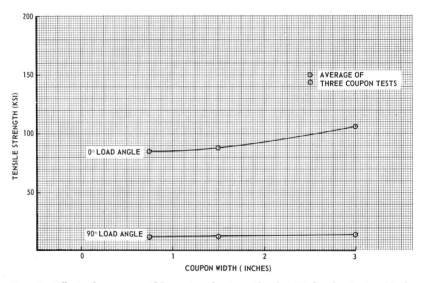

Fɪɢ. 8—Effect of coupon width on tensile strength of ±30-deg laminates (8-ply boron/SP 272).

±α-deg beam specimens over the coupon specimens for α greater than 30 is thought to be due to inaccuracies of the beam analysis becoming evident as the ultimate stress levels decrease.

Another laminate construction of common interest is that of a 0/90-deg orientation. Axial coupons, 1½ in. wide, of this construction were tested at varying load angles. The results are shown in Fig. 5. Note that the 45-deg load angle test for a 0/90-deg laminate is identical to a ±α-deg laminate test where α = 45. Inter-

Composite Materials

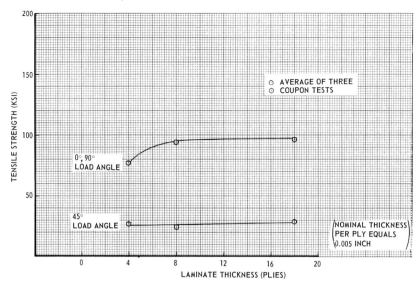

FIG. 9—Effect of coupon thickness on tensile strength of 0/90-deg laminates (1½-in.-width boron/SP 272).

mediate load angle strength values were calculated at every 5 deg using a computer program listed in Ref 8. This program solves the orthotropic strength relation equations presented in MIL-HDBK-17, [9], using the strength values obtained at three load angles: 0, 45, and 90 deg.

A similar study was accomplished for 1 in. wide, 0/±45-deg axial coupons using strength values obtained at the same three load angles: 0, 45, and 90 deg. The results are presented in Fig. 6.

An effect of axial coupon width study was performed on 0/90 and ±30-deg laminates. Figure 7 presents the results of the 0/90-deg laminate tests conducted at 0 and 45-deg load angles. Figure 8 presents the results for the ±30-deg laminate tests conducted at 0 and 90 deg. The only significant effect on strength occurred for the 0 deg loaded ±30-deg laminate. Apparently, the ±α-deg laminates for larger values of α exhibit sufficient strain capability and nonlinear stress-strain response to negate the use of a strength measurement as a means to differentiate the effects of coupon widths. The curves do indicate that the use of 1½-in.-wide specimens for obtaining strengths of 0/90 and ±α-deg laminates yields valid data.

Figure 9 presents the results of an effect of laminate thickness on the tensile strength of a 0/90-deg laminate tested at 0 and 45-deg load angles. The results indicate that 8-ply laminates provide nominal strength data. The 45-deg oriented laminate seems little influenced by the thickness; however, the 4-ply, 0 deg oriented results are reduced significantly. Apparently, a 4-ply tension coupon specimen does not have enough plies to transfer the loading uniformly in an efficient laminate construction.

TENSILE ELASTIC PROPERTIES

Typical stress-strain curves for ±α-deg constructions where α = 0, 30, 45, 60, and 90 are presented in Fig. 10. Figure 11 presents the companion typical stress-strain curves for 0/90 and 0/±α-deg constructions where α = 30, 45, and 60. All these curves present the axial and transverse strains for an axial tensile loading of 1½-in. wide coupons (1 in. wide for the 0/±α-deg coupons). The curves do

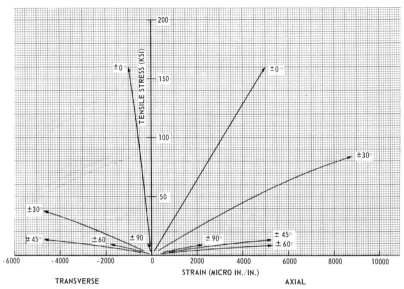

FIG. 10—Typical tension stress-strain response of $\pm\alpha$-deg laminates (8-ply boron/SP 272).

not extend to the failure load since strain readings seldom were accomplished at the moment of failure. Initial moduli and initial Poisson's ratios were measured for all the instrumented specimens. The following presents the average values for varying widths and thicknesses:

Laminate Orientation, deg	Initial Modulus, E_x	Poisson's Ratio, ν_{yx}
± 0	31.7×10^6 psi	0.20
± 30	12.0	1.25
± 45	2.9	0.85
± 60	2.3	0.30
± 90	3.7	0.02
0/90	16.5	0.05
0/± 30	17.4	0.96
0/± 45	12.8	0.84
0/± 60	12.4	0.27

Comparisons of the effects of beam and coupon test methods on the measured stress-strain response of 0, 0/90, ± 30, and 0/± 45-deg laminates are presented in Figs. 12–15, respectively. The 0 and ± 30-deg laminates were tested at 0 and 90-deg load angles. The 0/90-deg laminates were tested 0 and 45-deg load angles. The 0/± 45-deg laminates were tested at a 0-deg load angle only.

For strain sensitive tests such as the 90-deg load angle of the 0-deg laminates, the beam provides additional strain capability for the specimen. For large angles of α in the $\pm\alpha$-deg laminates, the beam tends to stiffen the axial stress-axial strain response. For a majority of the tests, the axial stress-transverse strain response also was stiffened by the beam specimens.

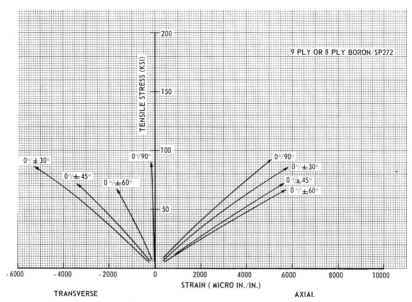

Fig. 11—Typical tension stress-strain response of 0/±α and 0/90-deg laminates.

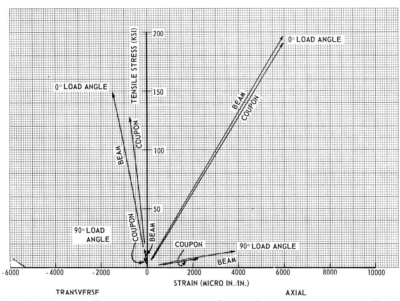

Fig. 12—Beam and coupon comparisons of typical stress-strain response for 0-deg laminates (6-ply boron/SP 272).

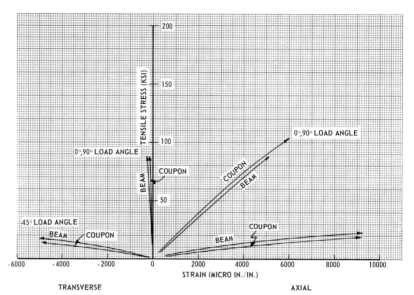

FIG. 13—Beam and coupon comparisons of typical stress-strain response for 0/90-deg laminates (8-ply boron/SP 272).

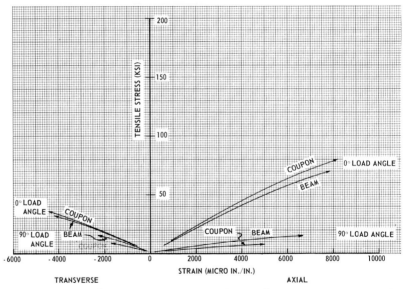

FIG. 14—Beam and coupon comparisons of typical stress-strain response for ±30-deg laminates (8-ply boron/SP 272).

CONCLUSIONS

A significant amount of test data has been compiled to provide a basis of comparisons of tensile strengths and elastic properties of angle-plied boron/epoxy laminates by axial coupon and long beam flexure test methods. The following can be concluded:

1. Present sandwich beam testing and analysis yield higher strength values for large angle (low-strength), angle-plied laminates than axial testing.

Composite Materials

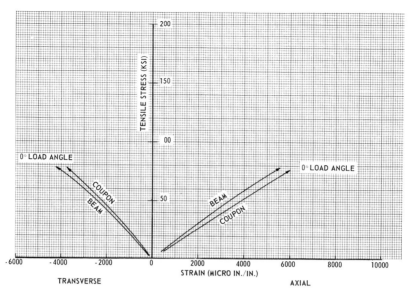

FIG. 15—Beam and coupon comparisons of typical stress-strain response for 0/±45-deg laminates (9-ply boron/SP 272).

2. Little effect of width of axial coupon tests was evident except for axial loading of a ±30-deg laminate. (However, all test widths were greater than ¾ in.)

3. Thin coupons of axially oriented, high-strength constructions yield lower-strength values than thicker laminates.

4. The beam test tends to stiffen the measured stress-strain response as compared to the axial coupon test. This is especially true for transverse strain measurements.

5. The beam test specimen allows more measured strain response than the axially loaded specimen.

References

[1] Lenoe, E. M., "Evaluation of Test Techniques for Advanced Composite Materials," Quarterly Report No. 5, AVCO Corporation Report AVSSD-0200-68-CR, 1968.

[2] Pagano, N. J. and Halpin. J. C., "Influence of End Constraint in the Testing of Anisotropic Bodies," *Journal of Composite Materials,* Vol. 2, 1968, pp. 18–31.

[3] Wu, E. M. and Thomas, R. L., "Off-Axis Test of a Composite," *Journal of Composite Materials,* Note to the Editor, Vol. 2, 1968, pp. 523–526.

[4] "Application of Advanced Fibrous Reinforced Composite Materials to Airframe Structures," Monthly Report No. 5, North American Aviation, Los Angeles Division Report NA-65-993-5, 1966.

[5] Dickerson, E. O. and DiMartino, B., "Off-Axis Strength and Testing of Filamentary Materials for Air-craft Application," *Advanced Fibrous Reinforced Composites,* Vol. 10, 10th National Symposium, Society of Aerospace Materials and Process Engineers, 1966.

[6] Waddoups, M. E. and Shockey, P. D., *Strength and Modulus Determination of Composite Materials with Sandwich Beam Tests,* General Dynamics/Ft. Worth Report FZM 4691, 1966.

[7] "Advanced Composites Aircraft Application Program–FY 1967 Final Report," AVCO Corporation Report R-2020, 1968.

[8] Lantz, R. B., "Effect of Load Angle on the Compressive Failure of Fiberglass/Epoxy Faced, Honeycomb Sandwich Structure," AVCO Corporation Report R-2031, 1968.

[9] "Plastics for Flight Vehicles, Part 1, Reinforced Plastics," MIL-17, Military Handbook, 1959.

Compression Testing of Aluminum-Boron Composites

N. R. ADSIT[1] AND J. D. FOREST[1]

Reference:

Adsit N. R. and Forest, J. D., "Compression Testing of Aluminum-Boron Composites," *Composite Materials: Testing and Design, ASTM STP 460,* American Society for Testing and Materials, 1969, pp. 108–121.

Abstract:

This paper presents methods for obtaining the mechanical behavior of aluminum-boron composites in compression. The property data obtained is compared with tensile data, and then used to predict the failure of simple structural elements. The analytical predictions are then confirmed by structural element tests.

Key Words:

composite materials, aluminum, boron, compression, testing, mechanical properties, evaluation, tests

The majority of aerospace structures are designed for compressive loadings—either pure axial compression, bending compression, or lateral collapsing pressure. Some types of structure, such as sandwich construction, are material strength limited. Stringer stiffened skin structure, on the other hand, is usually modulus limited to prevent local instability failures. In either type of application it is necessary to obtain an accurate compressive stress-strain relationship for the material so that modulus, tangent modulus, and limiting strengths are known.

Results of compression tests on isotropic, homogeneous materials usually are similar to results from tension tests. One objective of this program was to find if this similarity was true with anisotropic aluminum-boron (Al-B) composite material. Adequate methods of testing for compressive behavior first had to be developed. Another objective of this program was to use the compressive data generated to predict the behavior of simple structural elements. These analytical predictions then were verified by testing structural elements.

ALUMINUM-BORON SPECIMENS

The material used for this investigation was a composite of 6061 aluminum and boron filaments combined by hot press diffusion bonding [1].[2] The boron filaments are 4.0 mils in diameter and have an average tensile strength in excess of 400 ksi and a modulus of elasticity of over 55×10^6 psi. Several variations of material were tested. Two material thicknesses were used; four filament layer material nominally 0.020 in. thick, and 16 filament layer material nominally 0.080 in. thick. Except for the structural element tests, the material was prepared as 12 by 12-in. panels. Five different materials were studied:

Unidirectional (UD) layup with filament volume percentages (v/o) of

(a) 25 v/o
(b) 37 v/o
(c) 50 v/o

---

[1] Senior engineering metallurgist and senior design engineer, respectively, Convair Division of General Dynamics, San Diego, Calif. 92112.

[2] The italic numbers in brackets refer to the list of references appended to this paper.

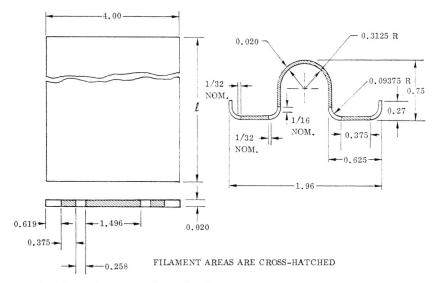

FIG. 1—Selectively reinforced panel and stringer.

Cross-ply (CP) layups of

    (*a*) 25 v/o each in 0 and 90 deg direction
    (*b*) 25 v/o each in +30 and −30 deg direction

The formed stiffeners were made from selectively reinforced panels (that is, there was 50 percent filament in some areas and no filament in other areas). The layup pattern is shown in Fig. 1.

In general, specimens were sheared out of the thin panels and then processed to final form by grinding or electric discharge machining (EDM). The thick materials could be sheared, but grinding was found to produce better specimens. In some cases it was necessary to bond tabs or ends onto the specimens.

TEST METHODS

The coupon compression tests on 0.080-in.-thick material were performed according to the ASTM Compressive Testing of Metallic Material at Room Temperature (E 9-61) with a Montgomery-Templin grip. This method was devised originally for testing thin sheet metal. The longitudinal composite specimens all failed at the ends by brooming. To prevent this from occurring, longer specimens were prepared with caps placed on each end of the specimen.

The 0.020-in.-thick material presented a more difficult testing task. The same method described above was tried first but proved totally inadequate. Longer specimens with tabs bonded onto the ends with EC 2216 epoxy adhesive then were prepared and the ends ground flat and parallel. The tabs on these specimens peeled off, and the composite failure was by end brooming.

A new type of specimen was then examined, as shown in Fig. 2. A specimen in the test fixture is shown in Fig. 3. This basic fixture was later modified by adding rods to support the specimen and prevent lateral buckling. The closely spaced rods can be seen in the upper view of Fig. 4 and under the specimen in the lower view of Fig. 4.

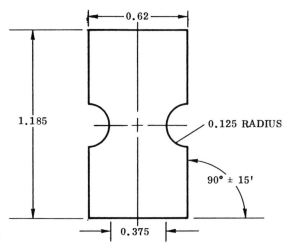

0.62

1.185

0.125 RADIUS

90° ± 15'

0.375

FIG. 2—Compression specimen for 0.020-in.-thick material.

ALL DIMENSIONS INCHES UNLESS OTHERWISE INDICATED

The compression tests were run on a Tinius Olsen testing machine. A compressometer was used to measure strain as directed in Test Method E 9-61 for the 0.080-in.-thick specimens. It was not possible to use a compressometer with the short 0.020-in.-thick specimens. In this case, a deflectometer was used to measure the relative head motion and the deflection of the testing machine (as measured by a head-to-head deflection run) subtracted. As a check of the test setup, a specimen was strain gaged and the strain measured on the SR-4 box. These data were in agreement with those measured by the deflectometer.

The specimen used for the crippling test is shown in Fig. 5. In order to prevent buckling and force the sheet to fail by crippling, split stainless steel tubes of ¾ in. diameter were clamped on one or both edges of the specimen during test to simulate simple supports. The tests were run in an Instron testing machine and the head deflection and load recorded. A head travel rate of 0.01 in./min was used.

The selectively reinforced aluminum-boron sheet material (Fig. 1) was formed into stiffeners and tested on an Instron testing machine [2]. The crosshead deflection and load were recorded at a head rate of 0.01 in./min.

FIG. 3—Test fixture for testing 0.020-in.-thick material. The compression specimen in the fixture shows a typical failure.

Composite Materials

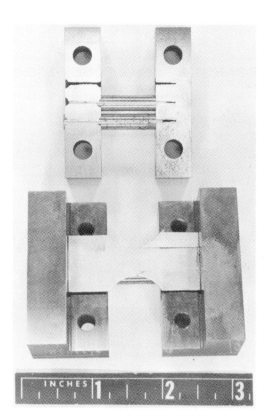

FIG. 4—Compression test fixture for 0.020-in.-thick composite. Rods provide lateral support.

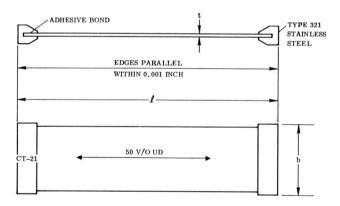

FIG. 5—Crippling test specimen geometry.

MATERIAL TEST RESULTS

The coupon compression test results for the 0.080-in.-thick material are given in Table 1. A typical stress-strain curve for a longitudinal specimen is shown in Fig. 6.

Those compression specimens tested so that the filaments were not oriented in the loading direction deformed by yielding and did not fail before they bottomed out on the fixture. Similar tension specimens do not yield, in most cases, but fail before a plastic deformation of 0.2 percent tensile strain is reached. The results of the specimens with end tabs are given in Table 2. Data from the companion specimens without end tabs are included for comparison. The mode of failure of these two types of specimens is shown in Fig. 7. The coupon compression test results for the 0.020-in.-thick material tested with the fixture shown in Fig. 4 are given in Table 3. Data for other specimen tests with and without reduced sections but, with lateral restraint, are given in Table 4.

The data given in Tables 1 and 2 are plotted in Figs. 8 and 9 along with tensile data [2] taken from the same or similar panels. The plots show that the strength and modulus of elasticity of 50 percent unidirectional material is very

Table 1—*Compression tests results on 0.080-in. thick aluminum-boron specimens.*

	No. of Tests	Ultimate Strength, ksi	Yield Strength, ksi	Modulus of Elasticity, 10^6 psi
50 v/o 0/90-deg cross ply:				
Longitudinal	5	80.0	...	21.9
Transverse	4	97.1	...	22.2
45 deg	3	...	15.2	23.7
50 v/o ± 30-deg cross ply:				
Longitudinal	5	42.1	...	28.8
Transverse	5	...	16.1	18.7
45 deg	3	54.4	...	24.7
50 v/o unidirectional:				
Longitudinal	5	183	...	34.6
Transverse	5	...	14.2	23.5
45 deg	3	...	16.9	25.1
25 v/o unidirectional:				
Longitudinal	5	126	...	21.9
Transverse	5	...	13.8	15.3
45 deg	3	...	14.8	18.2

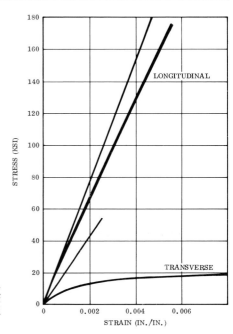

Fig. 6—Typical compressive stress-strain curve of 50 percent unidirectional material.

Table 2—*Compression tests on 0.080-in.-thick 50 v/o unidirectional composite.*

	Longitudinal	Transverse	45 deg
With end fixtures:			
Strength ultimate, ksi	184	. . .	. . .
Yield, ksi	. . .	13.7	16.8
Modulus of elasticity, 10^6 psi	31.2	24.9	22.8
No end fixtures:			
Strength ultimate, ksi	183	. . .	. . .
Yield, ksi	. . .	14.6	16.9
Modulus of elasticity, 10^6 psi	36.8	22.8	26.7

little different for tension or compression. These data also indicate that specimen thickness does not seem to affect the measured property. When plots of the strength and modulus of elasticity against orientation, and as a function of thickness and mode of testing, are made for the other material configurations, the compressive strength is higher but the modulus of elasticity is very nearly the same as for tension. These data taken together indicate that the compression strength and modulus are at least as high as the tensile strength.

The role of the aluminum matrix is substantially different in compression than in tension. In tension, the matrix is used to shear the load into the filaments and to shear the load around broken filaments. The matrix serves the same purpose in compression but also must stabilize the filament against buckling. As the filament content increases, the amount of matrix material between the filaments must decrease. This suggests that as the filament content of the composite is increased, the matrix may reach a point where it is less effective in preventing the filaments from buckling.

The difficulty in causing the material to fail in the center of the specimen, rather than brooming at the ends, makes it difficult to assess how much the compres-

Fig. 7—Compression test specimens showing failure modes with and without end caps on 50 percent unidirectional material. The specimen on the right (261-CL-2) has no end support and failed by brooming. The specimen on the left (261-CL-4) was tested with the end fixture as shown.

Table 3—*Compression test results on 0.020-in.-thick aluminum-boron specimens tested with no lateral support.*

	No. of Test	Ultimate Strength, ksi	Yield Strength, ksi	Modulus of Elasticity, 10^6 psi
50 v/o 0/90-deg cross ply:				
Longitudinal...........	5	143	...	25.1
Transverse............	5	88.2	...	18.8
45 deg................	3	73.9	46.8	22.7
50 v/o ± 30-deg cross ply:				
Longitudinal...........	5	128	...	31.2
Transverse............	5	43.4	32.6	23.5
45 deg................	3	99.6	55.0	23.0
50 v/o unidirectional:				
Longitudinal...........	11	165	...	30.6
Transverse............	5	23.8	21.6	18.7
45 deg................	3	39.1	20.9	16.7
37 v/o unidirectional:				
Longitudinal...........	5	155	...	20.2
Transverse............	5	22.8	17.4	14.8
25 v/o unidirectional:				
Longitudinal...........	5	122	...	23.4
Transverse............	5	20.5	18.7	14.7
45 deg................	3	34.8	24.5	19.7

Table 4—*Compression test results of 0.020-in.-thick material. Fixture provided lateral support.[a]*

Specimen No.	Shape of Specimen	Ultimate Strength, ksi	Yield Strength, ksi	Modulus of Elasticity
30-S-12...........	reduced	203	...	29.8
A30-S-12..........	reduced	150	...	39.4
30-S-8............	straight	182	...	30.5
A30-S-8...........	reduced	171	...	...
49-S-1............	straight	156	...	30.1

[a] All specimen 50 v/o unidirectional composite.

sive strength exceeds the tensile strength and whether it is higher at low-filament content. The fact that the specimens of the 0.020-in.-thick material gave results comparable to those obtained on the 0.080-in. specimens is significant. The 0.080-in. longitudinal specimens failed at the ends by brooming; however, those fitted with end caps showed no significant improvement in strength. The results from the tests on the 0.080-in. material are the same as the results of the 0.020-in.-thick material that failed in the center. One must also consider that these thin specimens could have buckled, but two factors suggest that this is not so. First, the section is reduced symmetrically to the center which is the maximum stress area. Thus, the length of specimen under maximum stress is very short. Also, since the ends are restrained to prevent buckling, we have a fixed end column in the test section which reduces the effective length to half of the actual length [3].

The possibility of lateral buckling was checked by making a fixture with support rods (Fig. 4). When straight sided and reduced section specimens were used, the measured strength and modulus are in substantial agreement with the 0.080-in.

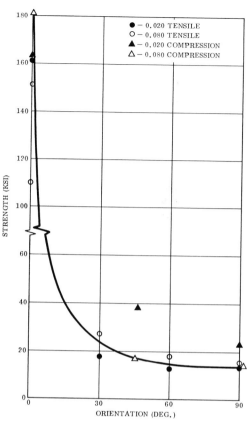

FIG. 8—Comparison of strength for four conditions of 50 volume percent unidirectional material.

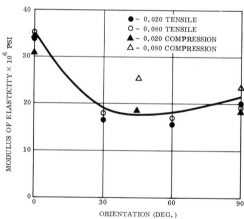

FIG. 9—Comparison of modulus of elasticity for four conditions of 50 volume percent unidirectional material.

specimens run with the Montgomery-Templin grip, and the short, thin specimens run with no lateral support.

One difference between tensile loading and compressive loading of aluminum-boron that is of some significance was noticed. When the filaments are oriented in directions other than the direction of load application, the specimens deform by yielding. Most of the tension testing performed by the authors [2] indicates very little plastic deformation in tension. The filaments exert a marked constraining effect upon the matrix. The transverse modulus of elasticity of unidirectional composites is about twice that of aluminum [2]. The effect of the filaments in the case of compressive loading is to subject the aluminum matrix to essentially hydrostatic pressure. In this state, the modulus is increased, and the matrix is capable of extensive plastic flow. In the case of tension, the matrix is in a state of triaxial tension which increases the modulus but restrains the matrix to little or no plastic deformation.

STRUCTURAL TEST RESULTS

ELEMENT CRIPPLING

The test results for the twelve specimens in this series are plotted in Fig. 10 for the one edge free case and in Fig. 11 for the no edge free case. The smooth curves drawn in the two figures are calculated values of the initial and ultimate crippling of flat elements with a modulus of 25.6×10^6 psi. Standard test coupons taken from the same panel as the crippling test specimens indicate that the reinforcement has an average modulus of 25.6×10^6 psi and an ultimate strength of about 100 ksi. Both the modulus and strength of this early experimental material are well below the tentative specification levels for 50 percent unidirectional aluminum-boron.

The theoretical initial buckling expression for an orthotropic plate having four fixed edges has been derived by Lekhnitskii [4]. A similar expression for the three edge free case has been derived by S. N. Dharmarajan for Convair. The resulting equation is transcendental in form as shown:

$$\beta \left(\alpha^2 - \mu_{xy} \frac{m^2 \pi^2}{a^2} \right)^2 \tan{(t\alpha b)} = \alpha \left(\beta^2 + \mu_{xy} \frac{m^2 \pi^2}{a^2} \right)^2 \tan{(\beta b)}$$

where α and β are:

$$\alpha = \left(\frac{m\pi}{a} \right) \sqrt{ \frac{D_{xy}}{D_y} + \sqrt{ \left(\frac{D_{xy}}{D_y} \right)^2 - \frac{D_x}{D_y} + \frac{N_x}{D_y} \left(\frac{a}{m\pi} \right)^2 } }$$

$$\beta = \left(\frac{m\pi}{a} \right) \sqrt{ -\frac{D_{xy}}{D_y} + \sqrt{ \left(\frac{D_{xy}}{D_y} \right)^2 - \frac{D_x}{D_y} + \frac{N_x}{D_y} \left(\frac{a}{m\pi} \right)^2 } }$$

The conventional nomenclature is used as explained in Ref 2. Dr. Dharmarajan also has derived expressions for the buckling of curved orthotropic elements which also are found in Ref. 2. Because of the trial-and-error solution involved in transcendental equations, the expressions for orthotropic element crippling have been programmed for computer solution. The curves of Figs. 10 and 11 were so derived.

The ultimate crippling behavior of either isotropic or orthotropic elements is very much more complex than the initial buckling behavior since inelastic, nonlinear strength properties are involved, as well as the conventional elastic properties. A successful approach for isotropic materials is a semiempirical ultimate strength method first described in Ref 5. This method was used by the authors, with appropriate modification of the Ramberg-Osgood expression defining the inelastic portion of the stress-strain curve, to produce the ultimate crippling curves of Figs. 10 and 11. The resulting relationship used is given below:

$$\left(\frac{F_{cc}}{\eta} \right)_{\text{Al-B}} = (E_c)_{\text{Al-B}} \left\{ \frac{F_{cc}}{9.7 \times 10^6} \left[1 + \frac{30}{7} \left(\frac{F_{cc}}{46,000} \right)^9 \right]^{1/2} \right\}_{\substack{2024-\text{T36} \\ \text{clad}}}$$

$$\eta = \sqrt{E_t / E}$$

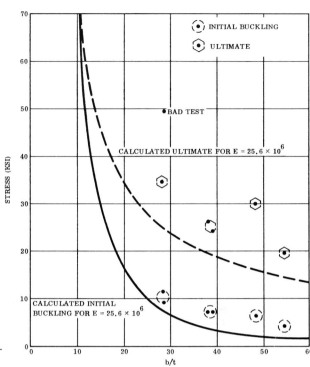

FIG. 10—Crippling tests—one edge free.

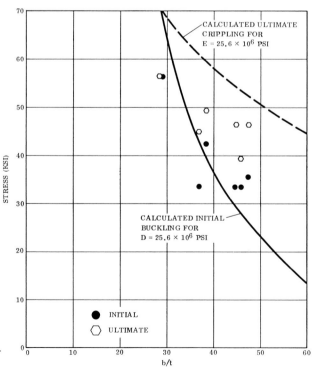

FIG. 11—Crippling tests—no edge free.

Table 5—*Stiffener test results.*

SPEC. NO.	SPEC. TYPE	HEAD RATE IN./MIN.	CHART SPEED IN./MIN.	LOAD AT FIRST VISIBLE BUCKLE (LB.)	BUCKLE LOAD BY DEFLECTION	MAX. LOAD AT FAILURE (LB.)
1	Hat	0.05	1.0		1060	1260*
2	Zee	0.01	1.0	1340	1180	1355
3	Hat	↑	1.0	500	500	2675
4	Zee		10.0	1000	1075	1265
5	Zee		1.0	1100	1060	1220
6	Hat		↑	1200	1200	1919*
7	Zee			1000	830	1065
8	Zee			900	800	920
9	RndHat			N.R.	4250	4400
10		↑		N.R.	3240	3600*
11				N.R.	N.R.	1320*
12				3000	3830	3920*
13		↓	↓	N.R.	3880	4130
14	RndHat	0.01	1.0	N.R.	N.R.	4120

*Specimens broomed at ends

SPEC. NO.	SPEC. TYPE	t	A	B	C	D	LENGTH (IN.)	Al–B AREA (IN.^{2})	TOTAL SPEC. AREA (IN.^{2})
1	Hat	0.030	0.97	1.125	0.60	0.66	0.95	0.0198	0.1237
2	Zee	0.030	0.47	1.125	0.49	0.28	1.84	0.0084	0.0588
3	Hat	0.030	1.02	1.110	0.56	0.66	3.57	0.0198	0.1228
4	Zee	0.030	0.50	1.110	0.41	0.33	1.93	0.0099	0.0545
5	Zee	0.029	0.50	0.80	0.44	0.36	1.08	0.01045	0.0445
6	Hat	0.029	0.85	0.83	0.40	0.46	3.11	0.01335	0.0883
7	Zee	0.029	0.50	0.82	0.28	0.25	2.54	0.00725	0.0428
8	Zee	0.029	0.52	0.82	0.35	0.28	1.28	0.0110	0.0454

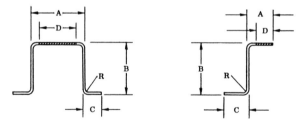

Flat hat & zee specimen geometry

Buckling and crippling tests are notoriously erratic, but it can be seen that the initial buckling values follow the theoretical curves for both edge restraint conditions quite well. The ultimate test values for the one-edge-free case are in reasonable agreement also with the lower bound calculated values. Little correlation was obtained in the no-edge free ultimate strentgh values. This is felt to be a result of inadequate edge support of the split tubes at the higher stress levels encountered in these tests. A more promising test method, which requires thicker material and larger specimens than were available in this series of tests, is one employing V-groove edge support blocks.

The importance of these tests, of course, is to verify the general method of calculating crippling behavior for orthotropic aluminum-boron, rather than to establish a totally empirical crippling criteria. The correlation obtained between theory and test was considered sufficient to proceed with confidence to the next level of instability testing—the behavior of formed stiffener elements in axial compression.

STIFFENER CRIPPLING

The results of the two series of stiffener tests are indicative of the improvement in basic material processing achieved over the past year. The initial test specimens (hat and zee sections with selectively reinforced flanges) were formed from material with a modulus of elasticity of 24×10^6 psi and an ultimate strength of 70 ksi in reinforced areas. The latter test specimens (round-top hat sections) were formed from material with a modulus of 32×10^6 psi and an ultimate strength of 160,000 psi.

The results of these tests are shown in Table 5. Several of the specimens failed by brooming of the ends, rather than by ultimate crippling. This type of failure is one of the exasperating features of testing the advanced, high-strength composites. It does not occur in all cases, however, so that useful test data were obtained from the tests.

In all the specimens tested, initial buckling occurred in the free flange first, and progressed slowly up the sides to the reinforced flange (or cap in the case of the hat sections). This behavior was expected, based on the b/t ratio and fixity of the various flanges. As in the flat element crippling tests, the value of the stiffener tests is in verifying the method of calculating stiffener crippling behavior rather than in establishing purely empirical values of crippling strength.

The method of prediction used is a summation of elements technique based on the crippling strength of each separate flange of the stiffener section. Values for initial and ultimate element buckling are based on the expressions described in the previous section. Figure 12 is a plot of flat plate crippling strengths for current specification value 50 percent unidirectional aluminum-boron having a modulus of 32×10^6 psi and an ultimate strength of 160,000 psi. A similar set of curves was also necessary for unreinforced 6061-F material to predict the behavior of the hat and zee sections with unreinforced flanges. Figure 13 plots these values, derived from experimental stress-strain behavior of the as-processed, unreinforced aluminum matrix.

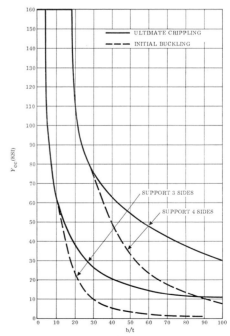

FIG. 12—Flat element crippling for 50 volume percent unidirectional aluminum-boron (6061-F).

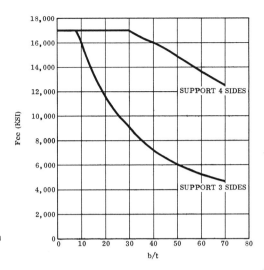

FIG. 13—6061-F aluminum crippling curves.

Table 6—*Comparison of theory and test stiffener crippling strengths.*

Specimen No.	Test Failing Load, lb	Calculated Failing Load, lb	Deviation, percent	Average Stress at Failure, ksi
1	1260	...	...	...
2	1355	1252	−8	...
3	2675	2610	−2	...
4	1265	1162	−8	...
5	1220	980	−19	...
6	1919	2370	−23	...
7	1065	995	−7	...
8	920	912	+9	...
9	4400	4432	−1	71.5
10	3600	4041	−12	61.0
11	1320	...	...	...
12	3920	4155	+6	59.0
13	4130	4170	+1	71.0
14	4120	4113	+1	86.3

Table 6 compares the analytical predictions with the experimental crippling behavior and also indicates the average stress in the stiffeners at ultimate failure. The area used in calculating average stress is the equivalent aluminum-boron area which consists of the actual aluminum-boron cross section and the reduced area of the unreinforced elements. This can be expressed mathematically as:

$$A_{\text{equivalent}} = A_{\text{Al-B}} + \frac{E_{\text{Al}}}{E_{\text{Al-B}}} (A_{\text{Al}})$$

As evidenced in Table 6, the average ultimate stress for good quality composite stiffeners is as high as 86 ksi, about 2½ times the strength of a well designed 7075-T6 aluminum stiffener of the same weight. In addition, the correlation between calculated and experimental behavior is excellent, especially for the later high-modulus and strength hat sections.

Composite Materials

CONCLUSIONS

Structural tests have been performed which verify analytical methods for predicting the crippling behavior of both flat elements (shown in Figs. 10 and 11) and formed stiffener shapes (shown in Table 6) of selectively reinforced, unidirectional aluminum boron. These methods require that accurate compressive stress-strain relationships for the composite laminates be determined by material testing.

An outgrowth of these tests is the realization that aluminum-boron exhibits considerable post-buckling strength similar to conventional isotrophic metals and that this added measure of strength can be predicted by available analytical tools. Note that the ultimate crippling strengths in Fig. 10 are three to four times the initial buckling strength. We conclude from this that aluminum-boron structures can be and, for efficiency should be, designed on an ultimate strength basis rather than on the no-buckling limit design approach normally used with polymer matrix composites.

It is evident from the data shown here that improvement in structural test methods for compression testing aluminum-boron is highly desirable. Larger flat elements with V-block edge supports may prove superior to the split tube supports used here. Stiffener elements with brazed doublers on each end or with potted ends could be used to prevent the brooming failures experienced in several tests of this type.

The modulus of elasticity of aluminum-boron composites is essentially the same in tension and compression at room temperature, as shown in Fig. 9. There does not seem to be an effect due to thickness. The compressive strength is at least that measured in tension at room temperature, as shown in Fig. 8.

The most common mode of failure of the longitudinal specimens is by brooming at the ends, and elaborate precautions must be taken to prevent this. Even then, the resulting strengths are not necessarily higher. The existing ASTM compression test method for metals, E 9-61, is an adequate starting point for aluminum-boron composites.

ACKNOWLEDGMENT

This work was partially supported by the Air Force Materials Laboratory under Contract No. F33615-67-C-1548, with Capt. W. J. Schulz as Air Force contract manager.

References

[1] Davis, L. W., "How Metal Matrix Composites Are Made," *Fiber-Strengthened Metallic Composites, ASTM STP 427*, American Society for Testing and Materials, 1967, pp. 69–90.

[2] "Evaluation of the Structural Behavior of Filament Reinforced Metal Matrix Composites," GDC-AVP-67-001, 5 volumes, General Dynamics Convair.

[3] Popov, E. P., *Mechanics of Materials*, Prentice-Hall, New York, 1952.

[4] Leknitskii, S. G., "Anisotropic Plates," University Microfilms, Ann Arbor, Mich.

[5] Cozzone, F. P. and Melcon, M. A., "Non-Dimensional Buckling Curves –Their Development and Application," *Journal of Aeronautical Sciences*, Oct. 1946.

Preliminary Evaluation of Test Standards for Boron Epoxy Laminates

**E. M. LENOE,[1] M. KNIGHT,[2]
AND C. SCHOENE[1]**

Reference:

Lenoe, E. M., Knight, M., and Schoene, C., "Preliminary Evaluation of Test Standards for Boron Epoxy Laminates," *Composite Materials: Testing and Design, ASTM STP 460,* American Society for Testing Materials, 1969, pp. 122–139.

Abstract:

This paper describes the first phase of a program of critical evaluation and development of test standards for advanced composite materials.

Emphasis is on flat laminates, and the work deals primarily with boron epoxy laminates fabricated using the 3M three-inch wide preimpregnated (prepreg) tape system. The results of approximately 1800 experiments are reported, and specific recommendations concerning specimen geometries and testing techniques are presented.

Four types of experiments were investigated, namely, tension, compression, flexure, and shear tests. Tension test variables examined were ply thickness, orientation of reinforcement, gage length and width, type of grip, specimen edge finish, loading rate, specimen alignment, and method of strain measurement. In compression testing, the controlling features investigated were number of plies, specimen width and length, unsupported gage length and end cap, and specimen cross-sectional geometry. The flexural experiments considered the effects of orientation of reinforcement, number of plies, load contact radii, span and type, and spacing of loads.

Interlaminar (short-beam) shear tests were considered, and the effects of span-to-depth ratio were determined. Torsion test data for rods of boron epoxy materials are discussed also with regard to the observed shear response of the resin system.

Key Words:

composite materials, boron, epoxy laminates, testing, tension, compression, flexure, shear, evaluation, tests

The purpose of this paper is to present the initial results of a program the objective of which is to review existing test techniques for advanced composites and to develop recommendations for standard text procedures.

Numerous aerospace companies have gained experience in the fabrication and testing of high-strength, high-modulus composites. The starting point for the program was a review of the more common existing techniques. In general, the types of coupon test procedures discussed here are not sufficient to obtain structural properties of advanced composites for design purposes. It also is evident that the fabrication of structures consisting of flat laminate elements will necessitate test procedures for quality control and hardware acceptance. Thus, it is imperative that test standards be developed to allow for purchase and delivery of structures built with such materials, and it is toward this objective that our efforts currently are aimed.

Composite materials composed of high-strength, high-modulus fibers placed in a resin matrix are evaluated ordinarily by experiments on relatively small specimens. The numerous defects—such as voids and nonuniformity in reinforcement spacing

---

[1] Group leader and senior engineer, respectively, Avco Space Systems Division, Avco Corp., Lowell, Mass. 01851.

[2] Project monitor, Air Force Materials Laboratory, Wright-Patterson Air Force Base, Ohio. 45433.

Table 1—*Summary of test plan (boron-epoxy).*

Type Test	Parameter
Tension............	specimen preparation
	tab application
	edge preparation
	tab material
	bond line
	strain rate
	specimen geometry
	0 and 90 deg fiber orientation
	cross and angle ply
Compression........	rectangular cross section
	specimen geometry
	fiber orientation
	circular cross section
	specimen geometry
Flexure............	specimen geometry
	load introduction
Shear	
Short beam.......	specimen geometry
	load introduction
Torsion..........	rods

and orientation—variabilities, and the inherent statistical nature of the moduli and strengths of constituents necessitate performing an extensive experimental evaluation of the specimen geometries and load introduction techniques. Establishing firm conclusions when dealing with large numbers of possible controlling influences would require an extraordinarily large series of experiments, particularly when the magnitude of the influences are small. Therefore, we seek only for indications of trends and to determine the relative importance of the controlling factors during the initial effort of the program. Thus, the number of replicas of each particular series of experiments rarely exceeded five, and the expected outcome of the work is the formulation of a detailed test plan to evaluate our tentative test standards.

This paper points out the kind, degree, and nature of controlling parameters such as specimen geometry and load introduction methods. Size effects, as well as the loading arrangement, are shown to have significant influence on observed strength and modulus values for the various stress states. The results serve to add emphasis to the importance of adopting uniform test standards for these unique reinforced plastics. Table 1 is a summary of the test plan.

The boron epoxy system was selected as the basic material for development of the recommendations. The particular laminates under discussion were fabricated using the 3-in.-wide boron epoxy preimpregnated (prepreg) tape system. Narmco and 3M were used as suppliers. The laminates were made by an autoclave pressure molding technique at 75 psi and following the recommended manufactures cure cycle. Flat laminates 9 by 9 in., with nominal 50 volume percent of boron reinforcement and with different numbers of plies, ranging from 6 to 100 were made. Most of the work was done on laminate with 0/90-deg reinforcement orientation.

TENSION

A variety of coupon specimen types are used to determine the tensile properties of advanced composites. They differ principally in size, shape, and loading technique. The straight-sided rectangular strip coupon has been used widely, and with care in testing provides acceptable test results. The specimens have tabs bonded to the ends by a lap-shear joint. Such tabs are load spreading devices and are used for introducing tension loads into the coupon. There are a number of factors which would conceivably influence the load-deformation response of the tensile coupon. Those included in this program are discussed with the data presented in the following sections.

INFLUENCE OF TAB APPLICATION AND SPECIMEN PREPARATION

It was concluded that no special edge or bond area surface treatment beyond normal care in cutting procedures and observation of manufacturer's recommendations for adhesive is required to produce satisfactory fiber glass tabs on unidirectional boron epoxy tension specimens.

INFLUENCE OF TAB MATERIAL

Most of the tension tests on advanced composite materials have utilized fiber glass tabs, bonded to the ends of the straight-sided test specimens. Modifications to this approach include the use of tapered metal insert tabs in order to achieve a more favorable load transfer at the specimen ends. The fiber glass tab has proven successful for certain specimen geometries and composite thicknesses. In general, for relatively thick specimens and at elevated temperatures, the tabs are inadequate, and the approach is to use localized heating. For this reason, we have tried several additional tab designs including:

1. Bonded aluminum tabs.
2. Titanium tabs bonded to a thin fiber glass layer which in turn is adhered to the composite.

The aluminum tab design was tested at temperatures of -67, 73, and $+375$ F. All failures occurred at the edge of the tabs. The tentative conclusions reached are that the aluminum, in the design used does not serve as an adequate tab material, and additional work will have to be done to determine if a satisfactory design can be found. The titanium tabs bonded to a thin fiber glass layer did not appear to offer significant advantage over the all fiber glass tabs. Added expense and complex preparation are debit features which suggest the use of a simpler fiber glass tab. The fiber glass tab appears adequate only for modest elevated temperatures up to the capabilities of the tab adhesives.

INFLUENCE OF BOND LINE

Specimens were tested to determine the influence of adhesion on a variety of bond lines. No significant variation was found in values for strength or modulus. From posttest visual inspection, it appears that an important factor in obtaining reproducible test data was the uniformity of the bond line, more so than the choice of adhesive.

INFLUENCE OF STRAIN RATE

Since the strain rate dependence of the boron epoxy will be influenced strongly by the resin materials, a detailed characterization of Narmco and 3M resin matrix was conducted. Tensile, compressive, and shear properties of the epoxy were determined at 73 and 350 F at strain rates of 0.005, 0.05, and 0.5 in./in./min. Thermal expansion, creep, and relaxation properties also were studied.

Considering the possible cumulative effects of time-dependent moduli, strengths, and strain to failure, when used in composites, it appears that the resin itself possibly will introduce more than 20 percent change over the range of two decades of strain. Such viscoelastic phenomena would be most evident in angle- and cross-ply composites, and very little influence would be observed in unidirectionally reinforced composites. Nonetheless, it is evidently desirable to adopt a uniform strain rate

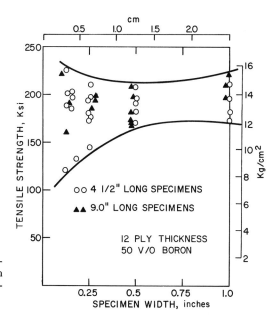

FIG. 1—Influence of specimen width and length on tensile strength of unidirectional boron epoxy.

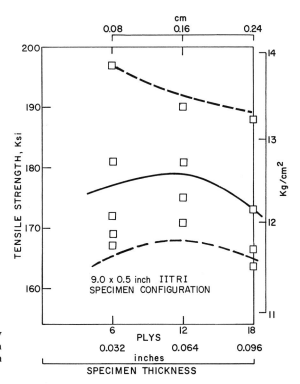

FIG. 2—Influence of ply thickness on tensile strength of unidirectional boron epoxy.

for general quality control procedures and in all other instances to take cognizance of time rate effects by matching the test conditions to the particular service applications.

INFLUENCE OF SPECIMEN WIDTH, LENGTH, AND THICKNESS ON UNIDIRECTIONALLY REINFORCED COMPOSITE

Figures 1 and 2 contain the results of an evaluation of geometric effects on performance of unidirectionally reinforced boron epoxy. In Fig. 1, the data for the short specimens seem to exhibit more scatter. We attribute this to the more exaggerated influence of nonuniform bond line and edge constraints. On the basis of the results, the narrower specimens in the range of 0.25 to 0.5 in. are recommended.

Ultimate strength measurements on 6, 12, and 18-ply composite specimens are represented in Fig. 2. The specimen configuration was 9 in. long and a ½ in. wide. The tensile coupon was similar to the IITRI specimen,[3] with the exception that the tabs were lengthened to 2½ in. The data suggest some possible decrease

[3] *Air Force Materials Laboratory Structural Design Guide for Advanced Composite Applications*, Wright-Patterson Air Force Base, Ohio.

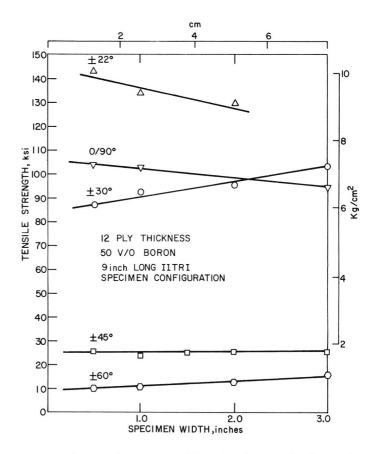

FIG. 3—Influence of specimen width on tensile strength of cross-ply and angle-ply boron epoxy.

in apparent strength for thicker specimens, and this suggests an upper bound on specimen thickness to be used.

INFLUENCE OF SPECIMEN WIDTH ON CROSS- AND ANGLE-PLY COMPOSITES

In this instance the coupons tested were 9 in. long and consisted of 12-plies. Figure 3 shows the influence of specimen width on tensile strength for the various angle-ply layups. The figure shows that strength trends from 10 percent decrease (0/90, ±22 deg) to greater than 20 percent increases (±60 deg) were observed with increasing width. Typical stress-strain response was determined for the ±22-deg angle ply for ½-in.-wide and 1-in.-wide specimens. Close similarity in behavior was observed for the two widths for the axial strain. Slight differences appeared for the transverse strains. The differences, however, are within typical experimental scatter.

Table 2—_Evaluation of tension specimens._

Variable Studied	Parameter	Remarks
1. Specimen preparation		
(a) Tab application	sanding versus chemical cleaning versus peel ply	no discernible influence on strength or modulus ($<10\%$ variation); therefore, use most economical procedure
(b) Machining....	diamond cut-off versus hand polishing (400 grit emery paper)	
2. Tab material........	aluminum, fiber glass, titanium outer cover on fiber glass	aluminum resulted predominantly in grip failures; the other materials were adequate up to post cure temperature of the adhesives
3. Bond line...........	adhesives: HT 424, AF 130, Epon 931, Eastman 910	no discernible influence, all adequate; Eastman 910 is recommended for room temperature due to thin, uniform bond line and resulting better specimen alignment
4. Strain rate effects....	strain rate sensitivity evaluation of resin (tension, compression, torsion at 73 and 350 F for $\dot\epsilon$ = 0.005, 0.05 and 0.5 in./in./min, and tension creep and relaxation tests)	$\sim$20% cumulative change in modulus, strength, and strain to failure; obviously, most important in angle-ply composites
5. Specimen geometry unidirectional composite; 0 and 90 deg	specimen width, length, and thickness (6, 12, and 18-plus; 4½ and 9.0 in. long, 0.25 to 1.0 in. wide)	(a) 0 deg—very little influence for ranges investigated; data scatter increases due to tab bond alignment problems (b) 90 deg—longer and wider coupons are apparently more sensitive to laminate bowing
6. Specimen geometry; cross and angle ply	specimen widths from ½ to 3.0 in. for 9-in.-long 12-ply laminates	depending on fiber orientation, strength changes of from minus 6% to plus 21% were observed for the range of widths tested

Recommendations—The IITRI type specimen should be used primarily for unidirectional laminates or laminates with the predominate reinforcement aligned with the load axis. Fiber glass tabs bonded with Eastman 910 are adequate for room temperature, HT 424 for elevated temperatures. Tests for plies ranging from 6 to 18 widths of ¼ to ½ in. and lengths shorter than 9.0 in. will give good results with proper tab bond alignment

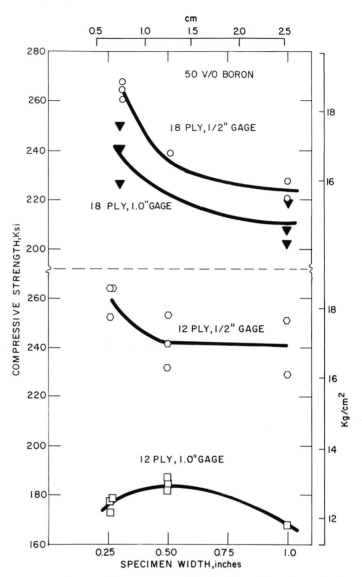

FIG. 4—Influence of specimen width, length, and number of plies on compressive strength of unidirectional boron epoxy.

Observed nominal stress-strain response for various other orientations also was studied. The 0/90-deg and ±22-deg composites exhibit fairly linear behavior while the ±30, ±45, and ±60-deg boron epoxy coupon specimens behave in an increasingly nonlinear manner. For angle plies larger than 22 deg the coupon specimens exhibit increasing shear deformation and fiber pullout.

The results of the work on the influence of variables on tensile strength are summarized in Table 2. The remarks are based on initial impressions of the influence of the variables and will serve as the basis for formulation of additional test plans during the program.

COMPRESSION TESTS

Two types of compression coupon specimens were evaluated: (1) rectangular configurations and (2) cylindrical specimens. Both types employed mild steel load platens which were recessed to fit around the composite. Tolerance and alignment of this fit were found to be extremely crucial to minimize scatter and obtain high-strength values.

The rounded steel load platens were designed to match smooth contoured surfaces in the compression fixture. Hardness of the load platens is an important factor as the boron filaments indent the steel. Indentation of fibers provides a degree of end fixity over and above the epoxy bond line effect but also limits reuse of the load platens as they soon become too uneven for good alignment. In the circular rods, for example, eccentricity of a few mils will introduce bending stresses greater

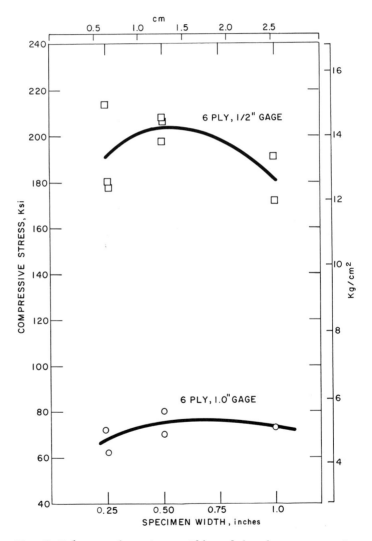

FIG. 5—Influence of specimen width and length on compressive strength of 6-ply unidirectional boron epoxy.

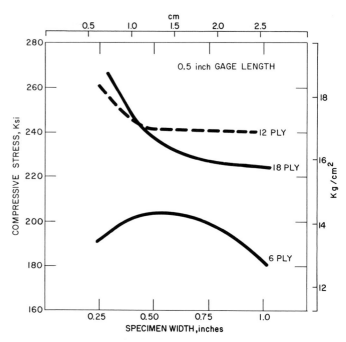

F<small>IG.</small> 6—Comparison of specimen width influence on compressive strength of 6, 12, and 18-ply boron epoxy.

than 60,000 psi. Parameters investigated for the rectangular cross sections were ply level, length, and width. Unidirectional fiber orientations were evaluated with loading parallel and perpendicular to the filaments. Results of the compressive experiments are shown in Figs. 4 through 10. Strength as a function of specimen width and gage length for 12 and 18-ply specimens appears in Fig. 4. The data show that the 1-in.-long, 18-ply composite is about 10 percent weaker than the ½-in.-long geometry. The corresponding reduction for the 12-ply material is approximately 30 percent. Another important result was that the ¼-in.-wide specimen generated higher values than the ½-in. or wider specimen.

Figure 5 shows that the 6-ply specimen gives much lower results. The influence of width is apparently random. Our conjecture is that alignment problems are too severe for the thin laminates, unless exceptional precautions are taken in fixture and specimen preparation.

Figure 6 is an attempt to summarize the data and suggests that this type of compression test be restricted to 12 to 18-ply materials and that the specimens be nominal ¼ in. widths and ½ in. lengths. Figure 7 demonstrates that bending eccentricities are larger in the 6 and 12 as compared to the 18-ply composite. These bending stresses are observable at fairly low load levels and probably are due to nonuniform indentation of boron into the steel platens, as well as overall misalignment.

Figures 8, 9, and 10 are the data obtained on the 90-deg specimens. The conclusion again is reached that the ¼-in.-wide and ½-in.-long geometry is most suited to this type of coupon test.

Compression experiments were completed also on circular rods with nominal 55 volume percent boron. The median value is 250,000 psi, but the highest strengths obtained ranged up to 358,000 psi.

Table 3 summarizes the results of the compression tests.

Table 3—*Evaluation of compression specimen coupons with recessed bonded steel load caps.*

Variable	Parameter	Remarks
Rectangular cross section, 0 and 90-deg fiber orientation	width, length, thickness ¼ to 1.0 in., 6, 12, and 18-plies	strength variations >100%, as a function of increased gage lengths
Circular cross section...	gage length (with constant ³⁄₁₆ in. diameter)	relatively small strength variations were observed; ultimate strength values considerably higher than for rectangular cross section

Recommendations—Avoid 6-ply and thinner material; use short narrow specimens ~ ¼ by ½ in. and take extra care in alignment of load train.

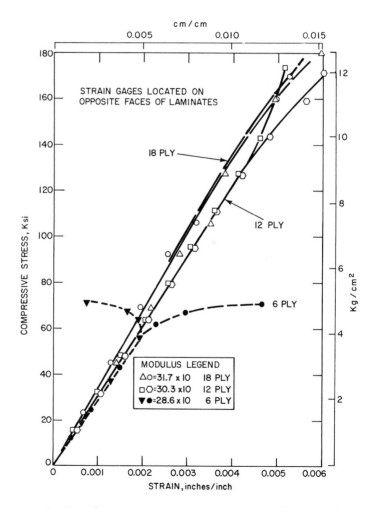

Fig. 7—Typical compressive stress-strain response of 6, 12, and 18-ply boron epoxy.

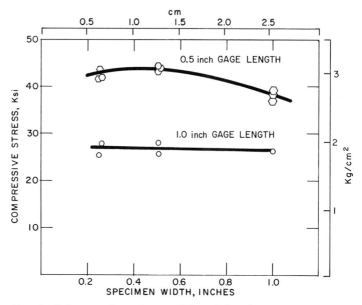

FIG. 8—Influence of specimen width and length on transverse compressive strength of 12-ply unidirectional boron epoxy.

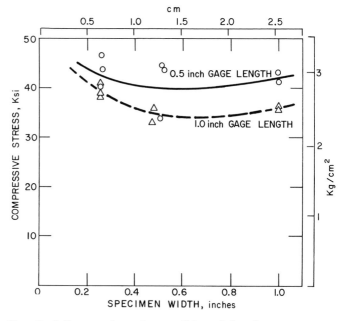

FIG. 9—Influence of specimen width and length on transverse compressive strength of 18-ply unidirectional boron epoxy.

FLEXURE TESTS

Experimental flexural strength and modulus values are influenced markedly by a number of variables including:
1. Fiber orientation and volume percent.
2. Specimen configuration.
3. Span-to-depth ratio.
4. Load-point contact radius and type (three versus four-point loads)

The variables considered in this program were specimen width, thickness, span-to-depth ratio, diameter of load contact points, and three versus four-point loads.

Boron epoxy flexure beams ordinarily fail on the tension side for selected span-to-depth ratios. In a 15-ply beam, for example, the first four layers on the tension side were fractured completely, and a small percentage of the next two also failed. The remaining filaments did not fracture. The distribution and percentage of fiber breaks does not coincide with simple beam theories for predicting flexural strength. Further posttest examinations of the beams of different widths currently are underway to establish the most likely failure modes.

Figures 11 through 14 present the parametric evaluation of beam width, number of plies, test temperature, support diameter, span, and type of load, while Table 4 summarizes the results of the experiments.

SHEAR TESTS

The interlaminar shear strength of composite materials often is determined by means of the so-called short-beam shear test method. In addition to quality control of production materials, the test is used quite often to compare different materials as well as to assess processing methods. In general, the test method is not appropriate for such usage as the experimental results are influenced markedly by a variety of factors including:
1. Constituent materials.
2. Specimen geometry: width, thickness, and length.
3. Fiber orientation.

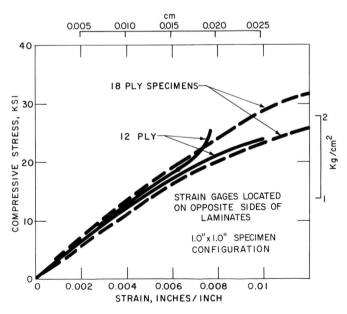

FIG. 10—Typical transverse compressive stress-strain response of 12 and 18-ply boron epoxy.

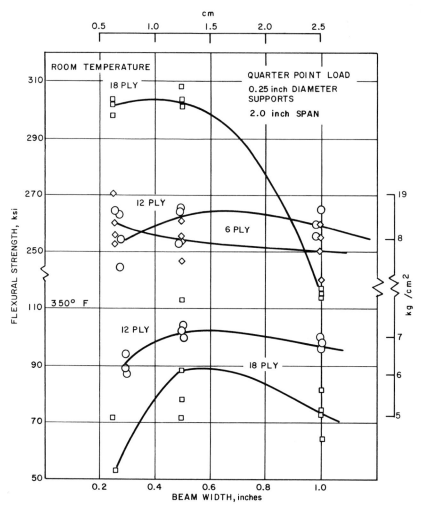

FIG. 11—Influence of beam width, number of plies, and test temperature on flexural strength (2.0-in. span) of unidirectional boron epoxy.

Table 4— *Evaluation of flexure specimens.*

Variable	Parameters	Remarks
Specimen geometry.....	width ($\frac{1}{4}$ to 1.0 in.) thickness (6, 12, and 18-ply)	widths $< \frac{1}{2}$ in. result in strength reductions effects are more pronounced at 350 deg
Loading..............	(a) Load contact diameter ($\frac{1}{8}$ and $\frac{1}{4}$ in. diameters)	at 350 deg, maximum strength variations were $\sim$20% and modulus variations $\sim$8%
	(b) span-to-depth ratio	most pronounced differences were at 350 deg and for 90-deg beams

Recommendations—Beam widths should be $> \frac{1}{2}$ in. and loading point diameters $\frac{1}{4}$ in. Span-to-depth ratios, etc. should be specified. Only one type of loading should be adopted to permit comparison of data. The test should be restricted to quality control usage.

4. Beam span.
5. Deflection stiffness and alignment of the test apparatus.
6. Nature of the load supports: diameter, shape, hardness, and modulus.

INFLUENCE OF BEAM WIDTH

From the results presented in Fig. 15, the data suggest that the wider beams appear to yield slightly lower nominal strengths. Figure 16 also shows a decrease in strength for the wider beams with a greater decrease associated with smaller diameter load supports.

INFLUENCE OF BEAM SUPPORT DIAMETER

Figure 16 demonstrates that the apparent shear strength measured using larger diameter load supports is higher on beams of various widths tested at 300 F.

INFLUENCE OF SPAN-TO-DEPTH RATIO

There has been a great deal of study of the effect of span-to-depth ratios necessary to achieve the desired interlaminar shear failure. Theoretically, this type of test should be satisfactory only so long as a pure shear failure occurs, and this is, of course, a difficult condition to attain. It is possible to solve the theoretical elasticity problem of determining the shear stresses in an orthotropic beam. The formula recently developed by Sattar and Kellog[4] of United Aircraft Corp., for instance, shows a significant effect of width-to-depth ratio on the magnitude and distribution of shear stresses in a short beam and provides approximate procedures for selecting various width-span-depth ratios to attempt to achieve only pure interlaminar shear failures.

[4] See p. 62.

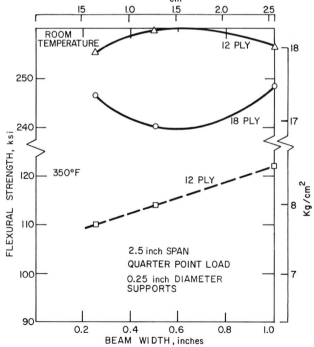

FIG. 12—Influence of beam width, number of plies, and test temperature on flexural strength (2.5-in. span) of unidirectional boron epoxy.

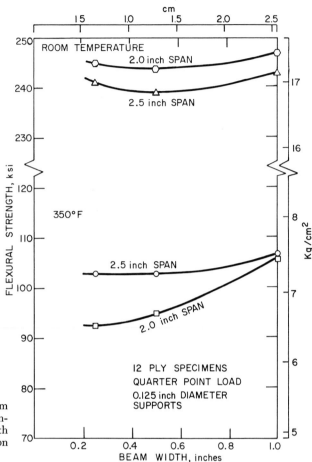

FIG. 13—Influence of beam width, span, and test temperature on flexural strength of unidirectional boron epoxy.

Table 5—*Evaluation of coupon shear tests.*

Variable	Parameters	Remarks
Short beam shear:		
Specimen geometry..	beam width and length	variations of $\sim$10% can be introduced by contact diameter
Loading............	load contact diameter and span-to-depth ratio	span-to-depth ratio increases drastically reduce apparent strength
Torsion tests:		
Uniaxially..........	$3/16$ in. diameter	strength comparable to resin properties are obtained; shear modulus can be measured

Recommendations—Short-Beam Shear: (for quality control only) Comparisons and materials judgments should be restricted to very similar components wherein close control of beam dimensions, resin content, and fiber orientations are maintained. Specimen dimensions should be approximately 0.25 by 0.60 in., and the majority of data generated to date has been on 15-ply. Thickness of the specimen must be reported. Comparisons of strength obtained with different ply levels do not appear to be justified. Suggested beam support diameter is 0.25 in. The support material type and hardness should be specified. Fiber orientations should be controlled closely and the test restricted to unidirectional directions.

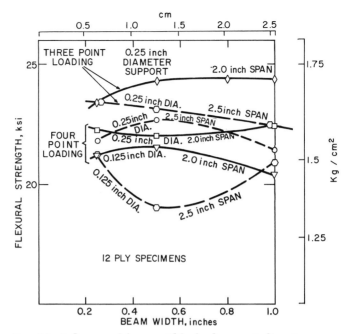

FIG. 14—Influence of beam width, load support diameter, span, and loading mode on transverse flexural strength of unidirectional boron epoxy.

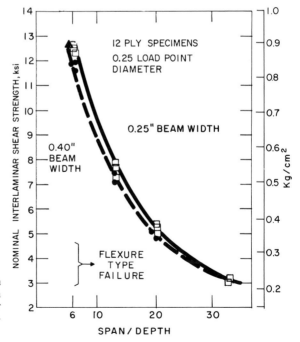

FIG. 15—Influence of span to depth and beam width on nominal interlaminar shear strength of unidirectional boron epoxy.

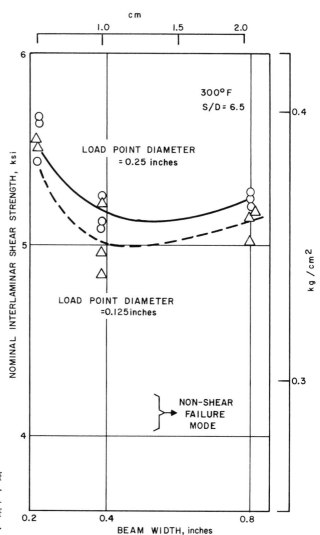

Fig. 16—Influence of beam width and load support diameter on interlaminar shear strength of unidirectional boron epoxy.

Such analysis, however, cannot account for a number of controlling factors, primarily due to the difficulty of obtaining an exact three dimensional elasticity solution for finite specimens under realistic load introduction conditions.

The load on the beam is neither a line load nor a uniformly distributed pressure under the beam supports. These load supports are ordinarily circular, and the actual load is difficult to assess.

In performing this test, it is well to recall that the experiment is designed supposedly to introduce a pure interlaminar shear. Interpretation of the test results requires examination of the fractured specimens to disclose whether a true "interlaminar" failure did occur or whether the reinforcement had been fractured. In general, this procedure requires an experienced or at least trained laboratory test technician. At elevated temperatures, in particular, care is required.

Coupon Shear Tests on Rods

Other methods of test can be used also to determine shearing properties of composites. In particular, attention is restricted to tests on small specimens. Experiments were completed for the boron epoxy using torsion on circular rods with unidirectional reinforcement parallel to the axis of twist (3.0 in. long by 0.180 in. diameter, 55 percent boron).

Figure 17 shows the typical test response.

Data from torsion experiments on rods are listed below:

Torsion tests on unidirectionally reinforced circular rods, 3/16 in. diameter.

Specimen	Maximum Shear Strength, psi
1	15.8×10^{-3}
2	15.2
3	15.4

Summary contents on shear testing are presented in Table 5.

ACKNOWLEDGMENTS

The authors wish to express their gratitude to P. Roy for his support in fabricating the composites used in this project; to E. Wagner, A. Hauze, and E. Winer for completing the experimental studies reported and to B. Ready for his preparation of the art work in this manuscript.

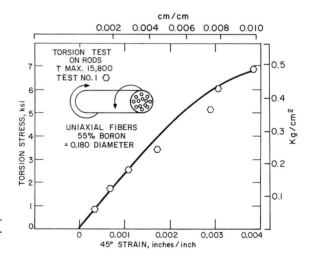

FIG. 17—Torsion stress versus 45-deg strain on circular boron epoxy rods.

Shear-Modulus Determination of Isotropic and Composite Materials

L. B. GRESZCZUK[1]

Reference:

Greszczuk, L. B., "Shear-Modulus Determination of Isotropic and Composite Materials," *Composite Materials: Testing and Design, ASTM STP 460*, American Society for Testing and Materials, 1969, pp. 140–149.

Abstract:

A new test technique that uses split rings to determine the shear modulus of isotropic and composite materials is described. In this test, two concentrated loads that are equal in magnitude but opposite in direction and act normal to the plane of the ring are applied at the point where the ring is split. Pertinent equations which relate the resultant ring deformation to the physical and mechanical properties are given. The out-of-plane deflection is predominantly due to shear deformation, which makes this test an ideal one for shear-modulus determination. Tests are conducted on steel and aluminum rings to demonstrate the accuracy of the new test method. The measured shear moduli are shown to agree within 2 percent of the values reported in the literature. The test then is used to measure the shear moduli of epoxy resin, unidirectional fiber glass composites, bidirectional fiber glass composites, unidirectional boron-epoxy composites, and unidirectional Thornel 40-epoxy composites. The experimental results are compared with theoretically predicted values, and a good agreement is found. Finally, a discussion of existing test method's for shear determination is presented, and the shear moduli obtained by various tests are compared.

Key Words:

testing, composit materials, shear modulus, ring test, fiber glass, boron, fiber composites, graphite composite, epoxy laminates, evaluation, tests

NOMENCLATURE

k_f — Volumetric filament content of a composite
k_r — Volumetric resin content of a composite
k_v — Volumetric void content of a composite
I — Moment of inertia, in.^{4}
J — Torsional property of a cross section, in.^{4}
R — Radius of the ring, in.
P — Load, lb
δ — Deflection, in.
ϵ — Strain, in./in.
σ — Stress, psi
m — Strain-stress ratio
E — Modulus of elasticity, psi
G — Shear modulus, psi

There are several test methods that are used to determine experimentally the shear moduli of isotropic and filamentary composite materials. These methods differ in complexity, the types of specimens that are employed, and the accuracy of the results that they yield. Two of the better known standard tests for shear-modu-

[1] Staff engineer, Advance Structures and Mechanical Department, McDonnell Douglas Astronautics Company—Western Division, Santa Monica, Calif. 90406.

lus determination are the plate-twist and the cylinder-torsion test. The plate-twist test, although deceptively simple to conduct, has caused problems when applied to composite materials. The cylinder-torsion, test, although it gives accurate results, is relatively expensive to conduct and requires some sophisticated test equipment.

Some of the other techniques that have been developed or used recently to determine the shear moduli of composites are the strain-rosette-instrumental tensile coupons that contain oriented filaments [1],[2] the solid-rod torsion test [2], and the Douglas ring test, which is the main topic of this paper. In addition to describing the Douglas ring test and the results it yields for shear modulus of isotropic and composite materials, this paper briefly discusses some of the other test methods for shear-modulus determination.

REVIEW OF EXISTING TEST METHODS

PLATE-TWIST TEST

Square, flat plates are used to obtain the shear modulus from the plate-twist test (Fig. 1a). Forces are applied at the four corners of the plate, and normal to its surface; two upward forces are applied at the ends of one diagonal, and two equal but downward forces at the ends of the other diagonal. Although the test itself is deceptively simple to conduct, the measured data may be difficult to interpret unless extreme care is taken in preparation of the specimens, in load-deflection measurements, and in the use of appropriate equations to convert the measured data into shear modulus. If the load-deflection equations that correspond to small deflection theory are to be used, the geometry of the plate specimen has to be such that, when loaded, the plate indeed does undergo small deflections. Otherwise, it may be necessary to use load-deformation equations of the large deflection theory, which, unfortunately, have not been published yet for plates made of filamentary materials. Furthermore, for certain types of composites it may be necessary to protect the corners of the plate, where the loads are to be applied, against excessive local deformations or crushing. The plates have to be perfectly flat. Any warpage of the plate during curing can have a significant influence on the plate deflection. These are but a few of the problems associated with the plate-twist test. The shear modulus obtained from this test is generally higher than the values obtained from other types of tests [2]. The underlying theory used to determine the shear modulus from the plate-twist test is given in numerous references, including Refs 3 and 4.

CYLINDER-TORSION TEST

Another test that has been used for the shear-modulus determination of composites is the cylinder-torsion test (Fig. 1b) [5]. This test yields reliable and accurate data. One drawback of the test is that it is relatively expensive and requires some sophisticated test equipment. Thin-walled cylinders are used in this test. Moreover, a torsion-test fixture is required.

[2] The italic numbers in brackets refer to the list of references appended to this paper.

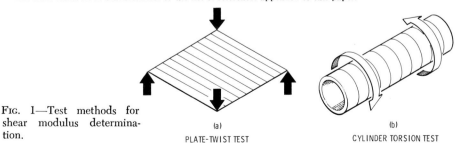

FIG. 1—Test methods for shear modulus determination.

(a)
PLATE-TWIST TEST

(b)
CYLINDER TORSION TEST

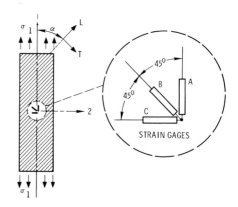

Fig. 2—Tensile coupon with oriented fibers.

Tensile Coupon with Oriented Fibers

Recently, the author developed a new test technique for measuring the shear modulus as well as other principal elastic constants of filamentary composites [1]. The technique employs strain rosette-instrumented tensile coupons that contain oriented filaments, as shown in Fig. 2. The following equation was given [1] for determining the principal shear modulus

$$G_{LT} = \frac{1}{2(m_A - m_C) + (2m_B - m_A - m_C)(\cot \alpha - \tan \alpha)} \quad \dots \dots (1)$$

where m's are the strain-stress ratios

$$m_A = \frac{\epsilon_A}{\sigma_1}, \qquad m_B = \frac{\epsilon_B}{\sigma_1}, \qquad m_C = \frac{\epsilon_C}{\sigma_1},$$

α is the fiber orientation angle; ϵ_A, ϵ_B, and ϵ_C are the three measured strains; and σ is the applied tensile or compressive stress. Thus, by conducting one simple tension test on a strain-rosette-instrumented coupon, one easily can obtain the principal shear modulus. Despite its simplicity, this test technique yields excellent results. For the case when $\alpha = 45$ deg, Eq 1 simplifies to

$$G_{LT} = \frac{1}{2(m_A - m_C)} \quad \dots \dots \dots \dots \dots \dots \dots \dots \dots (2)$$

Solid-Rod Torsion Test

Another test technique that yields reliable data is the solid-rod torsion test [2]. This test is similar to the cylinder-torsion test, except that a solid rod is used. To conduct this test, a torsion machine is required. If such a machine is not available, one either has to resort to some of the test methods described above to determine the shear modulus or use the Douglas ring test, which is described herein.

DOUGLAS RING TEST

General Considerations

If a split ring is subjected to two forces that are equal in magnitude but opposite in direction and are acting normal to the plane of the ring (Fig. 3), the resultant vertical deflection at the point of the application of loads is [6]

$$\delta = \frac{\pi P R^3}{EI} + \frac{3\pi R^3 P}{GJ} \quad \dots \dots \dots \dots \dots \dots \dots \dots \dots (3)$$

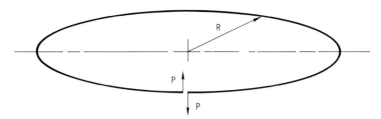

FIG. 3—Split ring subjected to out-of-plane loading.

The first term in Eq 3 represents deflection due to bending of the ring, while the second term represents the deflection due to torsion. The term J is the property of the ring cross section. For rings of rectangular cross sections, J may be expressed as

$$J = \beta ab^3 \dots\dots\dots\dots\dots\dots\dots\dots\dots\dots (4)$$

where β is a numerical factor which depends on the a/b ratio. Table 1 gives values of β for various a/b ratios.

Table 1—β parameter for rings of rectangular cross section.

a/b	1.0	1.2	1.5	1.75	2.0	2.5	3.0	4.0	5.0	6.0	8.0	10.0	∞
β	0.141	0.166	0.196	0.214	0.229	0.249	0.263	0.281	0.291	0.299	0.307	0.313	0.333

As the E/G ratio increases, the bending component of the total ring deflection $(\pi PR^3/EI)$, decreases, but the torsional component of the total ring deflection $(3\pi PR^3/JG)$ increases [6]. For example, for a ring of square cross section and made of material for which $E/G = 3$, 84 percent of the total ring deflection is due to torsion, and 16 percent is due to bending. For a material with $E/G = 10$, 94.7 percent of the total ring deflection is due to torsion, and only 5.3 percent is due to bending. The fact that the deflection is predominantly due to shear deformation makes this test an ideal one for shear-modulus determination and especially for shear-modulus determination of high-modulus filamentary composites.

If Eq 3 is solved for G, the following expression is obtained for its determination

$$G = \frac{3}{J\left(\dfrac{\delta}{P\pi R^3} - \dfrac{1}{EI}\right)} \dots\dots\dots\dots\dots\dots\dots\dots (5)$$

Thus, to determine the shear modulus from the proposed test, it is necessary to know the ring geometry, Young's modulus of the ring, and the load-deflection data for a ring subjected to out-of-plane loading.

There are several test methods for determining the Young's modulus of the ring; these are discussed in detail in Ref 6. For a split ring, the modulus can be determined by subjecting the ring to two diametrically opposite tensile forces that act in the plane of the ring and are applied at ±90 deg from where the ring is split. The Young's modulus is obtained from the following equation

$$E = \frac{\pi PR^3}{2I\delta_D} \dots\dots\dots\dots\dots\dots\dots\dots\dots\dots\dots\dots (6)$$

where δ_D is the change in ring diameter across the points where the loads are applied.

For rings made of filamentary composite materials, there is another method by which E can be evaluated. In the case of unidirectional composite rings, E can be computed from the following equation

$$E = E_f k_f + E_r k_r \left(1 - \frac{k_v}{k_r}\right) \quad\quad\quad\quad\quad (7)$$

where:

k_f = fiber volume fraction,
k_r = resin volume fraction,
k_v = void volume fraction, and
E_f and E_r = Young's moduli of fibers and matrix, respectively.

The accuracy of Eq 7 has been established by numerous investigators. Even if there were an error in E as computed by Eq 7, the error resulting from the use of the computed value of E in Eq 5 would be quite small, since for most composites the bending component of the total deflection is quite small. For example, for fiber glass ring $E/G = 6$. If there were a ± 10 percent error in the E computed from Eq 7, the resultant error in G would be only ± 1 percent. For composites with higher E/G ratios the error in G resulting from the possible error in the computed value of E would be less than 1 percent. Thus, the use of Eq 7 in combination with out-of-plane bending test is justified.

It is worth noting here that the out-of-plane ring bending test is essentially a torsion test performed without a torsion machine.

EXPERIMENTAL RESULTS

To establish the feasibility and the accuracy of the proposed test technique, tests were performed on rings made of isotropic and filamentary composite materials. In the case of isotropic materials, rings made of aluminum, steel, and epoxy resin were tested. The composite rings that were tested included: unidirectional glass-epoxy rings, bidirectional glass-epoxy rings, unidirectional boron-epoxy rings, and unidirectional Thornel 40-epoxy rings. The dimensions of the rings, their construction, and other pertinent data are shown in Table 2.

Table 2—Ring properties.

Specimen Designation	Material	Ring Width, in.	Ring Height, in.	Ring Radius, in.	Fiber Content by Volume, %	Void Content by Volume, %	Young's Modulus, $\times 10^{-6}$ psi
A-1	aluminum (6061-T651)	0.500	0.506	2.750	...	...	10[b]
S-1	steel (1020 CR)	0.500	0.496	2.755	...	...	29[b]
E-4	epoxy (828-1031 MNA/BDMA)	0.400	0.400	2.730	...	...	0.51[c]
F-1	fiber glass (unidirectional	0.406	0.406	3.140	69.4	0.10	8.75[d]
F-2	fiber glass (bidirectional)[a]	0.516	0.518	3.212	54.9	4.80	4.65[c]
F-3	fiber glass (bidirectional)[a]	0.463	0.463	3.137	58.4	1.15	4.86[c]
F-4	fiber glass (unidirectional)	0.490	0.488	3.115	51.0	5.75	6.32[c]
B-1	boron (unidirectional)	0.250	0.255	2.982	77.6	3.8	36.26[c]
T-1	Thornel 40 (unidirectional)	0.253	0.250	2.964	51.3	4.6	15.63[c]

[a] Dispersed construction: ⅔ hoops, ⅓ longitudinals.
[b] From Ref 7.
[c] Measured values.
[d] Calculated from Eq 7.

The test setup used to obtain the load deflection data from the split rings that were subjected to out-of-plane loading is shown in Fig. 4. To eliminate ring twisting during testing, the load was applied by means of vertically aligned U-grips, which passed through the hole drilled radially through the ring. A pulley arrangement was provided to counterbalance the ring weight, thus allowing the ring to be suspended in air in a horizontal position. The tensile load was applied through the U-grips. Some of the rings were tested on the Instron test machine and some on a Baldwin Universal testing machine in which the head travel was measured with a Wiedeman-Baldwin deflectometer. Some of the typical load-deflection curves that were obtained are shown in Figs. 5, 6, and 7. Since the cross-sectional areas of various rings differed, as shown in Table 2, the load-deflection curves are not necessarily comparable.

The experimental values of the shear moduli obtained from the Douglas ring test were determined from Eq 5 using the load-deflection data given in Figs. 5, 6, and 7 and the data given in Table 2. Table 3 shows the final results, as

Table 3—Comparison of measured shear moduli with theory and values given in literature.

Specimen Designation	Material	Measured Shear Modulus, $\times 10^{-6}$ psi	Shear Modulus Obtained from Literature, $\times 10^{-6}$ psi [7]	Predicted Shear Modulus, $\times 10^{-6}$ psi [8]	Error, %	Number of Tests (t) or Specimens (s) Tested
A-1........	aluminum	3.72	3.8	...	+2.15	2 (t)
S-1........	steel	10.80	11	...	+1.82	2 (t)
E-4........	epoxy-resin	0.188	...	...	...	4 (s)
F-1........	unidirectional fiber glass	1.342	...	1.351	+0.67	3 (t)
F-2........	bidirectional fiber glass	0.746	...	0.693	−7.10	2 (t)
F-3........	bidirectional fiber glass	0.907	...	0.911	+0.44	3 (t)
F-4........	unidirectional fiber glass	0.606	...	0.608	+0.33	2 (t)
B-1........	unidirectional boron-epoxy	3.051	...	2.910[a]	−4.62[a]	4 (t)
T-1........	unidirectional Thornel 40-epoxy	0.429	...	0.673[b] (0.444)[c]	+56.9[b] (+3.49)[c]	2 (t)

[a] Calculated by numerical integration; theory of Ref 8 not applicable if $k > 73$.
[b] Calculated assuming fiber to be isotropic and $G_f = 16 \times 10^6$ psi.
[c] Calculated assuming fiber to be anisotropic and $G_f = 1.2 \times 10^6$ psi [9].

well as the shear moduli reported in literature, and, in the case of filamentary composites, the theoretically predicted values.

For isotropic materials, the measured shear moduli show an excellent agreement with the values cited in the literature, thus confirming the accuracy of the ring-test technique. In the case of filamentary composites, the theoretical values of the shear moduli calculated from equation of Ref 8 show, in most cases, excellent correlation with the test data.

COMPARISON OF SHEAR MODULI OBTAINED
FROM VARIOUS TYPES OF TESTS

To further verify the accuracy of the Douglas ring test, a comparison has been made of the shear moduli of composites that were obtained by various test methods.

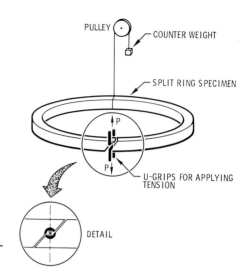

FIG. 4—Test setup for out-of-plane ring testing.

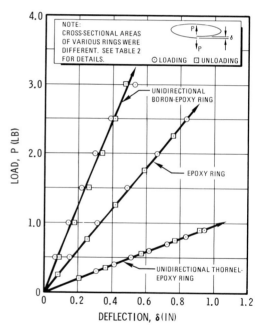

FIG. 5—Typical load-deflection curves for boron-epoxy, Thornel-epoxy, and epoxy rings.

The results are shown in Table 4. Results from the following types of tests are compared: plate-twist test, cylinder-torsion test, solid-rod torsion test, oriented-fiber-tension-coupon test, and Douglas ring test. From the discussion, it is quite apparent that the specimens used in different tests are of different size, shape, and geometry and obviously are fabricated differently. Consequently, there may be variations in the internal microstructure of the various types of specimens, including variations in fiber content, void content, and uniformity of fiber distribution. To make the comparison of test results meaningful, only the data that have been obtained for approximately the same fiber content are compared. In arriving at the average values shown in Table 4, only the data for specimens with $k = 69 \pm 1$ percent

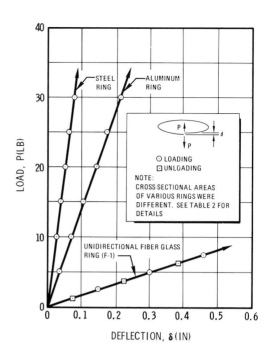

FIG. 6—Typical load-deflection curves for steel, aluminum, and fiber glass rings.

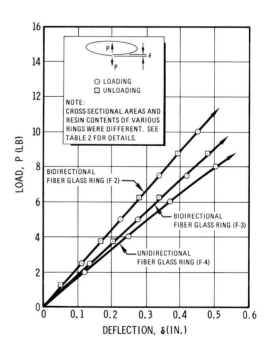

FIG. 7—Typical load-deflection curves for fiber glass rings.

were averaged. Not all of the references that are cited in Table 4 give a complete characterization of test specimens. It is expected that by correcting for, say, the void content, a much better correlation could have been obtained for the shear modulus that was obtained by the various test techniques. Nevertheless, even without normalizing the data, quite a good correlation exists between the shear moduli obtained from the various tests. The fact that the shear moduli obtained from the Douglas ring test agree quite well with the values obtained from other types of tests further substantiates the merit at this simple, inexpensive test.

Table 4—*Comparison of shear moduli of S-glass-epoxy composites determined by various test methods.*

Type of Test	Average Fiber Content (%)	Average Void Content (%)	Average Shear Modulus $\times 10^{-6}$ psi	Number of Tests	References
Plate Twist	---[a]	---	---	---	---
Cylinder Torsion	68.8	N. A.[b]	1.44	3	[5]
Solid-Rod Torsion	69.2	N. A.	1.15	4	[2]
Tensile Coupon With Oriented Fibers	68.9	1.6	1.27	14	[1]
Douglas Ring Test	69.4	0.1	1.35	3	Present Paper

[a] No experimental data could be found for a shear modulus of composite with k = 69 + 1 percent which was obtained from the plate-twist test. For k = 61.2 percent the shear modulus obtained from the plate-twist test was 1.52 x 10^{-6} psi [2].
[b] Not Available

CONCLUSIONS

In comparison to the existing test methods, it has been shown that the ring test for shear-modulus determination is simple, inexpensive, and yields accurate data. The accuracy of this new test method has been demonstrated by: (1) conducting tests on well known isotropic materials and comparing the measured results with the values given in literature, (2) comparing the measured shear moduli of composite materials with the values cited in open literature, and (3) comparing the measured values of shear moduli with the theoretically predicted results based on a previously verified theory. The test method applies equally well to isotropic and composite materials. The fact that circular rings are used makes this test an ideal one for shear-modulus determination of composites, since rings appear to be one of the simplest good-quality test specimens that can be fabricated from composites. Finally, the adaptation of the ring to shear-modulus determination complements the Naval Ordnance Laboratory (NOL) ring tests, which have been employed for such a variety of tests that they can be considered the keystone in composite material characterization.

ACKNOWLEDGMENT

Work presented herein was conducted under the sponsorship of the McDonnell Douglas Astronautics Company–Western Division under an Independent Research and Development Program, Account No. 81391-008.

References

[1] Greszczuk, L. B., "New Test Technique for Shear Modulus and Other Elastic Constants of Filamentary Composites," *U. S. Air Force–ASTM Symposium on Testing Techniques for Filament Reinforced Plastics*, Dayton, Ohio, Sept. 1966 (published as Technical Report AFML-TR-66-274, Sept. 1966).

[2] Adams, D. F. and Thomas, R. L., "Test Methods for the Determination of Unidirectional Composite Shear Properties," *Twelfth National Symposium*, Society of Aerospace Materials and Process Engineers, Aneheim, Calif. 10–12 Oct. 1967, p. AC-5.

[3] Witt, R. K., Hoppmann, W. H., II, and Buxbaum, R. S., "Determination of Elastic Constants of Orthotropic Materials With Special Reference to Laminates," *Bulletin*, American Society for Testing and Materials, No. 194, Dec. 1953, p. 53.

[4] Beckett, R. E., Dohrmann, R. J., and Ives, K. D., "An Experimental Method for Determining the Elastic Constants of Orthogonally Stiffened Plates," Meeting of the Society for Experimental Stress Analysis (SESA), New York, N. Y., 2 Nov. 1961.

[5] Feldman, A., Stang, D. A., and Tasi, J., "An Experimental Determination of Stiffness Properties of Thin-Shell Composite Cylinders," Meeting of the Society for Experimental Stress Analysis (SESA), Denver, Colo., 5–7 May 1965.

[6] Greszczuk, L. B., "Douglas Ring Test for Shear Modulus Determination of Isotropic and Composite Materials," *Proceedings, 23rd Annual Technical and Management Conference*, Society of the Plastics Industry, Washington, D. C., 6–9 Feb. 1968.

[7] "Metallic Materials and Elements for Aerospace Vehicle Structures," MIL-HDBK-5, Department of Defense, Washington, D. C., 1967.

[8] Greszczuk, L. B., "Theoretical and Experimental Studies on Properties and Behavior of Filamentary Composites," *Proceedings, 21st Annual Technical and Management Conference*, Society of the Plastics Industry, Chicago, Ill., 8–10 Feb. 1966.

[9] "Integrated Research on Carbon Composite Materials," Technical Report, AFML-TR-66-310, Part II, Dec. 1967, p. 269.

End Plugs for External Pressure Tests of Composite Cylinders

R. J. MILLER[1]

Reference:

Miller, R. J., "End Plugs for External Pressure Tests of Composite Cylinders," *Composite Materials: Testing and Design,* ASTM STP 460, American Society for Testing and Materials, 1969, pp. 150–159.

Abstract:

A method of reducing moment and shear loads which occur at end plugs during compression tests of composite ortho-tropic cylinders by external radial pressure is presented. The primary goal is to reduce the interlaminar shear stress to a permissible level and reduce the bending stresses of the cylinder to a negligible amount. The need to reduce the interlaminar shear stress is a result of the generally low shear strength of resin used in composite laminates.

Key Words:

composite materials, compression, cylinder, fiber glass, filament, interlaminar shear, pressure, epoxy laminates, stress, evaluation, tests

NOMENCLATURE

B_y Extensional stiffness in hoop direction of cylinder
D_x Bending stiffness in longitudinal direction of cylinder
E_g Modulus of elasticity of filament
E_r Modulus of elasticity of resin
K Percent filament by volume in laminate
l Half length of cylinder
M Bending moment (in.·lb)
P External pressure (psi)
Q Shear load (lb/in.)
R_m Mean radius of cylinder
t Cylinder wall thickness
y Deflection
μ_g Poisson's ratio of filament
μ_r Poisson's ratio of resin
μ_{yx} Poisson's ratio giving strain in the x-direction as a result of a stress in the y-direction
μ_{xy} Poisson's ratio giving strain in the y-direction as a result of a stress in the x-direction

SUBSCRIPTS

o Refers to value at origin or coordinate system
b Refers to value along interval $-b \leq x \leq 0$

The reduction of discontinuity end shears and moments is of great importance in the determination of maximum compressive strength of filament-reinforced plastic cylinders by external-pressure tests. This is caused primarily by the need to support the cylinder end against buckling and the low interlaminar shear strength of fila-ment-reinforced plastic laminates as compared with isotropic materials. In the past, when the discontinuity shear stresses were not considered a problem in the testing

---

[1] Engineer scientist specialist, Advance Structures and Mechanical Department, Research and Development Directorate, McDonnell Douglas Astronautics Company–Western Division, Santa Monica, Calif. 90406.

of isotropic materials, cylindrical end plugs were used to support the ends of the cylinders. This kind of support has been used extensively in testing fiber glass cylinders, but has resulted in many premature failures of the cylinders caused by shear at the ends of the cylinder. In such cases, the maximum compressive strength of the filament-reinforced plastic laminate has not been obtained. Another method of support utilizes flat plates that are pressed against the ends of the cylinder. This method is also unsatisfactory because it does not provide sufficient support at the ends, and the cylinder may fail prematurely by buckling as a ring.

This report deals with the design of contoured end plugs that will prevent high end discontinuity moment and shears and also maintain circular ends. Two separate designs are considered. The first allows the maximum shear and moment along the plug to be specified and the length and the contour of the plug to be obtained. In the second method, the moment and shear loads are mutually dependent; therefore, when the value for one is chosen, the other is fixed accordingly. The primary advantage of the second method is that the contour of the plug is a constant radius, whereas in the first method it is not, and manufacturing is thus more difficult.

REDUCTION OF BENDING MOMENT AND SHEAR AT END FIXTURE

It will be shown that the discontinuity moment and shear occurring at the end fixture during external-radial-pressure tests of filament-reinforced plastic cylinders can be reduced by the use of curved end plugs. A test fixture as shown in Fig. 1 will be designed. The contour of this fixture will allow the cylinder to have a predetermined shear and moment as specified by the contour of the plug when the cylinder deflects along the plug. It will be assumed that at the point where the cylinder leaves the plug, the boundary conditions will be the same as those for a fixed-end cylinder with an initial slope.

First, the moment in the cylinder at the point where it leaves the plug will be determined from the classical equation for the deformation of cylindrical shells (Eq 1) [1–6].[2]

$$\frac{d^4y}{dx^4} + 4\beta^4 y - \frac{P}{D_x} = 0 \dots \dots \dots \dots \dots \dots (1)$$

where:

$$\beta = \left(\frac{B_y}{4R_m^2 D_x}\right)^{1/4}$$

[2] The italic numbers in brackets refer to the list of references appended to this paper.

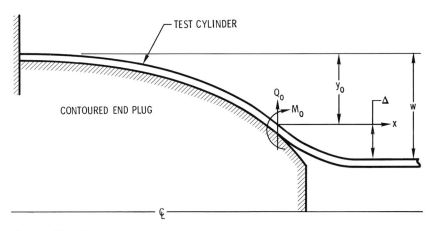

FIG. 1—Test fixture.

The general solution to this equation is

$$y = e^{\beta x}(C_1 \cos \beta x + C_2 \sin \beta x) + e^{-\beta x}(C_3 \cos \beta x + C_4 \sin \beta x)$$
$$+ \frac{PR_m^2}{B_y}(1 - \mu_{yx}\mu_{xy}) \dots\dots\dots\dots\dots\dots\dots\dots (2)$$

If it is assumed that the cylinder is long enough ($\beta l > 2\pi$) that the conditions at one end do not affect the other end, it can be concluded that the first term in Eq 2 must drop out as the local bending produced by the forces M_o and Q_o dies out rapidly as x increases. Therefore, $C_1 = C_2 = 0$, and Eq 3 is obtained:

$$y = e^{-\beta x}(C_3 \cos \beta x + C_4 \sin \beta x) + \frac{PR_m^2}{B_y}(1 - \mu_{yx}\mu_{xy}) \dots\dots\dots\dots (3)$$

The constants C_3 and C_4 must now be determined using the conditions:

$$\left. \begin{array}{c} M_x = M_o \\ Q_x = Q_o \end{array} \right\} x = 0 \dots\dots\dots\dots\dots\dots\dots\dots (4)$$

and the stress-displacement relations:

$$\left. \begin{array}{c} \dfrac{d^2 y}{dx^2} = -\dfrac{M_x}{D_x} \\[2mm] \dfrac{d^3 y}{dx^3} = -\dfrac{Q_x}{D_x} \end{array} \right\} \dots\dots\dots\dots\dots\dots\dots\dots (5)$$

Differentiating Eq 3 yields:

$$\frac{dy}{dx} = -\beta e^{-\beta x}[(C_3 - C_4) \cos \beta x + (C_3 + C_4) \sin \beta x] \dots\dots\dots\dots (6)$$

$$\frac{d^2 y}{dx^2} = -2\beta^2 e^{-\beta x}(C_4 \cos \beta x - C_3 \sin \beta x) \dots\dots\dots\dots\dots (7)$$

$$\frac{d^3 y}{dx^3} = 2\beta^3 e^{-\beta x}[(C_4 + C_3) \cos \beta x - (C_3 - C_4) \sin \beta x] \dots\dots\dots\dots (8)$$

Solving Eqs 5, 6, 7, and 8 at $x = 0$ yields:

$$C_3 = -\frac{1}{2\beta^3 D_x}(Q_o + \beta M_o)$$

$$C_4 = \frac{M_o}{2\beta^2 D_x}$$

Equations 3 and 6 can be solved simultaneously for M_o by substitution of the values for C_3 and C_4 and by letting $y' = y'_o$ and $y = y_o$ at $x = 0$. Therefore,

$$M_o = -(\beta\Delta - y'_o)2\beta D_x \dots\dots\dots\dots\dots\dots\dots\dots (9)$$

$$Q_o = 2\beta^2 D_x(2\beta\Delta - y'_o) \dots\dots\dots\dots\dots\dots\dots (10)$$

where:

$$\Delta = \frac{PR_m^2}{B_y}(1 - \mu_{yx}\mu_{xy}) - y_o$$

Equation 9 gives the moment in the cylinder at the point where it leaves the plug in terms of the properties of the cylinder, the external pressure, the slope, and the deflection of the plug at this point. By specifying the contour of the plug, it is possible to determine the moment, shear, slope, and deflection at any

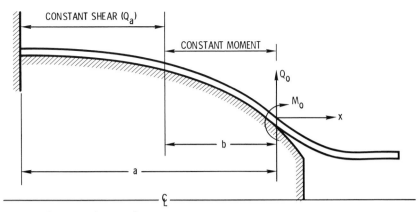

Fig. 2—Constant shear and moment.

point along its surface. This enables us to determine the exact point where these conditions satisfy Eq 9 and thus the exact point of departure of the cylinder from the plug.

There are two methods for specifying the contour of the plug.

METHOD 1: DISCONTINUITY REDUCTION

In Method 1, the plug contour is determined by the following equations:

$$y''' = \frac{Q_a}{D_x} = C \qquad [-a \leq x \leq -b \text{ (Fig. 2)}] \dots\dots\dots\dots\dots (11)$$

$$y_b'' = -\frac{M_o}{D_x} = B \qquad [-b \leq x \leq 0 \text{ (Fig. 2)}] \dots\dots\dots\dots (12)$$

Equation 11 results in a constant shear and increasing moment with increasing x, while Eq 12 results in zero shear and constant moment with increasing x, as shown in Fig. 2. Therefore, it is desirable to begin the initial contour of the plug using Eq 11, and, when the moment reaches the allowable level, to use Eq 12 to determine the contour.

Integrating Eq 11 and utilizing the conditions $y'' = 0$ at $x = -a$ gives:

$$y'' = C(x + a) \qquad (-a \leq x \leq -b) \dots\dots\dots\dots\dots\dots (13)$$

Integrating Eq 13 and utilizing the conditions that $y' = 0$ at $x = -a$ yields:

$$y' = \frac{C}{2}(x^2 + 2ax + a^2) \qquad (-a \leq x \leq -b) \dots\dots\dots\dots (14)$$

Solving for the deflection y by integrating Eq 14 and utilizing the conditions that $y = 0$ at $x = -a$ results in:

$$y = C\left(\frac{x^3}{6} + \frac{ax^2}{2} + \frac{a^2x}{2} + \frac{a^3}{6}\right) \qquad (-a \leq x \leq -b) \dots\dots\dots\dots (15)$$

The curvature for the interval $-b \leq x \leq 0$ is given by Eq 13 solved at $x = -b$:

$$y'' = C(-b + a) \dots\dots\dots\dots\dots\dots\dots\dots\dots\dots (16)$$

Integrating Eq 16 and using Eq 14 to determine the slope at $x = -b$ gives:

$$y' = C(a - b)x + \frac{C}{2}(a^2 - b^2) \qquad (-b \leq x \leq 0) \dots\dots\dots\dots (17)$$

Integrating Eq 17 and using Eq 15 to determine the deflection at $x = -b$, Eq 18 is obtained:

$$y = \frac{C}{2}(a-b)x^2 + \frac{C}{2}(a^3-b^2)x + \frac{C}{6}(a^2-b^3) \qquad (-b \le x \le 0) \dots \dots (18)$$

The slope (y_o') and deflection (y_o) at $x = 0$ can be obtained from Eqs 17 and 18, respectively:

$$y_o' = \frac{C}{2}(a^2-b^2) \dots \dots \dots \dots \dots \dots \dots \dots \dots (19)$$

$$y_o = \frac{C}{6}(a^3-b^3) \dots \dots \dots \dots \dots \dots \dots \dots \dots (20)$$

Assuming M_o and Q_a are given quantities, Eqs 9, 11, 12, 16, 19, and 20 can be solved simultaneously for plug length a. Substituting Eqs 11 and 12 into Eq 16 and solving for b yields:

$$b = a - \frac{M_o}{Q_a} \dots \dots \dots \dots \dots \dots \dots \dots \dots (21)$$

Substituting Eqs 11, 19, 20, and 21 into Eq 9 and solving for plug length a yields:

$$a = \left[-\left(\frac{-\beta M_o + 2Q_a}{2Q_a} \right) \pm \sqrt{\left(\frac{-\beta M_o + 2Q_a}{2Q_a} \right)^2} \right.$$
$$\left. - 2\beta \left(\frac{1}{2\beta} + \frac{D_x \beta P R_m^2}{B_y M_o}(1 - \mu_{yx}\mu_{xy}) + \frac{\beta M_o^2}{6Q_a^2} - \frac{M_o}{2Q_a} \right) \right] \frac{1}{\beta} \dots (22)$$

Therefore, by giving the allowable moment and shear, the length and shape of the required end plug can be determined. It is now possible to design an external-pressure filament-reinforced plastic specimen that will not fail because of shear loads and that will have low discontinuity stresses at the point where the cylinder leaves the plug. Therefore, even if a short test section is used between the points where the cylinder leaves the plugs, the influence of the discontinuity stress on the true compressive stress at the center of the cylinder will be negligible. Thus, the true compressive stress at failure can be calculated quite accurately without using strain gages by assuming a state of pure membrane stress exists.

The bending moment and shear at any point along the cylinder after it leaves the plug can be calculated from the following equations, where $x = 0$ at the point where the cylinder leaves the plug.

$$M_x = M_o[e^{-\beta x}(\cos \beta x + \sin \beta x)] + \frac{Q_o}{\beta}e^{-\beta x}\sin \beta x \qquad [0 \le x \le (l-a)] \dots (23)$$

$$Q_x = -2\beta M_o e^{-\beta x}\sin \beta x - Q_o e^{-\beta x}(\sin \beta x - \cos \beta x) \qquad [0 \le x \le (l-a) \quad (24)$$

METHOD 2: DISCONTINUITY REDUCTION

Because of the simplicity of the algebra and construction of the plug, nonlinear theory will be used to describe the surface of the plug; that is,

$$\frac{M}{D_x} = \frac{y''}{[1 + (y')^2]^{3/2}}$$

will be used instead of $\frac{M}{D_x} = y''$, which was used to derive Eq 9. In nonlinear theory,

Composite Materials

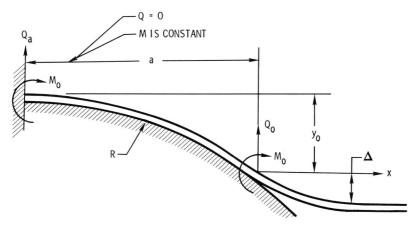

Fɪɢ. 3—Constant moment.

a constant moment is given by a constant radius of curvature. Using nonlinear theory to determine the surface contour, a plug, as shown in Fig. 3, will be designed. The shear at $x = -a$ is given by Q_a and will be equal to zero along the interval $-a \leq x \leq 0$; whereas, the moment will remain constant along the interval $-a \leq x \leq 0$ and then diminish gradually. The moment and shear along the interval $0 \leq x \leq (l - a)$ can be determined, using linear theory, from Eqs 23 and 24.

The point at which the cylinder leaves the plug must be determined. Equation 9 gives the moment at this point.

$$M_o = -(\beta\Delta - y_o')2\beta D_x \dots \dots \dots (9)$$

where:

$$M_o = -\frac{D_x}{R} \dots \dots \dots (25)$$

$$\Delta = w - y_o \dots \dots \dots 26)$$

$$y_o = R - (R^2 - a^2)^{1/2} \dots \dots \dots (27)$$

$$y_o' = a(R^2 - a^2)^{-1/2} \dots \dots \dots (28)$$

Therefore,

$$-M_o = [\beta(w - [R - (R^2 - a^2)^{1/2}]) - a(R^2 - x^2)^{-1/2}]2\beta D_x \dots \dots (29)$$

Let

$$C = \frac{-M_o}{2\beta D_x} - \beta(w - R)$$

Equation 29 then becomes

$$(\beta^2 R^2 - C^2)R^2 - 2\beta R^2 a + (C^2 - 2\beta^2 R^2 + 1)a^2 + 2\beta a^3 + \beta^2 a^4 = 0 \dots \dots (30)$$

If the maximum allowable moment is given, Eq 30 can be solved for a. The shear load at the beginning of the contoured surface is given by:

$$Q_a = 2\beta M_o \dots \dots \dots (31)$$

THEORETICAL RESULTS

Of the two methods presented for the design of contoured plugs, the second is the most practical because it allows the shortest plug for a given allowable moment or shear and a simpler method of fabrication. Theoretical results will be

Table 1—Cylinder properties.

D_z = 6567 lb·in.²	P = 15,000 psi	B = 3.3 in.
E_Y = 5.718 × 10⁶ psi	β = 1.613	A = 3 in.
E_g = 10 × 10⁶ psi	E_r = 0.5 × 10⁶ psi	K = 0.70
μ_g = 0.20	μ_r = 0.36	$t_1 = t_2 = t_3$ = 0.1 in.

obtained for the cylinder shown in Fig. 4 with the properties shown in Table 1. Figures 5 and 6 give an assembled view of the cylinders and plugs.

For the test fixture shown in Fig. 5, the maximum moment was found to be approximately 3000 in. lb/in., which is equivalent to approximately 190,000 psi bending stress in the outer filaments. The maximum shear was approximately 9000 lb/in., or 45,000 psi interlaminar shear stress, which is approximately five times the allowable for fiber glass-resin laminates currently in use. Figure 7 shows a

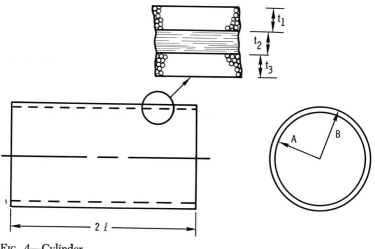

Fig. 4—Cylinder.

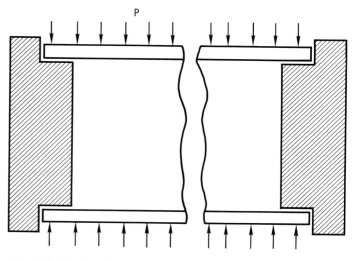

Fig. 5—Cylindrical plugs.

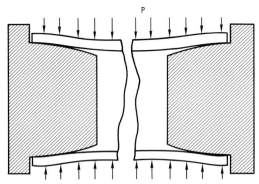

Fig. 6—Contoured plugs.

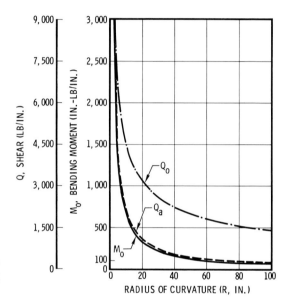

Fig. 7—Moment and shear values as a function of plug curvature (Method 2).

plot of moment and shear versus meridional radius of curvature. If the allowable interlaminar shear stress were assumed to be 9000 psi, the maximum shear force would have to be reduced to 1800 lb/in. or less. This would require a plug with a meridional radius of curvature of at least 60 in.

EXPERIMENTAL RESULTS

Plugs designed by the methods given in this paper were used in testing ring-stiffened fiber glass cylinders under external pressure [7]. Three such specimens were designed and tested without failure at or caused by the end plugs. The first specimen failed at 55.5 percent of design pressure as a result of bond failure between the rings and shell. In the next test, in which the bond strength between rings and shells was improved, the specimen failed at 95 percent of design pressure. The final specimen tested consisted of a shell cut from the same specimen as the second, but the ring design and spacing were improved. This specimen failed at 105 percent of design pressure. This indicated that the end plugs performed as expected and that failure was not caused by bending or interlaminar shear stresses. More information about the results of these tests can be obtained from

FIG. 8—Photograph of test fixture.

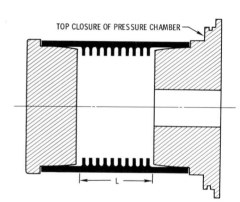

TOP CLOSURE OF PRESSURE CHAMBER

FIG. 9—Cross section of end fixtures and cylinder assembly.

L

Ref 7. Figure 8, taken from Ref 7, shows the plugs used. A cross section of the specimen and plug assembly is shown in Fig. 9.

The methods outlined in this paper were used also to design plugs to be used for external-pressure tests on monocoque fiber glass cylinders to obtain ultimate compression strength [8]. Table 2 gives the failure pressure of the cylinder tested along with the shear stresses using contoured plugs as well as the shear stresses that would have occurred if solid cylindrical plugs had been used. The cylinder was a hoop-long-hoop construction, as shown in Fig. 4, with dimensions and properties as shown in Table 2.

The high values of stress at failure given in Table 2 indicate that the cylinders failed close to the maximum compressive strength of E-glass-epoxy composites, again showing that the contoured plugs performed as expected. The conclusion was drawn in Ref 8 that the strain-gage data indicated that the cylinder may have buckled rather than failed in compression.

From Table 2, it is clear that the interlaminar bond shear stress that would occur if the plug was not contoured would cause failure at a pressure far below the actual failure pressure. It is also clear that the interlaminar shear stress using the contoured plug was reduced to 22 percent of that without a contoured plug. Additional experimental results using contoured plugs are given in Ref 9, where

Table 2—*Data on compression test of fiber glass epoxy cylinder.*

Failure Pressure, psi	Maximum Composite Stress,[a] psi	Bond Shear Stress Without Contoured Plug,[b] psi	Bond Shear Stress With Contoured Plug,[c] psi	Transverse Shear Force Without Contoured Plug,[b] lb/in.	Transverse Shear Force With Contoured Plug,[c] lb/in.
16 800	170 000	36 640	7 900	12 800	2 763

$$E_g = 10.5 \times 10^6 \text{ psi} \qquad t_1 = 0.130 \text{ in.}$$
$$E_r = 0.5 \times 10^6 \text{ psi} \qquad t_2 = 0.166 \text{ in.}$$
$$\mu_g = 0.2 \qquad\qquad\qquad t_3 = 0.118 \text{ in.}$$
$$\mu_r = 0.36 \qquad\qquad\quad R = 75 \text{ in.}$$

[a] Hoop stress at inner radius of cylinder.
[b] Denotes theoretical value if cylindrical plug was used.
[c] Denotes theoretical value when contoured plug was used.

layer stresses up to 210,000 psi were reported for composite cylinders fabricated using S-glass-epoxy laminates.

SUMMARY

The theory presented in this paper has been substantiated by results obtained from several experimental programs conducted to determine the compressive strength of fiber glass cylinders subjected to external pressure. In each of these programs, the use of contoured end plugs has prevented successfully the occurrence of high interlaminar shear stresses at the end plugs, therefore eliminating these stresses as a failure mode.

References

[1] Timoshenko, S., *Theory of Plates and Shells*, McGraw-Hill, New York, 1959.

[2] Hartog, Den, *Advanced Strength of Materials*, McGraw-Hill, New York, 1952.

[3] Roark, R. J., *Formulas for Stress and Strain*, McGraw-Hill, New York, 1954.

[4] Greszczuk, L. B., "Membrane Analysis Methods for Composite Structures," Douglas Report No. SM-41543, 28 March 1962.

[5] Miller, R. J., "Principal Elastic Moduli for Filament Reinforced Plastic," Douglas Report No. SM-47748, 23 Oct. 1964.

[6] Hetenyi, M., *Beams on Elastic Foundation*, The University of Michigan Press, Ann Arbor, 1946.

[7] Miller, R. J., Nebesar, R. J., and Schneider, M. H., "Design and Test of Ring-Stiffened Fiber Glass Cylinders Under External Pressure," Douglas Paper No. 3535, Feb. 1966.

[8] Jacobson, H. R., "Final Report, Optimum Construction of Reinforced Plastic Cylinders Subjected to High External Pressure," MDAC-WD Report No. SM-45871, March 1968.

[9] Stone, F. E., "Study of Residual Stresses in Thick Glass-Filament-Reinforced Laminates," Douglas Report No. SM-49252, Oct. 1965.

Strength Theories of Failure for Anisotropic Materials

B. E. KAMINSKI[1] AND R. B. LANTZ[2]

Reference:

Kaminski, B. E. and Lantz, R. B., "Strength Theories of Failure for Anisotropic Materials," *Composite Materials: Testing and Design, ASTM STP 460,* American Society for Testing and Materials, 1969, pp. 160–169.

Abstract:

A discussion of several failure criteria from their initial application to homogeneous, orthotropic materials to their extension to "quasi-homogeneous," anisotropic materials is presented. Also discussed are the assumptions, general limitations, and the physical and analytical significance of the respective "material constants." The basic criteria include the Lame'-Navier maximum stress theory, Henky-von Mises distortional energy theory, Tresca maximum shear stress theory, and the St. Venant maximum strain stream theory.

Key Words:

composite materials, mechanical properties, anisotropic, unidirectional, fiber composites, fiberglass, reinforced plastics, boron, epoxy laminates, evaluation, tests

Composites are gaining acceptance in the aerospace industry as a result of the quest for lightweight, efficient structure. However, being a new field of endeavor, analysis and design of structural composites suffer from the lack of a well-ordered body of knowledge. One of the problems facing a designer of anisotropic composite structures is to locate and utilize a failure theory valid for the particular composite material being investigated out of the myriad material matrix/reinforcement material combinations possible. Often the implicit assumptions of the failure criterion chosen or the existance of alternate criteria are unknown.

This paper traces the developments of several failure criteria as initially presented for application to homogeneous, orthotropic materials and their extension to "quasi-homogeneous," anisotropic materials. Such a generalization allows the use of a failure criterion for unidirectionally reinforced composites, a common basic ply of laminated materials. The assumptions, general limitations, and the physical and analytical significance of the respective "material constants" are discussed for several anisotropic adaptations of the basic classes of homogeneous, isotropic failure theories. These basic classes include the Lame'-Navier maximum stress theory, Henky-von Mises distortional energy theory, Tresca maximum shear stress theory, and the St. Venant maximum strain theory.

MAXIMUM STRESS THEORY

In 1920, Jenkin [1][3] proposed a modification of the Lame'-Navier maximum stress theory to predict the failure of an orthotropic material. He suggested that stresses on the material be resolved into stresses along the natural axes of the material. The strength of the material was said to be reached when any one of

[1] Structures engineer, General Dynamics Corp., Fort Worth, Tex. 76101; formerly, graduate student, Georgia Institute of Technology, Atlanta, Ga. 30332.
[2] Senior engineer, AVCO Aerostructures Division, Nashville, Tenn. 37202.
[3] The italic numbers in brackets refer to the list of references appended to this paper.

the stresses associated with these natural axes reached its maximum value. That is, failure occurs when either

$$\left.\begin{aligned}\sigma_1 &= F_1\\[2mm]\sigma_2 &= F_2\end{aligned}\right\} \dots\dots\dots\dots\dots\dots\dots\dots\dots\dots (1)$$

where F_1 and F_2 are defined as either tensile or compressive strengths in the 1 and 2 directions, respectively, as determined by uniaxial testing. Tensile strength values are positive, and compressive strength values are negative. The tensile or compressive strengths were used depending on the sign of the applied stress.

DISTORTIONAL ENERGY THEORY

Circa 1950, Norris [2,3] and Hill [4] independently developed their generalizations of the von Mises isotropic distortional energy yield criterion,

$$(\sigma_1 - \sigma_2)^2 + (\sigma_2 - \sigma_3)^2 + (\sigma_3 - \sigma_1)^2 + 6(\sigma_{12}^2 + \sigma_{23}^2 + \sigma_{31}^2) = 2\sigma_0^2 \dots\dots (2)$$

to account for the anisotropy of their respective problems.

To provide a failure criterion for plywood, Norris [2] proposed, in 1946, an "interaction" formula relating the stress components which would cause failure of the composite under applied load. The condition of failure was given to be

$$\frac{\sigma_1^2}{F_1^2} + \frac{\sigma_2^2}{F_2^2} + \frac{\sigma_{12}^2}{F_{12}^2} = 1 \dots\dots\dots\dots\dots\dots\dots\dots (3)$$

In that study, Norris observed that "this empirical formula fitted the experimental data so well that it seemed to point to some existing physical mechanism responsible for the strength of orthotropic materials. Its form suggested the Henky-von Mises theory."

Consequently in 1950, Norris [3] presented a strength theory for general orthotropic materials, assuming that the orthotropic material can be represented by an isotropic material containing regularly spaced voids in the shape of rectangular prisms (the resulting material resembles a waffle grid). Norris calculated the energy due to distortion for a set of orthogonal axes parallel to the cellular walls. In doing so, the following additional assumptions were made:

1. Each set of walls is subjected to a two-dimensional stress system.

2. If it is assumed that the walls do not buckle when they are stressed, the values of the stresses in the isotropic walls will be proportional to the values of the applied gross stresses.

3. Bending of the grid walls is neglected.

4. The distortional energy has just the value associated with failure of the isotropic material.

The theory then becomes a set of three von Mises-type equations, one for each set of walls:

$$\left.\begin{aligned}\frac{\sigma_1^2}{F_1^2} - \frac{\sigma_1\sigma_2}{F_1F_2} + \frac{\sigma_2^2}{F_2^2} + \frac{\sigma_{12}^2}{F_{12}^2} &= 1\\[3mm]\frac{\sigma_2^2}{F_3^2} - \frac{\sigma_2\sigma_3}{F_3F_1} + \frac{\sigma_3^2}{F_1^2} + \frac{\sigma_{23}^2}{F_{31}^2} &= 1\\[3mm]\frac{\sigma_3^2}{F_3^2} - \frac{\sigma_3\sigma_1}{F_3F_1} + \frac{\sigma_1^2}{F_1^2} + \frac{\sigma_{31}^2}{F_{31}^2} &= 1\end{aligned}\right\} \dots\dots\dots\dots (4)$$

Assuming plane stress, the equations reduced to

$$\left.\begin{array}{c} \dfrac{\sigma_1^2}{F_1^2} - \dfrac{\sigma_1\sigma_2}{F_1 F_2} + \dfrac{\sigma_2^2}{F_2^2} + \dfrac{\sigma_{12}^2}{F_{22}^2} = 1 \\[12pt] \dfrac{\sigma_2^2}{F_2^2} = 1 \\[12pt] \dfrac{\sigma_1^2}{F_1^2} = 1 \end{array}\right\} \quad \dots\dots\dots\dots\dots\dots (5)$$

Comparing the first of Eq 4 with the interaction formula, Eq 3, the two equations are identical with the exception of the cross term.

Note that if the material were such that the isotropic strength relations would hold,

$$F_1 = F_2 = F_3 = \sqrt{3}\,F_{12} = \sqrt{3}\,F_{23} = \sqrt{3}\,F_{31} = \sigma_0$$

then Eq 4 reduce to the von Mises isotropic yield criteria.

Hill [4] was concerned with the tendency of isotropic metals to exhibit certain anisotropic properties when undergoing metal working involving severe strains. Hill claimed he had a physical interpretation of von Mises' "plastic potential" which allowed him to generalize von Mises' yield criterion for application to anisotropic metals. By making an analogy, Hill selected a homogeneous quadratic in the stresses to represent the "plastic potential." It was assumed that:

1. The material possessed a natural set of orthogonal axes at every point.
2. The yield stress in tension for any direction was the same as that in compression.
3. From the symmetry imposed by Item 1, terms in which any one shear stress occurs linearly were rejected.
4. Superposition of hydrostatic pressure would not influence yielding.

The "plastic potential" or yield criterion has the form

$$2f(\sigma_{ij}) = A_1(\sigma_2 - \sigma_3)^2 + A_2(\sigma_3 - \sigma_1)^2 + A_3(\sigma_1 - \sigma_2)^2$$
$$+ 2A_4\sigma_{23}^2 + 2A_5\sigma_{31}^2 + 2A_6\sigma_{12}^2 = 1 \dots (6)$$

where:

$2f(\sigma_{ij}) =$ the plastic potential;
$2A_1 = (F_2)^{-2} + (F_3)^{-2} - (F_1)^{-2}$;
$2A_2 = (F_3)^{-2} + (F_1)^{-2} - (F_2)^{-2}$;
$2A_3 = (F_1)^{-2} + (F_2)^{-2} - (F_3)^{-3}$;
$2A_4 = (F_{23})^{-2}$;
$2A_5 = (F_{31})^{-2}$; and
$2A_6 = (F_{12})^{-2}$.

F_1, F_2, and F_3 are determined from either uniaxial tension or compression tests in the 1, 2, and 3 directions, respectively and F_{12}, F_{23}, F_{31} are determined from pure shear tests in the 12, 23, and 31 planes, respectively.

For the case of plane stress, Eq 6 reduces to

$$2f(\sigma_{ij}) = (A_2 + A_3)\sigma_1^2 - 2A_3\sigma_1\sigma_2 + (A_3 + A_1)\sigma_2^2 + 2A_6\sigma_{12}^2 = 1 \dots\dots (7)$$

Applying the isotropic strength relationships will reduce either Eq 6 or 7 to the isotropic von Mises criterion.

Marin [5] in 1957 attempted a further generalization of the von Mises theory to account for the difference in tensile and compressive yield strengths and the variation of strength with direction. In doing so, Marin limited his theory to principal stresses. For isotropic materials, the use of principal stresses has the advantage of eliminating one variable, namely, shear. However, for an anisotropic material, the shear variable is replaced with one of angular orientation due to the dependence of the mechanical properties on the direction of applied load. Thus, the use of principal axes fails to provide either a mathematical advantage or additional physical significance over the use of natural axes. It also has been proved by Greszczuk [6] that when dealing with anisotropic materials, axes of principal stress are not always coincident with axes of principal strain.

An application of Hill's criterion to unidirectional, fiber-reinforced composites was proposed by Azzi and Tsai [7] in 1965. Recognizing that in most real fiber-reinforced materials the fiber positions in the cross section are dispersed randomly and the material properties in directions transverse to the fiber direction are equal, they assumed that the material was transversely isotropic and set F_3 equal to F_2 in Hill's criterion. Thus, the failure criterion reduces to

$$\frac{(\sigma_1^2 - \sigma_1\sigma_2)}{F_1^2} + \frac{\sigma_2^2}{F_2^2} + \frac{\sigma_{12}^2}{F_{12}^2} = 1 \dots\dots\dots\dots\dots\dots(8)$$

Recently, a modification to Hill's failure surface as defined by Eq 6 has been proposed to take into account the differences between tensile and compressive strengths. In 1967, Hoffman [8] formulated a fracture condition which included linear terms of σ_1, σ_2, and σ_3. Thus, the failure criterion was defined as

$$C_1(\sigma_2 - \sigma_3)^2 + C_2(\sigma_3 - \sigma_1)^2 + C_3(\sigma_1 - \sigma_2)^2 + C_4\sigma_1$$
$$+ C_5\sigma_2 + C_6\sigma_3 + C_7\sigma_{23}^2 + C_8\sigma_{31}^2 + C_9\sigma_{12}^2 = 1..(9)$$

where:

$C_1 = \dfrac{1}{2}\left[\dfrac{1}{F_{t2}F_{c2}} + \dfrac{1}{F_{t3}F_{c3}} - \dfrac{1}{F_{t1}F_{c1}}\right]$;

C_2 and C_3 by permutation of 1, 2, 3;

$C_4 = \dfrac{1}{F_{t1}} - \dfrac{1}{F_{c1}}$;

C_5 and C_6 by permutation of 1, 2, 3;

$C_7 = (F_{23})^{-2}$;

C_8 and C_9 by permutation of 1, 2, 3;

and where c and t refer to compression and tension, respectively.

Assuming a state of plane stress and that the material is transversely isotropic, Eq 9 becomes

$$D_1(\sigma_1^2 - \sigma_1\sigma_2) + D_2\sigma_2^2 + D_3\sigma_1 + D_4\sigma_2 + D_5\sigma_{12}^2 = 1\dots\dots\dots\dots(10)$$

where:

$D_1 = (F_{t1}F_{c1})^{-1}$;
$D_2 = (F_{t2}F_{c2})^{-1}$;
$D_3 = (F_{c1} - F_{t1})/(F_{c1}F_{t1})$;
$D_4 = (F_{c2} - F_{t2})/(F_{c2}F_{t2})$; and
$D_5 = (F_{12})^{-2}$.

MAXIMUM SHEAR STRESS THEORY

Hu [9] in 1958 presented an anisotropic yield criterion based on a modification of the isotropic Tresca (maximum shear stress) criterion. Acknowledging the fact that the orientation of the principal stress axes is a factor in the formulation of a yield criterion for anisotropic materials, he limited the theory to the particular case where the principal stress axes and the natural axes are coincident. For strongly anisotropic materials, application of Hu's theory is restricted severely.

MAXIMUM STRAIN THEORY

An application of St. Venant's maximum strain theory to anisotropic materials was proposed by General Dynamics, Fort Worth Division, in 1966 [10]. They postulated that failure occurred when any one of the strains associated with the material axes of the material reaches its maximum value, that is,

$$\epsilon_1 = K_1$$
$$\epsilon_2 = K_2$$

and

$$\epsilon_{12} = K_{12}$$

where K_1 and K_2 are either the tensile or compressive ultimate strains, and K_{12} is the ultimate shear strain. The failure envelope easily can be found in terms of stresses as follows. Hooke's law for an orthotropic material is defined as:

$$\begin{vmatrix} \epsilon_1 \\ \epsilon_2 \\ \epsilon_{12} \end{vmatrix} = \begin{vmatrix} S_{11} & S_{12} & 0 \\ S_{12} & S_{22} & 0 \\ 0 & 0 & S_{66} \end{vmatrix} \begin{vmatrix} \sigma_1 \\ \sigma_2 \\ \sigma_{12} \end{vmatrix} \dots\dots\dots\dots\dots\dots (11)$$

where:

$$S_{11} = E_1^{-1};$$
$$S_{12} = -\nu_{12}E_1^{-1} = -\nu_{21}E_2^{-1};$$
$$S_{22} = E_2^{-1}; \text{ and}$$
$$S_{66} = G_{12}^{-1}.$$

The envelope of strains at failure will be bound by the stresses producing the strains K_1, K_2, and K_{12}. Since about the natural axes the normal and shear stress-strain relations are uncoupled, they may be investigated independently. By rewriting Eq 11 in terms of limiting strain, the following is obtained:

$$\left.\begin{aligned} \sigma_2 &= -\frac{1}{\nu_{12}}(K_1 E_1 - \sigma_1) \\ \sigma_2 &= E_2\left(K_2 + \frac{\nu_{12}}{E_1}\sigma_1\right) \\ \sigma_{12} &= K_{12}G_{12} \end{aligned}\right\} \dots\dots\dots\dots\dots\dots (12)$$

Using the relationships,

$$F_1 = E_1 K_1$$
$$F_2 = E_2 K_2$$

and

$$F_{12} = G_{12} K_{12}$$

Equations 12 reduced to:

$$\left.\begin{aligned} \sigma_2 &= \frac{-1}{\nu_{12}}(F_1 - \sigma_1) \\ \sigma_2 &= F_2 + \nu_{12}\frac{E_2}{E_1}\sigma_1 \\ \sigma_{12} &= F_{12} \end{aligned}\right\} \dots\dots\dots\dots\dots\dots (13)$$

The first two expressions of Eqs 12 can be shown to be of the same form as St. Venant's isotropic maximum strain criteria,

$$\left.\begin{aligned} \sigma_I - \nu\sigma_{II} &= \sigma_0 \\ \sigma_{II} - \nu\sigma_I &= \sigma_0 \end{aligned}\right\} \dots\dots\dots\dots\dots\dots (14)$$

Composite Materials

by substituting in the isotropic material properties:

$$\sigma_0 = F_1 = F_2, \qquad E_1 = E_2 = E, \qquad \nu_{21} = \nu_{12} = \nu$$

where σ_I and σ_{II} are principal stresses.

DISCUSSION

In the past it has been a general practice when comparing failure theories to normalize the axes, that is, the stress variables σ_1, σ_2, and σ_{12} and replaced by σ_1/F_1, σ_2/F_2, and σ_{12}/F_{12}. In doing so, an impression is generated that certain theories produce smooth failure surfaces across the σ_1 and σ_2 axes (in a constant σ_{12} plane) when in reality, this is not true.

Consider an ideal isotropic material with identical strengths in tension and compression. Shown in Fig. 1 are the theories presented here for the special case when σ_{12} is zero. Norris' interaction formula defines a circle. The theories of Azzi-Tsai and Hoffman are identical and define an ellipse. For Norris' failure theory, the first of Eq 4 defines an ellipse in first and third quadrants identical to that generated by Azzi-Tsai and Hoffman. However, when the first equation of Norris' failure theory is applied to the second and fourth quadrants, mirror images of the first and third quadrants are generated. The remaining two expressions of Eq 4 are equivalent to those of the maximum stress theory and enclose an area interior to that established by the first expression. Thus, Norris' failure theory reduces to that of maximum stress.

Figure 2 presents the failure theories for the almost isotropic Ph15-7Mo stainless steel where the surfaces were generated for each quadrant. A slight variation exists between the curves of Azzi-Tsai and Hoffman due to the fact that Hoffman's theory takes into account the different strengths of the material in tension and compression (170 ksi versus 179 ksi, respectively), and, therefore, defines only one ellipse for

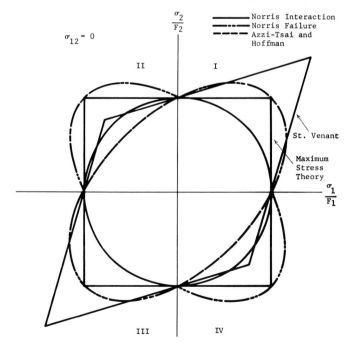

FIG. 1—Ideal material.

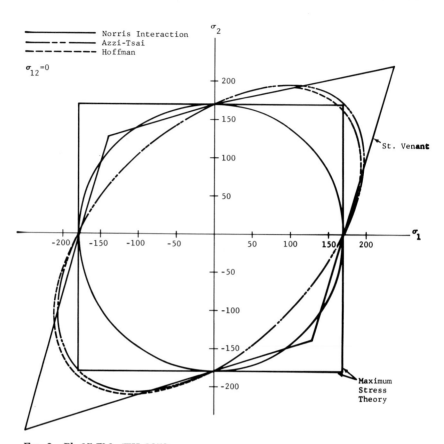

Fig. 2—Ph 15-7Mo/TH 1050.

the four quadrants. Again, Norris' failure theory is identical with that of maximum stress.

Now consider two fiber-reinforced, unidirectional materials, Scotchply Type 1002 [11] and Narmco 5505 [12]. Failure surfaces for each of these materials are shown in Figs. 3 and 4, respectively for $\sigma_{12} = 0$. The theories of Norris (interaction) and Azzi-Tsai produce surfaces, interior to the rectangle generated by the maximum stress theory, which are composed of four curved line segments. Each of these segments is defined by the equation of an ellipse. Although the surface composed of these segments is continuous, it may not be smooth as one might be lead to believe from the use of normalized axes. In particular, Azzi-Tsai's theory includes a cross term $(C\sigma_1\sigma_2)$ which, from analytical geometry, rotates the ellipse defined by each quadrant. Thus, the ellipses of any two adjacent quadrants are not necessarily normal to the separating natural axis, that is, tangents to the curved segments at the common axis intercept may not be collinear. For the two materials shown here, the angular rotation is in the neighborhood of 2 to 5 deg and varies for each quadrant depending upon the limiting strengths. A comparison of the surfaces generated by Azzi-Tsai's theory to that of Norris' interaction formula shows that the effect of the rotation is small. Hoffman's theory, which defines only one ellipse, is both continuous and smooth.

In Figs. 3 and 4, St. Venant's theory emphasizes the point that while a theory may be applicable to one material, it may not be totally satisfactory for another. In particular, the use of St. Venant's theory appears to present no problem when

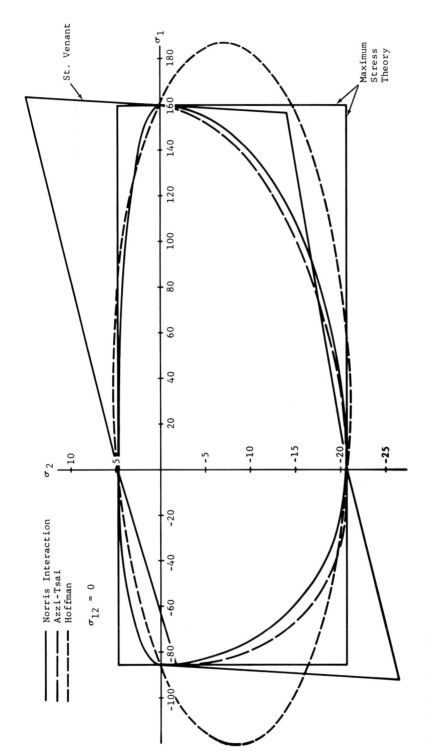

FIG. 3—Scotchply Type 1002.

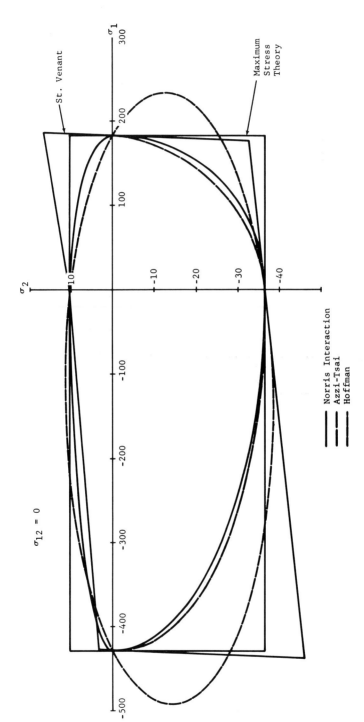

Fig. 4—Narmco 5505.

applied to boron. However, the theory predicts a uniaxial compressive failure $(-\sigma_1)$ for the fiber glass material at -63 ksi even though the value of -85.7 ksi was used to establish the criteria.

The theory predicted a transverse tensile failure under a uniaxial compressive load. Therefore, it would be theoretically impossible to determine experimentally the compressive strength of the unidirectional material in its primary direction. However, a close examination of Eqs 13 indicates that a small change in the value of Poisson's ratio has a strong effect on the theory. Since Poisson's ratio for the fiber glass material is known [11] to vary with load, it is probable that the above dilemma does not exist.

CONCLUSIONS

The assumptions, general limitations, and the physical and analytical significance of several failure criteria and their "material constants" have been discussed as applicable to composite materials. It is evident from the significant differences between the theories that final selection of a failure criterion for a given composite material can be substantiated only through experimental verification.

References

[1] Jenkins, C. F., "Report on Materials of Construction Used in Aircraft and Aircraft Engines," Great Britain Aeronautical Research Committee, 1920.

[2] Norris, C. B. and McKinnon, P. F., "Compression, Tension and Shear Tests on Pellow-Poplar Plywood Panels of Sizes That Do Not Buckle with Tests Made at Various Angles to the Face Grain," Forest Products Laboratory Report No. 1328, 1946.

[3] Norris, C. B., "Strength of Orthotropic Materials Subjected to Combined Stresses," Forest Products Laboratory Report No. 1816, 1950.

[4] Hill, R., "A Theory of the Yielding and Plastic Flow of Anisotropic Metals," Proceedings of the Royal Society, Series A, Vol. 193, 1948, pp. 281–297.

[5] Marin, J., "Theories of Strength for Combined Stresses and Non-isotropic Materials," Journal of the Aeronautical Sciences, Vol. 24, No. 4, 1957.

[6] Greszczuk, L. B., "Elastic Constants and Analysis Methods for Filament Wound Shell Struc-tures," Douglas Aircraft Company Report No. SM45849, Appendix A, 1964.

[7] Azzi, V. D. and Tsai, S. W., "Anisotropic Strength of Composites," Experimental Mechanics, Vol. 5, 1965, pp. 283–288.

[8] Hoffman, O., "The Brittle Strength of Orthotropic Materials," Journal of Composite Materials, Vol. 1, 1967, pp. 200–206.

[9] Hu, L. W., "Modified Tresca's Yield Condition and Associated Flow Rules for Anisotropic Materials and Application," Journal of the Franklin Institute, 1958.

[10] Waddoups, M. E., "Advanced Composite Material Mechanics for the Design and Stress Analyst," General Dynamics, Fort Worth Division Report FZM-4763, 1967.

[11] Davis, J. W. and Zurkowski, N. R., "Put the Strength and Stiffness Where You Need It," Reinforced Plastics Division, 3M Company.

[12] "Advanced Composites Aircraft Application Program–FY67 Final Report," AVCO Corporation Report R-2020, 1968.

Optimization of a Boron Filament Reinforced Composite Matrix

P. W. JUNEAU, JR.,[1]
L. H. SHENKER,[1] AND
V. N. SAFFIRE[1]

Reference:

Juneau, P. W., Jr., Shenker, L. H., and Saffire, V. N., "Optimization of a Boron Filament Reinforced Composite Matrix," *Composite Materials: Testing and Design, ASTM STP 460*, American Society for Testing and Materials, 1969, pp. 170–181.

Abstract:

Most polymeric matrix materials currently in use with boron filaments are based on epoxy resins having limited strength retention characteristics at elevated temperatures. In order to provide materials capable of fulfilling the requirements for a matrix having the optimum properties necessary for short time, elevated temperature composite application in the aerospace field, an intensive development program was undertaken.

An effective evaluation of resin systems for composite applications was undertaken by choosing a test configuration which imposed major stresses in the weak direction of the matrix and matrix-fiber interface. This testing concept made it necessary to compare materials having low flexural modulus and tensile values.

The evaluation configuration chosen for this study was a 9-ply boron-resin laminate, with the filaments oriented in the ±45 deg direction. Evaluation consisted of determining the flexural modulus of a 1-in.-wide by 4.5-in.-long plate over a temperature range up to 600 F, and observing the reduction in modulus as the temperature increased. One plate could thus be used to test the entire range of temperatures, and could then be reconfigured for a tension test, permitting maximum utilization of materials. In addition, through other laboratory evaluation techniques and fabrication studies, it was possible to evaluate a considerable number of resin formulations and modifications.

In general, it was found that toughness is the key property required of the resin matrix, rather than high modulus and high strength alone. Unmodified phenolic resins were found to be brittle and subject to crazing under thermal stress on cooldown from cure temperatures. Epoxies have relatively low heat distortion temperatures and are not necessarily compatible with heat shield materials cured simultaneously. A phenolic resin modified with epoxy and polyvinyl formal was found to have a combination of properties that approaches an optimum for a moderate elevated temperature (500 F) resin matrix for boron filaments.

Key Words:

composite materials, fiber composites, epoxy laminates, boron filaments, mechanical properties, evaluation, tests

The use of highly efficient structural composites, based on high-strength, high-modulus filaments such as boron, must be considered first from the point of view of the application, particularly when extremes of environment may be encountered. For example, where a boron reinforced composite is to be employed as the structural component of a re-entry vehicle, it is important to consider the interaction of the elements of the structural composite with those of the heat shield. The com-

---

[1] Supervising engineer, manager, plastics technology, and manager, Materials Laboratory, respectively, General Electric Co., King of Prussia, Pa. 19101.

patibility of these elements, both thermally and mechanically, is of the utmost importance in maintaining structural integrity of the vehicle as it encounters the forces generated by re-entry.[2]

Advanced high performance heat shields currently are made using thermally stable fibrous reinforcements bonded with a thermally table resin having good charring properties. The fibers are relatively dense, thermally conductive materials such as carbon, graphite, quartz, or silica. The resinous binder employed in these structures must be capable of forming mechanically satisfactory composites in conjunction with the reinforcing fibers. It also must hold the fibers in position while being exposed to high temperatures and high shear forces, and must decompose in an efficient way to form a mechanically stable char structure capable of reradiating some of the heat generated by re-entry into the earth's atmosphere and of holding fibers in position. It is desirable, therefore, to choose resin systems for heat shield applications from that group of materials capable of forming relatively large portions of carbon upon thermal decomposition, while giving off copious quantities of low molecular weight gases. The interaction of the gas with the char to form stable carbonaceous products is an added consideration. Figure 1 is a simplified sketch of this process. Upon examining this sketch, it is apparent that the use of a structural composite that is stable to high temperature will be advantageous in decreasing the thickness requirements, and thus the weight, of the heat shield.

DISCUSSION

Of the resins available, phenolic resins have been used extensively in applications where good thermal stability and mechanical properties are desired. They combine ease of application with versatility and good processing characteristics. Particularly in the field of heat shield technology, phenolic resins reinforced with nylon, glass,

[2] Juneau, P. W., Jr., "Thermally Stable Resin Matrices for Boron Composites" Major Status Report, *Advanced Composite Hardware*, Sept. 1966.

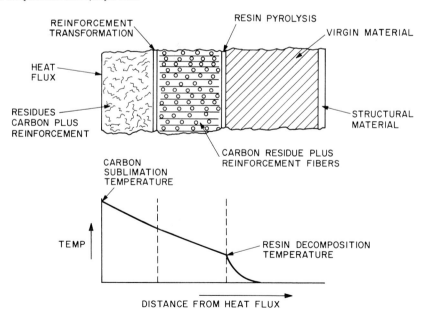

FIG. 1—Representation of multizone transformations in reinforced heat shield materials.

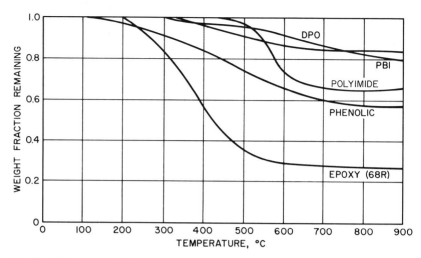

FIG. 2—TGA curves of five resins.

silica, and carbonaceous fibers have found wide application and utility. Figure 2 presents the results of thermogravimetric analyses (TGA) performed on a number of resins including, for comparison purposes, an epoxy resin resigned for structural applications. It is readily apparent from this graph that the resins displaying a combination of weight retention to relatively high temperatures and a good yield of char are those based on benzenoid or heterocyclic structures, such as the phenolic, polyimide, polybenzimidazole, and diphenyl oxide structures.

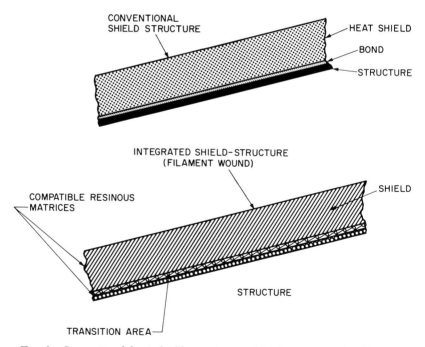

FIG. 3—Conventional heat shield structure combination compared with an integrated shield/structure composite.

Based on an assessment of charring characteristics, strength retention at elevated temperatures and processing characteristics, resins such as phenolic, polyimide, and diphenyl oxide appear to have the greatest potential. They can be used as ablation materials and, together with high-strength, high-modulus filaments, in structural applications of interest to re-entry vehicle technology.

The use of these advanced resins to form integrated shield/structure composites presents a number of potential advantages. Such composites would consist of a heat shield and a structure with a compatible resin system having optimum mechanical and thermal properties integrally fabricated and cured at one time. For example, the shield/structure bond problem would be eliminated, and it would be possible to operate at a higher heat shield backface temperature than is now feasible. The stuctural composite portion of the integrated assembly would have the same elevated temperature capability as the shield, since it uses a thermally compatible resin system. Since less heat shield insulation thickness would be needed than is necessary in the case of a bonded shield, a thinner, lighter weight thermostructural composite would result. The overall result would be a lighter weight, more easily produced system than the currently used aluminum structure-heat shield combinations. Figure 3 illustrates this concept, and compares the integrated structure with a commonly used adhesive bonded assembly. Note that an integral heat shield/structure could be operated at temperatures limited only by the mechanical properties of the resin.

The thermally stable resins previously mentioned are employed customarily with fibrous reinforcements of relatively small diameter, containing a moderate-to-high surface area. Since these are condensation resins, eliminating one molecule of water as each new chemical linkage is formed, careful processing must be used, even with the small diameter fibers, to produce parts free from voids. During cure of such composites, the escape of water of condensation is either by capillary action along the fibers, or by diffusion through the thickness of the material during molding or laminating. When condensation resins are employed as matrix materials with nonporous large diameter filaments, water of condensation has only limited access to escape by means of capillary paths, and increased emphasis must be placed on diffusion through the composite thickness. Careful control of the B-stage condition of the resin is necessary, so that a minimum amount of volatile materials may be present when the resin is processed to its final degree of cure. At the same time, the resin must retain adequate flow characteristics to knit together the layers of composite material and form a coherent structure.

Figure 4 shows the strength retention capabilities of several filaments, including

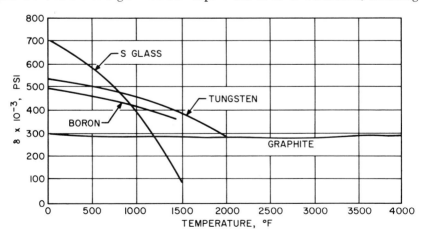

FIG. 4—Strength versus temperature properties of filamentary reinforcements.

boron and graphite, as a function of temperature, indicating their potential capability. In order to utilize to the fullest the outstanding thermal and physical properties of boron filaments, as well as the other advanced filaments becoming available, it is necessary to employ them with matrix materials that can retain their integrity during and after exposure to a wide range of environmental conditions.

The selection of a matrix material was guided by considerations of both resin properties and the requirements of the composite structure. Specifically, the resin selection criteria involved the ability to prepare impregnated boron tapes for the filament winding process selected for test specimen and prototype fabrication. Because the heat shield resin was preselected as a phenolic, the structural composite matrix was required to be compatible with condensation resins throughout the processing cycle. It also had to exhibit good strength retention characteristics and provide continuous load transfer at elevated temperatures. The use of a high-temperature epoxy resin was considered for this application, since this would simplify processing by eliminating both the shrinkage and volatile problems associated with phenolic or other condensation resins. However, the susceptibility of epoxy resin formulations to attack by moisture during the critical periods of the cure cycle, coupled with an upper temperature stability limit of about 400 F, limited the utility of these materials for this application.

The general class of phenolic resins was investigated next for system compatibility. These thermally stable resins are employed customarily with fibrous reinforcements of relatively small diameter, containing a moderate-to-high surface area. Since these are condensation resins, eliminating one molecule of water as each new chemical linkage is formed, particularly careful processing techniques must be used, even with the small diameter fibers, to produce parts free of voids.

The mechanical properties of most unmodified and unreinforced phenolic resins are poor, due to the highly cross-linked nature of the macromolecule.[3] This, together with the high degree of aromaticity induced by the high molar concentration of phenol and its derivatives, leads to a polymer that has brittle characteristics including high-modulus, low-tensile strength, and high-residual stress caused by curing. When used with large diameter filaments, they must fill in relatively large spaces between filaments. Cracking of the matrix then results due to shrinkage upon cure. Suitable modifications, however, such as the incorporation of other impact resistant resins into the basic phenolic, can overcome these deficiencies.[4,5]

Once the processability of the selected candidate resin system was established by making it into a 29 filament tape, it was further evaluated using two criteria: (a) machinability and (b) flexural modulus versus temperature of a ±45-deg oriented composite. Tapes were made using a number of phenolic resins, both in the unmodified form, and modified with either polyvinyl formal and epoxy resins, or both. A modified epoxy also was evaluated. Table 1 lists the different tape materials and gives typical properties. The tape was made into composites by preparing sheets of parallel oriented filaments by winding on a drum, then laying up 9-ply 8 by 8-in. plates and curing them in an autoclave. The ±45-deg oriented specimens were machined out of the plates using a carborundum wheel. Machining characteristics were noted—fraying and splintering of the fibers in the composite indicated a brittle matrix having poor adhesion to the boron filaments. Figure 5 compares the edges of specimens having a tough matrix with one having a brittle matrix. Note the extensive fraying and resin debonding shown by the brittle matrix composite compared to the smooth edge of the composite made with a tough matrix.

Initial investigations were carried out on unmodified resins. The rationale used

[3] Ferry, J. D., *Viscoelastic Properties of Polymers*, Wiley, New York, 1961.
[4] Achhammer, Tryon, and Kline, "Chemical Structure and Stability Relationships in Polymers," *Modern Plastics*, Vol. 37, No. 4, Dec. 1959.
[5] Madorsky, S. L. and Straus, S., "Stability of Thermoset Plastics at High Temperature," *Modern Plastics*, Vol. 38, No. 6, p. 134.

Fig. 5—Machined edge of boron composite ($\times 60$).

Table 1—*Tape properties.*

Resin	Comments	Resin Content, %	Volatiles, %	Gel Time at 160 C
A. Phenolic................	high-modulus brittle MilR9299	32.2	4.3	60 to 70 s
B. DPO....................	intermediate	34.1	4.5	65 to 75 s
C. Phenolic................	intermediate MilR9299	26.4	4.2	80 s
D. Phenolic................	intermediate MilR9299	36.5	4.8	55 to 60 s
E. Phenolic................	intermediate MilR9299	30.2	4.6	60 to 65 s
F. DPO....................	frangible MilR9299	37.3	3.4	65 to 70 s
G. Phenolic................	intermediate MilR9299	38.1	5.6	40 s
H. Phenolic................	intermediate MilR9299	41.0	6.4	110 s
I. (A + Formvar)..........	...	39.8	7.8	53 s
J. (B + Formvar)..........	...	34.2	6.7	35 s
K. (D + Formvar).........	...	41.2	7.6	10 s
L. (G + Formvar).........	...	37.1	7.1	36 s
M. (H + Formvar)..........	...	43.8	8.6	127 s
N. Commercial modified phenolic adhesive formulation	...	39.0	10.0	42 s
O. High-temperature epoxy...	...	34.4	none	9 min

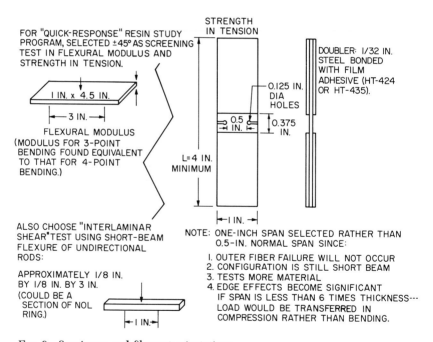

FIG. 6—Specimens and filament orientations.

in performing the mechanical tests is shown in Fig. 6. It was necessary to use a flexure modulus test in which the resin was preferentially stressed in order to determine rapidly the response of a number of candidate materials in a critical fashion. Figure 7 shows the flexural modulus properties of ±45-deg boron composites made with phenolic resins A and B. Note the rapid modulus decay with temperature.

Resins C, D, E, G, and H were evaluated also in the unmodified form, and showed similar properties, characteristically shown in Fig. 8. Resins G and H showed some promise based on modulus retention, but were excessively brittle, and were modified with polyvinyl formal. Test results on composites made from these modified materials, from a commercial modified phenolic, and from a modified epoxy, are shown in Fig. 9. These data show that the modulus retention characteristics of the resins

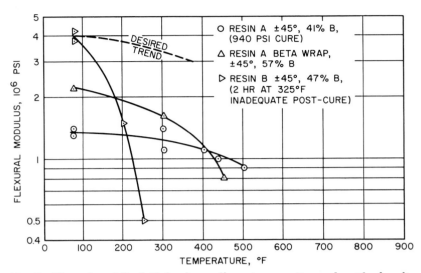

FIG. 7—Flexural moduli of 45-deg boron filament composites made with phenolic resins A and B.

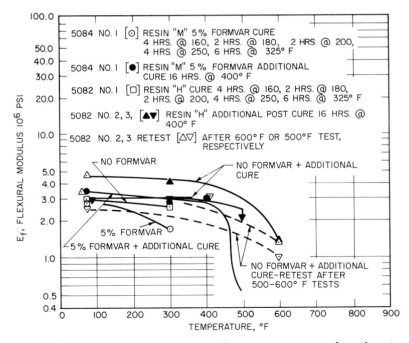

FIG. 8—Flexural moduli of 45-deg boron filament composites made with resins H and M.

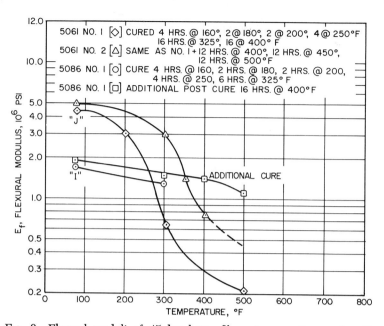

FIG. 9—Flexural moduli of 45-deg boron filament composites made with resins I and J.

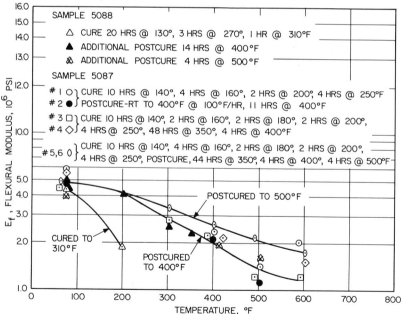

FIG. 10—Flexural moduli of 45-deg boron filament composites made with modified phenolic resin N.

were not improved by the addition of polyvinyl formal, but in some cases, the low temperature moduli were increased.

The commercial modified resin (an adhesive formulation consisting of a phenolic resin modified by the addition of epoxy and polyvinyl formal resins) showed good modulus retention at high temperatures and was selected for further evaluation.

A modified epoxy also was selected. Figure 10 shows the flexural modulus versus temperature of a ±45-deg, 9-ply boron composite using resin N. Note the relatively slow decay of modulus properties with temperature. In Fig. 11, the modulus versus temperature properties of ±45-deg boron filament composites made with resins N and O (a modified epoxy) are compared. The superiority of resin N is immediately

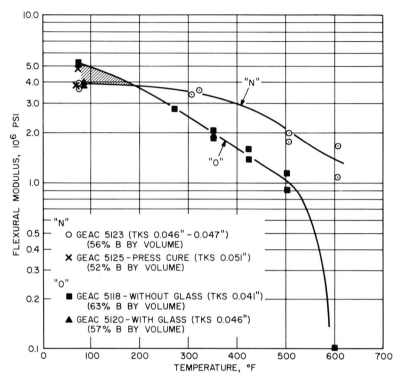

Fig. 11—Flexural moduli of 45-deg boron filament composites made with resins N and O.

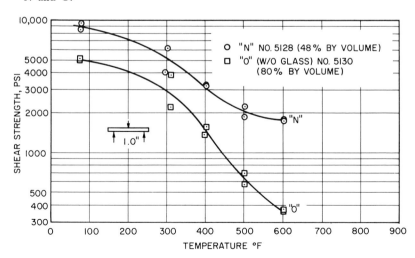

Fig. 12—Interlaminar shear strength, short-beam 0-deg flexural, boron composites made with resins N and O.

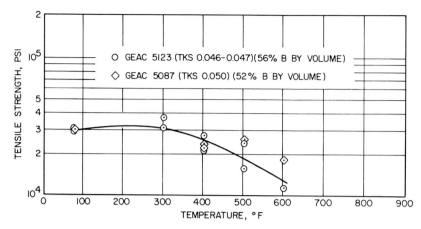

Fig. 13—Tensile strength of ±45-deg boron filament composite—modified phenolic resin N.

apparent at this point, resin N was selected as the principal candidate for application in structural composites, and further evaluation was carried out to confirm the validity of the selection. Figure 12 shows short-beam shear values obtained on unidirectional composites made with the two resins. Resin N was the superior material, and the 1.0-in.-span shear test was able to provide a clearer distinction between the two by de-emphasizing fiber effects. Figure 13 shows the tensile strength of ±45-deg composite made with resin N. This fiber orientation stresses the resin interface, and confirms the modulus data. Figure 14 shows the flexural and tensile moduli of 0/90-deg composites made of the two resins. This orientation is not resin sensitive, and does not readily distinguish between materials. Figure 15 shows the tensile strength properties of 0/90-deg composites. The superior adhesion of resin N is evident. The tensile strengths of unidirectional composites are shown in Fig. 16.

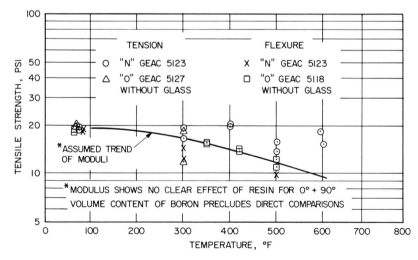

Fig. 14—Flexural and tensile modulus of 0-deg (±90 deg) boron filament composites.

Composite Materials

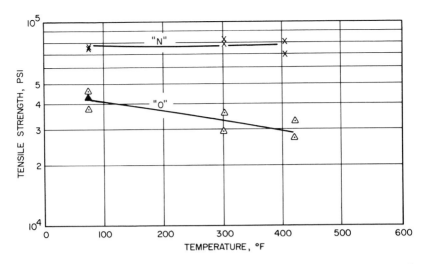

FIG. 15—Tensile strength of 0-deg (±90 deg) boron filament composites N and O.

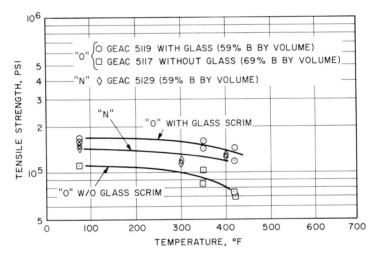

FIG. 16—Tensile strength of unidirectional boron filament composite.

CONCLUSIONS

An effective evaluation technique has been demonstrated that enables material choices to be made based on the assessment of a critical property. The use of ±45-deg oriented flexure modulus specimens provided a convenient index of resin performance and permitted rapid determination of the suitability of a number of candidate materials for a critical high-temperature application. The evaluation technique was further verified by laminate property tests in other orientations.

ACKNOWLEDGMENTS

This work was performed under Contract Number AF33(615)-5364 from the Air Force Materials Laboratory, Air Force Systems Command, Wright-Patterson Air Force Base, Ohio. The monitor for this program was W. J. Iller.

The authors are indebted to R. Kaufman and L. Cohen for their part in the performance of this work. Mr. Kaufman was responsible for specimen fabrication, and Mr. Cohen was responsible for the development of mechanical testing techniques.

Defining the Adhesion Characteristics in Advanced Composites

G. L. HANNA[1] AND
SAMUEL . STEINGISER[1]

Reference:

Hanna, G. L. and Steingiser, Samuel, "Defining the Adhesion Characteristics in Advanced Composites," *Composite Materials: Testing and Design, ASTM STP 460*, American Society for Testing and Materials, 1969, pp. 182–191.

Abstract:

The object of this study was to develop testing techniques that would permit a valid measure of the in-plane shear properties of advanced filament reinforced composites. The shear properties of glass and graphite reinforced epoxy materials were measured by short-beam shear and torsion tests. The specimen geometries and testing procedures necessary to make valid measures of shear properties were established for each of the tests and were found applicable to all materials studied.

The torsion tube and torsion rod tests are the best available methods to accurately measure shear modulus, G, and shear strength, τ_u, by a single

determination. The short-beam shear test at a span-to-depth ratio of 4 is capable of giving the same shear strength value as the torsion tests on composites only if a shear failure occurs. However, short-beam shear specimens of composites possessing good adhesion often fail in tension, and no correlation can be made between the calculated shear strengths obtained from short-beam shear testing and torsion testing. Torsion specimens then are required to measure shear strength.

The theory that the maximum interfacial shear strength that can be achieved in a composite is equal to the shear strength of the matrix material was examined. Torsion tests were conducted on unreinforced epoxy resin as well as on glass and graphite reinforced composites fabricated to achieve maximum adhesion. In no instance did the composite shear strength exceed that of the matrix resin (that is, about 14 ksi).

Key Words:

composite materials, fiber composites, mechanical properties, torsion shear, short-beam shear, graphite epoxy, glass-fibers, epoxy laminates, fractography, evaluation, tests

The rapid increase in the use of advanced composite materials for aerospace structural applications necessitates the availability of reliable design data. Consequently, there has been a flurry of activity to develop mechanical property test methods that are capable of producing the necessary data. Two of the most critical design parameters for composite materials are the shear strength and shear modulus. Several so-called shear tests have been developed that are simple and economical to conduct, but nonetheless impose other components of stress on the specimen in addition to the shear mode (for example, the short-beam shear and the notch-tension interlaminar shear tests). Other shear test techniques, such as the plate twist and "picture frame" tests, have been valuable but costly in both material and special fixturing requirements.

The torsion shear test has long been used to produce a pure state of shear in a hollow tube or cylindrical rod specimen. However, new test equipment had to be developed with the necessary sensitivity at the low torque ranges required to test composites. Now that this equipment is available, the torsion test is probably

---

[1] Group leader and scientist, respectively, Monsanto Research Corp., Dayton, Ohio. 45407. Mr. Hanna and Dr. Steingiser are personal members of ASTM.

the best available shear test,[2] and it can be conducted with the ease and rapidity of a tension test. As in the case of tension testing, both modulus and strength can be determined on a single specimen.

Although the material requirements to test in torsion are moderate, a test such as the short-beam shear test requires significantly less material and may be preferred in quality control programs or screening studies. However, in such an instance a correlation must be made between torsional and the selected screening test data on the material of interest.

The object of this program was to determine if a correlation could be made between horizontal short-beam shear and torsion shear data.

EXPERIMENTAL PROCEDURES

Thornel 40-epoxy flat laminates were prepared to examine the horizontal short-beam shear test method, and unidirectional 0.25 by 6-in.-long rods of E-glass and graphite reinforced epoxy were prepared to compare the applicability of the torsion shear and short-beam shear tests. Several types of graphite reinforcement were used in the 0.25-in.-diameter rod specimens—Thornel 40, Thornel 50, and Morganite Type I. Some Thornel yarns were heat treated by various techniques to improve the interlaminar shear strength of composites made from them.[3] The epoxy resin employed in all composites consisted of 27 parts of ZZL-0820 and 100 parts of ERL-2256. The composites were cured for 2 h at 180 F followed by 4 h at 300 F and 50 psi. The fiber volume content of each specimen was determined by gravimetric methods.

The Thornel 40-epoxy flat laminates were tested under short-beam flexural loading (Fig. 1) to define the minimum span-to-depth ratio (l/d) that could be employed to obtain an interlaminar failure without imposing severe compressive stresses under the loading nose. Specimens were tested at span-to-depth ratios between 3 and 14.

[2] Pagano, N. J., Air Force Materials Laboratory, Wright-Patterson AFB, Ohio, private communication.
[3] Steingiser, Samuel and Cass, R. A., "Graphite Fiber Reinforced Composites," Monsanto Research Corporation, Dayton, Ohio, AFML-TR-68-357, Part I, Nov. 1968.

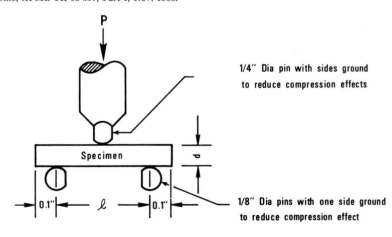

P

1/4" Dia pin with sides ground to reduce compression effects

Specimen

0.1" l 0.1"

1/8" Dia pins with one side ground to reduce compression effect

PROPOSED SPECIMEN GEOMETRY [3]

b, width = 0.25 in

d, depth = 0.125 to 0.25 in

l, span

FIG. 1—Short-beam shear test method.

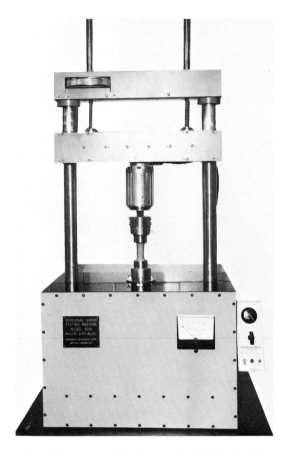

FIG. 2—Torsion shear tester
for rods and tubes.

All short-beam shear tests were conducted at a crosshead speed of 0.02 in./min. The apparent shear strength, S_H, measured by this test was calculated from the formula

$$S_H = \frac{0.75P}{bd}$$

where:

P = maximum load (lb),
b = specimen width (in.), and
d = specimen thickness (in.).

In most cases a short-beam shear specimen (0.215 by 0.125 by 0.7 in.) was prepared from the end of the 0.25-in.-diameter rod and tested at a span-to-depth ratio of 4. The remaining section of each rod was then tested on a torsion testing machine (shown in Fig. 2) designed, built, and marketed by Monsanto Research Corp. The torsion tester has a 6000 lb·in. torque capacity applied at a constant rate of twist, and it permits changes in specimen length without introducing spurious stresses during the test. Tests can be conducted at rates of twist from 15 to 150 deg/min in either direction, and 0.125 to 2.62-in.-diameter specimens can be accommodated. Collet type grips (Figs. 3 and 4) are used to achieve a high degree of concentricity (±0.2 percent) in loading. One grip is free to move axially to prevent the application of axial forces. Torsion tube specimens also may be cemented

Composite Materials

FIG. 3—Torsion shear rod specimen and collet gripping assembly.

FIG. 4—Torsion shear tube specimen and collet gripping assembly.

to conical seats if preferred. Two-element strain gages are applied to the surface of each specimen to obtain the shear strain measurements necessary to calculate shear modulus.

The graphite-epoxy specimens were tested at a twist rate of 50 deg/min, and the E-glass-epoxy composites were tested at 100 deg/min. The specimen gage length was always greater than 2.25 in. The torsion shear strengths, τ_u, were calculated from the formula

$$\tau_u = \frac{M_T D}{2J}$$

where:

M_T = maximum applied torsional moment (lb·in.),
D = rod diameter (in.), and
J = polar moment of inertia (in.) ($J = \pi D^4/32$ for a rod).

The torsional tangent shear modulus, G, is defined as

$$G = \frac{\Delta M_T L}{J \Delta\theta}$$

where:

L = gage length of the rod and
$\Delta\theta$ = change in angle of twist corresponding to the applied ΔM_T.

The failure mode was defined visually or microscopically for each short-beam shear and torsion test specimen. A scanning electron microscope was employed in many fractography studies.

Hanna and Steingiser Adhesion Characteristics   *185*

RESULTS AND DISCUSSION

The horizontal short-beam shear test has been employed for a number of years to measure the apparent shear strength of composite materials. However, the test has a serious deficiency in that a shear failure does not always occur. This difficulty perhaps most seriously plagued those who were engaged in boron-epoxy evaluation, and more recently those who were producing high shear strength (>8000 psi) graphite-epoxy composites. The question that arises from this test deficiency is, "If I obtain a tensile failure, what meaning does the strength value I calculate have, if any?" As we will show later, a short-beam shear test that yields a tensile failure is of no value to the designer. He does not know whether or not the strength calculated from such a test is less than or greater than the true shear strength, τ_u.

The horizontal short-beam shear test has not been standardized as yet. Many companies have established their own internal procedures, which have now formed the basis for a method to be written if the industry so desires. A standard specimen depth and width and test fixtures have been proposed,[4] as shown in Fig. 1. However, the span-to-depth ratio (l/d) still must be defined. Values of l/d between 3 and 8 have been employed for several materials with varying degrees of success. The initial objective of this program was to define a span-to-depth ratio for consideration as a standard geometry.

Short-beam shear specimens were prepared from two Thornel 40-epoxy composites and tested at span-to-depth ratios of 3, 4, 5, 6, 8, and 14. One composite was prepared from yarn that had been treated to increase the interlaminar shear strength. The other was fabricated from as-received yarn. As expected, high span-to-depth ratios resulted in tensile components of failure, and shear failures were obtained at the low l/d values (Fig. 5). Several other graphite-epoxy composites were tested at l/d values of 3, 4, and 6 to confirm the results shown in Fig. 5. These data are presented in Table 1. Failures were determined to be tensile or shear in nature by microscopic or visual examinations. Examples of the types

---

[4] "Proposed Tentative Method of Test for Apparent Horizontal Shear Strength of Flat Laminates," ASTM Committee D-20, Subcommittee XVIII Working Document, 15 May 1967.

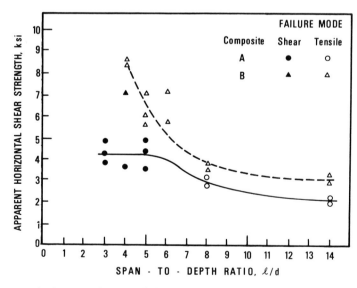

FIG. 5—Apparent horizontal shear strength versus span-to-depth ratio for Thornel 40-epoxy composites.

Table 1—*Effect of span-to-depth ratio on apparent horizontal shear strength of Thornel 40-epoxy 0-deg laminates.*

Specimen	Span-to-Depth Ratio, l/d	Average Apparent Horizontal Shear Strength, S_H, ksi	Failure Mode
A..................	4	8.5	shear
	5	5.7	tensile
B..................	4	6.9	shear
	5	5.8	tensile
C..................	4	8.5	tensile
	6	6.1	tensile
D..................	4	7.7	tensile
	6	5.4	tensile
F..................	3	7.0	shear
	4	6.9	shear
	6	6.0	tensile
G..................	3	7.1	shear
	4	7.0	shear
	6	6.3	tensile

of failure modes observed are shown in Fig. 6. The fractography of each type of failure was studied employing the scanning electron microscope. As shown in Fig. 7, a shear failure results in debonding of the fiber from the matrix. The fiber failures are a result of posttest sectioning to enable study of the failure mode. Fractography of complex failures revealed both debonding and constituent failures, and predominantly tensile failures were nearly flat.

From the data shown in Fig. 5 and Table 1, four significant conclusions can be drawn.

1. The maximum span-to-depth ratio at which shear failure will occur is dependent on the shear strength of the composite. The lower shear strength composite, made from the as-received Thornel 40, exhibited shear failures and a nearly constant apparent horizontal shear strength, S_H, (S_H ranged between 3600 and 5000 psi) at an $l/d \leq 8$. However, the high shear strength material, fabricated from the treated Thornel 40 yarn, failed in a shear mode only at an $l/d \leq 4$.

2. A span-to-depth ratio of 4 is the minimum that can be used without introducing severe compressive components in the specimen. Although a maximum load value was measured (slope of the load-deflection curve equals zero) at $l/d = 3$,

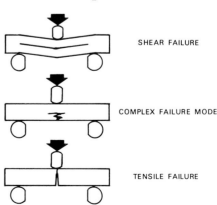

SHEAR FAILURE

COMPLEX FAILURE MODE

Fig. 6—Types of failures obtained in graphite-epoxy short-beam shear specimens.

TENSILE FAILURE

Hanna and Steingiser Adhesion Characteristics

Fig. 7—Scanning electron micrographs of Thornel 40-epoxy resin composites tested by the short-beam shear method (reduced 63 percent for reproduction).

the load subsequently increased rapidly in an asymptotic manner. Complete specimen failures occurred for $l/d \geq 4$. Although compressive stress components certainly exist at l/d ratios of 4 and greater, the ability exists to measure nearly equal values of S_H regardless of l/d, provided shear failures occur. Therefore, the adoption of $l/d = 4$ is suggested as a standard geometry that produces the best possibilities for a shear failure to occur in a material without introducing severe compressive stresses.

3. It appears that with sufficient experience in applying the horizontal short-beam shear test to a given material, the load-deflection curve can be employed to determine the failure mode instead of microscope analyses. It was found that graphite-epoxy composites fail abruptly without a significant decrease in the slope of the load-deflection curve (Fig. 8c) when the tensile failure mode occurs. Shear failures are denoted by a gradual decrease in the slope of the load-deflection curve to a zero value after which complete failure occurs (Fig. 8a).

4. Whenever a shear failure was obtained, noncatastrophic cracking initiated before maximum load was reached. As shown in Fig. 8a, the load at which cracking initiated (that is, the first event) corresponded to the point at which deviation from linearity occurred on the load-deflection curve. The cracking was observed by sharp audible reports as well as by microscopically examining specimens that were loaded to the point where the first audible report occurred. The point of shear crack initiation is of significantly more importance to the designer than the failure strength. One of the objectives of the designer is to prevent any degradation

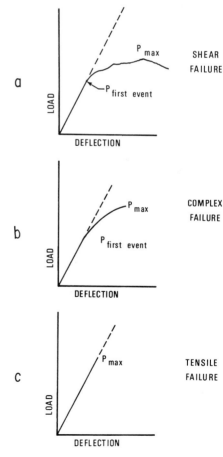

Fig. 8—Load-deflection curves for graphite-epoxy short-beam shear specimens.

Table 2—Comparison of shear data for 0-deg composites.

Material	Test Method	Fiber Volume, %	Shear Modulus, G, Mpsi	Shear Strength,[d] ksi	Failure Mode
E-Glass epoxy	TR[a]	58	1.17	13.0	shear
	TR[a]	58	1.22	12.5	shear
	SBS[b]	58	...	13.6	shear
Thornel 40-epoxy (untreated fiber)	TR[a]	60	0.68	4.80	shear
	TR[a]	60	0.62	4.20	shear
	SBS[b]	50	...	4.62	shear
Thornel 40-epoxy (treated fiber)	TR[a]	57	...	7.58	shear
	TR[a]	58	0.76	7.68	shear
	TR[a]	55	...	6.86	shear
	TR[a]	55	...	7.75	shear
	SBS[b]	57	...	11.4	tensile
Thornel 50-epoxy (untreated fiber)	TR[a]	55	0.66	5.10	shear
	TR[a]	55	0.61	5.22	shear
	SBS[b]	55	...	6.40	tensile
Thornel 50-epoxy (treated fiber)	TR[a]	53	...	8.12	shear
	TR[a]	54	...	9.02	shear
	TR[a]	42	0.74	7.50	shear
	TR[a]	42	0.71	7.40	shear
	SBS[b]	60	...	8.99	tensile
	SBS[b]	42	...	7.94	tensile
Morganite Type I (MO-B)-epoxy (treated fiber)	TR[a]	55	0.68	5.71	shear
	SBS[b]	55	...	6.40	tensile
Morganite Type I (MO-H)-epoxy (untreated fiber)	TR[a]	61	1.37	3.26	shear
	SBS[b]	61	...	3.25	shear
Morganite Type I-epoxy (treated fiber)	TR[a]	58	1.02	8.50	shear
	TR[a]	57	1.01	8.20	shear
	SBS[b]	58	...	8.7	shear
	SBS[b]	57	...	11.1	tensile
Epoxy resin[c]	TR[a]	0	0.25	14.1	tensile
	TR[a]	0	0.25	14.1	tensile
	TR[a]	0	0.25	14.1	tensile
	TR[a]	0	...	13.7	tensile
	TR[a]	0	...	13.7	tensile

[a] 0.25-in.-diameter molded torsion rod. Strain gages used for modulus determination.
[b] Short-beam shear test at $l/d = 4$.
[c] Resin—27 parts ZZL-0820/100 parts ERL-2256; Cure—72 h at room temperature plus 2 h at 150 C.
[d] These values represent true shear strength only when shear failure occurs.

of the material that would lead to a fatigue failure. Therefore, we believe that both the stress at crack initiation (first event) and the ultimate shear strength, S_H, should be reported from all data.

After defining the horizontal short-beam shear test method, it is still of paramount importance to determine the real value of short-beam shear data to the designer. The fact that a pure state of shear does not exist in the short-beam shear specimen is enough to make one skeptical of the data obtained. We therefore compared the torsional shear rod and the short-beam shear ($l/d = 4$) tests on several graphite and glass reinforced composites. The torsion shear test was selected since it imposes a pure state of shear on the composite and the true shear strength, τ_u, and shear modulus, G, can be defined. Although a torsion tube specimen presents a less severe shear gradient through the specimen than the torsion rod geometry, torsion rods (cast or machined), and torsion tubes have been shown to have equal ability to measure torsional properties of advanced composites at room temperature.[5]

A comparison of the data obtained by the two shear test methods is presented in Table 2. The torsion rod and short-beam shear tests produce very similar shear strength values if a shear failure is obtained in the short-beam shear specimen. The E-glass-epoxy and untreated Thornel 40-epoxy composite data demonstrate the similarity in shear strengths obtained. However, when short-beam shear specimens fail with tensile components (as was the case for the untreated Thornel 40- and both treated and untreated Thornel 50-epoxy composites) the torsion shear strengths, τ_u, were either less than or equal to the short-beam shear strengths. Therefore, short-beam shear tests that result in tensile failures cannot be interpreted as even a conservative estimate of the material's shear strength. The true shear strength may be significantly below the measured S_H value as was the case for treated Thornel 40-epoxy ($\tau_u = 7.58$ ksi, $S_H = 11.4$ ksi). Whenever a short-beam shear test results in any tensile components of failure the data should not be reported, and a test capable of producing shear failures (for example, torsion tube or torsion rod) should be substituted.

A few torsion rods also were prepared from the epoxy resin without reinforcement. The average torsional strength obtained on these rods was about 14.0 ksi (Table 2). It is interesting that the torsional strength of the epoxy resin appears to be the limiting value that can be obtained for a composite since in no instance does the shear strength of any composite exceed that of the pure epoxy.

CONCLUSIONS

The short-beam shear test can be employed as a screening test for determining the shear strength of advanced composites provided a shear failure does occur. The torsional shear and short-beam shear (at $l/d = 4$) strengths obtained for several glass and graphite reinforced epoxies were comparable provided the short-beam shear specimen did fail along the neutral axis (that is, plane of maximum shear stress).

However, the complex stress state present in the short-beam shear specimen prevents its use to develop design data, since tensile failures often occur and torsion rod or torsion tube tests should be conducted. The torsion test is also the best method to obtain both shear modulus and shear strength from a single test. With the recent availability of appropriate torsion testing machines for advanced composites, torsion testing for shear properties is as simple and economical to conduct as a short-beam shear test.

ACKNOWLEDGMENT

A portion of the data reported herein was obtained under Contract No. F33615-67-C-1728 which was administered by the Air Force Materials Laboratory, Plastics and Composites Branch (MANC), Directorate of Laboratories, Wright-Patterson Air Force Base, Ohio.

[5] Adams, D. F. and Thomas, R. L., "Test Methods for the Determination of Unidirectional Composite Shear Properties," *Advances in Structural Composites*, Society of Aerospace Material and Process Engineers, Vol. 12, 1967.

Fatigue Testing and Thermal-Mechanical Treatment Effects on Aluminum-Boron Composites

H. SHIMIZU[1] AND
J. F. DOLOWY, JR.[1]

Reference:

Shimizu, H. and Dolowy, J. F., Jr., "Fatigue Testing and Thermal Mechanical Treatment Effects on Aluminum-Boron Composites," *Composite Materials: Testing and Design*, ASTM STP 460, American Society for Testing and Materials, 1969, p. 192–202.

Abstract:

Fatigue strength data are presented in the form of constant lifetime curves over a short ratio range, principally for a tension/tension mode with a low minimum stress and a maximum stress closely approaching the full tensile strength of the material. A brief description of the complex stress distribution within the matrix and its consequences are given, and the effects of thermal/or mechanical treatments or both are discussed in terms of the stress state modifications and the effects on longitudinal tensile strength. Regions of high triaxial stress within the matrix results in a higher than predicted Young's modulus, with a fracture stress which approaches and occasionally exceeds the rule of mixtures at strain levels less than the strain capability of the filaments. Under these circumstances, it was suspected that the strength of the composite was limited by the early fracture initiation in the matrix. Thus, composite material strength should be optimized by decreasing the initial or residual matrix stress level to the point where fracture is initiated simultaneously in the matrix and the filament. Stress modifications and resultant strength improvements were accomplished by selective thermal and mechanical treatments to the composites after fabrication.

Key Words:

fiber composites, composite materials, aluminum, boron, reinforcement, fatigue, tensile, mechanical properties, heat treatment, evaluation, tests

Application studies of metal-matrix composite materials for aerospace propulsion structures were conducted at Marquardt which lead to an in-depth study of the mechanical behavior of the aluminum-boron composite system [1,2,3].[2] In this paper, attention is focused on the fatigue properties and the effects of thermal and mechanical treatments on tensile properties. The data presented are derived mainly from in-house efforts which supplemented the recent work conducted for the Air Force.[3]

At the outset of composite materials development, the initial evaluations usually involved simple tension specimen characterizations. As the materials development proceeded, the areas of application developed, and the need for practical engineering design information emerged. Among the important requirements were fatigue properties and, for the aluminum-boron composite system, the effects of heat or mechanical treatments.

This report deals with aluminum-boron composites with matrices of 1145 and

---

[1] Special members of the Advanced Technical Staff, Aerospace Products Division, The Marquardt Corp., Van Nuys, Calif. 91409. Mr. Shimizu is a personal member ASTM.
[2] The italic numbers in brackets refer to the list of references appended to this paper.
[3] Contract F33615-68-C-1165.

6061, with emphasis on the latter. The 6061 alloy is representative of a useful wrought, heat-treatable alloy, and was selected because of its formability (for composite fabricability by diffusion bonding) and brazability (for composite joining in high-temperature applications). These properties are not mutually exclusive, since both are associated intimately with and affected by the initial melting temperature (solidus) which, in turn, is controlled by the alloy composition.

The experimental aspect of this paper includes fatigue studies on uniaxial and biaxial aluminum-boron specimens (6061 and 1145 matrix alloys), both at room temperature and elevated temperatures. A Goodman diagram shows the payoffs obtainable with composites. A limited study of the effects of edge notches and center holes cut by electrical discharge machining (EDM) is presented also. The second area of study covers the effect of thermal and mechanical treatments on the properties of 6061 matrix composites, specifically the effects of T6 treatments and subsequent transverse cold rolling.

FATIGUE TEST METHODS

Material Fabrication and Specimen Preparation

The composite materials used were predominantly unidirectionally reinforced and fabricated at Marquardt, using the vacuum diffusion-bonding technique. The material is fabricated by preparing layups of 0.004-in.-diameter boron filament on aluminum sheets and temporarily held by an organic adhesive which escapes without residue before final pressing temperatures are reached. Selection of foil thickness and filament spacing are controlled by the requirements for reinforcement volume fraction. Foil thickness ranges generally from 0.002 to 0.006 in., and filament spacings are typically about 180/in., while the total thickness ranged from 0.018 to 0.050 in., corresponding to three- to ten-plies.

The actual consolidation of this material was conducted at temperatures exceeding 900 F and pressures of several thousand psi. From this it can be appreciated that the "as-fabricated" composite material has a large longitudinal residual stress resulting from the fabrication cycle, due to the large difference in the thermal expansion (or contraction) characteristics between the two constituents. The internal stress pattern is complicated by the lateral restraint imposed on the matrix by the closely spaced filaments. On application of an external tensile load, the difference in the Poisson ratios of the constituents ($v_m > v_f$) causes a further extension of the magnitude of the lateral stress field. These materials had tensile strength capabilities, as shown in Fig. 1; the number shown above each data band represents the number of test points included. The data consistency has improved greatly with time; at present 6061 aluminum-boron can be fabricated with a maximum variation from panel to panel of less than ± 15 percent of the average ultimate tensile strength.

The test specimens (Fig. 2) were prepared by three techniques: all the reduced-section specimens were cut by EDM techniques, while the rectangular specimens were either abrasively cut with a carbide cut-off wheel with coolant, or sheared with an ordinary paper cutter. In all cases, any rough edges were touched up with emery paper prior to testing.

Testing

The test technique used throughout this study was tension/tension and tension/compression in a standard Sonntag fatigue machine which uses a fixed initial extension on the test specimen with an inertia pendulum superimposing a cyclic load. The standard sheet-metal grips were used also with the composite specimen. Since it was not desirable to tighten the grips to the point of damaging filaments at the grip edges, and because some grip slippage was noted during the early tests as the cyclic load was applied, a sleeve of emery paper was placed (rough surface

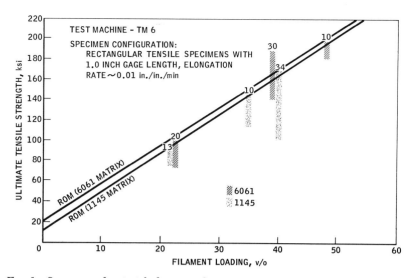

FIG. 1—Summary of uniaxial aluminum-boron strength.

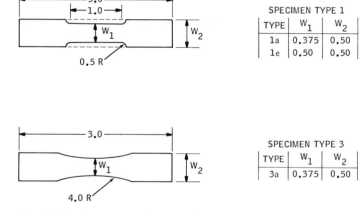

FIG. 2—Tension test specimen configurations.

toward the composite specimen) around each specimen. Soft aluminum tabs cemented to the grip area of each specimen also eliminated the grip-slippage problem; however, since the emery paper was simpler, the latter was used.

Because of the excessive material-property data scatter in the tensile data on early material, the data was plotted on basis of "percentage of average ultimate tensile strength" as well as on an absolute stress basis. On the percentage basis, it was possible to consolidate all filament volume fractions and various matrix alloys on a single chart (Fig. 3).

All the tension/tension fatigue tests were run with from 8 to 10 ksi minimum stress, with maximum stress (depending on material capabilities) of from 65 ksi for 22 volume percent to 140 ksi for 40 volume percent aluminum-boron. A limited number of tests was run from approximately the same maximum tensile stress level into a low compressive level of about 40 ksi; the possibility of specimen and gripping instability precluded higher compressive stresses. The tension/compres-

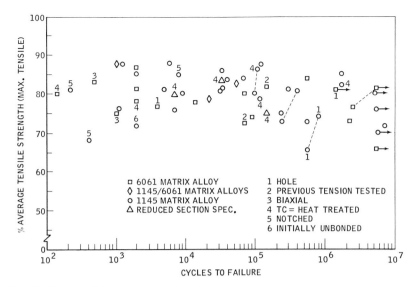

FIG. 3—Fatigue data summary, all materials.

sion fatigue data were essentially the same as had been obtained from the tension/tension tests.

The standard specimen was a rectangular, parallel-sided blank approximately 0.5 in. wide, and longer than 3.5 in. (Fig. 2). A series of reduced-section specimens showed similar results for uniaxial aluminum-boron, but longitudinal cracks initiated at the reduced section to radius transition point and propagated longitudinally into the grip section. However, this occurred early in the test life and appeared to have little effect on the fatigue properties of the remaining parallel-sided specimen. This phenomenon can be attributed to the relatively low shear strength of the matrix material and demonstrates the lesser efficiency of discontinuous filaments in transferring cyclic stresses to adjacent filaments. This phenomenon has been reported also in fiber glass experiments.[4] This occurrence also suggests that large plastic strains must be occurring in the matrix of a composite at points of major reinforcement discontinuities. For these reasons, notched tension tests in the longitudinal direction do not show severe strength degradation, and on properly notched fatigue specimens no effect has been noted.

Fatigue Test Results

The fatigue test results are presented as S/N curves for 22 and 40 volume percent 6061 matrix uniaxial specimens in Figs. 4 and 5, respectively. The corresponding amplitude ratio also is given as an auxiliary ordinate. For the 22 volume percent material, a narrow scatter band is seen with only one failed specimen below the lower bound, and this specimen was a remnant piece from a previous tension specimen. For 10^6 cycle life, stresses of approximately 60 ksi can be accommodated. For the 40 volume percent data, a greater amount of test data showed a similarly shaped scatter band with stress capability at 10^6 cycles of approximately 110 ksi. Figure 6 shows S/N data for 37 volume percent uniaxial 1145 aluminum-boron: here it can be seen that the low-cycle data have noticeably more scatter. However, at the high cycles, a very narrow band occurred. For 10^6 cycles the maximum stress capability is approximately 95 ksi.

---

[4] Personal communication from AVLABS, Ft. Eustis, Va.

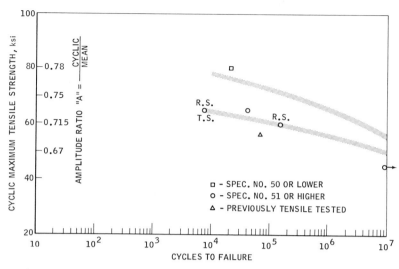

FIG. 4—Fatigue data for 6061 aluminum- 22 volume percent boron, uniaxial composites.

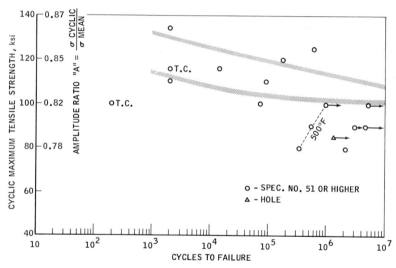

FIG. 5—Fatigue data for 6061 aluminum- 40 volume percent boron, uniaxial composites.

Figure 7 shows a Goodman diagram for uniaxial about 40 volume percent 6061 aluminum-boron. Curves for about 10^2, 10^4, and 10^6 cycles are included, and data for two conventional heat-treatable aluminum alloys also are shown. Similar data have been generated by several other investigators [4,5] on recent material, and [6,7] on early composite systems.

Figure 3, presented earlier, plotted all the data shown above, as well as 22 volume percent 1145/6061 data, and both 32 and 40 volume percent data for 1145 matrix. In all these cases, the use of percentage of average ultimate tensile strength on the ordinate of an S/N type curve proved to be an excellent normalizing technique, decreasing the data scatter significantly.

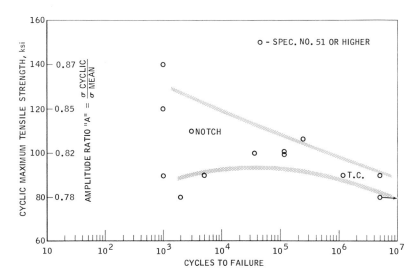

FIG. 6—Fatigue data for 1145 aluminum- 37 volume percent boron, uniaxial composites.

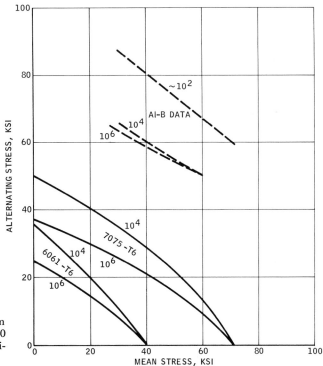

FIG. 7—Goodman diagram for 6061 aluminum- 40 volume percent boron, uniaxial composites.

Figure 8 shows the results of elevated-temperature fatigue tests and both edge-notched and "center-hole" specimen tests as an S/N curve. Interestingly, the elevated-temperature data show significant scatter with the 700 F point well above the 500 F points for 40 volume percent material. Previous tension tests have shown that a minimum in the temperature-versus-strength curve occurs about 500 F,

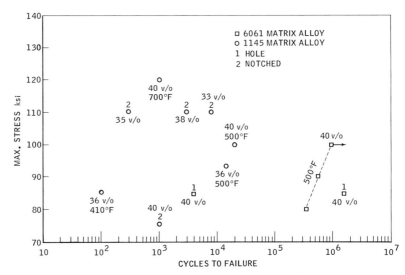

FIG. 8—Fatigue data for uniaxial specimens with notches or holes and for elevated temperatures.

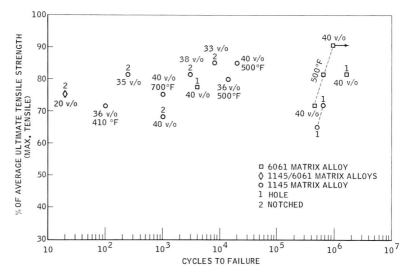

Fig. 9—Fatigue data for uniaxial specimens with notches or holes and for elevated temperatures (normalized).

with test points above 700 F approaching the room-temperature strength. The lowest point on this curve represents one of two specimens which was edge notched, using hand cutters, and possibly causing damage to the composite specimen at the notch root.

Figure 9 shows the same fatigue data plotted as percentage of average room-temperature ultimate tensile strength versus cycles to failure, and as was noted earlier, this data "normalizing" technique markedly decreases the data scatter. Both these curves tend to indicate little or no effect on the fatigue capabilities by either elevated temperatures or irregular specimen; also, the general trend (S/N curve)

Fig. 10—Nontypical fatigue fractures.

would be approximately a horizontal line from 100 cycles out to 2×10^6 at approximately 80 percent of average room-temperature ultimate strength.

The bulk of the test specimens exhibited a fracture similar to that seen in tension testing. The failure was catastrophic in nature, with little or no change in specimen stiffness prior to fracture; very few pullouts were seen (when they occurred, were only short lengths), and the plane of fracture would shift occasionally part of the way across or through the thickness of the specimen. Several "special" fracture types are shown in Figs. 10 and 11. The two specimens shown exhibited a delamination-type failure which was expected on a few unbounded specimens; however, both these specimens were felt, from tensile data, to be moderate-to-well bonded. The notched and "holed" specimens initially cracked longitudinally; however, when notched with hand cutters (diagonals), damage was incurred at the notch tip, causing the tensile-type failure shown in the bottom specimen of Fig. 11.

Thermal-Mechanical Treatment Effects

In order to understand the effects of thermal and mechanical treatments on composite material properties, it is necessary to know the initial conditions existing in the as-fabricated material. In earlier studies, it was noted that composite tensile fracture occurs before the specimens attain a strain level corresponding to the high-strength achievable by boron filaments. This was attributed to the complex stress state in the matrix, including regions of triaxial tension which caused low strain elastic failure initiation in the matrix. The stresses originate from two sources: the residual stresses resulting from the fabrication temperature cycle, and the stresses due to the imposed tension test loads. The stresses from the two sources tend to be cumulative. Thus if the residual stresses in the matrix can be relieved, the composite should then be capable of higher load-imposed stresses. Further, since the fibers are initially in longitudinal compression, the total composite strain should exceed the normal maximum strain capability of the filaments. Studies at Marquardt have shown that 6061 aluminum-boron uniaxial and biaxial composite tensile properties can be enhanced by heat treating the composite to approximate the standard 6061 aluminum T6 condition, and further by cold rolling the heat-treated specimens. Two mechanisms are immediately evident from these treatments. The solution heat-treatment and water-quenching process can be visualized as creat-

Fig. 11—Typical fatigue fracture for specimens with notches or holes.

ing a severe and indeterminate internal residual stress state in the composite, with the subsequent artificial aging process causing precipitation of intermetallic particles to harden and strengthen the aluminum, as well as allowing selective internal creep to relieve the residual stress. The mechanical cross rolling, however, achieves matrix strengthening by cold-working effects while simultaneously effecting an additional relief of the longitudinal residual stresses.

Data have been generated on several panels of 36 to 40 volume percent (total) aluminum-boron for 6061-T6 and 6061-T6 rolled to a 10 percent reduction in one or two passes. The data are summarized in Fig. 12 for the uniaxial material. The average as-fabricated strength is 147 ksi within a range extending from a minimum of 119 ksi to a maximum level of 190 ksi. For the T6 condition material, the average strength is 160 ksi with a spread ranging from 136 to 185 ksi, while for the few specimens T6 treated and then cold rolled, the average strength is approximately 175 ksi with a high of 200 ksi and a low of 154 ksi. Although the scatter was somewhat excessive, this was the result of some intentional variations of fabrication parameters and some minor modifications in the technique used to achieve the T6 condition. In addition, the variations of average filament strength (340 to 425 ksi) must be considered. If each panel is considered separately, the data scatter is less than ±10 percent of the average ultimate tensile strength.

By considering the average data points it is evident that heat treating to the T6 condition increases the longitudinal ultimate strength and decreases the scatter. A cross-rolling treatment of a T6 material further increases the average strength, but the number of data points is insufficient for effective appraisal of the data scatter.

For several uniaxial panels, tests were conducted in the transverse direction. The as-fabricated data spanned the range 12 to 16 ksi, while the data for the T6 and the T6-plus-rolled materials spanned the range of 14 to 20 ksi. It is felt

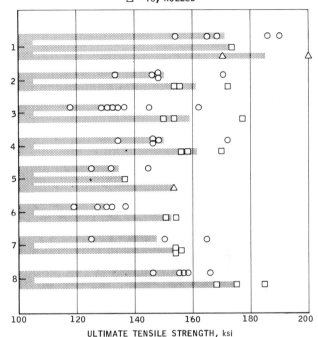

O AS FABRICATED
□ - T6
△ - T6, ROLLED

Fig. 12—Effect of thermal and mechanical treatments on strength of uniaxal 6061 aluminum- 40 volume percent boron composites.

ULTIMATE TENSILE STRENGTH, ksi

that these data are not representative of the true capability of the composite. This is due to the sensitivity of transverse tests to edge defects. Further, the T6-treated specimens exhibited a slight curvature which imposes bending stresses during testing, possibly contributing to the data depression.

A single panel of orthogonally biaxial aluminum-boron (28 and 14 volume percent) in the as-fabricated condition showed strengths of 90 ±5 ksi, while in the T6 condition, the strengths increased to 115 ±20 ksi, all tests being conducted in the direction of higher reinforcement.

CONCLUSIONS

The fatigue capabilities of uniaxial and biaxial aluminum-boron composite materials can be predicted as a function of the corresponding tensile strength. This study showed, for 10^6 cycle life, maximum fatigue stresses of approximately 75 percent of the average ultimate tensile strength can be expected (for "A" ratio of 0.67 to 0.99). Also, the effects of edge notches and center holes do not appear to decrease the efficiency of the adjacent composite material; however, in several cases, notches were placed in these specimens by hand with ordinary cutters. In these particular cases, twisting occurred, and visible damage existed at the point of the notch causing a lower-than-expected fatigue life.

Elevated temperatures did not affect markedly the fatigue behavior; however, the 400 and 500 F tests fell below the 700 F test which is similar to the influence of temperature on tension tests. The effects of thermomechanical treatments on fatigue results have been started, but data are not available as yet.

The effectiveness of the T6 condition on the uniaxial and biaxial tensile properties indicates strengthening of approximately 10 percent for the T6 and approximately 20 percent for the T6 rolled material over the as-fabricated material. The transverse data, although much less consistent than desired, also showed a marked strength increase. It strongly is suspected that much of the data scatter noted in the thermomechanical effects study and in the fatigue are created actually by variations in the constituent properties and variations in the bond quality of the basic composite. Initial studies at Marquardt have shown bond quality and fracture appearance to be directly correlatable with several properties of the composite.

References

[1] Shimizu, H. et al, "Metal-Matrix Composites Behavior and Aerospace Applications," Paper 670861, *Transactions*, Society of Automotive Engineers, 1968, pp. 2661–2668.

[2] Dolowy, J. F., Jr. et al, "Design and Testing of Composite Materials for Use in Ramjet Inlet Structures," Paper No. 68-1041, American Institute of Aeronautics and Astronautics, Philadelphia, Pa., Oct. 1968.

[3] Taylor, R. J. et al, "Application of Composite Materials to Ramjet Inlet Structures," AFML-TR-67-85, Vol. I, Air Force Materials Laboratory, May 1968.

[4] Antony, K. C. and Chang, W. H., "Mechanical Properties of A1-B Composites," *Transactions*, Amer-

ican Society for Metals, Vol. 61, 1968, pp. 550–558.

[5] Kreider, K. G. et al, "Plasma Sprayed Metal Matrix Fiber Reinforced Composites," AFML-TR-68-119, Air Force Materials Laboratory, July 1968.

[6] Toy, A., Atteridge, D. G., and Sinizer, D. I., "Development and Evaluation of the Diffusion Bonding Process as a Method to Produce Fibrous Reinforced Metal Matrix Composite Materials," AFML-TR-66-350, Air Force Materials Laboratory, Nov. 1966.

[7] Kreider, K. G. and Leverant, G. R., "Boron Fiber Metal Matrix Composites by Plasma Spraying," AFML-TR-66-219, Air Force Materials Laboratory, July 1966.

Mechanical Testing of Metal Matrix Composites

K. G. KREIDER[1]

Reference:

Kreider, K. G., "Mechanical Testing of Metal Matrix Composites," *Composite Materials: Testing and Design,* *ASTM STP 460,* American Society for Testing and Materials, 1969, pp. 203–214.

Abstract:

Testing techniques were developed for the mechanical testing of metal matrix, high-modulus filament composites. The testing of boron-aluminum composites was emphasized using both unidirectional reinforcement and multi-directional reinforcement. Tension, compression, shear, fatigue, and creep testing are discussed. Because of the anisotropy of the material, certain adaptations of standard tension testing techniques were developed in the gripping, specimen design, and extensometry.

The tension test at temperatures up to 500 C of materials with room-temperature tensile strengths in the reinforced direction of up to 200,000 psi and shear strengths in nonreinforced directions of 10,000 psi required specimens of modified geometry. These specimens had long grip lengths to prevent delaminating shear failures and only mildly reduced sections to prevent longitudinal shear from the shoulder into the grip. The gripping of the materials required abrasive paper or doublers to eliminate slippage in the grips without making indentations in the brittle boron fiber as occurs with serrated grips. Strain measurement was accomplished with both strain gages and clip on linear variable differential transducer extensometers. Both methods required special considerations due to anisotropy and the brittle nature of the fibers. Similarly, in compression and shear tests particular specimen designs and loading point constraints were found superior.

The recommended material information input and reportable information from the tests will be discussed. The input information is more complex than is the case for isotropic materials and includes the properties of the constituent phases of the composite, the geometry of their combination, and the thermal and mechanical history of the part which lead to internal stresses between the constituents.

The axial unidirectionally reinforced elastic properties of the composites can be predicted from the rule of mixtures, and the stress strain curve demonstrated three regions; elastic-elastic behavior, elastic-plastic behavior, and fiber breakage similar to that observed by McDanels et al. It was found that the rule of mixtures prediction of strength could be applied better if the statistical nature of the fiber strength and the critical reinforcing length were taken into consideration.

An analysis of the state of stress of the composite structures in the tests described is presented together with a discussion of the value and limitations of the tests in determining engineering properties.

Key Words:

composite materials, mechanical properties, boron, aluminum, evaluation, tests

This paper is a discussion of observations on the static and fatigue testing of metal matrix composites performed at United Aircraft Research Laboratories. Some

[1] Program manager, composite materials, United Aircraft Research Laboratories, East Hartford, Conn. 06108.

of the results of most of the tests described in this paper have been disclosed previously [1,2].[2] It was felt that the composites would be used primarily in applications where metals currently are being used, and efforts were made to use standard ASTM metal specimen tests whenever possible. In addition, it was recognized that certain properties should be determined which were particularly important in composites with a metal matrix such as aluminum and a high-modulus brittle fiber such as boron. Some problems were encountered in testing the composites using tests designed for isotropic ductile materials, and modifications in the testing procedure were made. The modifications were made with the intention that the same critical property be determined that is tested for in the standard test. Such tests as burst tests or bend tests which are not correlated as easily to standard basic tests for metals used in structural applications are not discussed. It is felt that, first, tests should be made in which the state of stress can be defined clearly. These tests also should provide a direct comparison with the properties of commonly used engineering metal alloys. The modifications employed relate primarily to the gripping of the specimen, reduction of secondary stresses such as shear stresses in a tension test, and techniques for increasing the precision of alignment and extensometry which appear to be more critical in the testing of composite materials.

It also should be emphasized that the specifications of the material to be tested received particular attention. With metal matrix composites the amount of each phase present (including porosity), the metallurgical condition and chemical composition of the phases, and information concerning residual stresses or thermal history during fabrication are important information.

TENSION TESTING

The tension testing of metal matrix composites such as boron-aluminum involves numerous problems not usually encountered in the testing of isotropic metals, which reflect the differences in properties between metals and composites. Important properties of composites which should be considered are: (1) the anisotropy of strength and elastic properties, (2) the brittle nature of the reinforcing phase which may be severely damaged by gripping forces or extensometer contacts, (3) the high-modulus and low-strain capacity of the reinforcement which magnify the problems of alignment and stress concentrations around the grips and in the zone of transition to the reduced section, and (4) the low interlaminar shear strength of the material which depends on the matrix and aggravates problems in gripping and load transfer.

In the testing of boron-aluminum sheet specimens a technique similar to that of the testing of sheet metal specimens was applied, and the same equipment was used as described in the ASTM handbook. However, several modifications in technique were made, particularly with respect to certain phases of the testing which were considered critical.

AXIAL

In axial testing (fiber parallel to load) of boron-aluminum it was found that the test specimen design described in ASTM Tension Testing of Metallic Materials E 8-61T was inadequate. The problem relates to the low shear strength of matrix and the high tensile strength of the matrix. This leads to two types of failure in shear parallel to the tension axis. In Fig. 1 the two shear failures are denoted by the shaded area in the grips which allows a shear pullout parallel to the grip faces and by a line through the gripped region parallel to the gage edge which separates the grip area shoulder from the central section. The first kind of failure can be eliminated by increasing the gripped area, thereby reducing the shear stress. The gripping area necessary to hold the specimen can be calculated from a force

---

[2] The italic numbers in brackets refer to the list of references appended to this paper.

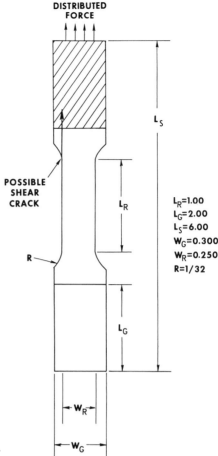

DISTRIBUTED
FORCE

POSSIBLE
SHEAR
CRACK

L_S

L_R

L_G

L_R=1.00
L_G=2.00
L_S=6.00
W_G=0.300
W_R=0.250
R=1/32

R

W_R

W_G

Fig. 1—Tension test specimen.

balance between tensile stress at fracture, σ_c, times the specimen cross-sectional area ($W_R t$) and the shear strength, τ_G, times the gripped area $2W_G L_G$.

$$2W_G L_G = \frac{\sigma_c W_R t}{\tau_G}$$

Since σ_c may be 200,000 psi, $t = 0.1$ in., and τ_G, 6000 psi, large gripped areas are necessary. The second type of failure, however, limits the useful grip width, W_G, with respect to the gage width, W_R (Fig. 1). The matrix shear strength. τ_s, necessary to prevent propagation of such cracks can be found by a force balance on either side of the crack. If the shear force is proportional to the ratio of the shoulder width (W_s) to the total width (W_g)

$$\tau_s l_g t = 2\tau_g W_g l_g \frac{W_s}{W_g}$$

(assuming a uniform friction force):

$$\tau_s l_g t = \sigma_c W_R t \frac{W_s}{2W_g}$$

or

$$\tau_s = \frac{\sigma_c(W_g - 2W_s)\dfrac{W_s}{W_g}}{l_g}$$

From the above considerations it can be concluded, therefore, that the grip area should be increased in length, but the difference between grip width and gage width should be reduced. It also will be noted that the gripping problems are more severe with thicker specimens. The specimen used most successfully was one with a grip length of 2½ in. and a gage width 80 percent of the grip width. This "mildly" reduced section also lowers the tensile stress concentration caused at the transition zone due to the change in transverse strain. Although fracture at the transition zone was common, so was central gage section fracture, and no difference in strength due to the location of the fracture was noted.

The gripping technique on the specimen is important. With sheet specimens compression grips were employed, and the compressive force was controlled carefully with screw clamps by controlling the torquing force. It is important to ensure the maximum friction force without crushing the specimen for the reasons given above with respect to gripping shear forces. However, file teeth or deeply separated grip faces were found to damage the fibers sufficiently to cause grip failures. It was found that a good compromise was to use knurled grips or abrasives with a roughness scale small in size with respect to the fiber diameter but coarse enough to prevent grip pullout. At high temperature, up to 500 C, copper liners were used which tended to bond to both the grip surface and the composite. This technique allowed tension testing at 500 C where the shear strength of aluminum is approximately 400 psi on specimens with 100,000 psi tensile strength.

The tension test specimens were mounted in a precision machined fixture and aligned in the grips using a ×10 microscope. The specimens were then transferred to the testing machine in a rigid fixture and mounted in a self-aligning loading train. This alignment, ±1 deg, is very critical since the strength drops precipitously at small angles (30 percent at 5 deg, Fig. 2).

The extensometry of the boron-aluminum specimens was performed assuming plane strain in the cross section allowing surface strain measurements. Both strain gages and clip on extensometers (linear variable differential transformer) were used to measure strain. It is extremely important to have both high sensitivity and high accuracy, since the strain to failure may be very low and the curve may have important characteristics which are informative about composite behavior. The axial stress-strain diagram in Fig. 3 has three regions, as described by McDanels et al [3], the elastic-elastic, elastic-plastic, and the region with fiber breakage. The stress-strain curve with transverse fibers (Fig. 3) also points to the need for sensitive and accurate extensometry. Strain gages have advantages in accuracy and the ability to measure strain at a given location. It is important, however, to measure axial strain on both sides of the specimen to determine bending movements present in the sheet specimen, particularly since the stress-strain curve is nearly elastic to, failure. Typically bending moment differential strains of 0.0002 were tolerated due to warpage of the test specimen and were accounted for. The flexibility of using rosettes for measuring strain in multiple directions is an added advantage of strain gages.

Clip on extensometers presented some problems due to the hard boron near the surface of the specimen which made slippage a problem compared with a ductile metal testing. The soft aluminum coupled with the light clipping forces used to ensure negligible damage to the specimen aggravates the problem of knife edge location precision. After comparison with strain gage measurements it was felt that the best accuracy that was obtained by this method was ±0.1 percent strain. More than one hundred axial tests were performed at temperatures up to

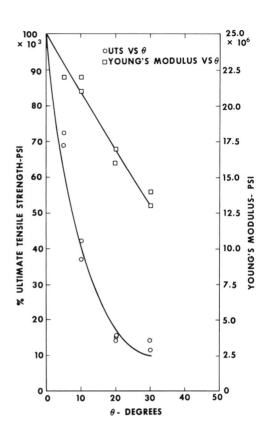

FIG. 2—Tensile properties as a function of angle between fiber axis and tensile axis.

500 C in this way and reported previously [2]. Values of up to 200,000 psi with 50 percent Borsic and aluminum alloy matrix were measured with most failures in the gage section.

An important observation in the testing and evaluation of boron-aluminum composites is the location and nature of the fracture. It was found that the electron scanning microscope was a useful tool in determining the exact nature of fracture. Because there are at least two phases present it is often important to report the location of the failure or whether the fiber, matrix, or interface has failed.

Certain properties of boron-aluminum composites should be well characterized when reporting the values of tension tests. These include the distribution in strength or the reinforcing phase which in turn leads to a distribution in strength of composite test specimens, the alignment and uniformity of distribution of the fibers, and the quality of the bonding of the composite. The mean deviation of strength values in boron-aluminum composites is greater than that usually found in standard wrought metal alloys. Typical results indicate that 80 percent of strength values fall within 10 percent of the average [2]. Although this mean deviation may be reduced by improving the relatively new material, information on the distribution of strengths or sample size of the tests is valuable. The alignment and distribution of the reinforcing phase can be checked with radiography. The radiographs in Fig. 4 illustrate this technique which is sensitive to the tungsten in the filament core. The alignment techniques previously described pertain to specimen alignment which assumes good filamentary alignment in the specimen. In the fabrication of boron-aluminum composites it is possible to fail to achieve complete interlaminar

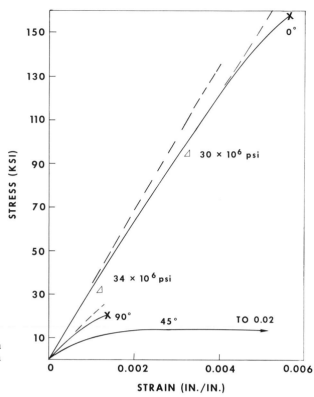

FIG. 3—Tensile stress strain curves for boron-aluminum composite.

STRAIN (IN./IN.)

bonding. The use of C-scan ultrasonic inspection is useful in determining the presence of such flaws. Figure 5 contains a photograph indicating deliberately placed flaws in a boron-aluminum composite together with a C-scan recording of such flaws. Although these flaws most drastically affect interlaminar shear strength they are also important in tensile properties.

Nonaxial

Nonaxial testing of the fiber composites has some different characteristic problems. In testing with the load perpendicular to the fibers the gripping problems are much simpler but the stress concentration problems are more severe due to splitting along the fibers analogous to the splitting of wood. With these specimens the edge surface condition is very critical, since the cutting of the ends of the fibers frequently leads to small cracks which are stress risers. It is important with these tests, also, to report the location of fracture. Figures 6a and b are electron scanning microscope pictures of transverse tension fractures in which the fibers failed longitudinally. A failure of this type leads to a stress-strain curve as depicted in Fig. 3. The testing of composites with acute angles between fiber and load also has some particular problems. The reaction of the anisotropic specimen to load leads to a torquing effect which makes interpretation more complex. This problem can be alleviated by testing balanced angle plies such as ±45 deg. The torsion and fiber rotation problem also can be minimized by the use of long gage lengths in specimens with very low strain to failure. In the testing of multidirectionally reinforced composites it is important to report the exact gage length-to-gage-width

A) ONE LAYER COMPOSITE   (B) THREE LAYER COMPOSITE

MAGNIFICATION: 5X

Fig. 4—Radiographs of boron-aluminum composites.

ratio, since a composite having fibers crossing the entire gage length would perform differently from one in which all fibers exited the edge.

SHEAR

Two techniques were used in testing shear properties. A standard sheet metal shear test specimen with slots was used to determine shear strength and is sketched in Fig. 7. It was felt that this test would give a conservative value of the shear strength parallel to the fibers due to stress concentrations compared with the three-point bend test or tube torsion tests. Similar precautions were employed with respect to alignment and surface condition to those used in axial tension testing. The speci-

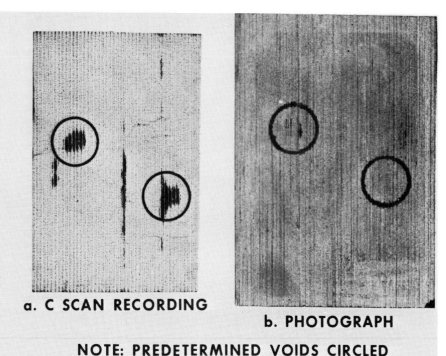

a. C SCAN RECORDING

b. PHOTOGRAPH

NOTE: PREDETERMINED VOIDS CIRCLED

Fig. 5—Ultrasonic inspection specimen.

men was ground with flats in the slots (Fig. 7) in order that a significant volume of the specimen was under nearly uniform shear stress. The shear strength also was measured by tension testing a specimen with the fibers 45 deg to the tensile axis. Under this kind of loading the specimen fails in the matrix parallel to the fibers, and the equation used to calculate the shear strength, τ, is as follows:

$$\tau = \frac{P}{A_o} \sin \theta \cos \theta^{2,6}$$

where:

P = load,

A = cross-sectional area, and

θ = angle between the applied load and the fiber axis.

The shear modulus was determined from this test by the method of Tsai [4]. This method employs the longitudinal modulus, E_{11}; the modulus at 90 deg to the fibers, E_{22}; and the initial slope in the test with the fibers 45 deg to the tensile axis. The apparent shear modulus also was determined by measuring the distortion in shear strain with strain gages in the slotted shear specimen placed on the zone of nearly constant shear stress. The two techniques indicated a shear modulus of 9.0 to 9.5×10^6 psi for the former method and 7.5 to 8.0×10^6 psi for the latter method for unidirectional composites with 50 percent boron and 50 percent aluminum.

COMPRESSION TESTS

Compression tests were performed on unsupported beams 0.20 by 0.25 and 0.75 in. long, as depicted in Fig. 8a. These specimens were loaded between tool steel

Composite Materials

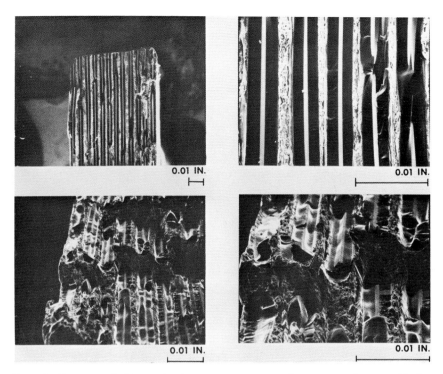

Fig. 6—Fracture of Borsic-aluminum composite specimen stressed perpendicular to fiber axis direction.

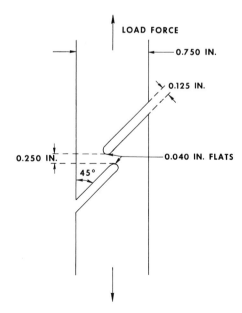

Fig. 7—Shear strength test specimen.

FIG. 8—(a) Borsic-aluminum compression specimen and (b) Borsic-aluminum impact specimen.

platens at a loading rate of 0.0025 in./min, and strain was measured with strain gages. It is felt that it is important that composites be tested in an unsupported manner so that no compressive loads are imposed in the X- or Z-direction (hydrostatic) which would lower the shear or tensile stresses which cause the failure. With the tool steel, the ends of the composite are embedded in the platens leading to a platen static friction coefficient of one. Unfortunately the severe depression in the steel caused high shear stresses on the composite at the edge and initiated the failure, Fig. 8a. The stress-strain curve was similar to the tensile stress-strain curve with the same modulus and matrix yielding starting at a high stress, and the compressive stress at failure was approximately 300,000 psi with a 50 percent boron composite.

IMPACT TESTS

Impact testing of composites should be performed according to standard metal testing procedures but should be performed at various critical loading angles with respect to the fiber. The observation of the fracture surface and crack characteristics is also important in these tests. Figure 8*b* is a photograph of a one-half size Charpy V-notch specimen after failure demonstrating the fibrous nature of the composite fracture. Two and one-half foot pounds were absorbed in this fracture.

FATIGUE TESTING

Both axial and cantilever bending tests have been performed on the boron-aluminum composite specimens. Axial fatigue testing was performed on a Tinius Olsen testing machine using a built-in program load control unit for low cycle testing, and a Sonntag universal fatigue testing machine was used for testing at 3600 rpm for high-cycle fatigue. Although the standard tension specimen was used in most tests, those tests which included high compressive stresses required a shorter gage length to prevent buckling. The grips used in tension testing were used also in fatigue testing and were found satisfactory. Only strain gages were used for extensometry because of inertial and fretting effects of clip-on mechanical extensometers. The stress strain cycle of the boron-aluminum composite changes considerably near the beginning and the end of a fatigue test, and strain measurement is important in understanding the behavior of the material. Results of these tests including Goodman diagrams have been reported previously [2].

Cantilever bending fatigue tests were made by fixing one end of the specimen in tension test type grips and vibrating the other end with a mechanical contactor. Since the bending moment is negligible at the contact, no failures were induced near the vibrator. Strain gages also were used for extensometry in these tests, together with the measurement of the deflection distance. This method allows the testing of thick specimens which have substantial interlaminar shear stresses and would have excessively high resonance frequencies and beam moduli for testing by a resonance or electromagnetic coupling technique. Typical Borsic-aluminum composites with 50 percent by volume fiber had 10^6 cycle lives at $\pm 120,000$-psi maximum stress.

It is felt that the axial fatigue testing is more useful for developing a basic understanding of the material since the stress state can be determined more easily and rigorously than in the bending tests. This is also true in static testing. The cantilever bending fatigue test is, however, very similar to certain service conditions for hardware and is an interesting test to analyze.

SUMMARY

It is felt that metal matrix composites such as boron-aluminum can be tested using tests based on standard ASTM tests used for metallic materials. Modifications in the tests are made to compensate for the anisotropic properties of the composite, the brittle nature of the reinforcing phase, and the low interlaminar shear strength of the composite compared to its tensile strength. These modifications, therefore, are made primarily in the gripping technique, reduction of secondary stresses, and in increasing the precision of alignment and extensometry.

References

[1] Kreider, K. G. and Leverant, G. R., "Services and Materials Necessary to Develop a Process to Produce Fibrous Reinforced Metal Matrix Composites," Final Report AFML-TR-66-219, Air Force Materials Laboratory, July 1966.

[2] Kreider, K. G., Schile, R. D., and Breinan, E. M., "Investigation of Plasma Sprayed Metal Matrix Fiber Reinforced Composites," Final Report AFML-TR-68-114, Air Force Materials Laboratory Aug. 1968.

[3] McDanels, D. L., Jech, R. W., and Weeton, J. W., "Stress Strain Behavior of Tungsten Fiber Reinforced Copper," NASA TN D-1881, National Aeronautics and Space Administration, Oct. 1963.

[4] Tsai, S., "A Test Method for the Determination of Shear Modulus and Shear Strength," Air Force Report AFML TR-66-372, Air Force Materials Laboratory, Jan. 1967.

Fatigue and Creep

Fatigue Fundamentals for Composite Materials

K. H. BOLLER[1]

Reference:

Boller, K. H., "Fatigue Fundamentals for Composite Materials," *Composite Materials: Testing and Design, ASTM STP 460,* American Society for Testing and Materials, 1969, pp. 217–235.

Abstract:

As composites are a combination of many organic and inorganic materials already studied, we draw on this information to guide us in strength evaluations, interpretation of results, and design aspects of the new materials. Basically, many micro- and macrostructural differences exist between the materials made of metals and those of reinforced plastics, sandwich constructions, and composites. The microstructure of the metals allows engineers to assume metals are isotropic and hence use simple formulas relating stress and strain. In contrast, composites are a mixture of dissimilar materials ranging from ductile, rubbery, weak, low-modulus materials to brittle, strong, high-modulus reinforcements. The result is a nonhomogeneous or anisotropic material involving complicated relations of stress and strain. If the dissimilar materials are arranged in layers and oriented in specific patterns, such as many reinforced plastics and sandwich constructions, the composite may be considered an orthotropic material and treated as such mathematically.

In addition to the various strength properties of the component materials, the composite also is affected by environmental factors, fabrication factors, and service factors. For example, temperature, moisture, various liquids, various gases, notches, imperfections, surface conditions, size, damping characteristics, adhesion, fastenings, cold working, orientation of crystals and fibers, and the rate of loading are all factors that affect the time and repetition of load, which is called fatigue. The effect of such factors generally cannot be predetermined mathematically, so that experimental evaluation must be made of each new composite.

In evaluating strength and related properties of the composites, experience with well-known materials provided the concepts for testing coupons, models, prototypes, and full-scale structures. Experimental evaluations include the effects of various environmental exposures, repeated and dead loads, various stress ranges, and as many mechanical tests as possible. Current trends are to use accelerated methods of test (such as the Prot, staircase, Probit, and rate process) to assay quickly the material's strength and life and to provide design criteria. Standard strength determinations provide stress-strain relations in tension, compression, and shear. Fatigue determinations provide curves of stress and strain magnitude versus usable life or time to failure. Superimposed on such graphs are the effects of environments in service. Just as there are many facets and ramifications in construction of composites, there are likewise many variables in the testing for establishing design criteria. If established procedures are not feasible, modifications are necessary to evaluate a new product and provide the assurance of safe design loads.

Key Words:

composite materials, mechanical properties, testing, fatigue, fiber composities, evaluations, tests

In the last 30 years, there has been a tremendous growth in the development and use of composite materials. Each increase in strength of a component material

[1] Engineer, Forest Products Laboratory, Forest Service, U. S. Department of Agriculture. The Laboratory is maintained at Madison, Wis., in cooperation with the University of Wisconsin. Personal member ASTM.

has triggered more growth. Just before World War II, the predominant reinforced composite used as a structural material was concrete reinforced with steel rods. During World War II, paper, cotton duck, and glass-reinforced plastic laminates were developed, and the British mosquito bomber using wood components became an outstanding example of the sandwich composite.

During the last few years high-strength fibers, such as E-glass, S-glass, boron, and carbon, were introduced as reinforcements in plastic matrices; now high-strength reinforcements are being used in metal and ceramic matrices. Each step in the development has produced stronger and stronger composites. To use them efficiently requires a knowledge of their strength and elastic characteristics, not only monotonic characteristics but the fatigue and time-dependent characteristics as well. The mixture of dissimilar materials has created evaluation problems that heretofore existed in only a few materials, wood and wood-base products, for example.

The root of the problem is misuse of relationships for isotropic materials on anisotropic composites. Previous generations have been developing, evaluating, and using homogeneous materials and applying isotropic material design criteria to their structures. Now, we are developing nonhomogeneous composites and need to apply anisotropic relationships, not isotropic ones, in the design criteria of structures using these materials. Utilizing the strength potential of these high-strength fibers presents a problem in fabricating and testing. The interaction and load transfer from one high-strength fiber to another, through the matrix, is not only a problem for the fabrication engineer but also for the evaluation engineer. He must have machines, fittings, and apparatus to transfer his loads into the net section of the coupon or the structural part. The nonhomogeneity of composites requires different techniques in evaluating the strength and time-dependent properties.

Because a great deal of information is available on the evaluation and use of homogeneous materials, we can draw heavily on this experience in arriving at "fatigue fundamentals for composites." We may outline facets of general testing and apply these to composites, such as: (1) microstructure, (2) the various applications or uses, (3) the nature of the strength properties, (4) how these are affected by service and environment, (5) various test methods, and (6) design aspects. Much of our current information on composites is based on experience with glass-reinforced plastics, typically highly orthotropic materials.

FATIGUE IN COMPOSITE MATERIALS

Assessing Time-Dependent Properties

From 1950 through 1966, about 5300 references on fatigue of metals have been cited in ASTM [1],[2] and several hundred additional references have been cited by Carden [2,3]. However, only about 100 references on fatigue of composites were cited during this same period [4]. Since then, a little more work has been done, but the field is still wide open for fatigue testing of composites.

The classic investigations on fatigue of metals began in 1860 when Wohler made his studies of railroad axles [5]. However, systematic investigations on strength of materials were begun much earlier. Galileo, as a result of tests on "tenacity" in about 1600, estimated that a rod of copper "4800 arms" long and suspended from one end would fail of its own weight. He also assumed that beams were stressed in tension over the whole section, and made calculations of strength on this basis. Mariotte in 1640 perceived that part of the section was in compression; still later, Coulomb located the neutral axis; and so on. Each generation has added to our knowledge of stress analysis.

Peterson, in 1950, reviewed some of the research since 1849 on the mechanism of fatigue [6]. This includes a discussion of early work concerning the nature

---

[2] The italic numbers in brackets refer to the list of references appended to this paper.

f fatigue, metallographic studies, work with single crystals and aggregates of a ew crystals, studies with X-ray methods, application of the electron microscope, esearch at elevated temperatures, and research with nonmetals into conditions governing intercrystalline and transcrystalline failure. He quotes many interesting facts concerning the discussion of fatigue 100 years ago. One such item states:

That a Mr. John Ramsbottom, Chief Engineer of the London and Great Northern Railroad, believes that a change takes place and comments, "A parallel case might be observed with reference to an ash stick which if doubled will break with a fibrous fracture but if subjected to vibration, however slight, running through it a great number of times, it will break in a different mode."

Here we have the first reference to fatigue failures of wood, a highly orthotropic material. Further, Peterson points out that the theories on the mechanism of fatigue were controversial. The question of crystallization was discussed a century ago but was not clarified completely for a long time. Older ideas were not discarded easily. It has been thought that magnetism was a factor; that steady load would cause deterioration of a member with time (load velocity would explain the mysterious service failures); and that repeated loading introduced permanent set. Even the term "fatigue" created a lot of discussion. Several items were thought to attribute plague us [7]. Peterson [6] also referred to work on a nonmetallic:

It is of interest to examine the conditions of failure in the material where slip and work hardening do not exist. Tests have been made of glass rods in bending under three conditions: (a) nonrotating, (b) rotating at 14 r.p.m., and (c) rotating at 10,000 r.p.m. For various stress levels, the average time to fracture was found to be approximately the same.

Thus we have an experiment showing that the effect of speed on a given material did not affect the fatigue life, but today we know that speed does affect composites.

From this background, I feel we can identify the basic principles of fatigue as: (a) fatigue failures are the result of repeated stresses and strains having various amplitudes and time functions (Fig. 1); (b) such alternating loads may be super-

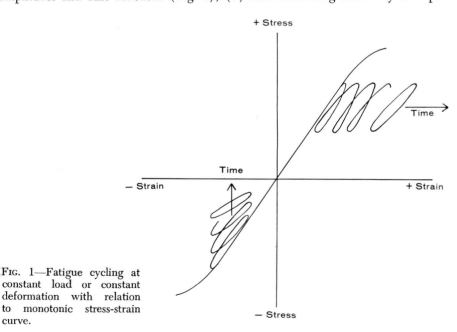

FIG. 1—Fatigue cycling at constant load or constant deformation with relation to monotonic stress-strain curve.

imposed on constant or variable mean loads; (c) these alternating loads may be at constant load or constant strain levels; (d) these alternating loads may be randomized completely; (e) the loads may be caused by temperature gradients, internal stress gradients, or other environmental conditions. Loads are applied in various forms, such as tension, compression, shear, or a combination of these. All materials are affected.

Another obvious conclusion is that the effect of time and environment must be evaluated by experiment and service to obtain empirical relationships of stress-strain-time-temperature, since pure theoretical relationships have not been agreed upon or verified.

STRUCTURE OF COMPOSITES

To better understand the fatigue fundamentals as applied to composite materials, we need to point out how composites differ from isotropic materials such as metals. We should start by reviewing the microscopic structure of a composite. It is usually a laminar or fibrous construction comprising a combination of complex dissimilar materials assembled and intimately fixed in relation to each other so as to use the properties of each to attain specific structural advantage for the whole assembly. The dissimilar materials range from ductile, rubbery, weak, low-modulus substances used as matrices, to brittle, strong, high-modulus fibers or flakes used as reinforcements. The result is definitely nonhomogeneous on a microscopic scale (Fig. 2). As a matter of fact, on this scale metals are not homogeneous either. In addition, the composite also may have the reinforcements arranged to give the product ultimate performance in specific directions. Mats or quasi-isotropic orientations of fibers can provide nearly equal strength in all directions in the plane of the laminate, while aligned reinforcements provide highly directional properties.

Reinforcement and matrix are scientifically proportioned for best performance and high strength-weight ratio. The ratio of reinforcement to matrix varies to conform to the requirements of the intended use of the composite material.

The sandwich construction, a special composite structure, uses a multidirectional composite as skin and a unidirectional separator as a core. A fundamental requirement of the components is that they must work together as a unit. Under load, the strain must be transferred from one strong element to another through the boundary interface and the weak matrix connecting the components. The result on a macroscopic scale is definitely a nonhomogeneous or anisotropic material requir-

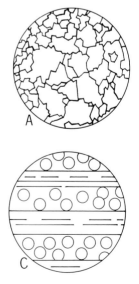
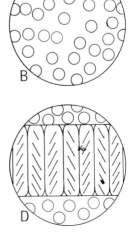

FIG. 2—Schematic microstructure of: A, metallic crystalline form; B, unidirectional reinforced composite; C, bidirectional composite; and D, sandwich.

Composite Materials

ing complicated relations of stress and strain, unlike those used in our classical equations for metals. If the dissimilar materials are arranged in layers or oriented in patterns as laminated reinforced plastics and sandwich constructions, the product may be considered an orthotropic composite and treated as such mathematically. To characterize the behavior of composites under load, a new discipline was born— micromechanics.

TESTING OF COMPOSITE MATERIALS

Our purpose in testing materials is to gather more information, so we have to ask pertinent questions. To what use are these composites subjected?

USES

Reinforced plastics are useful in many relatively low-temperature environments. Plastic matrix composites may be used up to 500 F. Many structural as well as nonstructural parts operate below this temperature, however, such as propulsion blades used for cooling towers, wind tunnels, and helicopters; pressure bottles, storage tanks, and pipes; boats; and many of the sundry parts of the outer surfaces of aircraft. A recent article in *Product Engineering* [8] cites the many uses of glass-reinforced plastics. More glass-fiber-reinforced plastics than ever before are being used in the new Boeing 737 jet. Bonded structures replace metal assemblies even in highly stressed areas. In the Boeing 737 jet, in the forward section, plastics are being used in the nose radome, in the nose-wheel well doors, in the body-to-wing fairings, and in the leading-edge enclosure panels. Aft, they are being used in the aileron tabs, body-to-wing fairings, lower body-to-wing fairings, hinged flap (inboard flap), wedge in the outboard midflap, and, of course, in the flap tracks. On the tail section, they have a vertical fin-tip assembly, rudder and trailing edge panels, fin-to-rudder fairings, elevator tab, tail cone, elevator and trailing edge, and finally, stabilizer trailing edge. Specifically, the article says:

Boeing engineers are making rudders, elevators, and control tabs of glass-reinforced plastics rather than assemblies of metal, eliminating much drilling, jigging, and riveting that aluminum assemblies require. A glass-reinforced plastic rudder on a test stand proved that it was strong enough. Test engineers subjected the plastic rudder to simulated flight conditions, vibrated it to measure fatigue endurance, and finally tried to tear it apart in the torsional test stand. It failed at 750 percent of its design load, but only because a metal spar buckled, shearing off metal rivets connecting the two parts.

This would indicate a need for more fundamental fatigue information on the quality of the glass-reinforced-plastic laminates used in this airplane. Similarly, the F-111 has many structural and nonstructural parts made of composites; approximately 40 percent are structural [9].

Structural applications in environments above 500 F, in highly stressed areas, and in corrosive environments of gases or liquids require composites with inorganic matrices, such as the metals and ceramics; these include motor parts and exhaust systems. In addition, metal matrix composites are being considered for such low-temperature applications as landing gears. Forged billets of large metal parts are a problem, according to Yaffee [10]. He says it is difficult to get larger and larger forging presses to manufacture billets strong enough to carry existing loads in our aircraft structures. He points out that some composite materials are being looked into by the Air Force Materials Laboratory for plasma-sprayed metal matrices around shaped continuous reinforcement fibers.

Another Air Force Materials Laboratory contract in this area was on diffusion-bonded filament-reinforced metal matrix composites. The objective of this program was the development of and evaluation of fabrication parameters and mechanical characteristics of four solid-state diffusion-bonded composite systems. Two systems used aluminum alloy matrices, one reinforced with beryllium filaments and the other with boron filaments. The two other composite systems are based on titanium alloy matrices, reinforced in one case with boron filaments and in the other with silicon carbide. Such composite materials show great promise for high-strength materials.

Monotonic Strength

To understand the fatigue fundamentals for composites, it is necessary to ask what is the strength and elastic range of the components. The monotonic strength properties of the components and of familiar composites are shown in Fig. 3. The kinds of materials that go into our composite are unlimited, and their range of strength and elastic properties is the widest known. The matrix, for example, may be a resin having a modulus of elasticity of about 250,000 psi or steel with 30 million psi. The reinforcements may be cotton duck having a modulus of about 500,000 psi or graphite with 68 million psi [11]. It is essential that a good interface bond exist between matrix and reinforcement so the weak matrix transfers the load from one fiber to another. This trait is the major difference between metals and composites, and therein lies a weakness of our composite.

It is fundamental that the monotonic strength properties are affected by the service environment. Both metal and plastic composites are affected by temperature. Elevated temperature usually causes degradation of both, but the metal ones withstand higher temperatures than do the plastics. Elevated temperature causes resins to melt, deploymerize, or oxidize. If the resin matrix was not cured fully at the time of fabrication, elevated temperature may cause initially some increase in strength before final degradation [12]. Hence, resin matrices may be unstable.

Moisture usually does not reduce the strength of metals as it does the strength of plastic composites. The resin matrix is relatively impervious to moisture, but moisture will weaken the interface between matrix and reinforcement and reduce strength. On the other hand, metals corrode in certain environments, and therefore plastics and composites are sometimes used to replace metals under such conditions.

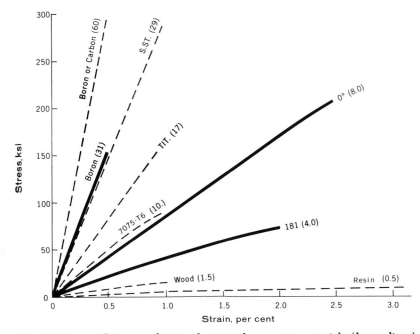

FIG. 3—Range of stress and strain for typical composite materials (heavy lines) and components (light lines).

Composite Materials

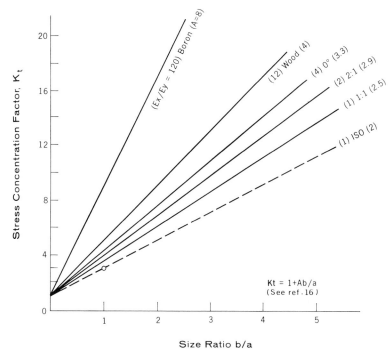

FIG. 4—Stress concentration factor, K_t, for infinitely wide plates.

The rate of load application affects the strength and stiffness properties of all composites. Those with the highest damping factor [13] are affected the most. Internal friction causes heating and strength reductions. However, the damping characteristic of composites [14] can be used to advantage to combat acoustical fatigue. Composites with a high damping factor also are good shock-absorbing materials.

Peterson [15] has cited monotonic stress concentration factors for metals and the effect of elastic and plastic relations in the vicinity of the concentration. The stress concentration factors due to notches are considerably higher for orthotropic composites than for isotropic materials [16] (Fig. 4).

FATIGUE DATA GENERATION

What laboratory test methods should we use to obtain the information required to design structures and yet simulate the service conditions? How should the data be presented? The answers are not simple. However, here are some guidelines.

In use, these various structural parts are subject to all stresses—tension, compression, shear, and combinations of these—the same as is basic for metal, and even more environment in space than our textbooks dreamed of. They usually are not subjected to a single load, but to a load repeated many times. The repetitions may be at constant stress, at constant strain, or combinations of these applied in a random sequence [17]. This time-dependent property—fatigue—is a service condition, and fatigue testing is simulated service testing at various alternating loads in alternating environments.

Fatigue characteristics are classified as a mechanical property, but which stresses and which stress ranges are we evaluating when we refer to the fatigue strength of a composite material? The service conditions of concern to one user of a composite, and simulated by him in a test program, are not necessarily the same service condition of the same composite for another user. They may have evaluated the same composite differently. Hence, they may generate different data having a different meaning, and thereby generate confusion and misinterpretation. A fundamental of fatigue testing is that the evaluation engineer should define completely the basic tests of his coupons, such as type of mechanical test, type of loading, range ratio, constants of test method, and specimen configuration, until such time as each of these variables can be standardized. The background for this standardization is ASTM STP 91 [18].

ASTM Manuals STP 91 [18] and STP 91A [19] are basic references which define our terms and make recommendations for testing and analysis. Basically, the quality engineer obtains information from coupons and models for reliable fundamental knowledge of the stress-cycle curve in compression, tension, and shear. Loadings should be obtained at several stress ratios plotting life diagrams, and then data obtained from models, prototypes, and structures with more specific service loads with each step. Most quality engineers are interested usually in the first phase, while designers are interested vitally in the succeeding phases.

The basic objective for determining the size and the shape of the coupons is to have the fittings and attachments transmit the forces to the net section, and the stresses should be as uniform as possible in that net section. The problem which confronts use of composites and our testing techniques is the transfer of loads to these strong fibers in the net section. We must grip the coupon or structural part so that the loads we think we have are transmitted through the body of the piece in question.

Specimen configurations vary over a wide range. The specimens from flat laminates have had sundry shapes in an effort to minimize stress concentrations due to test fixtures [20]. For tension tests, the ASTM Test for Tensile Properties of Plastics (D 638-68) includes several shapes. Boron laminates need tab ends on the tension specimen. Filament-wound structures are simulated by the Naval Ordnance Laboratory (NOL) ring test in tension, compression, or interlaminar shear. Compression specimen design of flat laminates requires an L/d ratio of 4:1, as well as clamps at the bearing ends. Thin laminates need lateral support, so various procedures to provide this support have been used. Compressive strengths usually are not equal to the tensile strengths, so that both properties need to be evaluated. For long-term creep and stress-rupture data, ASTM describes the dumbbell-shaped compression specimen. A rectangular specimen, if adequately clamped and machined, has been found satisfactory also. Structural parts have been both bonded and riveted to adjoining members. Flat laminates are machined to given sizes, while filament-wound reinforcements may have specimens in the form of cylinders or rings as models for evaluation of the composite. In testing sandwich coupons in compression, the loads must be balanced equally on both faces.

For fundamental fatigue information, the specimen should be subjected to either constant alternating stress or strain of various range ratios (minimum stress divided by the maximum stress). For a range ratio of -1.0 the Forest Products Laboratory has used ¼ by 1½ by 6-in. specimens, clamped at both ends for 2 in., leaving 2 in. unsupported.

There are four broad ranges of speed for fatigue testing: (1) acoustical, that is, high frequencies and usually low amplitude [21]; (2) low stress-high cycle life; (3) intermediate cycle life; and (4) high stress-low cycle life. These four broad ranges are the same as for metals testing. If the range ratio for any of the tests in these four broad ranges is $+1$, (maximum and minimum stresses are equal in value and sign and the alternating stress amplitude is equal to zero), a special case of fatigue testing is the result; that is, the mean stress is constant and we have a stress-rupture and creep evaluation. This is a limit on constant fatigue life diagrams.

The speed for evaluating the fatigue strength of metals, which have low damping characteristics, has been 1800 rpm and above; the speeds for plastics and plastic

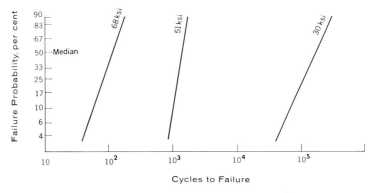

FIG. 5—Trend of life expectancy at three stress amplitudes.

components, which have high damping characteristics, have usually been 900 rpm and below. It is reasonable to expect that the metal composites will conform to the characteristics of metals and will be evaluated by metals procedures. For studying strain behavior and strain distribution, slow speeds have been found practical. In recent years, more emphasis has been placed on a random program of loads [17] to simulate service environments in fatigue testing of coupons, and especially structures.

Special test methods can accelerate the acquisition of fatigue data. Such methods as Prot's continually increasing amplitude [22,23], the staircase method [18], the Probit, and rate-process methods have been used with varying degrees of success.

No information-gathering test method can be used unless the data are reliable [24]. Both Freudenthal and Weibull [19] have made numerous studies to establish the statistical pattern of life expectancy. For example, the failure probability plots presented in Refs 19 and 25 show the life expectancy on a plot of failure probability versus the number of cycles to failure at various stress amplitudes (Fig. 5). The scatter varies from a fraction of a log increment at the high stress levels to greater than a log increment at the low stress levels.

The reliability of test data is a very important consideration when choosing data for use. Reliability not only involves the scatter of individual test points in a single series of tests of one batch, but in a broader sense also involves the life expectancy when the weakest individual member encounters the most severe service loading [26]. In this sense, the variability results from repeated reproduction of the material or structural parts. Fatigue tests on specimens from a single panel show the inherent variability of that material using a specific test method. More reliable data are obtained from different fabrications by different manufacturers under different evaluation conditions, so we have fundamental "round-robins" for evaluation of men, machines, and materials.

DATA PRESENTATION

The scope of the quality engineer usually is limited to the evaluation of his material. Within this scope, however, he should present the fatigue information to the designer in few relatively simple graphs, without sacrificing completeness of information on material, test method, specimen description, or previous history. Data should be presented with nomenclature consistent with ASTM STP 91 [18].

The simplest form of data presentation is the Wohler plot, which is the S-N curve plotted on semilog coordinates with stress as the ordinate and log of cycles to failure as the abscissa (Fig. 6). At ¼ cycle, the ultimate strength of the material under that particular condition is presented. Subsequent data points or curves show the effect of cyclic loads and environments. Additional lines may show confidence

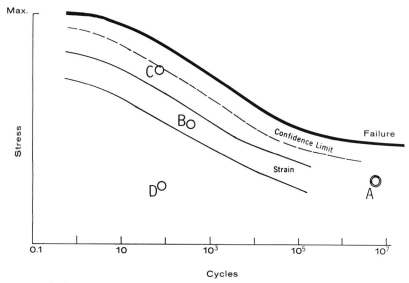

FIG. 6—Sample S-N curve.

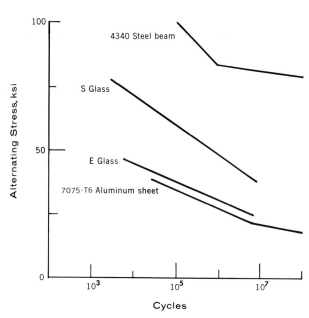

FIG. 7—Effect of materials.

limits or lines of constant strain. If such additional information is superimposed on typical S-N coordinates, it shows the effect of many variables, such as materials, environments, mean stress, and shapes [14,15,22,25,27–32] (Figs. 7–12). Another plot, similar to the Wohler plot, substitutes the strain for stress in the ordinate and then shows the cyclic effect on the abscissa.

Blatherwick and Olsen [33] and Peterson [15] have pointed out a reduction in stress concentration factor K_t for isotropic materials with an increasing number of cycles. Thus the factor is a function of fatigue life. Also, when the strain amplitude at the root of the notch was held constant, the maximum stress decreased

Composite Materials

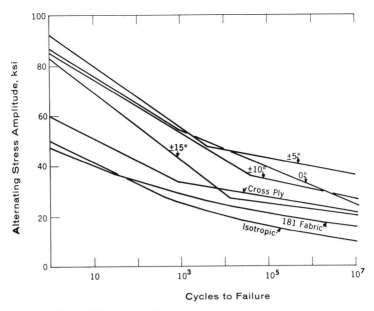

FIG. 8—Effect of fiber orientation.

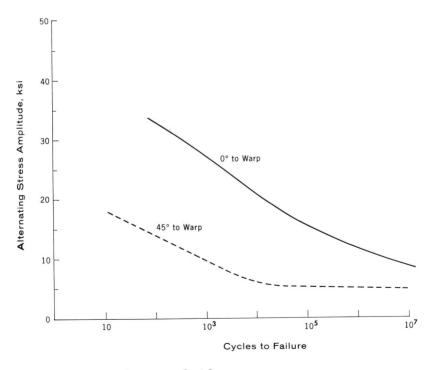

FIG. 9—Effect of loading at 0 and 45 deg to warp.

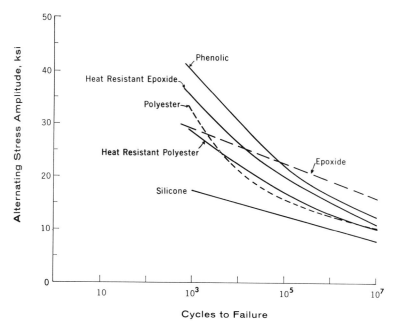

FIG. 10—Effect of resins.

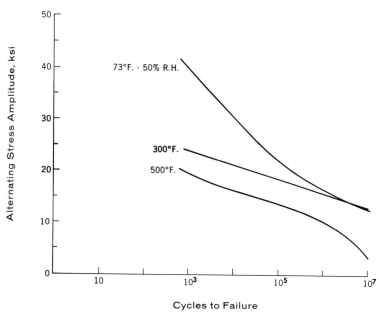

FIG. 11—Effect of temperature.

with increased fatigue cycles; when load amplitude was constant, the maximum stress amplitude decreased with fatigue cycles, despite an increasing strain amplitude. The K_t factor for composites also depends on elastic constants.

The anisotropic nature of composites causes a major difference between the stress distribution around notches in such composites as orthotropic laminates and

Composite Materials

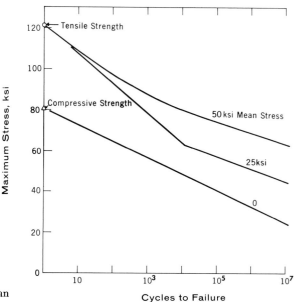

Fig. 12—Effect of mean
stress levels.

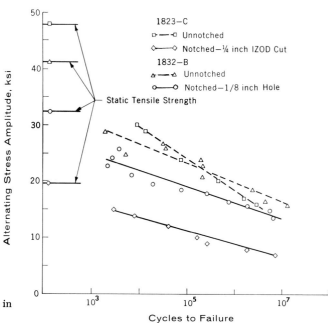

Fig. 13—Effect of notches in
fabric laminate.

that around notches of similar shape in isotropic materials. Recent work shows that the isotropic values for K_t need to be modified to include the orthotropic elastic constants [16]. This can be done by using Neuber's nomograph in Ref 18 for shape factor and adding to it the elastic correction. For example, a circular hole causes a K_t value of 3 in isotropic materials and quasi-isotropic laminates, while the K_t value in orthotropic laminates may be between 3 and 9 (Fig. 4).

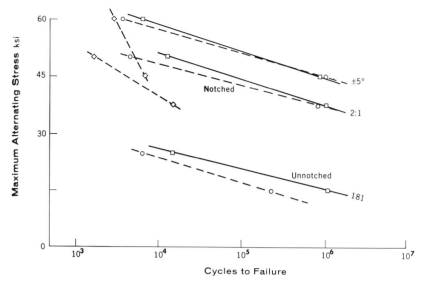

FIG. 14—Effect of notches in unwoven fiber laminates.

The fatigue stress concentration factor K_f, however, approaches 1 (Figs. 13 and 14). This phenomenon needs further exploration to determine the extent to which the internal micrscopic stress concentrations affect the macroscopic stress concentrations. The time-dependent characteristic and repetition of stresses evidently cause the peak stresses in critical locations to be equalized or supported by neighboring fibers, so that the K_f value is reduced to nearly 1.

The second most common method of data presentation is some form of constant-life diagram (Fig. 15). Grover [7] points out how the family of S-N curves at various mean stress levels are combined to show the life at various maximum stresses, minimum stresses, alternating stress amplitudes, mean stresses, or stress range ratios. It has been found [28] that, for plastic composites used in tension as well as compression, the scales for maximum stress and mean stress need to be extended into the negative region (Fig. 16). And since tensile strength is not necessarily equal to the compressive strength, the limit life lines at ¼ cycle are not at the same maximum stress or at the same mean stress level when zero amplitude exists [34] (Fig. 17). This form of life diagram also shows the stress-rupture information for the material.

A new concept in fatigue monitoring is made possible by the introduction of the S-N gage. This gage, instead of being made of an elastic metal and stressed well below its endurance limit, is of a ductile metal that deforms and stays deformed with each cycle of stress or strain; thus, its electrical resistance changes with each cycle. A measure of the percent change in resistance, if correlated with the percent of remaining life, is a measure of the fatigue characteristic of the structure. Work with this gage is being done on metals and should be applicable to metal composites. It may be possible to use S-N gage generated data to predict the remaining life in a highly stressed part. For example, a measurement of electrical resistance may be used under precalibrated conditions to be indicative of the remaining life of a stressed part. The application of S-N gages to resin composites is questionable because the operating strains of resin composites are greater than for the metals in the current S-N gages.

As fatigue characteristics are a function of so many variables, data presentation should include a good description of all factors that affect test results: a complete description of the material, including the percentage of each component, orientation

Composite Materials

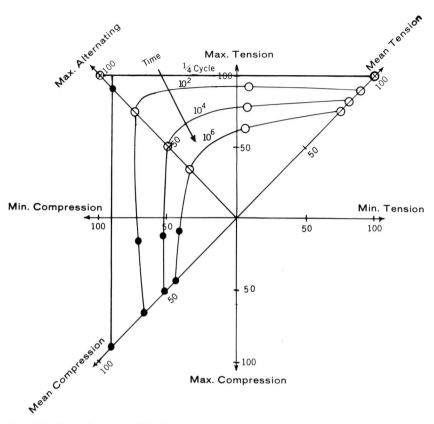

FIG. 15—Typical constant-life diagram.

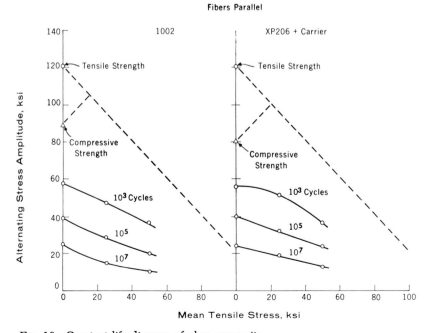

FIG. 16—Constant-life diagram of glass composite.

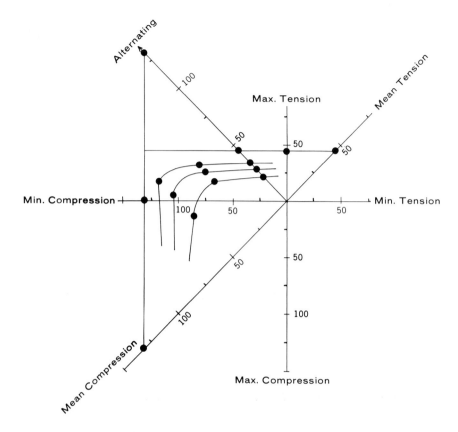

FIG. 17—Constant-life diagram of boron composite.

of components, and any previous history that might affect strength. Data also should include descriptions of the specimen size and shape, the test environment, speed of testing, and of course the stress and duration of that test [7,18]. Many times, data are meaningless because pertinent factors were not cited. Committees such as *ASTM E-9* [18,19] and Federal committees such as those for MIL-HDBK-5 [35] and MIL-HDBK-17 are outlining procedures and data reporting so that fatigue information has meaning.

DESIGN ASPECTS

The pros and cons of the application of fatigue data generated in the laboratory on small coupons to design are out of the scope of this presentation on fatigue fundamentals for composites. Nevertheless, it is fundamental that the quality engineer know how the data are used. Dolan [36] has cited the behavior of models of material versus structural mechanics, and factors and development of a design involving coupons, as well as a feedback from tests of structures. Nixon [26] has said he feels data from coupons are of little use. Rather, he thinks we need to know the capacity of our products in fatigue under actual conditions—not in failure. Nixon, nevertheless, recommends testing to failure in fatigue, saying that an endurance test that does not result in failure gives little information about the capability of the product.

Grover [7] points out two design approaches to prevent fatigue. One is the safe-life approach, that is, a stress level well below the average *S-N* curve or the 5 percent exclusion limit is used for the entire structure (stress level A or

B, for example, on Fig. 6). Examples of the utilization of safe-life approach are in the design of shafts, gears, and heavy equipment where weight is no problem; also, in equipment, the safe-life method is used by designers to plan for obsolescence. The second approach is the fail-safe approach, that is, if one part fails by fatigue, other adjacent parts will sustain the extra load until the structure can be repaired (stress levels C and D, respectively, on Fig. 6). In a fail-safe approach where weight is a factor, areas are reduced, but critical areas are watched for signs of damage. The S-N gage is a new concept for this monitoring. The feedback of service compliance to the quality engineer is of utmost importance for safer structures.

SUMMARY

This presentation has pointed out the basic principles of fatigue as an engineering phenomenon applicable to all materials and structures. Materials get tired; they wear out because loads are repeated again and again under changing conditions—superimposed mean loads, temperature gradients, stress concentrations, and environments. We need to know the capacity of our products under dynamic conditions, for fatigue is not just a one-time ultimate-strength test. Furthermore, there is a need for standardizing composite products so confidence and reliance may be achieved through tests and service records.

The basic concept of fatigue of composites is not different than that for metals, but rather more complicated. This is due principally to the anisotropic structure of composite materials. This anisotropy presents a problem in load transfer from one strong link to another through a weak link. This anisotropy should not be treated with simple mathematics but should be characterized by micromechanic relationships. Monotonically, the composites have a high notch sensitivity, greater than for isotropic materials. However, dynamically, they seem highly insensitive to notches. The multicontinuous load paths through fibers and matrices provide a number of fail-safe miniature structural elements that "iron out" stress concentrations. A high damping factor (causing internal heating) is a distinct characteristic which is used to advantage in some applications to reduce abnormal vibration and noise. These composites, however, especially those with a plastic matrix, are subject to variations in strength due to heat. Their chemical instability can cause unexpected trends in strength as a function of time.

Because composites are mixtures, the proportions can be arranged for maximum strength-weight ratio, and also for maximum strength in a given direction. Being nonhomogeneous, the usual isotropic strength relationships and formulas for design criteria are not applicable. Completely different anisotropic relationships are applicable for monotonic strength considerations, but for fatigue analysis the basic empirical S-N curves and constant-life diagrams need to be generated. Constant-life diagrams have a scale axis extended to include compressive stresses, and for composites the limiting constant-life lines at ¼ cycle for tension and compression are not necessarily equal.

The variables in a composite evaluation seem to be astronomical, but when properly understood they really are not. Each variable must be considered separately and weighed according to how it affects service life. Because of the complexity of the behavior of composite materials under dynamic stress, there is a serious need for data generated according to basic principles and reported precisely. When dealing with fatigue fundamentals that may affect human lives, proceed with caution. When in doubt, run a fatigue test of the product under simulated service loading.

References

[1] American Society for Testing and Materials, References on fatigue. ASTM STP 9-AA, 1950–54; ASTM STP 9-BB, 1955–1959; ASTM STP 9-L, 1960; ASTM STP 9-M, 1961–1962; ASTM

STP 9-N, 1963; ASTM STP 9-O, 1964; ASTM STP 9-P, 1965–1966, on microfiche.

[2] Carden, A. E., "Bibliography of the Literature on Multiaxial Stress Fatigue," report sponsored by NASA grant NSG-381, Research in the Aerospace Physical Science and Engineering, Report MH67-AEC-2, Aug. 1967.

[3] Carden, A. E., "Bibliography of the Literature on Thermal Fatigue," report sponsored by the NASA grant NSG-381, Report MH67-AEC-3, Department of Engineering Mechanics, College of Engineering, University of Alabama, 1967.

[4] U. S. Forest Products Laboratory, "Forest Products Laboratory List of Publications on Modified Woods, Paper-Base Laminates, and Reinforced Plastic Laminates," 66-031, Madison, Wis., 1966.

[5] Gibbons, C. H., "Materials Testing Machines," Baldwin Locomotives, Baldwin Locomotive Works, Philadelphia, Pa., Vol. 13, No. 1, April–July 1934.

[6] Peterson, R. E., "Discussions of a Century Ago Concerning the Nature of Fatigue and Review of Some of the Subsequent Researches Concerning the Mechanism of Fatigue," ASTM Bulletin. No. 164, American Society for Testing and Materials, Feb. 1950, p. 50.

[7] Grover, H. J., "Fatigue of Aircraft Structures," Naval Air Systems Command, Department of the Navy, NAVAIR 01-1A-13, 1966, U. S. Government Printing Office, Washington, D. C.

[8] "Reinforced Plastics Go Up, Up and Away in Airliners," Product Engineering, Vol. 39, No. 8, 1968, p. 97.

[9] Rosato, D. V., "Environmental Effects on Polymeric Materials," Materials, Vol. 2, Rosato, D. V. and Schwartz, R. T., eds., Interscience Publishers, New York, 1968, pp. 1229–2216.

[10] Yaffee, M. L., "Forming Holds Key to Future Structures," Aviation Week and Space Technology, Vol. 88, No. 18, 1968, p. 99.

[11] Simon, R. A. and Prasen, S. P., "Properties of Carbon Fiber Composites: Effect of Coating with Silicon Carbide," Modern Plastics, Vol. 45, No. 13, 1968, p. 227.

[12] Jurevic, W. G. and Rittenhouse, J. B., "Structural Plastics Application Handbook," AFML-TR-67-332, Wright-Patterson Air Force Base, Ohio, 1967.

[13] Lazan, B. J., "Damping Properties of Materials and Their Relationship to Resonant Fatigue Strength of Parts," Report R55GL129, General Electric Co., 11 Jan. 1955.

[14] Boller, K. H., "Resume of Fatigue Characteristics of Reinforced Plastic Laminates Subjected to Axial Loading," Proceedings of 10th Sagamore Army Materials Research Conference, Syracuse University Press, Syracuse, N. Y., 1964, p. 325.

[15] Peterson, R. E., "Engineering and Design Aspects," Materials Research and Standards, American Society for Testing and Materials, Vol. 3, No. 2, 1963, p. 122.

[16] Boller, K. H., "Effect of Notches on Fatigue Strength of Composite Materials," AFML-TR- (in press), Wright-Patterson Air Force Base, Ohio.

[17] Swanson, R. S., "Random Load Fatigue Testing—A State of the Art Survey," Materials Research and Standards, American Society for Testing and Materials, Vol. 8, No. 4, 1968, p. 11.

[18] Manual on Fatigue Testing, ASTM STP 91, American Society for Testing and Materials, 1949.

[19] A Guide for Fatigue Testing and the Statistical Analysis of Fatigue Data," Supplement to Manual on Fatigue Testing, ASTM STP 91A, American Society for Testing and Materials, 1963.

[20] Hofer, K. E., Jr., "An Investigation of the Fatigue and Creep Properties of Glass Reinforced Plastics for Primary Aircraft Structures," Illinois Institute of Technology, Chicago, Ill., 1968.

[21] Trapp, W. J., "A Review of Acoustical Fatigue," Proceedings of 10th Sagamore Army Materials Research Conference, Syracuse University Press, Syracuse, N. Y., 1964, p. 261.

[22] Boller, K. H., "Application of Prot Test Method to Stress-Rup-

ture Curves of Glass-Reinforced Plastic Laminates," Report 2118, Forest Products Laboratory, 1958.

[23] Prot, E. M., "Fatigue Testing Under Progressive Loading," (translated by Ward, E. J., Captain USAF), WADC-TR-52-148, Wright-Patterson Air Force Base, Ohio, 1952.

[24] "F-111 Fatigue Failure Solution Expected," Aviation Week and Space Technology, Vol. 89, No. 12, 1968, p. 39.

[25] Boller, K. H., "Effect of Single-Step Change in Stress on Fatigue Life of Plastic Laminates Reinforced with Unwoven 'E' Glass Fibers," AFML-TR-66-220, Wright-Patterson Air Force Base, Ohio, 1966.

[26] Nixon, Frank, "Testing for Satisfactory Life," Relation of Testing and Service Performance, ASTM STP 423, American Society for Testing and Materials, 1967, pp. 43–60.

[27] Boller, K. H., Werren, Fred, and Kimball, K. E., "Fatigue Tests of Glass-Fabric-Base Laminates Subjected to Axial Loading," Report 1823 and Supplements A, B, and C, Forest Products Laboratory, 1952.

[28] Boller, K. H., "Fatigue Properties of Various Glass-Fiber-Reinforced Plastic Laminates," WADC Technical Report 55-389, Wright-Patterson Air Force Base, Ohio, 1956.

[29] Boller, K. H., "Fatigue Properties of Plastic Laminates Reinforced with Unwoven Glass Fibers," ASD-TDR-62-464, Wright-Patterson Air Force Base, Ohio, 1962.

[30] Boller, K. H., "Fatigue Strength of Plastic Laminates Reinforced with Unwoven 'S' Glass Fibers," AFML-TR-64-403, Wright-Patterson Air Force Base, Ohio, 1965.

[31] Stevens, G. H., "Fatigue Strength of Phenolic Laminates from 1 Cycle to 10 Million Cycles of Repeated Load," U. S. Forest Service Research Note FPL-027, Forest Products Laboratory, Madison, Wis.

[32] Stevens, G. H. and Boller, K. H., "Effect of Type of Reinforcement on Fatigue Properties of Plastic Laminates," WADC-TR-59-27, Wright-Patterson Air Force Base, Ohio, 1959.

[33] Blatherwick, A. A. and Olsen, B. K., "Stress Redistribution in Notched Specimens During Fatigue Cycling," Experimental Mechanics, Vol. 8, No. 8, 1968, p. 356.

[34] Wilson, F. M. and Lane, E. K., "Research on Resin-Impregnated, Collimated Boron Filaments and Improved High-Modulus, High-Strength Filaments and Composites," AFML-TR-65-382, Wright-Patterson Air Force Base, Ohio, 1965, p. 101.

[35] Moon, D. P. and Hyler, W. S., of Battelle Memorial Institute, "Guidelines of Presentation of Data for MIL-HDBK-5," AFML-TR-66-386, Wright-Patterson Air Force Base, Ohio.

[36] Dolan, T. J., "Models of the Fatigue Process," Proceedings of 10th Sagamore Army Materials Research Conference. Syracuse University Press, Syracuse, N. Y., 1964, p. 1.

Creep and Fatigue Behavior of Unidirectional and Cross-Plied Composites

I. J. TOTH[1]

Reference:

Toth, I. J., "Creep and Fatigue Behavior of Unidirectional and Cross-Plied Composites," *Composite Materials: Testing and Design, ASTM STP 460*, American Society for Testing and Materials, 1969, pp. 236–253.

Abstract:

The tensile, fatigue, creep, and stress-rupture behavior of unidirectional, unidirectional off-axis, and cross-plied composites of 25 volume percent boron filaments in a matrix of 6061 aluminum has been investigated. The tensile investigation has shown the composites to be of reasonably high quality. The fatigue strength of boron-reinforced 6061 aluminum was observed to be superior to those of boron-reinforced 1100 and 2024 aluminum composites. The failure mechanism is discussed in terms of buildup of stress concentration in the matrix to stress levels capable of fracturing proximate filaments. The effect of off-axis loading on the fatigue behavior was found to be similar to that observed on their tensile properties. Cross plying has changed the fracture mode from predominately matrix shear failure to failure involving filament fracture.

The excellent creep and stress-rupture properties of boron filaments were utilized fully in the unidirectional composites when tested in the reinforcement direction. The off-axis creep and stress-rupture behavior of the unidirectional composite was characterized by a rapid reduction of properties with increasing off-axis orientation. Cross plying changed the mode of creep failure for many composite configurations from shear to tensile creep and improved the off-axis stress-rupture properties compared to corresponding unidirectional composites. In general, the creep behavior of cross-plied composites was characterized by a large degree of filament rotation towards the load axis.

Key Words:

composite materials, mechanical properties, fatigue, creep, stress rupture, boron fiber reinforcement, aluminum, fiber composites. fiber orientation studies, evaluation, tests

Composites are of interest in present day technology because they offer the opportunity to "tailor make" a material with a combination of properties that is not available in any single material. Little work has been done, however, on the efficient tailoring of the laminate to specific strength and stiffness requirements. Satisfactory performance in static tension tests is used frequently as the criterion for further development of a composite system for use as a structural or semistructural material. However, in actual service, the failure of a component is determined frequently either by its response to alternating or fluctuating stresses, that is, fatigue, or its behavior under sustained load at elevated temperatures, that is, creep. Therefore, prior to the acceptance of a metal-matrix composite system as a structural material for many applications, it is necessary to evaluate thoroughly the fatigue and creep characteristics of that particular system. In this paper, attention will be focused on the tension-tension fatigue and axial creep behavior of unidirectional (U-D) and cross-plied (C-P) composites of boron reinforced 6061 aluminum.

[1] Principal engineer, TRW Equipment Laboratories, TRW Inc., Cleveland, Ohio. 44117.

The experimental procedures, results, and discussions presented here are a condensation of recent work performed on an Air Force contract [1],[2] and hence for more detail the given reference should be consulted.

EXPERIMENTAL PROCEDURE

All of the composites were prepared at a constant boron filament content of 25 nominal volume percent by diffusion bonding a stackup of ten-ply wound monotape of boron-6061 (T-O) aluminum. The various types of U-D and C-P 25 volume percent boron-6061 aluminum composites in this investigation were all in the as pressed condition and are shown in Table 1.

Table 1—*Types of 25 volume percent boron-6061 aluminum investigated.*

Composite Type		Loading Direction, deg
Unidirectional (U-D)	A	0
	B	5
	C	20
	D	45
±5 deg Cross ply (C-P)	E	0
	F	15
	G	25
±45-deg Cross ply (C-P)	H	0
	T	45

In the production of C-P composites, alternate plies at $\pm\theta$ deg to the common axis were used. The vacuum diffusion bonding parameters were 975 F for 1 h under 6700 psi.

Tension, fatigue, and creep specimens were prepared to the respective configurations shown in Figs. 1 and 2. Tensile properties were determined on an Instron

[2] The italic numbers in brackets refer to the list of references appended to this paper.

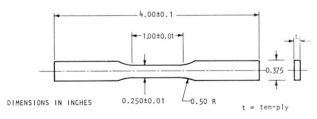

FIG. 1—Tension and creep specimen geometry.

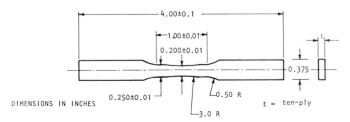

FIG. 2—Fatigue specimen configuration.

Toth on Unidirectional and Cross-Plied Composites

testing machine. Fatigue testing was performed in a standard Baldwin Sonntag unit using tension-tension loading with a ratio of static to dynamic load of 3:2. All creep and stress-rupture tests were performed in air at temperatures of 400 and 600 F in Arcweld machines under constant axial loads. Extensometers of the linear variable differential transformer (LVDT) type were used to continuously monitor the creep strain.

Subsequent to mechanical tests, selected specimens were examined metallographically and radiographically to identify failure mode.

RESULTS AND DISCUSSION

TENSILE

The room-temperature off-axis tensile values are presented in Fig. 3. The experimental strength values show good agreement with the predicted strength curve which was calculated from Tsai's equation [2]. The tensile results of the 0-deg composite indicated a calculated filament stress at failure approximately 390,000 psi, and a rule of mixture modulus. These values characterize the composites to be of a reasonably high quality in the light of the present state of the art in aluminum-boron (Al-B) composite fabrication.

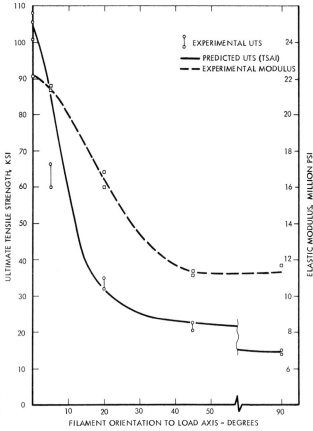

FIG. 3—Correlation of off-axis tensile values with predicted curve using Tsai's analyses.

Composite Materials

FATIGUE BEHAVIOR OF COMPOSITES

In Fig. 4, the present S/N curve of a 0-deg U-D boron-6061 aluminum composite is compared to that of boron-1100 aluminum and boron-2024 aluminum. These composites can compare more meaningfully when the results are considered in terms of the ratios of endurance limit (at 10^7 cycles) to ultimate tensile strength. These ratios were obtained as 0.63, 0.76, and 0.42 for 6061 aluminum 25 volume percent boron, 1100 aluminum 25 volume percent boron, and 2024 aluminum 20 volume percent boron, respectively.

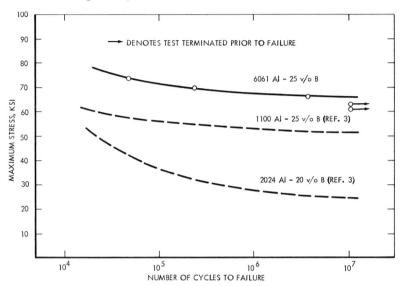

FIG. 4—Effect of matrix on the fatigue property of boron reinforced unidirectional composites.

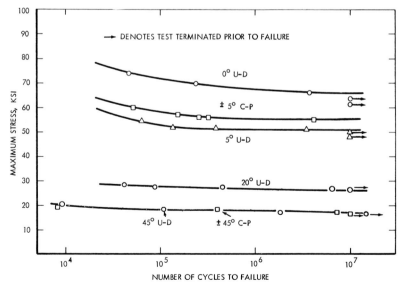

FIG. 5—Fatigue data for unidirectional U-D and C-P 6061 aluminum 25 volume percent boron composites.

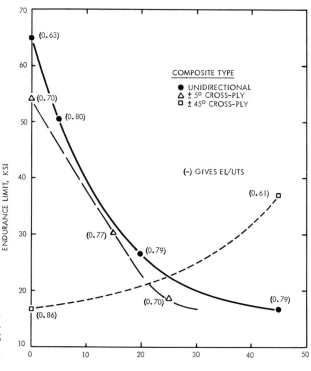

FIG. 6—Comparison of endurance limits of unidirectional and cross-plied 25 volume percent boron 6061 aluminum composites.

The results of the fatigue investigation of some unidirectional and cross-plied specimens are presented for comparison in Fig. 5. It can be seen that the fatigue behavior of these composites are characterized by a rather flat S/N curve. In comparing the unidirectional and the cross-plied composites at 5 and 45-deg filament orientations to the load axis, it is evident that at 25 volume percent reinforcement the anticipated higher than normal stress field at filament crossover points have no detrimental effect on composite fatigue behavior.

The effect of off-axis loading on the endurance limits (10^7 cycles) with respect to filament direction is shown for both U-D and C-P composites in Fig. 6. The ratios of endurance limit to ultimate tensile strength are given in parentheses. These ratios show no significant trend with filament orientation.

The fatigue fracture surfaces of two typical cross-plied composites are compared with their corresponding unidirectional partners in Fig. 7. It can be seen clearly that even at 5-deg filament orientation, the fracture mode of the unidirectional material contains some areas of shear failure, whereas the cross-plied material fails entirely in tension. The fracture paths of some fatigue-tested specimens are shown in Fig. 8 in direct relation to filament orientation. Specimen Type A (0-deg U-D), shows a typical fatigue fracture path where filament fracture is the controlling failure mechanism. Specimen Type G (±5-deg C-P) was tested at 25 deg to the common axis. This specimen together with the 20-deg U-D specimen (not shown here) indicates that even at a filament orientation of 20 deg the failure mode is basically matrix fatigue and shear failure. In the case of cross-plied materials, however, the fracture path must run through one half of the filaments in the specimen cross section. Specimen Type H (±45-deg C-P) shows, adjacent to the fracture surface, some filament rotation towards the load axis. Specimen

Fig. 7—Comparing fatigue fracture surfaces of (a) 5-deg U-D, (b) ±5-deg C-P, (c) 45-deg U-D, and (d) ±45-deg C-P.

Type D (45-deg U-D) indicates failure of the matrix parallel to the filament direction.

In an attempt to find differences in the mechanism of filament fracture under various test conditions, the fracture surface of a number of tension, fatigue, and creep tested specimens, having lower than 20-deg filament orientation to the load axis, were examined metallographically focusing on the filaments themselves. These

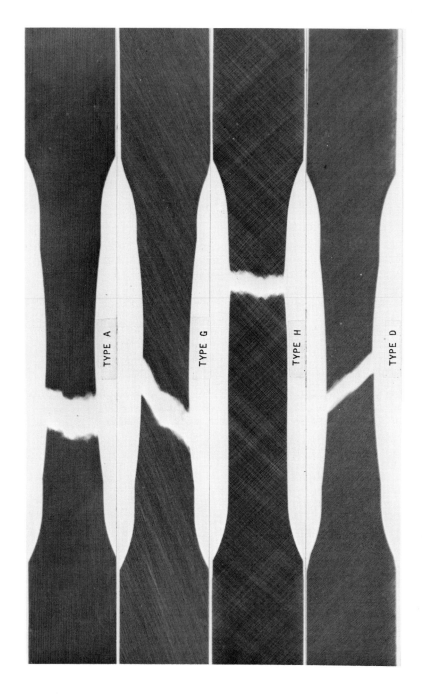

Fig. 8—Radiography of some fatigue tested 25 volume percent boron-6061 aluminum composites.

are shown in Fig. 9 as having origins of fracture at: the filament surface, the boron-tungsten core interface, an inclusion in the boron, and the site of a pre-existing crack.

It was observed that filament fracture mode Type A is found mainly in fatigue, Type B is found mainly in creep, and Type C is found mainly in tension tested composites. However, the difference in their occurrence is too small, as well as the fact that their distribution over the fracture surface is too random to allow the conclusion that the filaments really fracture by a different mechanism under the different testing conditions. The metallographic investigation of these composite fracture surfaces, therefore, suggests that under the conditions studied, the filaments have fractured basically in a tensile manner. But this then implies that boron filaments are not susceptible to fatigue damage. This implication, in fact, is substantiated by the results of some fatigue tests carried out at the Air Force Materials Laboratory on boron filaments.[3] Similar results were obtained in terms of fracture stress levels under both cyclic and normal tensile loading conditions.

If the filaments are not susceptible to fatigue damage and if they control the fatigue behavior of the composite, a horizontal S/N curve may be anticipated. However, if it is assumed that the matrix relaxes in cyclic loading so that its contribution to fatigue strength can be disregarded completely, the endurance limit for the 25 volume percent boron-6061 aluminum (0-deg U-D) composite should still be 97,000 psi (which is 25 percent of the filament strength utilization found in tensile). The actual endurance limit of this composite, however, is only 65,000 psi. Therefore it is obvious that the matrix has a significant effect on the fatigue behavior of this composite even when tested parallel to the filament direction.

In the light of the above discussion, it is pertinent to consider the fatigue behavior of the unreinforced matrix. The endurance limit at 10^7 cycles of the 6061 aluminum matrix, using an R factor[4] of 0.2, is about 16,000 psi in the T-O condition, and about 30,000 psi in the T-6 condition [4]. For this investigation, the matrix has not been heat treated, and, hence, it is assumed to be in the T-O condition. Actually, the ultimate tensile strength of an 11-layer diffusion bonded (under similar conditions used for the composite) 6061 foil was found to be 22,000 psi, which is somewhat higher than expected for annealed material. Nevertheless, using the endurance limit of 16,000 psi and the tensile strength of 22,000 psi gives an endurance/tensile ratio of about 0.73 for the matrix material. It is interesting to note that this value is very close to the ratios shown in Fig. 6 for the composite materials. The closeness of these ratios suggests that the matrix controls the fatigue behavior of the composite.

Some interesting observations were presented recently by Cooper [5] on the fracture toughness of metal-matrix composites. Most relevant was the observations that in a composite of copper reinforced with brittle tungsten filament, cracking occurs only in the filaments. The mechanism of failure postulated is that the crack propagating in the filament is arrested at the filament-matrix interface. However, as a result of redistribution of the load in the composite, the stress concentration in the matrix at the point of filament fracture is sufficient to cause fracture of the next filament immediately ahead of the existing line of filament breaks. Thus, the crack advances through the composite by bridging the matrix and breaking only filaments as it propagates. This mechanism is directly applicable to the situation arising during the fatigue of boron reinforced aluminum composites, but there is one important exception. The exception is that the boron filaments fracture not by a fatigue mechanism, and, hence, as shown in Fig. 10, the filament breaks are not lined up as in the copper-tungsten system.

In Fig. 11, the photomicrograph of a fatigue tested 0-deg U-D composite shows that the matrix can work harden to the extent that, in certain areas, it even reaches its fatigue limit. The fact that very few such matrix fatigue cracks are observed

---

[3] Shimmin, K. D., Air Force Materials Laboratory, private communication.
[4] $R = \sigma\ \text{min}/\sigma\ \text{max}$.

Toth on Unidirectional and Cross-Plied Composites

243

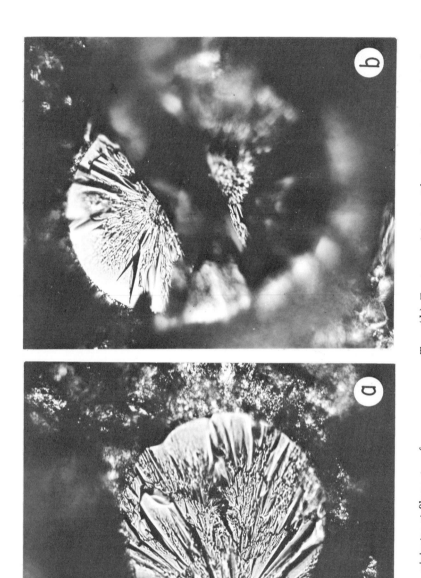

Type (a) Fracture originates at filament surface.   Type (b) Fractures originates at boron/tungsten core interface.

Fɪɢ. 9—Typical boron filament fracture surface in a 6061 aluminum matrix (×500).

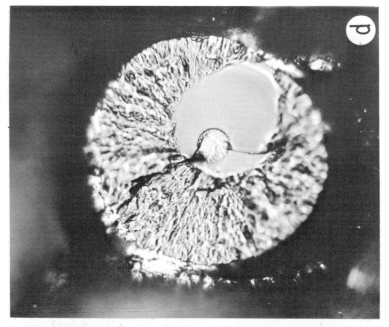

Type (*c*) Fracture originates at an inclusion.
FIG. 9—Continued.

Type (*d*) Fracture originates at site of pre-existing crack.

245

Fig. 10—Photomicrograph of a 0-deg U-D composite adjacent to fatigue fracture (×100).

Fig. 11—Photomicrograph of a fatigue tested 0-deg U-D composite showing that the matrix at certain areas has reached its fatigue limit (×250).

in the composites indicates that such an extent of matrix fatigue hardening is not necessary to build up sufficient stress concentration to fracture proximate filaments.

In summary of the above discussions, it is suggested tentatively that failure, for a 0 deg or low angle 25 volume percent boron-aluminum composite under tension-tension cyclic loading occurs in the following manner:

1. Initial filament fracture arises from basically three sources: (a) filament breaks developed during composite fabrication, (b) filament fragmentation near the surface during specimen preparation by grinding and, (c) initial filament breaks occurring at points of weakness during actual fatigue loading.

2. Stress concentrations in the matrix resulting from above filament breaks grow in magnitude with time under cyclic loading as a result of matrix work hardening and some fatigue damage.

3. Cracks initiate at matrix-filament interface at site of proximate filaments when magnitude of stress concentration is high enough. The stress involved are higher for matrices having higher work hardening rate and hence result in earlier crack initiation.

4. The nucleated cracks propagate by a shear mechanism along matrix-filament interface to a point where the existing stress level exceeds the filament tensile fracture stress and results in filament fracture.

This failure mode of fracturing proximate filaments can continue in a localized plane until the effective composite area in that plane is reduced enough for composite tensile failure to occur.

This suggested failure mode is based on the presently available experimental data. It is felt, however, that in order to determine the exact mechanism of composite fatigue failure more experimental work is required. For example, there is a need to determine:

(a) degree of matrix fatigue damage expecially adjacent to the reinforcement where a submicroscopic crack can produce filament failure,

(b) strain to fracture especially for off-axis and cross-plied composites, and

(c) effect of filament packing density or preferably interfilament separation distance on composite fatigue behavior.

Furthermore, residual strength measurement and metallographic examination of specimens suspended after fatigue testing at selected loads just prior to predicted failure is also necessary in studying the failure mechanism in detail.

CREEP AND STRESS-RUPTURE PROPERTIES

In Fig. 12, the stress-rupture behavior of 45-deg U-D and ±45-deg C-P composites are compared. It can be seen that at 600 F, the 100-h stress-rupture strength of the cross-plied composite is about three times greater than the corresponding unidirectional one.

The 100-h stress-rupture strength for the unidirectional composite as a function of filament orientation is shown in Fig. 13. The very rapid reduction in stress-rupture properties with filament orientation is discouraging from a practical application point of view. The 100-h stress-rupture strength over ultimate tensile strength ratios are decreasing with both increasing temperature and filament orientation. Thus, it is evident that for many applications, such as jet engine compressor blades, aluminum-matrix composite need to be strengthened in the off-axis directions.

The effect of applied stress on minimum creep rate for some of the composite types are shown in Fig. 14. In general, the trend of the creep behavior is consistent with that observed for homogeneous materials. It is interesting to note the superior creep behavior of the ±45-deg C-P composite as compared to the 45-deg U-D one, especially at the higher temperature.

The superior stress-rupture strength and minimum creep rate of the ±45-deg C-P material as compared to the 45-deg U-D composite is somewhat misleading without considering their total creep behavior. A comparison of typical creep curves

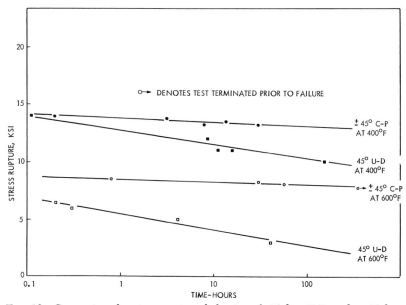

Fig. 12—Comparing the stress-rupture behavior of 45-deg U-D and ±45-deg C-P composites.

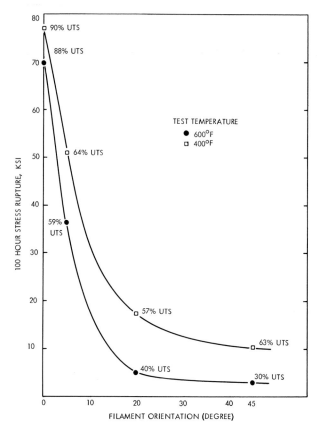

Fig. 13—100-h stress-rupture as a function of filament orientation of 25 volume percent boron-6061 aluminum composites.

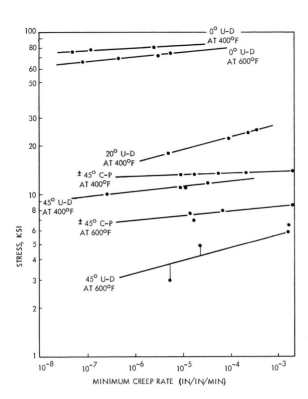

FIG. 14—Effect of stress on minimum creep rate.

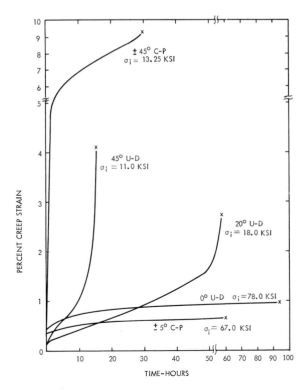

FIG. 15—Creep curves of some U-D and C-P 25 volume percent boron-6061 aluminum at 400 F.

249

FIG. 16—Comparing creep fracture surfaces of (a) 0-deg U-D, (b) 20-deg U-D, (c) 45-deg U-D, and (d) ±45-deg C-P composites were tested at 400 F.

of some unidirectional and cross-plied composites is presented for 400 F in Fig. 15. Unfortunately, these creep curves cannot be compared directly because differing original stresses (σ) were used in order to give stress-rupture results in reasonable times. It is, however, evident in comparing the curves of the 45-deg U-D and ±45-deg C-P composites that they behave quite differently. All of the latter type of composites were characterized by an unusually large primary or first-stage creep.

In comparing the fracture surfaces of some of the creep-tested composites, in Fig. 16, it is evident that cross plying changes the fracture mode. Failure of the 45-deg U-D composite takes place in the matrix parallel to the filament direction

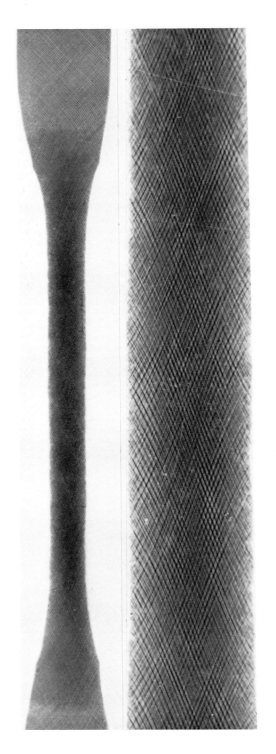

FIG. 17—Radiograph of Type H (±45-deg C-P) composite after creep testing at 600 F showing filament rotation towards load axis. Test was terminated without fracture at 51.3 percent creep elongation (*top*) (×3.4) and (*bottom*) (×11).

by a shear-controlled process. The quantitative interpretation of this cannot be attempted since there are no data available on the time-dependent shear behavior of the matrix. In case of the off-axis unidirectional composite, the creep elongation is usually not uniform over the measured gage length because of unbalanced filament rotation toward the load axis causes some specimen misalignment. In case of the ±45-deg C-P composite, the previously operative easy shear path is reinforced, in fact, with one half of the filaments. Thus, for failure to occur, one half of the filaments must break. Prior to these breaks, however, this composite is capable of extensive creep elongation in a uniform manner presumably due to the balanced nature of the cross plying. This takes place, as shown in Fig. 17, by filament rotation toward the load axis. The degree of this rotation is such that the original ±45-deg filament orientation is reduced to ±21 deg. Consequently, it is questionable whether the creep properties of the 45-deg U-D composite are realistically comparable to those of the ±45-deg C-P composite

The creep curve for the 0-deg U-D composite, illustrated in Fig. 15, has three characteristic regions: (1) initial strain on loading according to simple composite theory, (2) quasi first-stage creep where further elastic straining of the filaments take place as the matrix relaxes, and (3) secondary stage creep region which probably is controlled by creep of the filaments. No accelerated third stage creep was observed.

Using the tungsten-filament-reinforced copper model composite system, McDaniels et al [6] recently formulated an equation for the prediction of creep behavior of 0-deg U-D composites based on the creep properties of the components. In order to use this prediction, and also to postulate a mechanism to explain quantitatively the low creep rates observed in the 0-deg U-D 25 volume percent boron-6061 aluminum composites, a knowledge of the creep behavior of boron filaments is required.

At present, the only available creep data on boron filaments were observed at a temperature of 1000 F. According to Metcalfe et al [7] at a temperature of 1000 F and a stress level of 264 ksi, the creep rate of the boron decreased from approximately 4.2×10^6 in./in./min at 1 h to 0.45×10^{-6} in./in./min at 15 h to a steady-state creep rate of 0.17×10^{-6} in./in./min after 65 h. These data, though obtained at a higher temperature than used for the testing of 25 volume percent boron-6061 aluminum, still can be compared with otherwise similarly obtained composite creep data, at least, to observe whether the general trend would permit the suggestion that the creep of the boron filaments control the steady-state creep of the composite. Consider the 0-deg U-D composite at 600 F at a stress level of 65 ksi which indicates an extrapolated rupture life of over 1000 h.

Assuming that in the first stage of the creep process the matrix relaxes and its contribution to overall composite strength is negligible, then the composite stress level of 65 ksi is equivalent to a calculated filament stress of 260 ksi. This stress level is consistent with that used above for the creep testing of boron filaments. The minimum creep rate of this composite at 600 F was 8.7×10^{-8} in./in./min. In comparing this creep rate to 1.7×10^{-7} in./in./min which is obtained in the boron filament at 1000 F, it seems reasonable that the second stage creep of the 0-deg U-D composite is controlled by the creep behavior of the boron filaments.

CONCLUSION

The tensile, fatigue, creep, and stress-rupture behavior of 25 volume percent boron filament reinforced 6061 aluminum has been investigated. The tensile investigation has shown that the composites fabricated as part of this program were reproducible with reasonably high quality.

It has been demonstrated that the tension-tension fatigue strength of the matrix can be increased significantly in the direction of the reinforcing filaments. The fatigue strength of boron reinforced 6061 aluminum was observed to be superior to those of boron reinforced 1100 and 2024 aluminum composites.

Since it appears that the filaments are not susceptible to fatigue damage, it was concluded that the behavior of the matrix under cyclic loading controls the fatigue behavior of the composite. A failure mechanism is suggested that involves slight fatigue damage and a buildup of stress concentration in the matrix, resulting mostly from points of initial filament failures, to a stress level capable of fracturing proximate filaments.

The effect of off-axis loading on the fatigue behavior of boron reinforced 6061 aluminum composites was found to be similar to that observed on their tensile properties. Cross plying has produced similar fatigue results to corresponding unidirectional composite, although it changed the fracture mode from predominately matrix shear failure to failure involving filament fracture.

The excellent creep and stress-rupture properties of boron filaments were utilized fully in the 0-deg U-D 25 volume percent boron-6061 aluminum composite. The creep behavior of this composite is characterized by three stages: (1) an initial elastic stress region, (2) a rather rapid quasi-first stage creep where the matrix relaxes, and (3) a very low strain rate secondary stage which appears to be controlled by the creep of filaments. Accelerated third stage creep was not observed. The 100-h stress-rupture life of the 0-deg U-D composite at 600 F was about 70,000 psi. The indications are that this can be doubled by increasing the filament loading to 50 volume percent.

The off-axis creep and stress-rupture behavior of the unidirectional composite is characterized by a rapid reduction in properties with increasing angle of filament orientation. Even at 5-deg orientation, the failure mode was predominately of matrix shear failure. These results indicated that for many applications, such as jet engine compressor blades, the aluminum-matrix composites will have to be strengthened in the off-axis directions by cross plying.

Cross plying has changed the fracture mode and improved the stress-rupture properties of correspondingly oriented unidirectional composites. The ±45-deg C-P composite is characterized by a surprisingly extensive creep elongation which is accompanied by a large degree of rotational alignment of filaments towards the specimen load axis.

ACKNOWLEDGMENTS

This work has made possible through the support of the Air Force Materials Laboratory, Wright-Patterson Air Force Base, Ohio, under Contract AF 33(615)-3062. The cooperation of K. D. Shimmin, Air Force project engineer, is acknowledged.

References

[1] Toth, I. J., "Time Dependent Mechanical Behavior of Composite Materials," Annual Technical Report on Contract AF 33(615)-3062, TRW Inc., Cleveland, Ohio, Nov. 1968.

[2] Tsai, S. W., "Mechanics of Composite Materials," AFML-TR-66-148, Air Force Materials Laboratory, Wright Patterson Air Force Base, Ohio, June 1966.

[3] Morris, A. W. H., "An Exploratory Investigation of the Time Dependent Mechanical Behavior of Composite Materials," report on Contract AF 33(615)-3062, TRW Inc., Cleveland, Ohio, Aug. 1967.

[4] "Metallic Materials and Elements for Aerospace Vehicle Structures,"
MIL-HDBK-5A, Department of Defense, Washington, D. C., 1966.

[5] Cooper, G. A., "The Potential of Fiber Reinforcement," Metals and Materials, April 1967.

[6] McDanels, D. L., Signorelli, R. A., and Weeton, J. W., "Analysis of Stress-Rupture and Creep Properties of Tungsten-Fiber-Reinforced Copper Composites," NASA TN-D-4173, National Aeronautics and Space Administration, Sept. 1967.

[7] Metcalfe, A. G. and Rose, F. K., "Testing of Thin Gage Materials," AFML-TR-68-64, Air Force Materials Laboratory, Wright-Patterson Air Force Base, Ohio, 1968.

Creep and Rupture of Graphite-Epoxy Composites

F. Y. SOLIMAN[1]

Reference:

Soliman, F. Y., "Creep and Rupture of Graphite-Epoxy Composites," *Composite Materials: Testing and Design, ASTM STP 460*, American Society for Testing and Materials, 1969, pp. 254–270.

Abstract:

Long-term loading can have a significant effect on the performance of aerospace composite structures. The size of this effect depends on the type of composite used, the fiber orientation within a particular composite, and the imposed stress and temperature levels. Test data for graphite-epoxy composite under undirectional state of stress and controlled temperature environment are presented. The behavior of specimens of different lamina orientation tested under identical conditions was compared. The stress situation and temperature level are the most influential parameters on the creep deformation of the material, being greatest where the reinforcing fibers are not aligned in the direction of the load application. A relation between time to rupture, temperature, and stress level was shown and compared to other composite systems. The results show the detrimental effects of stress level, time, and temperature on the strength and deformation. Based on these data, the behavior under long-term loading could be extrapolated from reasonably short-time tests composite materials, fiber composites,

Key Words:

composite materials, fiber composites, creep, creep testing, creep rupture, creep recovery, rupture mechanism, fiber-reinforced materials, graphite composite, mechanical properties, evaluation, tests

Advanced composites made of boron or graphite filaments are being used in structural applications in several types of aircraft and helicopters. They usually offer a weight saving of 20 percent or more over the current technology and enable the designer to tailor specific composites that meet the needs of his particular structure. The behavior of these structures has been investigated by many workers, who have predicted the response of these materials due to certain types of loads. However, little work has been done to evaluate the carrying capacity and dimensional stability of composites under service environment for an expected life time.

Although data on short-term properties provide a basis for predicting the potential and behavior of a composite system, such information is not accurate enough to forecast behavior during actual service. This behavior may differ very significantly if the system experiences an elevated temperature level above the basic design level. Creep and creep rupture represents one the long-term characteristics of structural materials which cannot be obtained easily from short-time tests.

It is well known that creep under a constant stress is a time-dependent phenomenon which is reflected in the form of a continuous increase in deformation as time progresses. The effect of this time sensitivity on the dimensional stability of a structural material plays a significant role in the usefulness of a particular material system for primary load-carrying structural members. The creep phenomenon in conventional materials has been the field of investigation for many pioneers who evaluated it both theoretically as well as experimentally. Creep in composites is a complicated phenomenon which is manifested by the apparent anisotropy and

---

[1] Scientist, Aerospace Sciences Laboratories, Lockheed-Georgia Co., Marietta, Ga. 30060.

the local heterogenity of the material. It has been shown by many investigators [1–3][2] that some fiber-reinforced composites are time sensitive. Steel [1] examined the creep behavior of glass-fiber-reinforced composites and showed their time sensitivity under loads was much less than their ultimate capacities. Findley [2] investigated the behavior of other types of composites under creep condition to predict the reliability of composite structures.

This paper presents some qualitative data concerning the creep behavior of an advanced composite system. It is intended to demonstrate the effect of different material parameters, such as stress levels, fiber orientations, and temperature on the creep deformation. Some theoretical considerations about the behavior of this composite system under creep conditions are developed. Analytical models of creep behavior are necessary tools in designing composite structures in aircraft, where a part of the loading is a function of the deformation response.

The following sections present the composite system considered, available experimental results, and a summary of conclusions and observations. The creep rupture mechanism is outlined and discussed briefly.

COMPOSITE MATERIAL

The experimental work was carried out on a graphite-fiber-reinforced composite. The graphite fiber was Thornel 50 with a Young's modulus of 47.4×10^6 psi and ultimate tensile strength of 275×10^3 psi. The matrix was ERL-2256 epoxy resin with 19.2 parts per hundred resin (phr) MPDA curing agent. Thirty percent by weight acetone was added to the resin bath during filament winding to adjust the viscosity.

The composite laminae were prepared by filament winding. The volume percent of the Thornel fibers was about 55 percent. Each lamina was B-staged by heating for 15 min at 95 C, followed by 25 min at 105 C.

Laminates were formed by hand layup of the B-staged laminae in the desired orientation. The laminates were press cured in the following cycle:

30 min at 70 C and 90 psi
30 min at 100 C and 180 psi
60 min at 160 C and 180 psi

The laminates were removed after the mold cooled, and they were post-cured for 120 min at 160 C. The resulting laminates were representative of typical graphite composite fabricated in a laboratory environment.

Two specimen configurations were cut from the laminates: the first measured 6 by 1 in. and the second measured 5 by 1 in.

EXPERIMENTAL PROCEDURES

Bending tests were conducted for (0 deg, 90 deg) laminated composite, and tension tests were performed for off-diagonal laminates.

FOUR-POINT BENDING TESTS

In testing advanced composites, flexure tests are considered the simplest and most economical to perform. These tests were conducted using a test fixture with two-point loading and two simple supports. The midspan of the specimen (dimension 1 by 1 in.) is under constant bending moment throughout its cross section. In all cases, to eliminate interlaminar srear failure, the tested span of the specimens was greater than 32 times the specimen thickness. The flexure specimen configura-

---

[2] The italic numbers in brackets refer to the list of references appended to this paper.

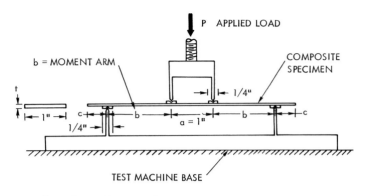

Fig. 1—Flexural test fixture and composite specimen.

tion and the loading-support jig are shown in Fig. 1. No end tabs were attached to the specimens.

TENSION TESTS

For off-angle, laminated, composite specimens, flexure tests present a risk in terms of the accuracy of measurements and the difficulties in interpreting the test results. Thus, a constant tensile stress condition was considered. However, the introduction of a uniform tensile stress condition throughout the specimen cross section is not an easy task. The conventional methods of testing have been shown [3] to be inadequate.

Figure 2 shows an alternative configuration of composite specimen and method of attachment. The specimen is attached to the testing machine by two pin-ended supports. This eliminates some of the end constraints that otherwise could be imposed by the machine jaws on the ends of the tested specimens. In this type of specimen, only uniaxial loads are transmitted to the end tabs. These uniaxial end loads are distributed throughout the specimen cross section, and a reasonably uniform state of stress is introduced.

The following data were produced from the bending and tension tests:

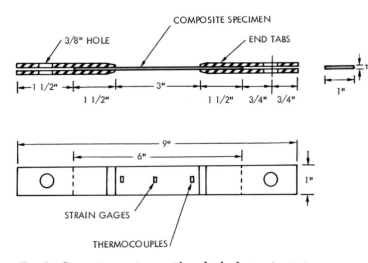

Fig. 2—Composite specimen with end tabs for tension test.

Short-Time Properties of the Composite

Several specimens were obtained from each laminate and tested under controlled environmental conditions. These data are used as the basis for evaluating the strength and stiffness of the composite.

Creep and Recovery or Creep and Stress-Rupture Data

Specimens were prepared from every laminate. Conditioning of the test specimens in terms of temperature exposure and humidity was controlled so that a uniform environment is sustained throughout the specimen before and during loading. Specimens were subjected to stress levels between 64 and 95 percent of the ultimate short-time strength. The time span of each tension test did not exceed 500 continuous hours, whereas the time span of the flexure test was more than 3000 hours. At the end of the testing period, unfailed specimens were removed and tested for strength retention. Recovery data for some unfailed specimens were recorded also after unloading.

Strength Retention Data

All unfailed specimens were tested statically for degradation in strength or stiffness after loading for 500 h. These short-time tests were conducted under the same environmental conditions as in the creep tests.

Straining of the tested composite specimens was monitored by means of foil gages mounted on the midspan of the specimens. Transverse and longitudinal gages were used to indicate the principal strains. Strains versus time were recorded automatically, whereas the applied stresses were controlled manually. The strain-time records were generally smooth, indicating continuous deformation up to the termination of the test or the failure of the specimen. Thermocouples were mounted at different locations on each specimen to control its temperature distribution and temperature variations. The temperature environment was maintained constant with a maximum variation of ±5 F throughout the tests.

Although the intent of this investigation was not to generate statistical results, most of the experiments were conducted at least twice in order to decrease the scatter of the test data and to ensure reproducibility.

EXPERIMENTAL RESULTS

The creep deformation of a graphite composite system under constant load is shown in Figs. 3, 4, and 5. Figure 3 illustrates the strain-time results for (0 deg, 90 deg) laminates under bending stresses. Figures 4 and 5 are for (30 deg, −60 deg) and (45 deg, −45 deg) laminates under uniaxial tensile stresses. In each of these figures, two stages of creep are evident: primary stage and secondary or steady-state creep. In the first stage, the creep rate is decreasing continuously; in the second stage there is a slight continuing decrease in the strain rate. This represents the minimum creep rate in the overall domain of loading time. It is also useful to note that, on application of the load, all specimens indicated an instantaneous strain which is a function of the applied stress level as well as the temperature condition. The third stage of creep or the tertiary stage, which did not appear in the previous figures due to the short loading time span, is demonstrated in Fig. 6. This stage is accompanied by a continuous increase in the creep rate as time progresses, and it terminates with the rupture of the test specimen.

Upon the removal of the external stresses from the unfailed specimens after a designed loading time, no complete strain recovery was noticed. Figure 7 illustrates the recovery strain versus time for (30 deg, −60 deg) laminates. After a period equal to 10 percent of the loading time, a state of equilibrium was reached, and 30 percent of the creep strain was unrecovered. The mechanical aftereffect thus

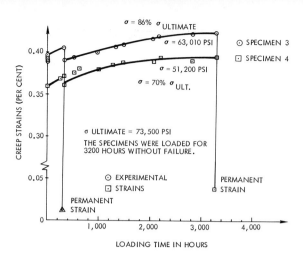

Fig. 3—Tensile creep strain for (0 deg, 90 deg) laminated composite specimen under constant bending moment and room temperature.

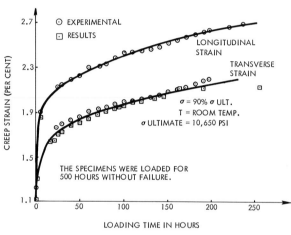

Fig. 4—Tension creep tests for (30 deg, −60 deg) laminated composite specimens under constant load (90 percent ultimate) and room temperature.

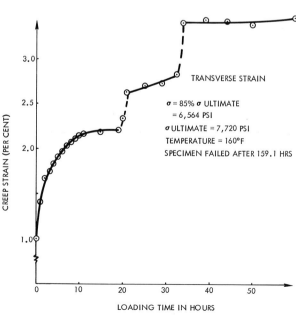

Fig. 5—Tension creep tests for (45 deg, −45 deg) laminated composite specimen under constant load (85 percent ultimate) and 160 F.

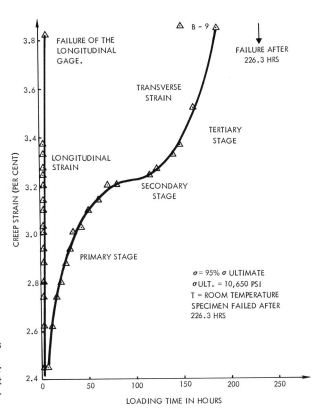

FIG. 6—Tension creep tests for (30 deg, −60 deg) laminated composite under constant load (95 percent ultimate) and room temperature.

Figure 6 labels:

3.8 — FAILURE OF THE LONGITUDINAL GAGE.

B - 9

FAILURE AFTER 226.3 HRS

TRANSVERSE STRAIN

TERTIARY STAGE

LONGITUDINAL STRAIN

SECONDARY STAGE

PRIMARY STAGE

σ = 95% σ ULTIMATE
σULT. = 10,650 PSI
T = ROOM TEMPERATURE
SPECIMEN FAILED AFTER 226.3 HRS

CREEP STRAIN (PER CENT)

LOADING TIME IN HOURS

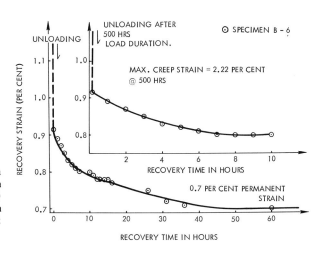

FIG. 7—Recovery strain after 500 h load duration for (30 deg, −60 deg) laminate tested at room temperature and 90 percent ultimate.

Figure 7 labels:

UNLOADING

UNLOADING AFTER 500 HRS LOAD DURATION.

⊙ SPECIMEN B - 6

MAX. CREEP STRAIN = 2.22 PER CENT @ 500 HRS

RECOVERY STRAIN (PER CENT)

RECOVERY TIME IN HOURS

0.7 PER CENT PERMANENT STRAIN

was recovered fully, part of it instantly and the rest after a sufficient length of time. The irrecoverable creep strain is attributed mainly to a permanent deformation in the matrix phase of the composite and a relative motion between the fibers and the matrix. The percent of the irrecoverable creep strain depends greatly on the orientation of the laminae, for example, in Fig. 3 the irrecoverable strain is less than 10 percent of the accumulated creep strain during loading, and it is

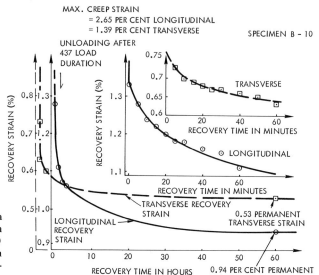

MAX. CREEP STRAIN
= 2.65 PER CENT LONGITUDINAL
= 1.39 PER CENT TRANSVERSE
SPECIMEN B - 10

FIG. 8—Recovery strain after 436-h load duration for (30 deg, −60 deg) laminate tested at room temperature and 85 percent ultimate.

reached instantaneously upon load removal. Figure 8 represents the strain recovery for (30 deg, −60 deg) laminate.

The extent of internal changes in the tested composite material is a direct function of the applied stress level, the temperature of the environment, and time. Thus, different shapes of creep curves are obtained if any one of the previous parameters change. The laminae orientation is also a major parameter that controls the behavior of the composite. These parameters are discussed separately below.

EFFECT OF FIBER ORIENTATION

If the fibers are in the direction of the load, most of the creep strain can be attributed to the deformation characteristics of the reinforcing fibers. The function of the matrix in this case is of a binding nature; the closer the fibers are aligned to the load direction, the less creep strain would be accumulated. Under tensile stresses, if a percentage of the fibers was in the direction of the load as in Fig. 3, the accumulated creep strain did not exceed 7 to 10 percent of the instantaneous strain after a loading period of more than 3000 h. No failure occurred in the specimens, and, upon removal of the load, most of the strain was recovered instantaneously, and only a small permanent strain was obtained. The situation is different, however, when the laminae are not aligned in the direction of the load. This is reflected not only in the magnitude of the observed instantaneous strain but also in the creep rate as a function of time during the different creep stages. In Fig. 4, where the fibers are aligned at (30 deg, −60 deg) to the direction of the load, the longitudinal as well as the transverse strains increased at least threefold as compared with that of (0 deg, 90 deg) laminates. Also, the accumulated creep strain during the secondary stage of creep was higher than the instantaneous strain. This shows that the inelastic deformation of the organic matrix phase of the composite system is more pronounced in this fiber orientation. A case of equilibrium was not reached even after a reasonable period of time. The same observation holds for the (45 deg, −45 deg) laminates, as presented in Fig. 5. In this case a state of equilibrium was reached sooner than in the case of (30 deg, −60 deg) laminates. Figure 9 represents a comparison of the creep srrain of two laminaes with different laminae orientations (30 deg, −60 deg) and (45 deg, −45 deg), tested under identical tensile stress and temperature levels. Here the time

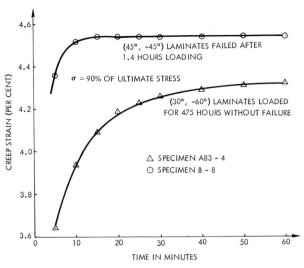

FIG. 9—The effect of fiber orientation on the creep strain at 160 F and 90 percent ultimate stress.

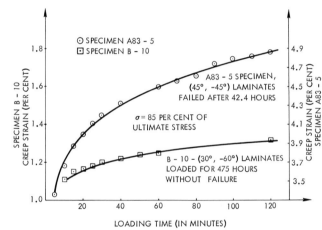

FIG. 10—The effect of fiber orientation on the creep strain at room temperature and 85 percent ultimate stress.

to reach equilibrium is clearly evident. This figure shows the superiority of one fiber orientation (30 deg, −60 deg) over the other (45 deg, −45 deg) in terms of creep strain under this type of loading, but the time to reach equilibrium is different in both cases. The test results shown in Fig. 10 are for the same laminae orientation as previously used, but the test was conducted at room temperature and lower stress level.

Again, the (30 deg, −60 deg) laminae did not accumulate creep strain as high as other laminates, even if they were tested under identical conditions. The creep strain in the (45 deg, −45 deg) laminates was at least threefold as that of (30 deg, −60 deg) laminates when both specimens were tested under identical percentage of stress level from their ultimate strength.

EFFECT OF STRESS LEVEL

The externally applied stress level on a composite specimen plays a significant role in determining its initial strain, as well as the creep strain rate during the loading period. This effect varies according to laminae orientation, which, in turn, is reflected in the nonlinearity of the stress-strain relation of these materials under static conditions. In unidirectional and bidirectional laminates under uniaxial loads, the increase in the instantaneous strain is expected to be a linear function of the load, since most of the load is carried by the fibers. Also, the creep rate during the secondary stage of creep can be assumed constant. This is shown in Fig. 3 for two different stress levels over a large time domain.

For different fiber orientation, the effect of the nonlinearity in the stress-strain relation is more pronounced and is rflected in the increase of both the instantaneous strain as well as the rate of creep during the secondary stage. Figure 11 represents the effect of the stress level on the creep strain for (30 deg, −60 deg) laminates tested under tensile stresses at room temperature. This effect is more evident in case of (45 deg, −45 deg) laminates, as shown in Fig. 12.

The rise of test temperature helps to increase the creep deformation of fiber reinforced organic composites. The introduction of this thermal energy is reflected in the severe nonlinearity of the material as attributed to the matrix deformation. As shown in Fig. 13 for (45 deg, −45 deg) laminates tested under two different levels of stress but identical temperature conditions, the increase in the stress level from 85 to 90 percent of the ultimate stress increases the initial strain by twofold. It also affects the life time of the tested specimens. This is shown in Figs. 14 and 15 for two different laminate systems, (30 deg, −60 deg) and (45 deg, −45 deg), respectively. tested under different stress levels and temperature conditions. These experimental data indicate a trend of a linear relationship between the level of the applied stress and the logarithm of the fracture time at a specified

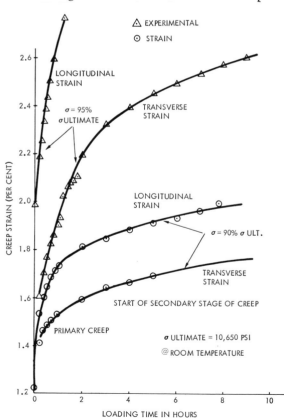

Fig. 11—The effect of stress level on creep strain for (30 deg, −60 deg) laminates at room temperature.

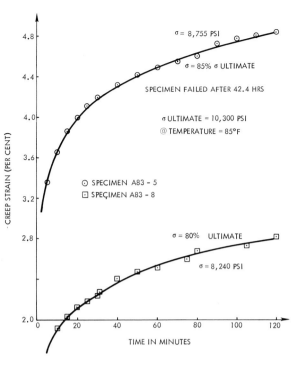

FIG. 12—The effect of stress level on creep strain for (45 deg, −45 deg) laminates at room temperature.

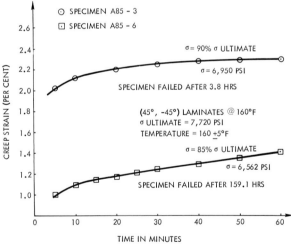

FIG. 13—The effect of stress level on creep strain for (45 deg, −45 deg) laminates at 160 F.

temperature level. Such a relation has been established for different metallic materials and was observed clearly in glass-fiber-reinforced composites [3]. Based on these qualitative data, it seems that a tentative relation of this form still holds for more advanced composites such as that considered here.

Using the information shown in these figures, it is possible to relate the fracture time of such composites to the energy level introduced into the system. The energy level is a direct function of the applied stress and the temperature imposed on the system. (This concept will be treated in greater detail in another forthcoming publication.)

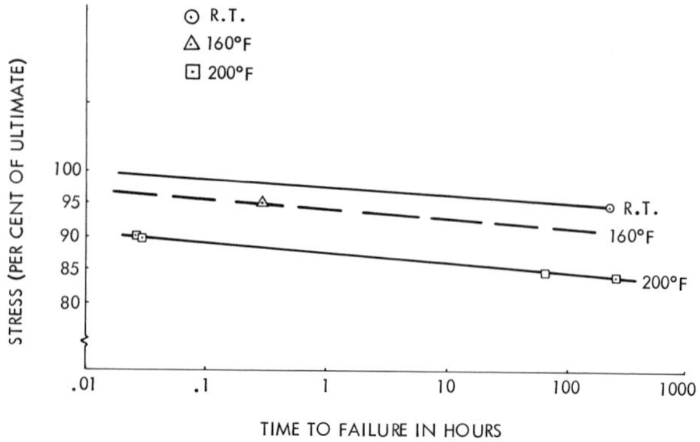

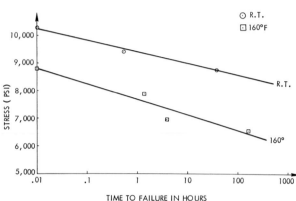

FIG. 14—The effect of stress level on the creep rupture time for (30 deg, —60 deg) laminates tested at different temperatures.

FIG. 15—The effect of stress level on the creep rupture time for (45 deg, —45 deg) laminates tested at different temperatures.

EFFECT OF TEMPERATURE LEVEL

Thermal effects in organic matrix materials is an important factor in determining the deformational characteristics of the resulting composite, as well as its survival time. The behavior of a composite at room temperature and low stress level (below 50 percent of ultimate) is very nearly in accordance with the elastic theory. At higher temperatures, however, and for high stress levels, the behavior of the composite becomes highly time dependent.

As the temperature of a creep test rises from 85 to 160 F and for stress levels higher than or equal to 70 percent of ultimate, the creep strain increases significantly, and thus the lifetime of the specimen decreases. This is shown clearly in Fig. 16 for (30 deg, —60 deg) laminated composite tested under 90 percent of its ultimate strength. Figure 17 illustrates the effect of temperature on the creep strain of the same laminate system tested at 85 percent of its ultimate. These figures indicate a considerable increase in the creep strain, especially in the first few hours after loading.

Thermal effects also are well reflected in the decrease of the creep rupture time of the two tested laminates. This is shown in Figs. 14 and 15 for (30 deg, —60 deg) and (45 deg, —45 deg) laminates. At the same stress level, a considerable decrease in the lifetime of both laminates is observed clearly. This can be attributed to the change in the character of the resin system as the temperature is elevated. The rise of the test temperature increases the ductility of the organic matrix and thus increases the rate of straining of the overall composite at the different stages of creep. It also results in accelerating the creep rupture time of the composite,

Composite Materials

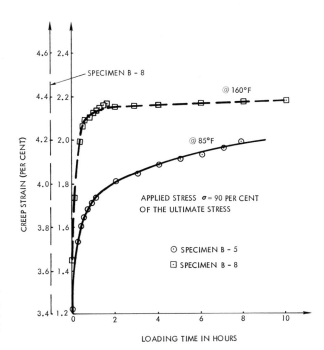

FIG. 16—The effect of temperature on the creep strain of (30 deg, −60 deg) laminates stressed at 90 percent of its ultimate.

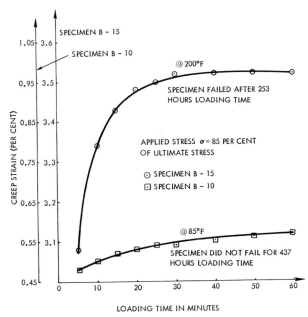

FIG. 17—The effect of temperature on the creep strain of (30 deg, −60 deg) laminates stressed at 85 percent of its ultimate.

especially at high stress levels. From observation of the fractured specimens at high temperature, a ductile fracture was more dominating.

As the temperature was elevated to 250 F, the life of the specimens was significantly short. All specimens tested under this temperature level or higher (300 F) did not survive more than 2 min. This temperature is in the neighborhood of the resin's transition temperature (about 320 F), at which its character changes from a solid state to a viscous rubbery state, thus reducing the bond stresses between the fibers and the matrix. The fractured specimens clearly demonstrated a state of fibers pullout accompanied by catastrophic failure of the overall specimen. This observation was more pronounced in the (45 deg, −45 deg) laminates, since the specimens did not carry the applied loads for more than a few seconds. The effect was consistent with most of the tested specimens.

As the temperature decreased, the ductility of the resin system decreased, and a brittle fracture occurred through crazing of the resin.

CREEP AND RUPTURE MECHANISMS

The creep and creep-rupture mechanism in a uniaxially stressed composite depends on many factors, such as the ratio of the moduli (E_f/E_m), volume fraction (V_f/V_m), and the fiber orientation with respect to the direction of the applied load. Two types of mechanisms can be introduced, depending on the constituents of the considered composite:

1. Fiber creep and rupture mechanism.
2. Matrix creep and rupture mechanism.

For a better understanding, let us first consider the stress distribution in the fibers and matrix of a uniaxially stressed composite.

The stress distribution in both fibers and matrix has been worked out by many investigators concerned with fiber-reinforced composites. It was stated that the matrix material in a uniaxially stressed composite transfers the internal stresses to the fibers in the form of shear stress at the fiber-matrix interface. The greater stress concentration of stress transfer occurs near the ends of the fibers. Tyson and Davies [7] concluded that the different states of stresses in the matrix and the fibers can be categorized as follows:

(a) Near the ends of the fibers the state of stress in the matrix material is nearly pure shear.

(b) At the ends of the fiber, the state of stress in the matrix is a combination of considerable tensile stresses acting parallel to the fiber axis, in addition to shear stresses at fiber-matrix interface.

(c) At the central portion of the fiber, the matrix is hardly stressed, and all the internal stresses are carried by the stiffer fibers. The fibers are highly stressed along most of its length to within a small distance from the ends of a fiber (5 to 10 times the fiber diameter). This is due to the high ratio between the modulus of the reinforcing fibers as compared to the modulus of the organic matrix.

These states of stresses for both fibers and matrix in a uniaxially stressed composite are schematically shown in Fig. 18. If the applied stress is less than the composite ultimate, then the orientation of the fibers with respect to the direction of the applied stress will be the determining factor for the creep and creep rupture mechanism.

Two types of laminae orientations will be considered.

A. Unidirectional and bidirectional laminae, where at least some of the reinforcing fibers are parallel to the load. In this case, the creep and creep rupture of the fibers is the main governing mechanism.

B. Off-angle laminates, where the fibers are neither parallel or perpendicular to the load. The creep and creep rupture mechanism of the matrix is the controlling mechanism.

CASE A. UNIDIRECTIONAL AND BIDIRECTIONAL LAMINATES

In laminated composites with 0 deg or (0 deg, 90 deg) fiber orientation, the continuous fibers in the load direction extend throughout the specimen. Thus, in the central span of a uniaxially stressed composite, most of the external load is supported by the unidirectional reinforcing continuous fibers, and no major contribu-

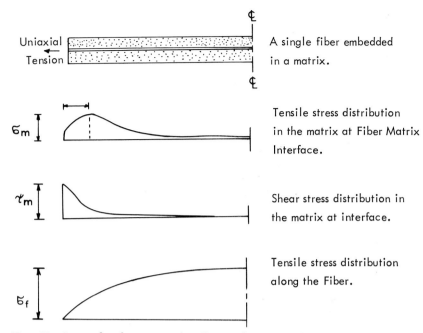

Uniaxial Tension

A single fiber embedded in a matrix.

σ_m

Tensile stress distribution in the matrix at Fiber Matrix Interface.

τ_m

Shear stress distribution in the matrix at interface.

σ_f

Tensile stress distribution along the Fiber.

FIG. 18—Stress distribution in the fiber and matrix of a uniaxially stressed composite.

tion is attributed to the matrix material. The creep strain of the overall composite is thus controlled by the creep of the fibers. High-strength, high-stiffness fibers are not very sensitive to long-time loading, especially if the applied load is less than 80 percent of its ultimate strength. A major creep strain cannot be expected as compared with the instantaneous elastic strain. Theoretically, the creep strain would approach a value equal to the instantaneous strain after infinite time.

If the applied stresses are less than the ultimate stresses of the composite, a case of equilibrium would result, and no catastrophic failure would occur in the specimen. However, if the externally applied stress is near the ultimate stress of the composite material (90 percent or more), the ultimate stress in any particular fiber may be reached after a period of time. This is due to the formation or existence of defects in a particular fiber or a group of fibers and the rise in the fiber internal stresses due to matrix relaxation. This is governed by the statistical distribution of fiber strength along its length [9]. The matrix adjacent to the fractured fiber has to carry this sudden load that previously was carried by the broken fiber.

In an organic composite, where $E_m \ll E_f$, the newly imposed tensile stress will exceed the ultimate stress of the matrix, and a crack will begin at the location of the fractured fiber. The propagation of this crack depends greatly on the degree of dutility of the matrix used in the composite. This ductility determines the ability of the resin system to absorb this suddenly released energy. If the matrix sustains this additional energy, then it will distribute it and transfer it to the other fibers in the immedite vicinity of the fractured fiber. In this case, the load of the broken fibers will be shared by other unbroken fibers adjacent to the broken one. The resulting tensile stress may exceed the ultimate strength of the adjacent fibers, and a gradual rupture of additional fibers would occur.

As was pointed out in Ref 6, if the matrix is ductile and the initiation of matrix cracks is resisted, a shear failure in the bond between the matrix and fiber interface is likely to begin near the end of the broken fiber. The propagation of the bond failure depends on the shear strength of the matrix material. If the strength of the matrix is greater than the newly imposed shear stress, a state of equilibrium would result, and no catastrophic failure would occur. In this case, the broken fibers are considered discontinuous. If the bond failure continues throughout the broken fiber, the composite specimen will have internal defects which greatly affect its stability and lifetime.

The introduction of additional stresses in the remainder of the fibers accelerates the failure process of the remaining fibers. This results in a tertiary stage of creep and eventual rupture of the composite fibers. After the failure of sufficient fibers, the matrix material can fail instantaneously or carry the load for a very short period of time, thus extending any existing tertiary creep stage.

A similar observation has been reported in testing unidirectional boron-reinforced composite under uniaxial tensile stresses. The creep characteristics of the boron fibers are also insignificant if the fibers are stressed along its axis.

The situation, however, is different in case of glass-reinforced composites. It has been reported by Steel [1] that, at stress levels which represent a fraction of the short-term potential strength of the composite, three stages of creep occurred and a significant amount of creep strain was observed. This was demonstrated for different types of glass-fiber composites under different environments.

CASE B. OFF-ANGLE LAMINATES

In case of off-angle reinforced composites, where the fibers are neither parallel nor perpendicular to the direction of the applied load, the fibers are considered to be discontinuous. Due to the application of the external loads, tensile stresses will exist in the matrix and the fibers. In addition, the matrix material is under shear stresses near the ends of the fiber at specimen edges or other discontinuities. Thus the fibers are considered to be under uniaxial tension, whereas the matrix is under shear and tensile stresses. Tensile stresses in the matrix and fibers cause creep deformation if the load is sustained for a reasonable length of time. In advanced composites with organic matrix, the creep of the matrix is usually greater than that of the fibers. Therefore, under creep conditions, relative shear stresses between the fibers and the matrix would be created and increased as the creep of the matrix progresses. Since the matrix at the fiber ends is subject to a combination of normal and shear stresses, its deformation would be much larger than that at the middle of the fiber. This is magnified by any unevenness in the applied stresses due to the method of testing or any external factors such as edge residual stresses due to cutting of the specimens.

Thus creep rupture starts at the edges of the specimen and progresses towards the center of the tested specimen. Two types of rupture mechanisms can be defined, depending on the volume percent of the constituents.

1. If the volume percent of the fibers in a composite is high ($V_f = 50$ percent or better), the spacing between fibers is small compared with the diameter of the fiber, and a bond failure is the most probable mechanism. The bond failure will occur at the ends of a fiber and will propagate towards the center.

A complete failure of an interface bond along a filament eliminates the carrying capability of the lamina where the fiber is placed. This, in turn, reduces the overall cross-sectional area of the laminate and increases the stress intensity in every remaining lamina. If the originally applied stress level is close to the ultimate stress of the laminate (80 percent or more), a rapid failure occurs in the measuring fibers and matrix.

If the applied stress is less than 50 percent of the ultimate, a new condition of equilibrium between the remaining laminae will result, and no failure will occur. In case more than 50 percent of the original number of laminae fail simultaneously upon loading, a catastrophic failure in the tested specimen results. This creep rupture mechanism has been observed clearly for two different laminae orientations.

In this type of creep failure, one of the laminae will have fiber-matrix bond failure, and the rest of the laminae fibers are either pulled out or broken due to tensile stresses in the fibers.

2. If the volume percent of the fibers is less than 50 percent, the spacing between any two consecutive fibers is usually larger than the diameter of the fiber. A matrix failure also will occur in this case, but not necessarily at the fiber-matrix interface. Failure between two consecutive fibers would occur through inelastic deformation in the matrix which is accompanied by a crack initiation and propagation parallel to the fibers in the outer lamina. This eliminates the carrying ability of this lamina, and the load will be redistributed among the remaining laminae.

The rest of the creep rutpure mechanism is similar to that stated in Item 1.

CONCLUSIONS

Time-dependent effects have been observed clearly in stressed graphite composites. The degree to which this effect can influence the design with these types of comparison depends greatly on the particular state of fiber orientation with respect to the direction of the applied load. The following conclusions and observations are for biaxially reinforced laminates, where the effects of different parameters such as fiber orientation, stress level, and test temperature are summarized briefly.

1. In all tested specimens, at least two stages of creep were observed easily: primary stage and steady state.

2. The tertiary stage of creep was noticed in specimens stressed under stress levels equal to 95 percent of its ultimate.

3. After 500 h of loading time, all unbroken specimens indicated some irrecoverable creep strain, which can be attributed to a permanent deformation in the matrix phase of the composite.

4. Laminae orientation effect: (a) The fiber orientation in the composite is the most influential factor in determining the creep characteristics of the composite system. If fibers are in the load direction, the accumulated creep strain is very small as compared with the instantaneous strain, and a negligible permanent strain will remain after unloading. (b) For (30 deg, −60 deg) laminates, a significant increase in the longitudinal and transverse strains was observed. The creep strain is even more pronounced in the case of the (45 deg, −45 deg) laminates.

5. Stress-level effects (a) The laminae orientation has a pronounced effect on the nonlinearity of the stress-strain relation of these composites. In uniaxial or biaxial laminae, where some of the fibers are aligned in the direction of the load, a linear stress-strain relation is well expected. In (30 deg, −60 deg) and (45 deg, −45 deg) laminates, the nonlinearity of the load-strain relation was obvious. Any external thermal energy supplied to the tested specimen magnified the degree of nonlinearity in the constitutive relations of this composite. Any slight increase in the stress level affects the creep and instanteous strains significantly. (b) Based on these limited experimental results, one can observe a linear relationship between the level of the applied stress and the logarithm of the fracture time. This relation holds for different temperature levels, at least for the three levels tested: 85, 160, and 200 F. At every specified temperature level, the linear relationship between the stress level and the fracture time has a slope different from that at another temperature. Such a relation has been shown to hold also for glass-fiber-reinforced composites. This relation suggests a dependence of the fracture time under constant load and temperature on the energy level in the system. The concept can be used to extrapolate the creep rupture time of this composite at low stress levels from the results obtained at elevated temperature and high stresses. A significant testing time can be eliminated if enough data are generated to guarantee the existence of this relation and to evaluate the different necessary parameters.

6. Temperature effect: (a) The rise in test temperature changes the character of the bonding matrix in the composite and thus improves its ductility. This, in

turn, increases the rate of straining of the overall composite and decreases its creep rupture time. Also, at temperatures higher than 85 F, the fracture time decreases significantly. (*b*) As the temperature decreases, the ductility of the resin system decreases, and a brittle fracture occurs through crazing of the resin system or failure of fibers.

7. Creep rupture mechanism: As expected, the dominating creep rupture mechanism depends on the laminae orientation, temperature, and aspect ratio of the reinforcing fibers, as well as the ductility of the bonding matrix. It is influenced also by the volume percentage of the constituents and the state of stress in the matrix material during the course of loading.

OBSERVATIONS

Failure strain under long-term loading is greater than that obtained during short-time testing for strength and stiffness.

If a sufficiently long time is allocated for a specimen to fail under a low stress level, the failure strain is expected to be higher than that obtained under a high stress level.

At a low temperature level, the scatter of the data was wider than that for high-temperature testing under the same stress level. This can be attributed to the ductility of the resin system.

ACKNOWLEDGMENTS

This study was performed under a Lockheed-Georgia Co. research program concerned with composite structures.

The author wishes to express his thanks to R. Waugh for supplying the laminates, J. Lowry for conducting the tests, and J. H. Sams, III, for his encouragement during the preparation of the paper. The author's conclusions are based on these qualitative experimental results. More data, however, are needed before a quantitative result could be obtained.

References

[1] Steel, D. J., "The Creep and Stress-Rupture of Reinforced Plastics," *Transactions of the Journal of Plastics,* Oct. 1965, pp. 161–167.

[2] Findley, W. N., "Creep and Relaxation of Plastics," *Machine Design,* May 1960, pp. 205–208.

[3] Pagano, N. and Halpin, J. "Influence of End Constraint in the Testing of Anisotropic Bodies," *Journal of Composite Materials,* Vol. 2, No. 1, Jan. 1968, pp. 18–31.

[4] DeSilva, A. R. T. "A Theoretical Analysis of Creep in Fibre Reinforced Composites," *Journal of the Mechanics and Physics of Solids,* Vol. 16, 1968, pp. 169–186.

[5] Dean, A. V., "The Reinforcement of Nickel-Base Alloys with High-Strength Tungsten Wires," *Journal of the Institute of Metals,* Vol. 95, March 1967, pp. 79–86.

[6] Mullin, J., Berry, J. M., and Gatti, A., "Some Fundamental Fracture Mechanisms Applicable to Advanced Filament Reinforced Composites," *Journal of Composite Materials,* Vol. 2, No. 1, 1968, pp. 82–103.

[7] Tyson, W. R. and Davies, G. I., "A Photoelastic Study of the Shear Stresses Associated with the Transfer of Stress During Fiber Reinforcement," *British Journal of Applied Physics,* Vol. 16, 1965, pp. 199–205.

[8] Baer, E., *Engineering Design for Plastics,* Reinhold, New York, 1964.

[9] Coleman, B. D., "A Stochastic Process Model for Mechanical Breakdown," *Transactions of the Society of Rheology,* Vol. 1, 1957, pp. 153–168.

Characterizing Strength of Unidirectional Composites

O. ISHAI[1] AND
R. E. LAVENGOOD[2]

Reference:

Ishai, O. and Lavengood, R. E., "Characterizing Strength of Unidirectional Composites," *Composite Materials: Testing and Design*, ASTM STP 460, American Society for Testing and Materials, 1969, pp. 271–281.

Abstract:

This study was made to determine the effect of test variables on the strength of unidirectional composites. Specimens of glass fiber reinforced epoxy of varying geometry were tested in tension and flexure at several strain rates. Both experimental and theoretical results show that specimen geometry strongly influences strength measurements for off-axis specimens having a small length-to-width ratio. This is a result of nonsymmetrical anisotropy in the specimens which causes twisting in flexure tests and shear coupling in tension tests. For slender specimens, having a free length-to-width ratio greater than six, both strength and stiffness measurements are almost unaffected by the boundary conditions of the test. A normalization procedure was used successfully to relate the strength of an off-axis composite to its transverse strength. While changes in loading mode, strain rate, and specimen geometry may produce significant changes in absolute strength values, normalized strengths proved to be essentially invariant.

Key Words:

composite materials, fiber, composites, mechanical properties, glass fibers, epoxy laminates, evaluation, tests

The way in which test results for composites are interpreted generally depends on whether the objective is a qualitative comparison of materials or a study of the absolute mechanical properties. Qualitative comparisons are fairly straightforward since it is usually only necessary to keep the test conditions constant throughout the testing program. Determination of absolute values may be more difficult but is often necessary, for instance, when relating basic constituent properties to macromechanical behavior or providing data for structural design. Of course, it is implicit in both types of testing that procedures, as well as methods of interpretation, must be consistent and reliable.

One of the requirements for reliability is that the measured properties be independent of the geometry of the specimen and loading mode. Unfortunately, this is often not the case with anisotropic materials. As a result, experimental data obtained from different sources but for the same composite system frequently show marked differences. Indeed, much of the data on composite properties in the literature must be considered questionable due to attempts to test composites as if they were isotropic and homogeneous.

In recognition of this problem, considerable effort has been spent in the Monsanto/Washington University ONR/ARPA Project on "High Performance Composites" on systematic characterization of composites and on determining the effect

[1] Visiting associate professor, Washington University, St. Louis, Mo., on leave from the Technion, Israel Institute of Technology. Personal member ASTM.
[2] Materials engineer, Monsanto Co., St. Louis, Mo. 63166.

of test variables on composite properties. Particular emphasis has been placed on testing polymeric composites which are reinforced unidirectionally with continuous and discontinuous fibers [1–3].[3] As a result, a substantial amount of data has been accumulated on the relations between ultimate stress and such variables as fiber orientation, matrix properties, strain rate, and temperature.

The generality of these relationships has been checked in different testing methods and compared with existing theories in the field of composite mechanics. Results have verified that improper experimental procedure or interpretation can lead to significant errors. The primary objective of this work is to point up some of the major factors which may interfere with strength characterization and lead to inaccurate analysis. In addition, this paper also attempts to show that with proper test procedures and data reduction, the effects of these factors can be reduced significantly, often to a negligible level.

EXPERIMENTAL PROCEDURE

Specimens were cut from unidirectionally reinforced filament wound plates of E-glass and epoxy. The roving used in preparing the plates, PPG 1062-T-6, was treated with an epoxy compatible coupler by the manufacturer. Two different epoxy matrices were used; the first, Epon-828 plus 20 parts per hundred resin (phr) Shell Curing Agent Z, has a linear stress-strain relationship and will be referred to as a "brittle matrix" in order to simplify notation, while the second system, Epon-815 plus 67 phr Versamide-140 has a distinct yield point preceding fracture will be referred to as a "ductile matrix." This notational system is based on the

[3] The italic numbers in brackets refer to the list of references appended to this paper.

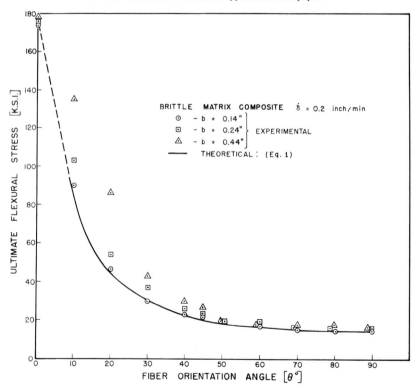

FIG. 1—The effect of fiber orientation on flexural strength for brittle matrix composite beams of different geometries.

deformational behavior of the unfilled resins and does not refer to the characteristics of the composites.

A diamond saw was used to cut the specimens. In all cases, the angle of the fibers relative to the axis of the specimen was controlled to ±1 deg. Flexure specimens were 3 in. long (for a 2-in. span test), ⅛ in. thick and had widths ranging from ⅛ to ½ in. All tension specimens were 6 in. long with a 3 in. gage length and had glass cloth tabs bonded to each end. Specimens with an orientation angle greater than 30 deg were ⅛ in. thick and shaped according to ASTM D 638-64T. The others were straight-sided strips ½₀ in. thick and ¼ in. wide. Tension and flexure tests were run on an Instron tester at constant crosshead speeds ranging from 0.0002 to 2.0 in./min. Test temperature was 23 ± 1 C with a relative humidity of 50 ± 5 percent.

RESULTS AND DISCUSSIONS

The influence of specimen geometry and mode of loading on strength and stiffness measurements was determined as a function of fiber orientation. Regardless of

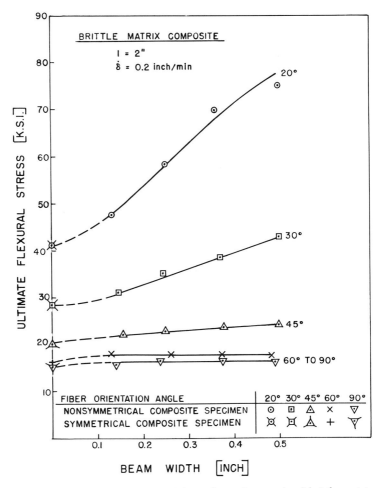

FIG. 2—The effect of beam width on flexural strength of brittle matrix composites at different fiber orientations.

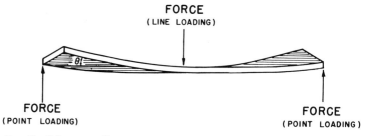

FORCE
(LINE LOADING)

FORCE
(POINT LOADING)

FORCE
(POINT LOADING)

FIG. 3—Schematic illustration of a nonsymmetrical wide beam twisting under flexural loading.

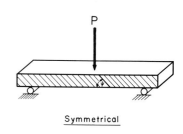

P

Symmetrical

FIG. 4—Schematic illustration of symmetrical and nonsymmetrical beam positions for flexural loading of slender specimens.

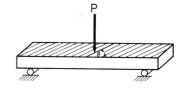

P

Non-Symmetrical

the test conditions involved, flexural and tensile strengths, of course, are strongly dependent on fiber orientation [2–5]. Both are maximum at $\theta = 0$ deg and decrease rapidly in the orientation range of 0 deg $< \theta <$ 10 deg. Part of this abrupt decrease results from a change in failure mode. In the vicinity of $\theta = 0$ deg, the composites fail by tensile fracture of the fibers, while for $\theta \geq 10$ deg an interfiber cleavage failure occurs. The decrease in strength with increasing orientation angle is illustrated in Fig. 1.

SPECIMEN GEOMETRY

Flexural strength[4] also decreases significantly with the beam width in the off-axis orientation range of 10 deg $< \theta <$ 45 deg (Fig. 1). This effect, illustrated in Fig. 2, is more pronounced for brittle matrix composites and is maximum at $\theta = 20$ deg for both systems. This dependence of strength on beam width results from the lack of symmetry in the beam with respect to the loading axis. This gives rise to a twisting distortion (Fig. 3) and results in combined torsional and bending moments in the midsection of the beam.

Beams were also tested in which the fibers were symmetrical with the loading axis. This was achieved by using a series of the ⅛ by ⅛-in. beams and simply rotating each beam 90 deg about its axis (Fig. 4). The beams in this series cor-

---

[4] Here, flexural strength is calculated by the simple beam theory.

respond to nonsymmetrical beams of zero width in that there is no twisting moment. In Fig. 2, it is shown that the curves for strength versus width for nonsymmetrical beams can be extrapolated to the symmetrical case.

A simplified analysis of the resulting state of stress on the failure plane is given in the Appendix. This analysis shows that the stresses in the midsection of the beam increase with decreasing specimen width and predicts the effect to be greatest in the orientation region of 10 deg $< \theta <$ 30 deg. This agrees well with experimental results.

The effect of specimen geometry on tensile strength also was investigated. In this case a nonuniform state of stress might result from shear coupling which induces shearing forces and bending couples at the ends of the specimen [6]. The magnitude of this effect is strongly dependent on the length-to-width ratio as well as the modular anisotropy ratio and orientation angle of the specimen [7,8]. However, the extent and seriousness of the effect on testing has not been determined previously for the type of specimens used here. Interestingly, with length-to-width ratios of six and twelve and orientation angles of 45 and 60 deg, no significant variation in strength was found. This is reasonable considering that the anisotropy ratio of this composite is less than four.

MODE OF LOADING

Tension Testing

According to anisotropic elasticity theory, the effects of nonsymmetrical anisotropy should be the greatest when a fixed gripping system is used. Such gripping induces local shear forces and bending moments by constraining shear distortions. The possibility of reducing the shear coupling effect by allowing more rotational freedom at the grips was investigated by testing nonsymmetrical tension specimens ($l/d \geq 6$) at various angles with both fixed grips and hinged grips (Fig. 5a and b). The

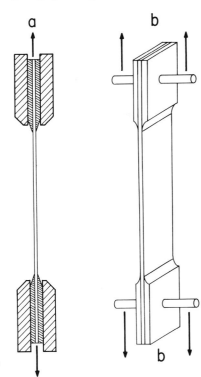

(a) Fixed grips.
(b) Hinged grips.

Fig. 5—Illustration of tension composite specimens used to investigate the effect of boundary conditions on shear coupling effects.

Table 1—*The effect of boundary gripping conditions on tensile strength of off-axis unidirectional composites.* ($1/d \geq 6$), *tensile strength in ksi.*

Type of Matrix	Gripping Conditions	Fiber Orientation Angle θ			
		60 deg	45 deg	30 deg	20 deg
Brittle..........	fixed	8.5	11.0	15.5	25.2
Brittle..........	hinged	8.1	10.7	16.7	24.0
Ductile..........	fixed	8.9	9.6	10.7	18.2
Ductile..........	hinged	8.7	9.6	11.3	17.7

results, tabulated in Table 1, show very small, probably insignificant, differences for the two gripping techniques.

Flexure Testing

The interpretation of flexural strength data is always questionable even for conventional isotropic homogeneous materials. No direct relationship is available to predict tensile strength values from ultimate flexure moment and vice versa. In many cases, however, flexure tests are the only practical method for mechanical characterization; thus, understanding the effect of basic variables such as orientation and strain rate on flexural parameters is of practical importance.

In a previous study [2], unidirectional composite strength versus fiber orientation was shown to be consistent with a failure criterion based on the maximum tensile traction acting on the interfiber failure plane. Only one material parameter is needed to define this relationship in the orientation range where brittle interfiber failure occurs. The most convenient parameter is transverse strength. Thus, strength at a specific angle can be normalized as follows:

$$\sigma_u(\theta)/\sigma_u 90 \text{ deg} = M_u(\theta)/M_u 90 \text{ deg} = 1/\sin\theta \dots\dots\dots\dots\dots (1)$$

where:

$\sigma_u(\theta)$ = ultimate uniaxial tensile stress acting on the composite externally,
$M_u(\theta)$ = ultimate maximum bending moment acting on composite beam,
$\sigma_u 90$ deg = tensile strength of transverse specimen ($\theta = 90$ deg),
$M_u 90$ deg = ultimate flexural moment of transverse specimen, and
θ = fiber orientation angle.

While the absolute values of tensile and flexural strength vary by a factor of about two, the normalized strengths are the same (Fig. 6). This illustrates that the normalized strengths, are not affected by the testing technique or loading mode, provided the interfiber tensile failure mode prevails.

The general applicability of this failure criterion and normalization technique was also checked with data available in the literature. Figure 7 shows that Eq 1 is not restricted to glass-epoxy composites and can be used equally well with metal matrix composites tested in uniaxial tension.

CONCLUSIONS

The following conclusions are based on experimental data for epoxy-glass composite specimens of various geometries which were tested under different loading conditions.

1. In off-axis flexure tests, specimen geometry strongly influences apparent strength. This is attributed to twisting moments, induced by shear coupling effects, which are superimposed on the bending moments. This effect can be reduced

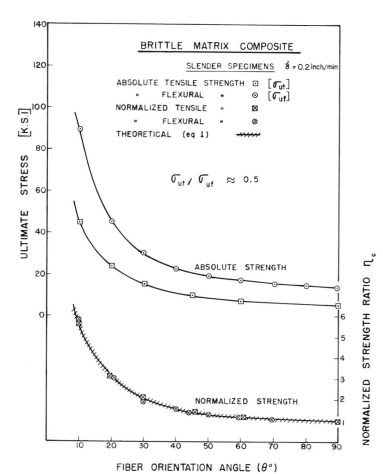

FIBER ORIENTATION ANGLE $(\theta°)$

FIG. 6—The absolute and normalized strength versus orientation relationships for the brittle matrix composite in flexural and tensile loading modes.

by increasing the free length-to-width ratio and is negligible for slender beams.

2. Boundary conditions of a test may have a significant effect on stress and strain distribution within a tensile composite, however. It was found that the ultimate tensile stress of slender specimens with low longitudinal-to-transverse modular ratio is practically the same with both fixed and hinged gripping conditions.

3. Generally speaking, by careful design of testing procedures, one can reliably characterize composites in flexure and tension tests if the specimens are slender. In this case even flexural data of nonsymmetrical off-axis specimens can yield fundamental relationships which are useful for predicting strength in other loading modes or test conditions.

ACKNOWLEDGMENTS

The authors wish to express their gratitude to T. L. Tolbert for assistance in the preparation of this paper and to J. J. Cornell and C. N. Rasnick for assistance in the experimental phase of this study.

The work described in this paper is part of the research conducted by the Monsanto/Washington University Association sponsored by the Advanced Re-

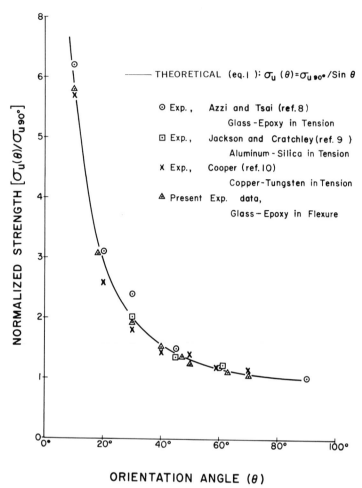

$$\text{THEORETICAL (eq.1): } \sigma_u(\theta) = \sigma_{u\,90°}/\sin\theta$$

⊙ Exp.,   Azzi and Tsai (ref.8)
            Glass-Epoxy in Tension

⊡ Exp.,   Jackson and Cratchley(ref. 9)
            Aluminum-Silica in Tension

X Exp.,   Cooper (ref.10)
            Copper-Tungsten in Tension

△ Present Exp. data,
            Glass-Epoxy in Flexure

NORMALIZED STRENGTH $[\sigma_u(\theta)/\sigma_{u\,90°}]$

ORIENTATION ANGLE (θ)

Fig. 7—An illustration of the general utility of Eq 1.

search Projects Agency, Department of Defense, under Office of Naval Research contract N00014-67-C-0218.

Appendix I

ANALYSIS OF ULTIMATE STRESSES FOR NONSYMMETRICAL UNIDIRECTIONAL COMPOSITE BEAMS

Consider an off-axis unidirectional composite beam, positioned nonsymmetrically with respect to the beam's major axis, X. The dimensions l, b, and h are span, width, and thickness, respectively; and the fibers are aligned at an angle θ to the major axis (Fig. 8).

In three-point flexural loading, when the beam twists as described earlier, the reaction at the supports, R, is a distance e from the center line. The moment

278   *Composite Materials*

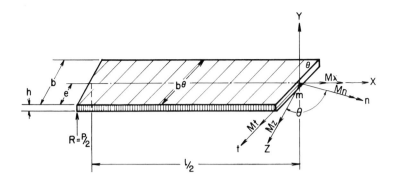

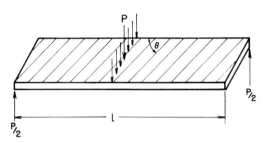

FIG. 8—Illustration of non-symmetrical flexure specimen and the internal moments acting on the failure plane.

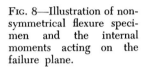

vestors acting at the midspan of the beam along X and Z axis are M_x and M_z and are related to the force P as follows:

$$\left.\begin{array}{l} M_x = Pe/2 \\ M_z = Pl/4 \end{array}\right\} \quad \dots\dots\dots\dots\dots\dots\dots\dots\dots\dots (2)$$

Their tangential and perpendicular projections on the fiber orientation plane, M_t and M_n, can be superimposed as follows:

$$\left.\begin{array}{l} M_t = M_z \sin \theta - M_x \cos \theta \\ M_n = M_z \cos \theta + M_x \sin \theta \end{array}\right\} \quad \dots\dots\dots\dots\dots\dots\dots\dots (3)$$

Assuming uniform stress distribution along the fiber orientation plane[5] (t taxis), the two major stresses, normal and shear, can be related to the bending moment M_t and the torque M_n. The critical combination of the two major stresses which initiates failure is assumed to be located at the lower midpoint, m, at the midsection of the beam (Fig. 8 (*top*)). The maximum values of the major normal stress, σ_θ, and the shear stress, τ_θ, at this point can be derived using the simple beam theory as follows:

$$\sigma_\theta = 6M_t/b_\theta h^2 \dots\dots\dots\dots\dots\dots\dots\dots\dots\dots\dots (4)$$
$$\tau_\theta = M_n/\alpha b_\theta h \dots\dots\dots\dots\dots\dots\dots\dots\dots\dots\dots (5)$$

where $b_\theta = b/\sin \theta$ is the beam width along fiber orientation and α is the torque constant for rectangular bars.

Substituting Eqs 2 and 3 into 4 and 5 gives:

$$\sigma_\theta = 6P/b_\theta h[(l/4) \sin \theta - (e/2) \cos \theta] \dots\dots\dots\dots\dots\dots (6)$$
$$\tau_\theta = P/\alpha b_\theta h[(l/4) \cos \theta + (e/2) \sin \theta] \dots\dots\dots\dots\dots\dots (7)$$

The failure criterion for unidirectional composites discussed in the text, implies that brittle interfiber failure initiates when the maximum tensile traction[6] exceeds

---

[5] This assumption becomes less justified in case of wide beams and small off-axis angles ($\theta < 30$ deg) where the moment vector M_z is a function of the coordinate t. However, this deviation from uniformity is small for slender beams ($l/b > 10$).

[6] The resultant of the normal and shear stresses.

Ishai and Lavengood on Unidirectional Composites 279

Table 2—Comparison between theoretical and experimental factor C for nonsymmetrical and symmetrical cases.

Fiber orientation θ.	20 deg	20 deg	30 deg	30 deg	45 deg	45 deg	60 deg	60 deg	90 deg
Beam's width $b = 2e$	0.5 in.	0.14 in.	0.5 in.	0.14 in.	0.5 in.	0.14 in.	0.5 in.	0.14 in.	$e = 0$
Actual beam's width $b\theta$.	1.46 in.	0.41 in.	1.0 in.	0.28 in.	0.71 in.	0.20 in.	0.58 in.	0.16 in.	0.5
Torque constant α..	0.33	0.27	0.33	0.25	-0.295	0.235	0.285	0.22	0.33
Theoretical C	1.92	1.46	1.79	1.33	1.40	1.17	1.17	1.06	1
Experimental C (brittle)	1.98	1.20	1.54	1.11	1.20	1.10	1.09	1.06	1.05
Experimental (ductile)	2.12	1.10	1.80	1.10	1.31	1.10	1.10	1.03	1.06

a specific limit, S_u, which is a constant material parameter for a given composite system [3]. According to this concept the following relationship exists at failure:

$$S_u^2 = \sigma_{u\theta}^2 + \tau_{u\theta}^2 = \text{constant} \quad \text{.....................(8)}$$

For the symmetrical beam:

$$S_u^2 = \sigma_u^2(\theta) \sin \theta = [1.5P_u^0 l/bh^2]^2 \sin^2 \theta \quad \text{..................(9a)}$$

whereas for the nonsymmetrical case

$$S_u^2 = [(6P_u/b_\theta h^2)(l \sin \theta/4 - e \cos \theta/2)]^2 + [(P_u/\alpha b_\theta h^2)(l \cos \theta/4 + e \sin \theta/2)]^2 \text{..}(9b)$$

In these equations, $\sigma_u(\theta)$ is the maximum composite tensile stress at failure calculated according to the simple beam theory, and P_u^0 and P_u are the ultimate central loads for the symmetrical and nonsymmetrical cases, respectively. The ratio of the measured ultimate flexural load of a nonsymmetrical beam to that expected for the symmetrical case can be defined as the correction factor C. This is a measure of the error caused by using the simple beam theory to calcuate the ultimate stress of nonsymmetrical beams. The correction factor C can be derived by equating Eqs 8 and 9 and rearranging:

$$1/C^2 = (P_u^0/P_u)^2 = [\sin \theta - 2(e/l) \cos \theta]^2 + \tfrac{1}{36}\alpha^2[\cos \theta + 2(e/l) \sin \theta] \quad (10)$$

Comparison of the theoretical values of C calculated with Eq 10 and experimental C values obtained from Fig. 2 are given in Table 2. Here, it is assumed, based on observation, that $e = b/2$. It is apparent that the general trends shown in Figs. 1 and 2 are confirmed by the theoretical analysis.

References

[1] Anderson, R. M. and Lavengood, R. E., "Variables Affecting Strength and Modulus of Short Fiber Composites," Journal of the Society of Plastics Engineers, Vol. 24, 1968, p. 20.

[2] Ishai, O., Anderson, R. M., and Lavengood, R. E., "Failure-Time Characteristics of Continuous Unidirectional Glass Epoxy Composites in Flexure," Monsanto/Washington University Association Interim Report, Dec. 1967, AD 827065.

[3] Ishai, O. and Lavengood, R. E., "Tensile Characteristics of Dis-

continuous Unidirectional Glass-Epoxy Composites," 24th Annual Technical Conference, 1969, Reinforced Plastics/Composites Division, Society of the Plastics Industry, Section 11-F.

[4] Chen, P. E. and Lin, J. M., "Transverse Properties of Fibrous Composites," Monsanto/Washington University Association, AD 840592.

[5] Ishai, O., "The Effect of Temperature on the Delayed Yield and Failure of 'Plasticized' Epoxy Resin," Polymer Engineer-

ing and Science, Vol. 9, March 1969, p. 131.

[6] Tsai, S. W., "Mechanics of Composites," Technical Report AFML-TR-66-149, Part I, Air Force Materials Laboratory, 1966.

[7] Pagano, N. J. and Halpin, J. C., "Influence of End Constraint in the Testing of Anisotropic Bodies," *Journal of Composite Materials,* Vol. 2, 1968, p. 18.

[8] Azzi, D. and Tsai, S. W., *Experimental Mechanics,* Vol. 5, 1965, p. 283.

[9] Jackson, P. W. and Crathley, D., *Journal of the Mechanics and Physics of Solids,* Vol. 15, 1967, p. 279.

[10] Cooper, G. A., *Journal of the Mechanics and Physics of Solids,* Vol. 14, 1966, p. 103.

Design and Application

Practical Influence of Fibrous Reinforced Composites in Aircraft Structural Design

H. C. SCHJELDERUP[1] AND B. H. JONES[1]

Reference:

Schjelderup, H. C. and Jones, B. H., "Practical Influence of Fibrous Reinforced Composites in Aircraft Structural Design," *Composite Materials: Testing and Design, ASTM STP 460*, American Society for Testing and Materials, 1969, pp. 285–306.

Abstract:

Practical design problems which have become evident as a result of current DOD-industry composite development programs are discussed.

Design problems related to residual stresses between matrix and filaments are reviewed, and the significance of thermal effects on the interface between filament and matrix is out-lined. Local failure and strength properties, as influenced by filament diameter, and the ever-present problem of interlaminar shear are studied.

The concept of balanced and minimum gage laminates is considered together with its interaction with design. The influence of design criteria, allowables, and joints on composite structures design is discussed. Detail design problems related to thermoelasticity and fatigue highlight some of the potential design problems which are unique to composites.

As experience in the design of composite structures is gained, many requirements for the future evolve. The requirements are basically a need for uniformity of practice in order for composites to become cost competitive with conventional materials. Suggestions and recommendations for future standards in composite design are presented.

Key Words:

composite materials, fiber composites, reinforced plastics, joints (junctions), aircraft structure, design, stress, tests, evolution

Great progress in the design of high-modulus filament composite structures has been made in the last few years as a result of a combined DOD-industry effort. However, with this progress some of the initial potential has been reduced by the influence of practical problems. In spite of these influences many successful structural components demonstrating significant weight savings have been designed and tested; among these components are the following:

1. General Dynamics F-111 horizontal tail.
2. Grumman FB-111 wing tip extension.
3. McDonnell-Douglas F-4 rudder and A-4 flaps.

Composite structures are more difficult to design than their conventional metallic counterparts, but the effort required to achieve success is well worthwhile. Experienced aircraft structural designers will agree that once the initial configuration and structural materials are determined, the foundation of a successful final design lies in the detailing. This means that the designer must appreciate the fundamentals of composite materials mechanics if the final design is to be successful.

The objective of this paper is to identify some of the different aspects of composite design and to bring them to the attention of the designer. Simple design formulas

[1] Director, Structures/Materials Research, and senior engineer/scientist, respectively, Douglas Aircraft Co. McDonnell Douglas Corp., Long Beach, Calif. 90801.

and data are not readily available for high-modulus composites. Much of the design data on composites is complicated and requires considerable study to master its applications.

The answers to all of the problems that concern designers are not available. However, these problems may be identified and discussed even through experimental data are not available to support some observations.

This paper begins by discussing the high-modulus composite laminate itself at a micromechanical scale in order to build the foundation for a discussion of some of the important design considerations relevant to composites. The fundamental problem of assembling the basic laminates into a semihomogeneous material then is reviewed in terms of balanced construction and minimum gage. A discussion of detail design problems related to joints, thermoelasticity, and fatigue highlight some of the potential design problems with composites. Finally, a few of the more important design areas requiring future development are summarized.

SOME MICROMECHANICAL INFLUENCES ON DESIGN

The objective of a completely integrated design procedure is to provide response predictions for multilaminated composites from the properties of the filament and matrix constituents. Composite fundamental strength properties are defined conveniently in terms of five principal strength: longitudinal tensile strength (F_{L_t}), longitudinal compressive strength (F_{L_c}), transverse tensile strength (F_{T_t}), transverse compressive strength (F_{T_c}), and the longitudinal transverse shear strength (F_{LT}).

LONGITUDINAL TENSILE AND COMPRESSIVE STRENGTH

Of these principal strengths, only one, longitudinal tension, can be predicted with any degree of accuracy [1].[2] The problem of predicting the longitudinal compressive strength has received a great deal of attention but has not been solved satisfactorily. Predictions of the compressive strength of glass-reinforced composites have been generally in error by a factor of three or more. Rosen [2] analyzed local buckling for two modes of local fiber interaction using an energy method. Two possible modes of failure were postulated:

1. In-phase (shear mode).
2. Out-of-phase (extensional).

These failures result in the two following critical stresses, the smaller of these dictating failure:

$$\sigma_c = 2\nu_f \left[\frac{\nu_f E_m E_f}{3(1 - \nu_f)} \right]^{1/2} \dots\dots\dots\dots\dots\dots\dots\dots (1)$$

$$\sigma_c = \frac{G_m}{1 - \nu_f} \dots\dots\dots\dots\dots\dots\dots\dots\dots\dots\dots\dots\dots\dots (2)$$

For most usable filament fractions, the extensional mode may be considered to predominate. From the foregoing equations it may be observed that, if the theoretical approach is justified, the diameter of the filament has no direct influence on compressive strength. It may be observed that, according to Eqs 1 and 2, the compressive strengths of boron and graphite composites should be approximately the same if the filament moduli are equal. This is not currently the case (Fig. 1).

The reasons for this inconsistency are varied. First, it seems certain that the mechanism of compressive failure is connected strongly to filament-matrix debonding rather than just matrix yielding. The experimental correlations between compressive strength and interlaminar shear strength (Fig. 2) significantly support the debonding hypothesis.

---

[2] The italic numbers in brackets refer to the list of references appended to this paper.

Filament Collimation and Compressive Strength

Another predominating factor on compressive strength may be the lack of collimation of the filaments. Composite materials, such as boron and certain types of S-glass, which utilize highly collimated filaments show remarkable compressive strength. For instance, boron can sustain up to 440,000 psi [3] with the filaments oriented in the 0-deg direction, and S-glass having a modulus some four times less than boron has shown compressive strengths up to 200,000 psi [2]. The use of graphite filaments, which in some cases have significant amounts of twist built into them, may limit the compressive strength attainable with this material.

Recently [4] the problem of buckling of a wire embedded in an elastic matrix has been considered together with the beam column behavior of an initially nonstraight wire embedded in a matrix. The type of result obtained is shown in Fig. 3 and clearly demonstrates the dependence of collimation on compressive response. It may be concluded that only for extremely small values (less than 0.0001) of the shear-modulus ratio (G_m/G_f) is there any possibility of an initially straight

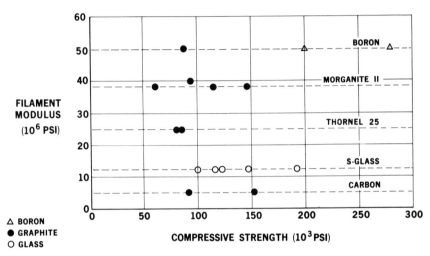

Fig. 1—Variation of composite compressive strength.

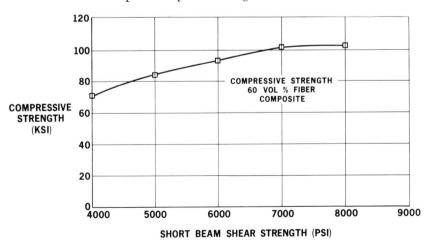

Fig. 2—Strength relationship of 0-deg graphite composite.

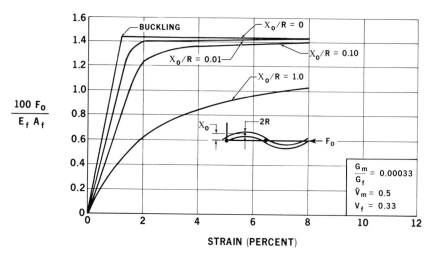

FIG. 3—Effect of filament collimation on compressive behavior of composite.

reinforcing wire buckling within the matrix. For most graphite/epoxy composites, the shear-modulus ratio is about 0.027. The observed differences in compressive and tensile behavior of certain filament reinforced composites may be explained also in terms of the beam-column behavior of the initially crooked reinforcement.

Another approach to the problem of the effect of twist on composite properties was made by Whitney [5]. This analysis indicated that the elastic modulus would not be affected significantly until the number of turns per inch were in excess of five. A consideration of the effect of twist on tensile strength suggested that the potential strength was likely to be reduced significantly if the number of twists per inch exceeded seven.

TRANSVERSE TENSION AND COMPRESSION

The problem of predicting the transverse plastic moduli of filamentary composites has been investigated extensively, and it is not intended to dwell on the results here (for a detailed bibliography, see Ref 6). However, although results derived on the assumption of idealized filament arrays are useful for design purposes, the distinctly nonuniform arrangements of filaments found in all practical composites suggest that a better approach would use results which bound the moduli, as advocated by Hashin and Rosen [7].

Analytical determinations of transverse strength have not attracted so much attention mainly because of the extreme complexity of the problem. It is apparent that, only when the factors which influence strength have been analyzed, will it be possible to direct research toward improving this property. The difficulties arise because of:

1. The complex nature of the stress field around the filaments.
2. The obvious difficulties of defining the exact mechanism which precipitates failure under the complex stress field.

At this time, a number of fundamental characteristics concerning the strength behavior of filamentary composites remains to be established. The extent to which behavior under transverse tension is related to that under transverse compression is obscure. Shu and Rosen [8] used limit analysis methods of plasticity and suggested that the matrix yield strength was a lower bound on strength. This is attained infrequently in many composites, but the correlation may be well obscured by the influence of voids. Obviously, a great deal of interesting work can be conducted in this area.

Composite Materials

Despite the significance of the longitudinal shear strength in design, prediction of this quantity has received little attention. The comments made above concerning the difficulties of achieving a rigorous solution to the transverse loading case are equally valid here. Shu and Rosen [8] also have considered this problem using limit-load methods of plasticity, but the bounds generally are separated too widely to be currently of value in design. The analysis also assumes an ideally plastic matrix which may be restrictive in many stituations.

Interlaminar Shear Characteristics and Joint Design

While the results of interlaminar shear tests are generally only of value for qualitative purposes, there is one area where the results are particularly relevant for design, and this occurs in the analysis of a composite joint.

An analytical determination of the interlaminar shear behavior of a general composite consisting of multioriented plies has not been made. Foye [9], however, considered a procedure for determining the longitudinal-transverse shear behavior of an assumed square array, for a matrix material described by the Ramberg-Osgood equation. The method employed a nine-node discrete element technique. From the analysis, the significant fact emerged that the inelastic elements exhibited no evidence of unloading with increasing stress level. While the results obtained reflected all the characteristics of the assumed bulk behavior of the matrix material, no comparison was made with experiment. The relevance of an analysis of longitudinal-transverse, stress-strain behavior to interlaminar shear arises from the equivalence of such behavior for the two modes if the array is completely random or symmetrically ordered and the filaments are parallel to the direction of the shear stresses, that is, the material is orthotropic (Fig. 4). For such a case, it is reasonable to assume that

$$G_{YZ} = G_{XZ} \neq G_{YX}$$

The preceding equation suggests that, for the particular case of a unidirectionally oriented composite, more precise data could be obtained from a longitudinal-trans-

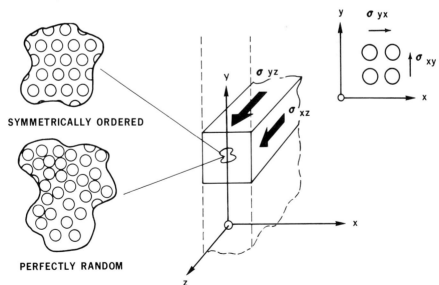

SYMMETRICALLY ORDERED

PERFECTLY RANDOM

FIG. 4—Shearing modes in unidirectional composite.

Schjelderup and Jones on Aircraft Design

verse shear test than the more conventional short-beam shear test. Such a procedure would not be applicable if the filaments show any type of preferential spacing with direction as frequently occurs in tape-laying operations or where a "scrim cloth" exists.

INFLUENCE OF THERMAL EFFECTS

Thermal residual stresses may be set up in the filament and matrix if differences in the coefficients of thermal expansion exist between the two and temperature changes occur after solidification of the matrix, either during fabrication or when the composite is in service. Generally it is considered that the stresses which are set up from the volumetric changes which occur during matrix solidification are insignificant because of the relaxation mechanism present in polymeric [10] and metal matrix systems. A measure of the micromechanical stresses which are set at the interface between the filament and the matrix is indicated in Fig. 5. It appears that the micromechanical stresses which are likely to be set up in graphite, boron, or glass composites utilizing epoxy matrices do not differ significantly. This is due to the fact that the coefficient of thermal expansion of the matrix system is so much greater than that for the filament. The influence of filament volume fraction on micromechanical residual stresses can be observed also.

SOME MACROMECHANICAL INFLUENCES ON DESIGN

In designing a composite sheet or plate, it is necessary to define the following parameters:
1. Orientation of the layers (to produce the desired elastic modulus).
2. Number of layers (to produce desired strength).
The number of layers is influenced strongly by the thickness per layer but should account also for the necessity of having to attain a "balanced configuration." The objective is to prevent distortion of the laminate both in its own plane and perpendicular to it under thermal or mechanical loading.

BALANCED AND UNBALANCED CONSTRUCTION

For in-plane balance the requirement is simply that any one layer has a mirror image with reference to the line of action of any external load which acts in the plane of the sheet. Under such conditions, neither mechanical nor thermal

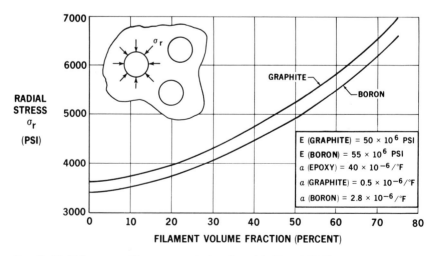

FIG. 5—Radial stress at filament-matrix interface ($\triangle$ T = 250 F).

loading produces shear distortion in the laminate, since the tendency of one layer to undergo thermal shear strain is counteracted exactly by the other layer [11].

The requirements for ensuring that no bending or warping in the sheet occur due to thermal or mechanical effects generally are more complicated to assess. Consider, for example, the four-ply system represented diagrammatically in Fig. 6. It is evident that under tension or compression the components of the forces in the filaments produce an out-of-balance couple which causes warping. If, however, the layers are arranged as shown in Fig. 7, then balance can be obtained for tensile, compressive, or shear loading. Under bending, however, the sheet would warp, since an unbalanced couple is induced. It can be shown that the minimum number of layers necessary to achieve balance in tension, compression, shear, and bending is seven.

This requirement creates particular problems in structures which are weight critical, and where minimum-gage situations are encountered, as in aircraft skins. Currently there is some difficulty in obtaining certain types of graphite-epoxy composite

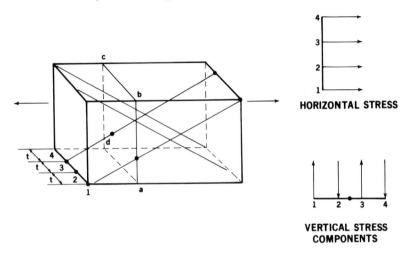

FIG. 6—Four-ply laminate; unbalanced in tension.

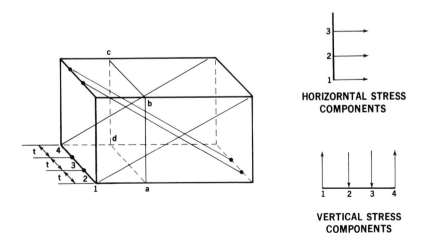

FIG. 7—Four-ply laminate; balanced in tension.

material in thicknesses below 0.08 in. per ply. This may result in skin thicknesses of 0.056 in. being utilized to obtain the necessary balanced properties, whereas, from elastic property and strength considerations, maybe half this thickness would be feasible. There is thus a need to be able to obtain composite material in a variety of thicknesses which permits the designer to achieve "balanced conditions" under minimum gage and allows optimum fabrication procedures to be attained.

While there may be good reasons for using completely balanced construction, this does not imply that unbalanced construction is unacceptable. The Douglas A-4 flap project utilized a skin having four plies arranged in the following manner: +45, −45, −45, and +45 deg. This was balanced for tension, compression, and shear but not for bending. Since the amount of bending that the skins experienced was relatively small compared to the predominant shear loading, this was not considered to be a serious problem.

The objections to utilizing unbalanced construction stem mainly from the difficulties of adequate designing with a material which would possibly induce severe coupling effects and also from the difficulties of obtaining meaningful design data to initiate the design.

Equally serious problems would arise at the manufacturing level where differential angular contractions would produce thermal warping which might be difficult to eliminate [12].

It would appear that, to achieve truly optimum design with composites, the possibility of using "unbalanced" laminates would arise. At this time, the problems of analyzing and testing such materials are not being solved. An analytical discussion of the magnitude of the difficulties which are encountered in even the tension testing of off-angle composites is given in Ref 13.

DESIGN PROBLEMS

The design procedure obviously must account for the total operational environment. Thus the design should include such considerations as: production feasibility, costs, maintenance, repairability, and a host of other problems. Here it is intended to comment only on a few topics which influence the overall efficiency of the design in the light of the foregoing discussion. The macromechanics of joint design will be considered at length, since the joining of composites is one of the most important problems facing designers today. An example of a typical thermoelastic problem will be cited together with some observations on the nature of fatigue.

Composite Joints

Because the first significant applications of fibrous reinforced composites were structures designed and fabricated using filament winding, much optimism has resulted from the belief that other designs in composites would have a small number of joints and be fabricated from a few basic integral components. Attempts to design aircraft structures and meet these goals have not been entirely successful, and the importance of determining design information on both mechanical and bonded joints of fibrous reinforced composites is obvious. In fact, as more and more component designs are studied, especially with consideration of the practical constraints of fabrication and tooling requirements, many composite designs appear conventional. The joining of high-strength and stiffness composites should account for the anisotropic nature of the material. Inherent weakness in the composite limit certain strength allowables. These aspects will be considered further in the following paragraphs.

Joint Design Considerations

To provide a better understanding about the performance and design of fibrous reinforced composites, a number of research programs have been sponsored by the Air Force. One of these is being carried out by the Douglas Aircraft Company for the Air Force Flight Dynamics Laboratory with the objectives of investigating strength trends and failure modes, evaluating design concepts, identifying promising joint concepts, and indicating weight-efficient joints. Published data from this program and company sponsored research will be used to support some observations which follow.

All results obtained to date are for tension coupons designed to simulate various practical joint design concepts as illustrated in Figs. 8 and 9. These types of tests, although simple in design, become the source of valuable strength allowable data to be used for the design and development of complex joints. With the current high costs of composite material structures, any tests of a complex nature are more expensive than the resulting data can justify. For example, practically every test point illustrated in this paper has resulted from the average of five coupon tests at a cost in excess of one thousand dollars.

Although the Air Force sponsored program is considerably broader in scope and considers other materials, fiber pattern, etc., the test data will be reported for boron filaments with a single, balanced (in tension) filament pattern. This

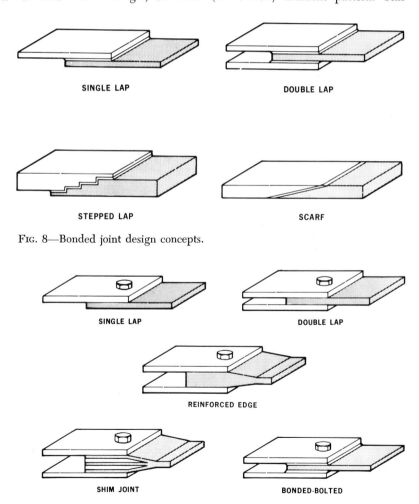

SINGLE LAP

DOUBLE LAP

STEPPED LAP

SCARF

Fig. 8—Bonded joint design concepts.

SINGLE LAP

DOUBLE LAP

REINFORCED EDGE

SHIM JOINT

BONDED-BOLTED

Fig. 9—Bolted joint design concepts.

Schjelderup and Jones on Aircraft Design   **293**

basic pattern consists of eight layers with 50 percent of the filament in the direction of loading and 50 percent at plus and minus 45 deg (0, +45, −45, 0, 0, −45, +45, 0). All test specimens were fabricated using 3-in. Narmco 5505 preimpregnated (prepreg) tape producing a laminate with a volume fraction from 51 to 59 percent boron and minimal voids. All bonded joints were fabricated using Shell 951 adhesive.

The Influence of the Adhesive

The choice of Shell 951 adhesive for bonding resulted from an early study in which the performance of various adhesive systems were evaluated. Ring shear tests were run and the adhesive strain and strength characteristics determined. As a result, two adhesives were considered in greater detail, 3M AF-130 and Shell 951. Typical stress strain curves are shown in Fig. 10. The essential differences between these two occurred in strength and ductility. In considering a choice between the two it is necessary to account for the interplay of the composite matrix material and the interlaminar shear strength. Linear analysis of bonded double lap joints will show large strains at the ends of the joint compared to an average throughout the joint. Either an increase in lap length or reduction in adherend stiffness Et will increase this strain concentration at the ends. Bonded joint failure for this simple type joint will occur when either the stresses and strains at the end of the joint exceed the capability of the matrix adhesive interface or stresses and strains induced in the laminate exceed the interlaminar strength of the composite.

Because the shear strains at the ends of the joint are large, without plastic flow at this interface, large stresses result and an early joint failure occurs. For example, the average shear stress developed by the 3M AF-130 in a ½-in. double lap test was little more than ten percent of its ultimate strength. In contrast to the brittle 3M AF-130 results, Fig. 11 shows the maximum average adhesive shear stress developed in various joints and lap lengths by the ductile Shell 951. The low values of adhesive shear stress for the ½-in. scarf joint is a result of the larger scarf angle which causes the adhesive to be loaded significantly in tension as well as shear. The loss in performance in the single lap joint is attributed to a similar cause. For the rest of the test data, average adhesive shear stress values are close to the adhesive strength. The ability to develop these high average adhesive stresses

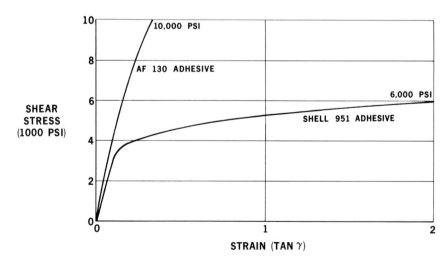

FIG. 10—Stress-strain curves from adhesive torsion ring test.

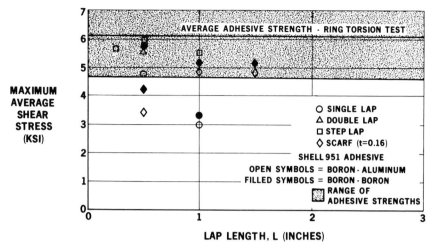

FIG. 11—Maximum average adhesive shear stress at failure.

compared to its ultimate strength shows the Shell 951 elastic properties to be optional for the range of thickness, stiffness, and lap lengths investigated.

Bonded Joint Performance

The performance of the various bonded joint design concepts is shown in Fig. 12. Figure 13 shows the joint strengths developed for single and double lap joints. Both joints for short lap lengths come close to developing the strength of the adhesive prior to failure. For the single lap, as the lap length increases, the effect of eccentricity is apparent because the joint fails to develop the adhesive shear strength and data scatter is large. However, within the bounds of the test data available, the full joint strength is developed, and tension failure in the adherend is the predominating mode of failure. The double lap joint performance shows little deviation from that expected from the adhesive until the adherend strength is developed.

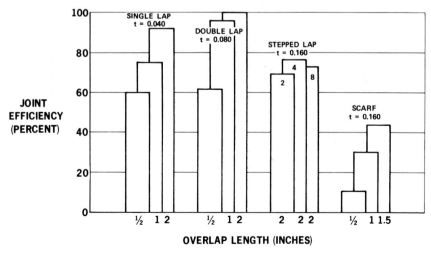

FIG. 12—A comparison of bonded joint strength efficiencies.

Schjelderup and Jones on Aircraft Design

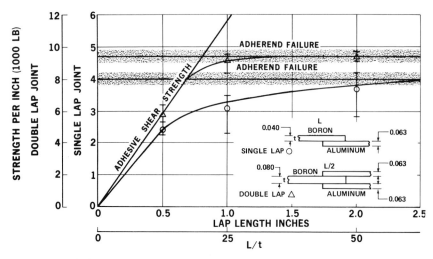

FIG. 13—Bonded joint performance.

The stepped lap joint is shown to be inefficient; however, this is not strictly true as the joint was strength limited by the equally thick aluminum adherend. A small increase in lap length would produce an efficient boron-to-boron joint. Because the adhesive system was so effective, multiple stepped laps beyond two did not improve the joint performance. A less ductile adhesive would have shown considerably more difference between a two-step and eight-step joint.

The scarf joint shows poor performance limited in this design application by the adhesive strength. To reach full efficiency either requires a greater adhesive strength or an extremely acute scarf angle. However, this type of joint can be made efficient with little weight penalty by increasing the adherent thickness at the joint to produce sufficient lap length at a practical scarf angle for the Shell 951 adhesive.

Failure Modes in Bonded Joints

Various modes of failure occurred for the joint designs studied, including cohesive failure of the adhesive, failure at the resin/adhesive interface, interlaminar shear of the laminate, and tension of the adherend. Figures 14, 15, and 16 illustrate these failure modes; however, considering the average values of adhesive shear stresses developed, it is apparent that effective adhesive joints can be designed.

The fact that some of the failures showed evidence of interlaminar shear of the adherend and resin failure may indicate that adhesives with higher basic strengths will perform no better, or possibly worse, than that studied, even if ductile.

Bolted Joint Performance

Double lap bolted joints strengths are compared in Fig. 17. A 100 percent efficient joint is indicated by the adherend strength value. Both edge-reinforced joints could be designed to full efficiency by increasing the reinforcing. However, a change in the design of the steel reinforced joint may be required to increase its efficiency, as some failures were occurring at the end of the tapered steel shims where the composite section was the same thickness as the basic adherend. Figure 18 shows shear-out stresses and tension stresses developed in various composite joints. The tension stresses are the average net stress developed in the section through the bolt. Although there appears to be a definite transition from shear-out to tension failures at an (e/D) ratio of four, the reason is probably the limited range of W/D. Specimens were designed not to fail due to excessive bearing stresses.

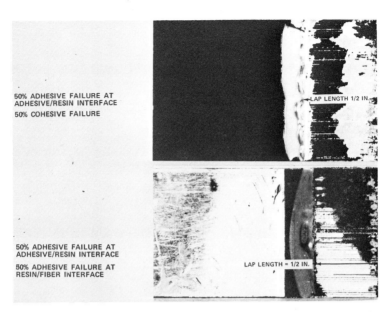

FIG. 14—Interfacial failures in simple lap adhesive joints ($L/\tau = 12.5$).

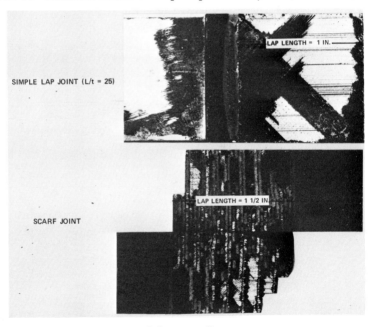

FIG. 15—Interlaminar shear failures in adhesive joints.

Failure Modes in Bolted Joints

Failures in the bolted composite joints appear conventional (Figs. 19 and 20), although, in comparison to metal joints, the composite stress allowables require greater edge distances and more area buildup to develop full efficiency. The greater edge distance is required because of the inherent weakness in composites in shear. To improve joint efficiency requires more area buildup because composite joints

Fig. 16—Adherend tension failure in simple lap adhesive ($L/\tau = 25$ to 50).

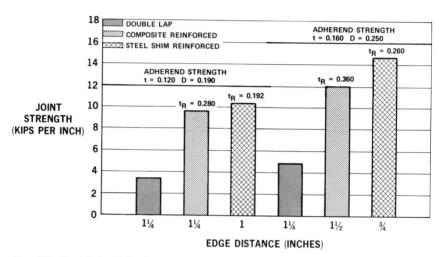

Fig. 17—Double lap bolted joint strength comparisons.

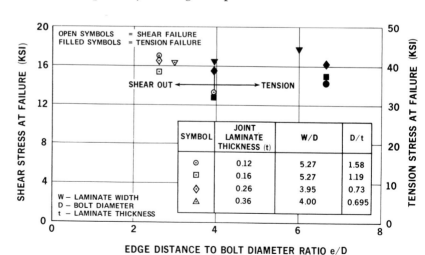

Fig. 18—Double lap bolted joint laminate failures.

Composite Materials

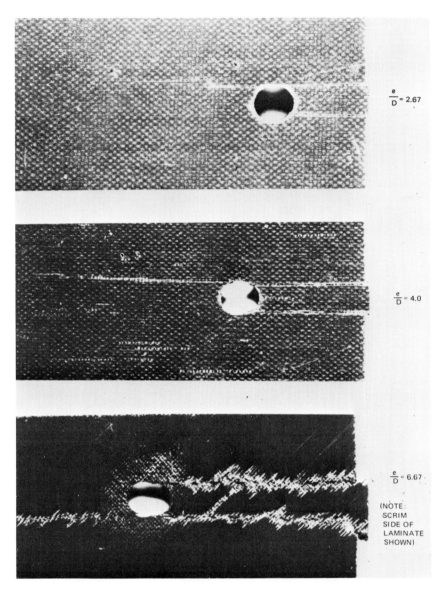

$\frac{e}{D} = 2.67$

$\frac{e}{D} = 4.0$

$\frac{e}{D} = 6.67$

(NOTE:
SCRIM
SIDE OF
LAMINATE
SHOWN)

FIG. 19—Shear-out failures of bolted joints ($D = \frac{3}{16}$).

are failing at average tension stresses considerably less than the basic adherend. One interesting characteristic regarding the failure of these bolted double lap joints is the small dispersion of data for a group of specimens; yet, both shear-out (Fig. 19) and tension shear-out (Fig. 20) failures were observed. This might be interpreted as indicating that the initiating failure mechanism was the same in both cases, although subsequent fracture occurred in different ways. Figure 21 shows the average shear stress at failure versus edge distance for these same tests. Considering the uniformity of the test data and observing that the tension stress at failure is consistent, it may be concluded that tension was the basic failure mechanism.

Schjelderup and Jones on Aircraft Design   **299**

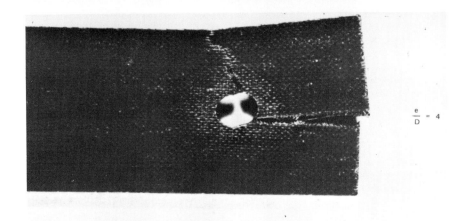

FIG. 20—Combined tension and shear-out failure of bolted joints $(D = \frac{3}{16})$.

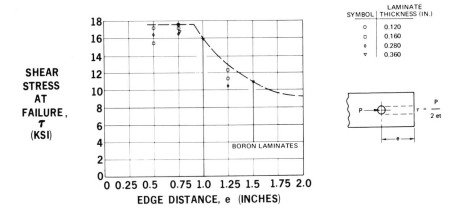

FIG. 21—Bolted joint shear stress at failure.

Composite Materials

Composite Joint Performance

The preceding has shown that, with proper design, efficient joints are feasible for high-modulus, fiber-reinforced composites. Figure 22 shows a static weight efficiency comparison of bonded and bolted joints. Although it may be unfair to compare bonded and bolted joints on a one-to-one basis, the figure does show that bonding appears considerably more efficient than bolting for permanent joints. The greatest cause for a weight penalty in the bolted joint is the low tension strength through the bolt and the low shear-out strength, increasing both thickness and lap distances, respectively.

The preceding discussion has been restricted to observations regarding static strength, but fatigue should be considered also. Lap bonded joints are subject to high stress concentrations, which, even with ductile joining materials, may have significant effect on fatigue life and influence some of the design conclusion.

THERMOELASTIC EFFECTS IN COMPOSITES

As mentioned previously, a design consideration that cannot be overlooked in high-modulus composite structures is that related to thermoelasticity. Many composite structures will be required to mate or interface with structures of other materials. In this respect graphite will present the greatest challenge with its infinitesimal coefficient of thermal expansion. For example, in a typical lap joint between aluminum and graphite sheets of equivalent stiffness, residual stresses on cool-down after bonding at 300 F could approach 21,000 psi, tension, in the aluminum and 63,000 psi, compression, in the graphite. For steel and graphite, stresses could be 29,000 psi in each. These stresses are very significant and illustrate the scope of the thermal problem that must be considered. Bonding stresses can be reduced by using lower-temperature cure adhesives, generally at the expense of strength and reliability. Undoubtedly stress relaxation will reduce values. However, even within the temperature ranges experienced in normal subsonic flight, stresses of considerable magnitude will be caused unless design details reduce constraints.

A Design Example

During installation of the Douglas A-4 boron flap on the aircraft, the continuous aluminum hinge would not match because of residual deformation caused during

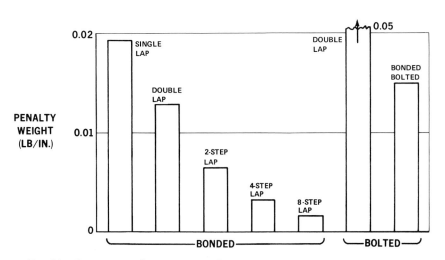

Fig. 22—Comparison of joint weight efficiency.

bonding of the hinge to the boron skin. The problem was obviated by removing material from the flap hinge segments, the amount removed being proportional to the distance from the hinge center.

A more severe thermal design problem exists with the equivalent A-4 graphite flap. The problem is extreme in the sense that the option does not exist to redesign the trailing edge of the wing. This current design difficulty is illustrative of the typical thermoelastic design considerations which arise. The problem has two parts:

1. The residual deformation and stresses resulting from bonding.
2. The mismatch resulting from the change of temperatures during normal flight operations.

Two design alternatives exist: bond the continuous hinge and provide clearances, as shown in Fig. 23, or cut the hinge into short lengths and provide clearances, as shown in Fig. 24. Clearances are shown for the ends of the hinge. These reduce linearly toward the center of the flap.

The advantage of the continuous hinge is that assembly tooling is simpler and residual bonding thermal stresses will occur only at the two ends, as shown in Fig. 25. On the other hand, because the aluminum hinge attaching the flap to the aircraft is bonded to the edge of the flap and not on the neutral axis, the stresses locked into the flap after curing cause the flap to be curved. The operating temperatures normally experienced by the aircraft will increase the effect. This misalignment with the mating hinge in the aircraft will cause assembly difficulties and induce fatigue loads in the hinge assembly during flight operation.

The segmented hinge will require more complicated tooling, and residual thermal stresses still will occur at each end of the hinge segment (Fig. 25). These stresses will be as severe as those of the continuous hinge, unless the segment is very short; however, the clearances are somewhat less. The loads set up in the hinge by the enforced straightening during installation on the aircraft are only marginally less than those for the continuous hinge and will occur in each segment. The one obvious merit of the segmented hinge is built-in fail safety. The hinge cannot unbond from end to end, as it can with a continuous hinge once failure is initiated.

The advantages of a moderate length segmented hinge over the continuous hinge are not outstanding. The potential assembly difficulty with the segmented hinge

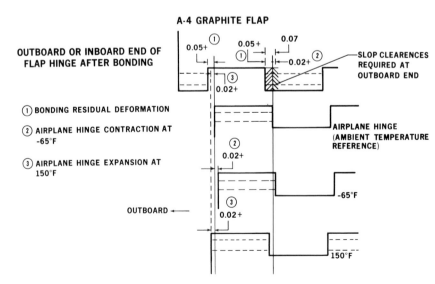

FIG. 23—Thermal interference; continuous hinge.

Composite Materials

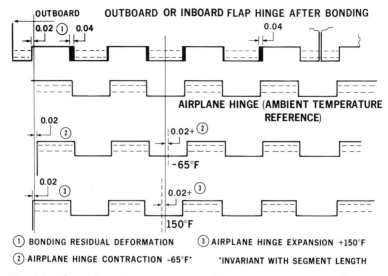

OUTBOARD   OUTBOARD OR INBOARD FLAP HINGE AFTER BONDING

0.02 ①   0.04   0.04

AIRPLANE HINGE (AMBIENT TEMPERATURE REFERENCE)

0.02   ②   0.02+ ②

-65°F

0.02   ③   0.02+ ③

150°F

① BONDING RESIDUAL DEFORMATION   ③ AIRPLANE HINGE EXPANSION +150°F

② AIRPLANE HINGE CONTRACTION -65°F*   *INVARIANT WITH SEGMENT LENGTH

Fig. 24—Thermal interference; segmented hinge.

caused by small tooling errors and thermal deformations outweight other advantages. Therefore, a continuous hinge is recommended for this design application.

Fatigue in Composite Materials

An area of possible concern in the design of high-modulus composite structures is compressive fatigue. This concern may at first appear unfounded, as a number of investigations have successfully fatigue-tested specimens to run out at maximum tension stresses as high as 80 percent ultimate. The concern is compressive failure under fatigue caused by progressive crazing of the matrix to the point that it may no longer stabilize the filament in compression. As mentioned earlier there is a strong relationship between interlaminar shear strength and compressive strength. Tension fatigue tests at high-stress values have shown definite evidence of crazing without failure.

A Design Example

Although inconclusive, this problem has been indicated by results from the Douglas Aircraft A-4 boron flap program.

The basic design of the flap skin is shown in Fig. 26. Design limit loads are mainly torsion with the skin loaded as shown. Thus, during fatigue cycles two filaments are loaded in tension, and the perpendicular filaments are loaded in compression. Two tests performed on a full-scale flap structure provided evidence in support of the hypothesis mentioned above. After 8000 cycles from zero load to 72 percent design limit load, the first flap failed statically at 181 percent design load. The mode of failure was compressive crippling of the lower skin. A second flap was loaded to 244 percent design limit load without failure. Subsequently, this flap was fatigue tested in a similar manner, except that the new fatigue load was the design limit load. This flap failed in an identical mode after 730 cycles (Fig. 27). The deduction is that, in the first case, crazing caused by the fatigue test had not progressed enough to precipitate a fatigue failure but had progressed sufficiently to reduce the ultimate strength. In the second case, the initial static

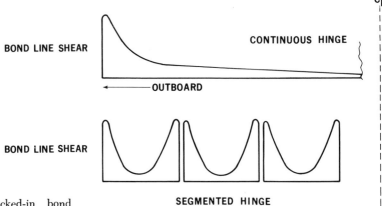

BOND LINE SHEAR

CONTINUOUS HINGE

←——— OUTBOARD

BOND LINE SHEAR

FIG. 25—Locked-in bond line stresses.

SEGMENTED HINGE

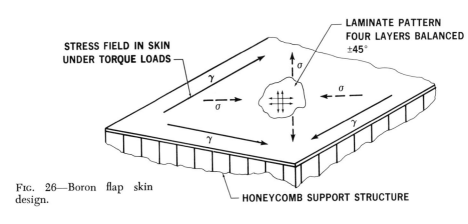

STRESS FIELD IN SKIN UNDER TORQUE LOADS

LAMINATE PATTERN FOUR LAYERS BALANCED ±45°

FIG. 26—Boron flap skin design.

HONEYCOMB SUPPORT STRUCTURE

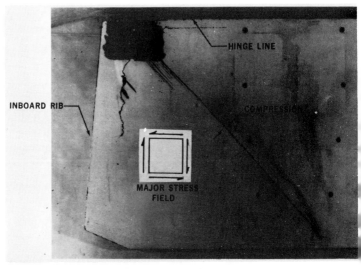

FIG. 27—Failure in upper skin fatigue test; Flap No. 2.

load had initiated crazing which progressed under the fatigue load to the point where the matrix could no longer stabilize the skin. This type of failure is particularly insidious from a design standpoint because a structure loaded predominantly in tension may have significant strength margin in tension but be degraded such that secondary compressive loads will initiate a failure.

How this problem will affect composite design is difficult to predict, but is does place considerable emphasis on the need for thorough development investigations in proving a design. Such an influence as balanced filaments may have significant effect on fatigue life. Fatigue evaluation must include spectrum loading because the load interaction in composites may even be stronger than metals. Intermittent compressive cycles may be required in a composite fatigue test.

CONCLUSIONS AND FUTURE REQUIREMENTS

The many research and development programs now in progress are providing the design experience and material performance data required for reliable structures of fibrous reinforced composites. However, because of the many variables in composites design, if economics is considered, a time comes when particular material systems must become standards such that material strength allowable data can become documented in design manuals. Then, using the current laminate strength analysis, extrapolation for various laminate designs may be made without the need for additional tests to determine design stress allowables.

Once a satisfactory material system is determined, the problem is to design a satisfactory industry-wide test program. In some respects this is being done today by the Air Force laboratories. However, it would be of value to devise and plan what tests of what patterns will provide sufficient data to allow interpolation rather than extrapolation to other conditions. The joining program, described in this paper, is a start towards this goal. The resin system, volume fraction, etc., were controlled within bounds common to many other programs. If others have confidence in this data, then tests do not have to be repeated for new designs.

Many fundamental problems unique to composites are still to be resolved. Some are obscure and must be deferred until composites play a stronger role in application. Others should be investigated now as part of the material development programs underway.

At this time design practice should use balanced laminate construction whenever possible, particularly in minimum gage situations. Fortunately, awareness of the problem is probably of more value than quantitative results from an expensive research program.

Continued effort is required in the evaluation of joints. For bonded joints, a standard test must be devised that may be used to determine the residual stress condition resulting from the bonding of a composite to other materials. Current analytical methods should give good estimates of other thermoelastic stress conditions but cannot account for residual stresses which occur during bonding.

Spectrum fatigue tests for multioriented filament laminates are needed to evaluate the significance of the fatigue interaction problem. Do sporadic high loading conditions strongly interact to cause only failure at lower stress levels? Are composites fatigue sensitive in compression? These are just a few of the many questions that remain to be answered before confident designs can be completed.

References

[1] Rosen, B. W., "Tensile Failures in Fibrous Composites," *Journal*, American Institute of Aeronautics and Astronautics, Vol. 2, No. 11, 1964.

[2] Rosen, B. W., "Mechanics of Composite Strengthening Fiber Composite Materials," American Society for Metals, 1964.

[3] Rogers, C. W., "Structural Air-

frame Application of Advanced Composite Materials," reports generated under contract AF 33(615)-5150, 1966/1967.

[4] Herrman, L. R., Mason, W. E., and Chan, S. T. R., "Response of Reinforcing Wires to Compressive States of Stress," *Journal of Composite Materials,* Vol. 1, No. 3, 1967.

[5] Whitney, J. M., "Geometrical Effects of Filament Twist on the Modulus and Strength of Graphite Fiber Reinforced Composites," AFML Technical Report TR-66-131, Air Force Materials Laboratory, 1966.

[6] Noyes, J. V. and Jones, B. H., "Crazing and Yielding of Reinforced Composites," AFML Technical Report TR-68-51, Air Force Materials Laboratory, March, 1968.

[7] Hashin, Z. V. and Rosen, B. W., "The Elastic Moduli of Fiber Reinforced Materials," *Journal of Applied Mechanics,* Vol. 31, 1964, p. 223.

[8] Shu, L. S. and Rosen, B. W.,

"Strength of Fiber Reinforced Composites by Limit Analysis Methods," *Journal of Composite Materials,* Vol. 1, No. 4, 1956967, p. 366.

[9] Foye, R. et al, "Advanced Design Concepts for Composite Structures," Quarterly Progress Report No. 3, Contract No. AF 33(615)-5150, 1967.

[10] McGarry, F. J., "Resin Cracking in Composites," *The Chemical Engineer,* London, Oct. 1964.

[11] Greszczuk, L. B., "Thermoelastic Considerations for Filamentary Structures," *Proceedings,* 20th Conference, Society of the Plastics Industry, Section 5C, 1965, p. 1.

[12] Gresham, H. E. and Hannah, C. G., "Reinforced Plastics for Jet Lift Engines," *Journal of the Royal Aeronautical Society,* London, Vol. 71, May, 1967, p. 355.

[13] Pagano, N. and Halpin, J., "Influence of End Constraint on the Testing of Anisotropic Bodies," *Journal of Composite Materials,* Vol. 2, No. 1, Jan. 1968, p. 18.

Determination of Design Allowables for Composite Materials

B. H. JONES[1]

Reference:

Jones, B. H., "Determination of Design Allowables for Composite Materials," *Composite Materials: Testing and Design, ASTM STP 460*, American Society for Testing and Materials, 1969, pp. 307–320.

Abstract:

The problems of establishing design allowables and factors of safety for filamentary composites are considered. A method is recommended for determining design data when small sample sizes are employed to establish basic laminate properties. It is advocated that the traditional factor of safety of 1.5 should not be employed without some reference to the coefficient of variation of the strength data. The concept of reliability is introduced. Methods for predicting design allowables under multiaxial stress are discussed in general terms.

The influence of lamina failure is considered together with means for predicting this. Results presented indicate that a linear analysis, based on the distortional energy criterion, can predict progressive lamina failure adequately for design purposes. A description of the phenomenon of crazing is included, and it is suggested that this is generally due to initial lamina failures. The paper concludes with some comments on the effects of testing and processing variables.

Key Words:

design allowables, safety factor, fiber composites, reliability, laminates, failure, evaluation, tests

The failure characteristics of composite materials are frequently more difficult to quantify for design purposes than are most homogeneous materials. Because the assessment of structural integrity involves defining the probability of failure, rather than whether the part will fail, the difficulty of quantifying failure characteristics of composites has made verification of strength criteria unsatisfactory apart from any *a priori* assumptions about failure mechanisms.

Data concerning composite materials generally exhibit appreciable scatter, particularly when samples from different sources are compared. This arises from the number of variables that influence the response of the material. From a statistical point of view, the problems of defining these influences are complex because:

1. Different failure and response mechanisms operate in each major material direction.

2. Even when the statistical distribution functions defining strength in each material direction are the same, the function parameters normally differ.

Complete definition of the variability of composites would prove to be an extremely time-consuming task. It is relevant to recall that complete statistical characterization of metals under all loading situations has yet to be attained, although these materials have been in use for a considerable time.

A great deal of attention is normally devoted to defining the initiation of laminate failure (first failure), because this is frequently deemed to be the ultimate effective load-carrying state. Of equal concern is the phenomenon of crazing, since this is considered to be critical to the ability of the laminate to withstand adverse environments.

---

[1] Formally senior engineer/scientist, Douglas Aircraft Co., McDonnell Douglas Corp., Long Beach, Calif. 90801; now with Aeronautical Research Associates of Princeton Inc., Princeton, N.J. 08540.

DESIGN ALLOWABLES

Regardless of the material used, the determination of design allowables for structural purposes involves essentially the same requirements. In all cases the inherent load-carrying ability of a structure must be such that the limit load can be sustained without excessive deformation and without a significant amount of permanent set. Of equal importance is the need to ensure that the structure is capable of withstanding the operational spectrum of loading and environment for a period considerably in excess of the specified service life. These requirements do not refer to the characteristics of the material being employed; nevertheless, they are the foundations upon which the design allowables must be established. It is the unique phenomenological behavior of laminated composite materials that makes the determination of allowables for optimum design a challenge.

Consistent criteria currently being investigated are aimed at permitting a rational consideration of the problem. The so-called semiempirical approach [1]² appears to be the most general, because it is based on the characteristics of the uniaxial lamina. Analytical methods can then be used to predict the behavior for the multidirectional situation. This method has obvious advantages over determining the behavior of specific orientations on an individual basis, particularly when an optimum design is desired.

DETERMINATION OF STRESS LIMITS

In general, the design limit load (DLL) stresses are chosen so that no lamina within the laminated structure is subjected to a stress exceeding certain limits, defined with reference to the monolayer properties. This represents an upper bound on the DLL stresses. There is also the general stipulation that the DLL should not produce a stress in any lamina in excess of two thirds of the ultimate load. That is, the following two conditions must be satisfied:

1. DLL stress < (ultimate/1.5), and
2. DLL stress < (characteristic stress × C)

where $C = 1$ or $1/1.15$ [1].

The characteristic stress is either proportional limit or some arbitrarily defined stress (two thirds of ultimate for transverse tension, stress at which the secant modulus has decreased to some arbitrary value, etc.). Additional requirements may exist also for such factors as fatigue life and residual strength. The rationale for such an approach to design allowables is naturally conservative, as it should be in view of the restricted amount of data available. It may be argued that considering ultimate load at the point where a particular lamina fails is unduly conservative. However, the formation of a significant crack in a metallic structure generally would mark the end of its useful load-bearing life. Hence, a similar criterion in laminated structures is reasonable, particularly because the crack would have deleterious effects on fatigue resistance and other strength factors.

DETERMINATION OF UNIAXIAL DESIGN ALLOWABLES

If the semiempirical approach is being employed, then it is necessary to determine data on the stress-strain behavior of the material to define moduli, proportional limit, ultimate stresses, etc. Because data are obtained from samples of finite size, it is necessary to account for variations in sample standard deviations and means for their relation to the true population by specifying confidence levels. Current practice [2] for metallic materials generally is to define a value above which at least 99 percent of the population is expected to fall with a confidence level of 95 percent (A-value). For composite materials, more realistic figures are 90 and 95 percent, respectively (B-value). If the design allowable is X_{QA} and the mean

---

² The italic numbers in brackets refer to the list of references appended to this paper.

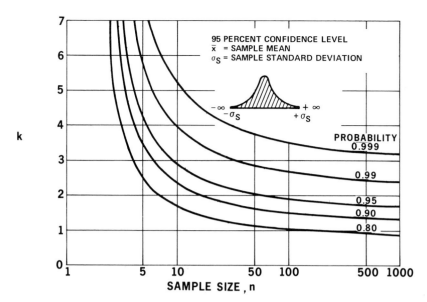

FIG. 1—One-sided tolerance limits and sample size for the normal distribution.

and standard deviations for the sample are $\bar{X}_S$ and σ_S respectively, then

$$X_{AQ} = \bar{X}_S - k\sigma_s \quad \ldots\ldots\ldots\ldots\ldots\ldots\ldots\ldots(1)$$

where k is the one-sided tolerance factor for the normal distribution at some particular confidence level and probability (Fig. 1). Employing an approach of this type tends to overconservatism, because Eq 1 involves the use of sample statistics rather than population values. Because the standard deviation is not convenient for purposes of comparing different population characteristics, Eq 1 will be written as

$$\frac{X_{AQ}}{\bar{X}} = 1 - k\gamma_S \quad \ldots\ldots\ldots\ldots\ldots\ldots\ldots\ldots(2)$$

where γ_S is the coefficient of variation of the sample.

Data in [1] and other documents indicate that, in general, tests are carried out on relatively small samples (fewer than five). It is obvious that considerable reductions in X_{AQ} must occur if the coefficients of variation currently characteristic of composites are present in the data. Therefore, it is suggested that the penalty for using small sample sizes is too high. From the relatively scarce data available, a survey indicated that under typical processing conditions certain coefficients of variation were characteristic of particular loading modes (Table 1).

An analysis was then carried out to determine the relationship between X_{AQ} and $\bar{X}$, assuming the population (rather than the sample) coefficient of variation (γ) was known. This resulted in the following expression [3]:

$$\frac{X_{AQ}}{\bar{X}} = \frac{1 + \gamma K_{1-Q}}{1 + \dfrac{\gamma K_{1-C}}{\sqrt{n}}} = b \quad \ldots\ldots\ldots\ldots\ldots\ldots\ldots(3)$$

where K_{1-Q} and K_{1-C} are the values of standard normally distributed variates exceeded by $(1 - Q)$ and $(1 - C)$ percentages of the population, respectively. For the A and B allowables referred to earlier, $Q = 0.01$ and 0.10, respectively. For a confidence level $C = 0.95$, it will be found that with $n = 3$, $b = 0.7$ for

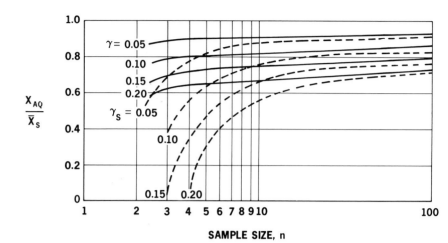

Fig. 2—Design allowables based on population and sample coefficients of variation.

Table 1—*Characteristic coefficients of variation for various loading modes.*

LOAD	COEFFICIENT OF VARIATION, γ	
	RANGE	AVERAGE
TENSION		
LONGITUDINAL	0.04 - 0.12	0.10
TRANSVERSE	0.01 - 0.20	0.11
COMPRESSION		
LONGITUDINAL	0.08 - 0.16	0.12
TRANSVERSE	0.05 - 0.11	0.08
SHEAR		
IN-PLANE	0.02 - 0.08	0.06
INTERLAMINAR	0.02 - 0.08	0.05
FLEXURE		
LONGITUDINAL	0.01 - 0.06	0.03
TRANSVERSE	0.01 - 0.02	0.08
FILAMENTS	0.06 - 0.19	0.12

an A allowable. When $n = 10$, $b = 0.73$, and when $n = 100$, $b = 0.76$. Thus, the proposed approach shows only a slight sensitivity to sample size.

A comparison of the approaches based on Eqs 2 and 3 is shown in Fig. 2 for assumed values of coefficients of variation γ and γ_s. The higher curve in each case is that which occurs from assuming the population coefficient of variation to be known. Thus, a higher design allowable results.

Therefore, it is suggested that the possibility of using characteristic coefficients of variation be investigated for tests carried out on small samples. This will lead to realistic designs having a historically meaningful base and eliminate some of the *ad hoc* testing that generally is carried out. The same approach could be employed on bonded joints, bolted joints, bonded-bolted joints, etc.

DESIGN ALLOWABLES IN COMBINED LOADING

If the semiempirical approach is used to determine design allowables, then for either unidirectional or multidirectional loading of the oriented laminate, a combined stress condition will exist in the laminae. There are a number of theories currently used to predict failure under such conditions, using maximum stress, maximum strain, or distortional energy criteria. In each case it is necessary to define monolayer properties in terms of principal strengths. Current practice is to employ the principal strength values evaluated, using the approach described earlier for design allowables defined in terms of safety factors. The type of relationship that results from using the distortional energy criterion of failure is shown in Fig. 3 for a boron-epoxy composite. B allowables were assumed at a factor of safety of 1.5. Ho [4] has considered the probability of failure of anisotropic materials under multiaxial stress, assuming a Weibull distribution and the maximum stress criterion.

Under combined tension, compression and shear, it is suggested that the total risk of rupture is

$$B = B_x + B_y' + B_{xy} = \int_v K \left[\frac{\sigma_x - \sigma_u}{\sigma_0} \right]^m dV + \int_v K' \left[\frac{\sigma_y' - \sigma_u'}{\sigma_0'} \right]^{m'} dV$$

$$+ \int_v K'' \left[\frac{\sigma_{xy} - \sigma_u''}{\sigma_0''} \right]^{m''} dV$$

$$\sigma_x \geq \sigma_u \geq 0$$
$$\sigma_y' \geq \sigma_u' \leq 0 \text{ (compression)}$$
$$|\sigma_{xy}| \geq |\sigma_u''| \geq 0$$

where:

B = total risk of rupture,
B_x = risk of rupture due to uniaxial tension
B_y' = risk of rupture due to uniaxial compression,
B_{xy} = risk of rupture due to shear,
$\sigma_x, \sigma_y', \sigma_{xy}$ = stresses in tension, compression, shear,
$\sigma_u, \sigma_u' \sigma_u''$ = threshold stresses,
$\sigma_0 \sigma_0', \sigma_0''$ = characteristic stresses,
m = material flaw intensity,
K = coefficient, and
V = volume.

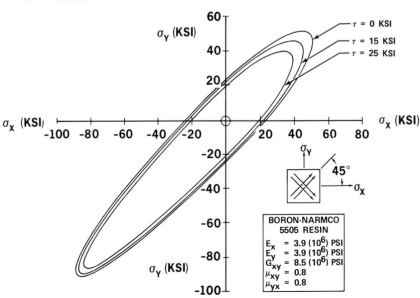

FIG. 3—Failure envelope for boron-epoxy laminated composite.

Jones on Determination of Design Allowables

The type of failure surfaces that would be obtained are shown diagrammatically in Fig. 4 for various probabilities of failure.

The objections to this approach are that no confidence levels may be conveniently incorporated in the analysis, and interaction effects are neglected. If, however, the maximum stress or strain criteria for failure in composite laminates are accepted, then the technique may prove valuable.

FACTOR OF SAFETY

Most metallic aircraft structures are designed with a factor of safety of 1.50. This figure, which is the ratio between ultimate and yield strengths, was developed for such materials as the following:

1. Steels with and without yield point.
2. Aluminum alloys (clad and unclad).
3. Magnesium alloys.

Because determination of a safety factor is, to a significant extent, heuristic, it is worth examining the concept in some detail to determine if it is logical to assume the same value for composite materials.

EXAMINATION OF THE CONCEPT

A factor of safety may be defined as

$$F_S = \frac{\text{ultimate load of structure}}{\text{design limit load of structure}}$$

if X_{UQ} is the ultimate strength related to the Q percentile, and X_{DL} is the applied (design limit) stress, then F_S may be written in terms of stress as

$$F_S = \frac{X_{UQ}}{X_{DL}} \dots \dots \dots (4)$$

If K_R is the value of a standard normally distributed variate exceeded by R percent of the poulation, then

$$K_R = \frac{X_{DL} - \bar{X}}{\sigma} \dots \dots \dots (5)$$

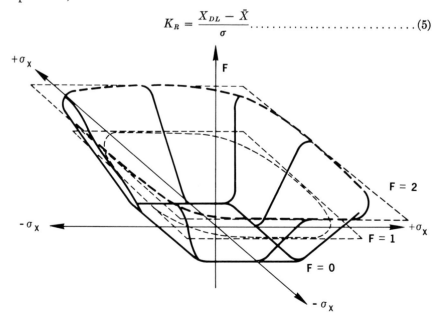

FIG. 4—Probability-of-failure surface (maximum stress criteria).

where $\bar{X} =$ mean and $\sigma =$ standard deviation.
Writing

$$X_{DL} = b\bar{X}_S \dots\dots\dots\dots\dots\dots\dots\dots\dots (6)$$

and

$$\gamma = \frac{\sigma}{\bar{X}}$$

then Eq 5 becomes

$$K_R = \frac{b - 1}{\gamma} \dots\dots\dots\dots\dots\dots\dots\dots (7)$$

Similarly, if K_{1-Q} denotes the value of a standard normally distributed variate exceeded by $(1 - Q)$ percent of the distribution, and if

$$X_{UQ} = d\bar{X} \dots\dots\dots\dots\dots\dots\dots\dots (8)$$

then

$$K_{1-Q} = \frac{d - 1}{\gamma} \dots\dots\dots\dots\dots\dots\dots\dots (9)$$

From Eqs 4, 6, 7, 8, and 9 is obtained

$$F_S = \frac{1 + \gamma K_{1-Q}}{1 + \gamma K_R} \dots\dots\dots\dots\dots\dots\dots (10)$$

The quantity R is the reliability level, and it specifies the probability that the strength exceeds the applied stress. That is,

$$R = pr(X > X_{DL})$$

where $X =$ random strength variable.

APPLICATION TO COMPOSITES

Using Eq 10 it becomes possible to determine the value of the reliability (R) attained in conventional structure when, say, $Q = 0.01$ (99 percent probability). If this is known, then the value of F_S to be used in composite structures to achieve the same reliability may be derived.

Figure 5 shows the factors of safety to be used in structures exhibiting various coefficients of variation and having the same reliability as a conventional structure designed to a 1.5 factor of safety. The assumed coefficient of variation in the conventional structure is assumed to be 0.03, a figure typical of aluminum [5].

It is evident from Fig. 5 that the factor of safety increases rapidly if the coefficient of variation of the data is greater than 0.007. To attain an A allowable status, factors of safety some 25 percent greater than those used for B allowables must be employed for coefficients of variation typical of the better class of composites.

This analysis puts the concept of factor of safety on a rational basis. It is evident that the value of 1.5 has meaning only when related to the concept of reliability. The designer should not use traditional factors of safety without careful appraisal of the data.

EFFECT OF DISTRIBUTION FUNCTION

A normal distribution of data has been assumed in the foregoing discussion. Table 2 indicates the manner in which the distribution function influences F_S for a constant value of $Q = 0.01$ and $\gamma = 0.12$. This choice of γ, which is typical

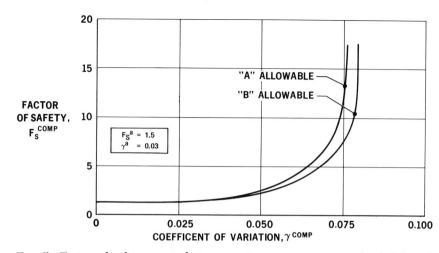

FIG. 5—Factor of safety required in composite structures to equal reliability of aluminum structures.

of composite data, does not produce the same degree of reliability as that implied by Fig. 5.

For the purpose of comparison, an old and a new design has been assumed, the original factor of safety in each case being 1.5. The results indicate that the normal distribution will tend to give a more conservative safety factor. This is worthy of consideration when analyzing data for design application.

INFLUENCE OF THE LAMINATE FAILURE MECHANISM ON DESIGN

As mentioned previously, current design practice takes particular account of both the yield and rupture points of the unidirectional layer.

Experimental evidence concerning the presence of a yield point in a unidirectional layer under tension is conflicting, seemingly being dependent on the method of testing. Generally, however, the linear-elastic filaments produce a linear stress-strain response almost up to failure, and slight nonlinearities may become evident as progressive filament failure occurs. In compression, response is generally similar.

Table 2—Influence of distribution on factor of safety.

DISTRIBUTION	QUANTITY	ORIGINAL VALUE	NEW VALUE
NORMAL	γ	0.12	0.15
	Q	0.01	0.01
	F_s	1.50	1.85
LOG NORMAL	γ	0.12	0.15
	Q	0.01	0.01
	F_s	1.50	1.65
WEIBULL	γ	0.12	0.15
	Q	0.01	0.01
	F_s	1.50	1.66

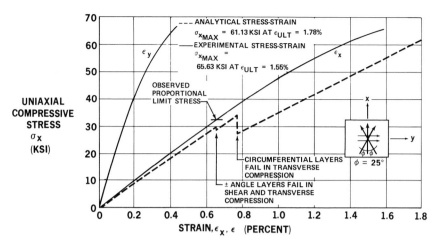

FIG. 6—Comparison of experimental and analytical stress-strain characteristics, uniaxial compression.

For transverse tension and compression, the frequently nonlinear behavior of the matrix becomes evident, as is the case in in-plane or interlaminar shear.

The behavior of a multilaminated composite reflects the characteristics of the constituent layers to a significant extent. It is considered that pronounced changes in the stiffness of the laminated composite under load are caused by fracturing of individual lamina. The load at which this occurs is of great interest to the designer, because a failed layer reduces the structural integrity of the structure under fatigue and adverse environment (for example, humidity).

To ascertain if this initial failure point could be predicted for a multioriented composite, a failure hypothesis based on the distortional energy criteria was used for predicting response. The details of this have been described in Ref 6 and will not be dealt with at length here. The equation used was suggested by Tsai [7] and based on Hill's method for describing continued deformation in anisotropic materials. It took the form

$$\left(\frac{\sigma_L}{F_L}\right)^2 - \left(\frac{\sigma_L \sigma_T}{F_L^2}\right) + \left(\frac{\sigma_T}{F_T}\right)^2 + \left(\frac{\sigma_{LT}}{F_{LT}}\right)^2 = 1$$

where F_L and F_T are the unidirectional strengths in the longitudinal and transverse directions and F_{LT} is the in-plane shear strength. The corresponding applied stresses are σ_L, σ_T, and σ_{LT}. The criterion above may be applied conveniently to multi-directionally oriented composites under any loading situation by imposing suitable strain ratios. This facilitates computation significantly.

Figure 6 is a comparison of the theoretical and experimental stress-strain responses of a glass-epoxy composite under compressive loading. The experimental curve shows a clear proportional limit at about the point where the circumferential and angle plies are predicted to fail. The stepwise decrease in load-carrying ability of the laminate arises from the failure of the individual layers. The principal reason why the experimental and theoretical curves do not coincide is that the testing machine is not a perfect straining device and tends to obliterate such small effects. The material was also assumed to be linear in its behavior—any departure from this would naturally influence the correlation. For most orientations, however, this assumption is not too restrictive. Comparison with a theoretical prediction using a nonlinear analysis and the maximum strain criterion is shown in Fig. 7 for a multidirectional boron-epoxy composite in tension. Further support to the general validity of the approach is given on Tables 3 and 4, which show the results of a number of compression tests on glass-epoxy, boron-epoxy, and boron-epoxy-whisker composites.

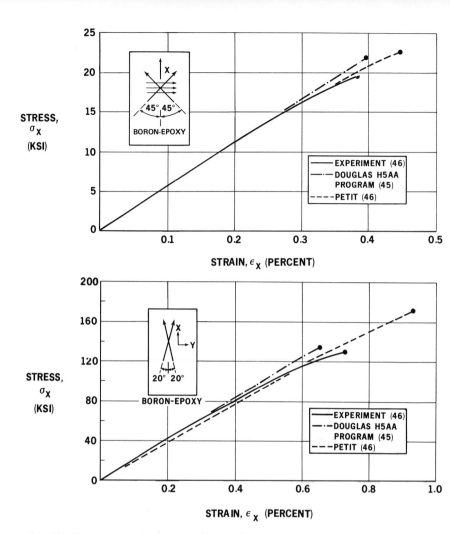

FIG. 7—Comparison of theories for predicting composite stress-strain behavior.

In summary then, it appears reasonable to conclude that, using either linear or nonlinear analysis, progressive degradation of a multilaminated composite can be predicted to a degree acceptable for design purposes. It is evident that some failures are more significant than others depending on the orientation of the lamina with respect to the loading direction. Whether these failures can be tolerated within the service life of the component will obviously depend on many factors. Fatigue life would be reduced seriously by the presence of such failures, as would the resistance to adverse environments.

THE PHENOMENON OF CRAZING

The phenomenon known as crazing became prominent during the hydrostatic proof-testing of filament-wound cylindrical pressure vessels [8]. The term crazing is applied generally to both the reduced optical transparency that frequently accompanies the application of high tensile stresses (particularly in the transparent thermo-

Composite Materials

Table 3—*Test versus theory comparison: compression of multidirectional composite.*

SPECIMEN GROUP	EXPERIMENTAL RESULTS				ANALYTICAL RESULTS				
	AVERAGE PROP LIMIT STRESS (PSI)	AVERAGE ULTIMATE STRESS (PSI)	AVERAGE E_x (10^6 PSI)	AVERAGE μ_{xy}	1ST FAILURE[a] STRESS (PSI)	2ND FAILURE[b] STRESS (PSI)	ULTIMATE STRESS (PSI)	E_x (10^6 PSI)	μ_{xy}
UC-45-1, -2, -3	23,170	52,200	4.89	0.350	18,230	23,570	47,810	3.99	0.261
UC-45-4, -6	24,680	34,520	2.89	0.359	20,580	26,100	30,340	3.17	0.266
UC-45-5, -7	26,260	35,070	2.95	0.243	19,380	20,670	25,340	2.89	0.265
UC-25-1, -2, -3	32,500	65,160	5.07	0.254	30,100	33,290	59,730	4.64	0.232

[a] ± ANGLE PLIES GOING OUT OF ACTION DUE TO LONGITUDINAL SHEAR AND TRANSVERSE COMPRESSION STRESSES

[b] CIRCUMFERENTIAL PLIES GOING OUT OF ACTION DUE TO TRANSVERSE COMPRESSION STRESSES ONLY

Table 4—*Test versus theory comparison: compression of multidirectional composite.*

CONFIGURATION	EXPERIMENTAL RESULTS		ANALYTICAL RESULTS	
	PROPORTIONAL LIMIT STRESS	ULTIMATE STRESS	FIRST FAILURE STRESS	ULTIMATE STRESS
BORON-EPOXY	NOT DEFINED	105.6	95.2	134.8
	NOT DEFINED	100.9	95.2	134.8
BORON-EPOXY + WHISKERS	65	170.3	96.2	134.8
	82	124.2	96.2	134.8
	NOT DEFINED	165.3	96.2	134.8

plastic polymers) and the formation of small cracks in thermoplastic and thermosetting materials. The first effect is caused by a change in the refractive index of the material at a local level and the second by the application of mechanical stress to incipient failure areas such as:

1. Micro- and macrovoids.
2. Impurities.
3. Giant molecule groups.
4. Areas where molecular chains are oriented in the direction of local stresses.

The localized alteration in refractive index may disappear on load removal, provided no plastic deformation has occurred, whereas crazing cracks are still evident subsequent to load removal. It is this latter type of crazing that will be discussed here, due to the increase of permeability with which it is associated and the effects it has on fatigue life, creep rupture strength, and ultimate strength.

As a consequence of the testing program and the strength analysis described in Ref 6, it became evident that crazing was not an independent phenomenon but a predictable, structural one based on transverse or shear failure. Crazing was not observed in the monolayer tests for any loading condition investigated. However, it was observed to occur in certain layers of multidirectionally oriented composites.

The mechanism by which crazing and yielding are considered to occur is as follows:

1. Initial yielding, characterized by the formation of a "knee" in the stress-strain curve, occurs when the first, localized, substantial failure takes place in whatever layer is loaded most critically. This is almost invariably a layer subjected to shear or transverse loading. Straining then takes place at a reduced stiffness. The layer that fails first continues to experience the gross composite strain, with additional fractures occuring throughout the layer at whatever range of strains is statistically characteristic of the material.

2. The subsequent failure of other layers may produce secondary "knees," with additional reductions in overall stiffness, although such effects may be masked during normal testing due to the nature of the loading machine.

3. The reduction in fatigue life, creep rupture strength, etc., observed after crazing is probably due to increased stresses in the layers remaining intact and to mechanical uncoupling, the latter effect predominating.

INFLUENCE OF PROCESSING AND TESTING

It should be emphasized that the data obtained from tests on composite materials must be assessed carefully for effects created by processing and testing variables. These include:

PROCESSING

1. Number, size, shape, and position of voids.
2. Variation of interface bonding strength.
3. Variation of interfilament spacing.
4. Variation of material properties from micro- to macrolevel.

TESTING

1. Influence of test specimen configuration.
2. Size effects.
3. Strain-rate effects.
4. Time-dependent effects.

The influence of voids is particularly significant on transverse tensile strength, in-plane and interlaminar shear strength, and longitudinal compressive strength. As has been reported, 10 percent voids can reduce transverse strength by as much as 60 percent [9]; compressive strength can be reduced by about 30 percent with 5 percent voids [10]; and interlaminar shear strength can decrease by 40 percent with 4 percent voids [11]. The effect of voids on the elastic properties of the matrix is indicated in Fig. 8 using a number of theoretical approaches.

The test specimen is very important in establishing meaningful design data, particularly in terms of specimen width and specimen configuration for multidrectional composites. In the main, tubular specimens provided more reliable data, although the specimens are clearly more expensive to produce.

CONCLUSIONS

The foregoing discussion has suggested the following:

1. Where design allowables are derived from data obtained from a small sample, it may be more satisfactory to employ a coefficient of variation characteristic of a particular loading situation rather than that exhibited by the sample.

2. The use of a factor of safety of 1.5 in all situations is questionable in view of the historical reason for using this figure. It is advocated that a factor of safety related to the concept of reliability be used.

3. Changes in the stiffness response of a multioriented composite are due to laminae rupturing. These changes generally can be predicted satisfactorily, using

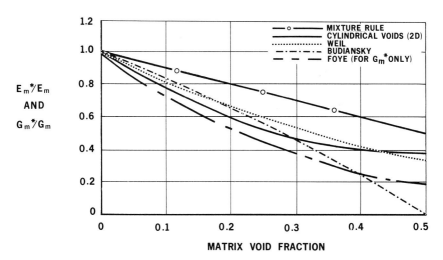

FIG. 8—Effect of voids on elastic properties of matrix.

suitable failure criteria and either linear on nonlinear analysis.

4. The phenomenon of crazing is associated with initial failure of a critically loaded lamina. It, too, can be satisfactorily predicted.

5. Data obtained from design purposes sould be reviewed carefully for the effects of processing and testing influences.

ACKNOWLEDGMENTS

Work described herein was carried out as part of a Douglas Aircraft Company Independent Research and Development (IRAD) project.

The author wishes to acknowledge the assistance of Howard Leve of the Structural Mechanics Section. Dr. Leve suggested the statistical analysis methods that were utilized.

References

[1] "Structural Design Guide for Advanced Composite Applications," Air Force Materials Laboratory (AFML), 1968.

[2] "Metallic Materials and Elements for Aerospace Vehicle Structures," Military Handbook, MIL-HDBK-5A, 1966.

[3] Leve, H., Contribution to "Advanced Composite Airframe Structures," Douglas Aircraft Company Report 3551T, Sept. 1967.

[4] Ho, J. Y. L., "Statistical Aspects of Failure," AFML-TR-66-310, Pt. I. Air Force Materials Laboratory, 1966, p. 74.

[5] Moon, D. P., Shinn, D. A., and Hyler, W. S., "Use of Statistical Considerations in Establishing Design Allowables for Military Handbook 5," 5th Reliability and Maintainability Conference, American Society of Mechanical Engineers, Society of Automotive Engineers, and American Institute of Aeronautics and Astronautics, 1966.

[6] Noyes, J. V. and Jones, B. H., "Crazing and Yielding of Reinforced Composites," AFML-TR-68-51, Air Force Materials Laboratory, 1968.

[7] Tsai, S. W., "Strength Characteristics of Composite Materials," NASA Report CR-224, National Aeronautics and Space Administration, 1965.

[8] Epstein, G. and Bandaruk, W., "The Crazing Phenomenon and Its Effects in Filament-Wound Pressure Vessels," Proceedings, 19th Conference, Society of the Plastics Industry, Section 19D, 1964.

[9] Brelant, S., *International Conference on Mechanics of Composite Materials*, Philadelphia, 1967, sponsored by the General Electric Co., to be published.

[10] Foye, R. L., "Compression Strength of Unidirectional Composites," Paper No. 66-143, American Institute of Aeronautics and Astronautics, 1966.

[11] Greszczuk, L. B., "Effect of Voids on Strength Properties of Filamentary Composites," *Proceedings*, 22nd Conference, Society of the Plastics Industry, Section 20A, 1967.

Characterization of Graphite Fiber/Resin Matrix Composites

R. A. ELKIN,[1] G. FUST,[2]
AND D. P. HANLEY[2]

Reference:

Elkin, R. A., Fust, G., and Hanley, D. P., "Characterization of Graphite Fiber/Resin Matrix Composites," *Composite Materials: Testing and Design, ASTM STP 460*, American Society for Testing and Materials, 1969, pp. 321–335.

Abstract:

The strength properties of a unidirectional lamina are essential inputs to the design of efficient composite airframe structural elements. To achieve the reliability required for such structures, these properties must be determined experimentally. A program was undertaken to characterize the static tensile, compressive, and in-plane shear strengths of Thornel-50, Morganite Type II, and HMG-50 unidirectional graphite fiber/resin matrix laminates.

Test results show that the Morganite fiber composites have significantly higher compressive, shear, and trans-verse tensile strengths than Thornel or HMG fiber composites. However, for longitudinal tension, the Thornel laminates have a slight strength advantage over the other two materials. Calculated strength envelopes for a typical airframe component laminate are presented as a convenient means of material comparison.

Correlations of laminate strengths with fiber content are shown, and optimum fiber contents for Thornel-50 laminates are estimated. Test methods used are reported, and differences in observed failure modes with the various materials are discussed. Test panel fabrication procedures, materials acceptance, and nondestructive inspection techniques are reviewed also. Composite strengths plotted as a function of qualitative ultrasonic inspection ratings show a promising correlation.

Key Words:

graphite, fiber composites, test methods, ultrasonics, strength, mechanical properties, fabrication, inspection, laminates, tension, compression, shear, resin, processes, unidirectional, longitudinal, transverse, failure mode, evaluation, tests

Development and availability of high-modulus, continuous graphite fibers both in this country and abroad has prompted considerable interest in the composite materials field. The aerospace industry with its demanding structural applications, in particular, has been stimulated strongly by the properties available with these materials. Notable among these properties are high stiffness and strength, low density, good fatigue and damping characteristics, ease of fabrication, and low-cost potential.

There are difficulties, however, in assessing various graphite fiber composites because of inadequate or conflicting data from material suppliers, and differences in fabrication and test procedures by the end-products users [1–4].[3] It was, therefore, a major objective of this work at Bell Aerosystems Company to compare several types of graphite composites through an integrated characterization program. Both material and process variations were considered, and emphasis was placed on static strength property measurements of unidirectional laminates. In the course of this

[1] Senior structures engineer, structural systems department, Textron's Bell Aerosystems Company, Buffalo, N. Y, 14240 (presently staff engineer, Rohr Corp., Riverside, Calif. 92502).
[2] Senior materials engineer, Nonmetallic Materials Laboratory, and chief, composite structures, Structural Systems Department, respectively, Textron's Bell Aerosystems Co., Buffalo, N. Y. 14240.
[3] The italic numbers in brackets refer to the list of references appended to this paper.

321

program 25 graphite fiber panels were fabricated and over 500 tests were performed on tension, compression, and shear specimens cut from these panels. Test methods for fiber glass and boron composites were found to be not entirely adequate for determining properties of graphite fiber materials. Consequently, a number of specimens were used in the development of test procedures. This paper presents the experimental results obtained, the test methods employed, and the fabrication and inspection techniques used. Data are presented also which indicate correlations between composite strengths and fiber content, and strength as a function of a qualitative nondestructive testing (NDT) rating factor. Optimum fiber contents are estimated for general applications, and failure envelopes for a typical airframe component laminate are given.

EXPERIMENTAL RESULTS

Analysis methods for determining elastic and strength properties of a multidirectional laminate from those measured on unidirectional lamina have been shown reasonably adequate [5]. Moreover, the methods of micromechanics are adequate for predicting the elastic constants but deficient in predicting strength properties of unidirectional lamina from the constituent fiber and matrix properties. The requirements for experimentally determined properties, therefore, are reduced to those of unidirectional lamina strengths. These properties have been determined for three graphite fiber composite materials: Thornel-50 (T-50),[4] Morganite Type II (M-II),[5] and HMG-50.[6] The impregnated resin systems used were USP's 798[7] with the HMG-50 and most of the T-50 panels and Narmco's 5605[8] with the M-II and one of the T-50 panels.

TEST DATA

Experimental results are shown in Fig. 1. Results are average ultimate strengths measured from a number of panel tests. Each upper or lower bound shown in Fig. 1 represents the average strength of four to six specimens cut from a single panel. The T-50 and M-II data were obtained from tests conducted on eight and three panels, respectively, while the HMG-50 data represent tests conducted on only two panels, hence, the reduced scatter for HMG-50, Fig. 1. In total, the data of Fig. 1 represent over 300 individual tests. (Additional specimens, which were used in test method development, are not included in this figure.) The ranges of the coefficients of variation of the test data are given in Table 1.

The test results given in Fig. 1 show that four out of the five strength properties are significantly higher for M-II than the other materials. For longitudinal tension, however, T-50 shows a slight advantage (approximately 12 percent) over the M-II or HMG-50 materials. Based upon reported fiber tensile strengths (Table 2), however, the M-II composite has the potential of exceeding the strength of the other materials in longitudinal tension also.

STRENGTH/PROCESS CORRELATIONS

Analysis of the T-50 strength data in conjunction with the processing methods (Table 3), and panel fiber and void contents (Table 4) indicate a number of trends. Figure 2 shows the tensile strengths of unidirectional T-50 composites as functions of fiber volume. Comparative data, Ref. 6, are included also. Longitudinal tensile strength tends to peak between 60 and 65 percent fiber, while the transverse tensile strength drops sharply at fiber volumes greater than 55 percent. The exceptionally low transverse tensile strengths, however, cannot be explained strictly on a fiber content basis. The possibility of differences in the resin system from one

---

[4] Union Carbide Corp., Carbon Products Division, New York, N. Y.
[5] Morganite Research and Development Ltd., London, England.
[6] HITCO, Materials Division, Gardena, Calif.
[7] U. S. Polymeric, Inc., Santa Anna, Calif.
[8] Whittaker Corp., Narmco Materials Division, Costa Mesa, Calif.

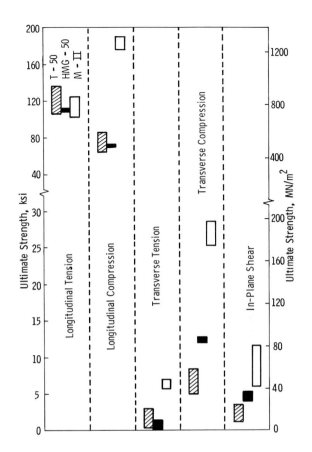

FIG. 1—Experimental results of unidirectional graphite fiber composite tests.

Table 1—Coefficients of variation, percent.

Material	Tension		Compression		Shear
	Longitudinal	Transverse	Longitudinal	Transverse	In-Plane
T-50	3 to 11	7 to 50	13 to 32	13 to 24	3 to 10
M-II	2 to 7	13 to 20	10 to 12	5 to 6	23 to 35
HMG-50	6 to 15	23 to 35	7 to 10	23 to 29	22 to 26

Table 2—Fiber properties.

Fiber	Modulus of Elasticity,[a] psi (GN/m^2)	Density,[a] g/cm^3	Fiber Tensile Strength,[b] ksi (MN/m^2)
T-50	50×10^6 (340)	1.63	250 (1720)
M-II	35 to 45×10^6 (240 to 310)	1.74	350 to 450 (2410 to 3100)
HMG-50	50×10^6 (340)	1.8	280 (1930)

[a] Vendor data.
[b] Ref 1.

Elkin et al on Graphite Composites

Table 3—Processing methods.

Process Schedule	Vacuum	Nominal Autoclave Pressure, psi (MN/m²)	Nominal Time/Temperature,[a] h/deg F (deg C)
A..........	no	200 (1.38)	1.0/250 (121)
		200 (1.38)	2.0/350 (177)
B..........	yes	200 (1.38)	1.0/250 (121)
		200 (1.38)	2.0/350 (177)
C..........	yes	200 (1.38)	0.03/200 (93)
		0	0.10/200 (93)
		200 (1.38)	1.0/250 (121)
		200 (1.38)	3.0/350 (177)
D..........	yes	60 (0.41)	1.0/250 (121)
		60 (0.41)	2.0/350 (177)
E..........	yes	0	1.0/250 (121)
		0	2.0/350 (177)
F..........	yes	0	1.0/250 (121)
		0	4.0/300 (149)
G..........	yes	0	24.0/275 (139)
H..........	no	1500 (10.34)[b]	1.0/250 (121)
		1500 (10.34)	3.0/350 (177)

[a] For all processes: cooled to less than or equal to 150 F (66 C) before panel removal.
[b] Cured in platen press.

Table 4—Panel fiber and void contents.

Material	Fabrication Method[a]	Fiber Content, volume %	Void Content, volume %	Density, g/cm³
T-50..............	A	70.3	2.6	1.49
	B	69.5	1.6	1.50
	D	66.1	3.0	1.47
	E	62.9	2.1	1.47
	F	62.1	2.6	1.46
	G	62.7	2.8	1.46
	H	58.1[b]	0.7[b]	1.47[b]
M-II..............	A	64.7	less than 0.5	1.58
	B	65.3	less than 0.5	1.58
	C	65.0	less than 0.5	1.58
HMG-50...........	A	63.2	2.0	1.58
	B	50.3	3.3	1.48

[a] See Table 3 for process description.
[b] Average of two panels.

lot of material to another presently is being investigated. Although the trend was not as clear as with tensile strength, longitudinal compressive strength showed a peak at a fiber volume of about 60 percent. The shear strength, which showed a tendency to drop above 50 to 55 percent fiber, was similar to the transverse tensile strength; that is, the strength dropped off rather than peaked out.

Several additional correlations were noted. Shear strength tended to decrease with increased void content, and void content decreased with increased cure pressure. Other process variables (see section on Inspection and Fabrication Procedures) prevented a complete correlation of fiber content and cure pressure for all panels. However, for the vac-bag/autoclave cured panels increased cure pressure resulted in increased fiber content.

Composite Materials

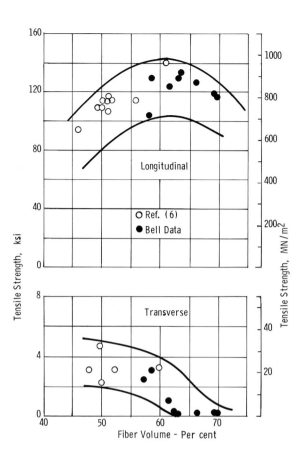

Fig. 2—Tensile properties of unidirectional T-50 composites as a function of fiber volume.

STRENGTH/INSPECTION CORRELATION

Because of the uniformity within the panels it was not possible to identify radiographically defects or material variations. In one instance foreign material, a small metal chip, was found in one panel.

After completion of the property measurement portion of the effort the ultrasonic inspection records for selected panels were reviewed. Each specimen of a particular type within a specific panel was assigned a qualitative ultrasonic rating based upon the size and location of regions which attenuated the signal. In reviewing the ultrasonic inspection records the best and worst specimens of a particular group were identified first and were given arbitrary ratings of 0 and 10, respectively. The remaining specimens in a given group were then listed in order of decreasing quality with no attempt made to quantitatively establish differences other than their general quality level as indicated by the amount of area which attenuated the sound. These ratings were established without reference to the mechanical properties associated with the particular specimens. For plotting purposes these intermediate quality specimens were located at equal intervals between 0 and 10, the best and worst ratings. In instances where several specimens were rated best, the average strength of the best rated specimens was used as the basis for non-dimensionalizing. The measured strength of each specimen then was divided by this average strength to obtain a normalized strength parameter. This was plotted as a function of the qualitative ultrasonic rating.

Correlation of the ultrasonic inspection results with longitudinal compressive strength is presented in Fig. 3. There appears to be a trend of decreasing strength

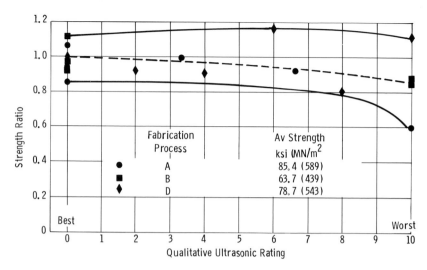

FIG. 3—Correlation of longitudinal compressive strength with ultrasonic inspection results.

with increasing area of material which attenuates sound to a greater extent than most of the material. Thus, it appears that ultrasonic inspection can be used to identify areas of material which will have low compressive strengths. No significant correlation was found between longitudinal tensile strength and the ultrasonic quality rating. With respect to the transverse compressive strength, the scatter band was quite wide so that the nature of the trend was difficult to identify positively. Similarly, the limited number of notch-shear specimens containing regions of high signal attenuation were quite small which made trends difficult to establish for this property. The limited data, however, did suggest a possible relationship between notch-shear strength and ultasonic rating. The high degree of variability in the transverse tensile data precluded establishment of any trends whatsoever.

MATERIAL COMPARISONS

Figure 4 shows the failure envelopes for the three materials for a typical laminate constructed with 60 percent of the thickness with fibers at 0 deg and 40 percent

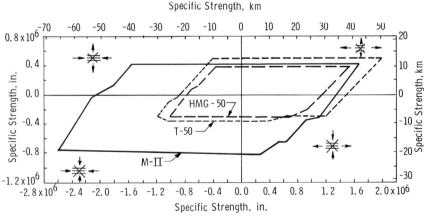

FIG. 4—Specific strength failure envelopes—60 percent (0 deg), 40 percent (±45 deg) laminate.

Table 5—*Ultimate lamina unidirectional strengths.*

| Material | Tension, ksi (MN/m^2) | | Compression, ksi (MN/m^2) | | Shear, ksi (MN/m^2) | Density, g/cm^3 |
	Longitudinal	Transverse	Longitudinal	Transverse	In-Plane	Composite
T-50............	135 (931)	3.0 (21)	80.0 (552)	8.0 (55)	3.5 (24)	1.49
M-II...........	120 (827)	6.5 (45)	185.0 (1276)	27.5 (190)	10.0 (69)	1.58
HMG-50........	110 (758)	2.0 (14)	70.0 (483)	12.5 (86)	5.0 (35)	1.58

at ±45 deg. The envelopes for the three materials are plotted on a specific strength basis using ultimate strength values shown in Table 5. The effect of the superior strength properties of M-II can be seen clearly.

The modulus of elasticity of both the T-50 and HMG-50 fibers is 25 percent greater than that of the M-II (Table 2). Therefore, in stability critical applications the T-50 and HMG-50 composites should show an advantage over the M-II. Figure 5 gives the weight per unit area versus compressive load on a long, simply supported orthotropic plate. At low load levels, where for minimum weight stability governs, both the T-50 and the HMG-50 are lighter than the M-II. At higher loadings more plies of material are required (each step shown in Fig. 5 represents the addition of one ply), and the increased thickness results in a strength critical rather than a stability critical plate. Since the slope of the strength critical segment of the plot is considerably steeper, a crossover occurs and the M-II, with higher compressive strength, becomes more efficient.

FAILURE MODES

Typical failure modes of longitudinal tension and compression specimens are shown in Fig. 6. The three tension specimens are T-50, HMG-50, and M-II pictured

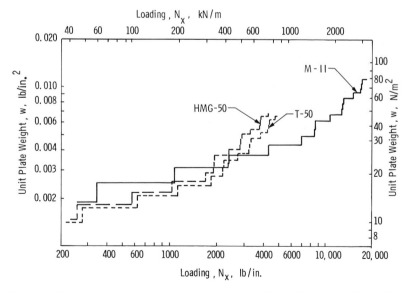

FIG. 5—Weight comparison-long simply supported 1.250–in. (31.75 mm)–wide plate under edgewise compression.

FIG. 6—Typical failure modes: longitudinal tension and compression specimens.

from top to bottom of the figure. The lowermost (compression) specimen is T-50. The T-50 tension specimens failed from end to end splintering apart into small pieces. The HMG-50 specimens failed in a similar manner, but broke into slightly larger pieces. A different failure mode was exhibited by the M-II specimens, these failing almost straight across with virtually no splintering. It appeared that the yarn bundles in the T-50 specimens failed progressively at weak points. The resin matrix was unable to adequately transfer load from a highly strained yarn to a yarn with lower strain due to the relatively low shear strength. The M-II specimens exhibited a greater capability to distribute internal loads. Consequently the M-II specimens failed at the weakest cross section, rather than progressively failing individual yarns or yarn bundles. The HMG-50 specimens having an intermediate shear strength between M-II and T-50 were able to transmit load more effectively than the T-50 specimens but not as well as the M-II specimens. Thus, the HMG-50 specimens failed in a manner somewhat similar to the T-50 specimens, but with larger bundles of yarn remaining intact.

No significant differences in the longitudinal compressive failure mode for the three materials were noted; hence, only one such specimen (T-50) is shown in Fig. 6. Failure modes typical of all three materials for transverse tension and transverse compression are shown in Fig. 7. A T-50 tension specimen is shown at the upper left and a M-II compression specimen at the upper right. Notch-shear specimens also are shown in this figure. The notch-shear specimens exhibited two types of failure. All of the T-50 and M-II specimens and those from one of the HMG-50 panels failed in a shear mode (Fig. 7, center). A number of the HMG-50 specimens from one panel, however, failed predominantly in a compressive mode (Fig. 7, bottom) with failure originating from the edge of the hole. Shear strengths of the two HMG-50 panels were not appreciably different (Fig. 1) irrespective of the different failure modes.

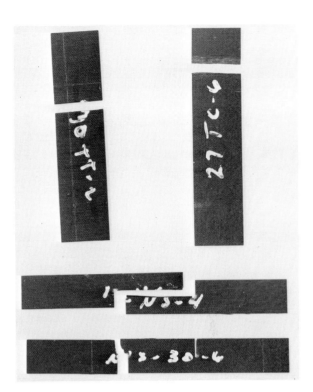

FIG. 7—Typical failure modes: transverse tension and compression and notch-shear specimens.

TEST METHODS

Test technology is lacking for unidirectional graphite composites with high strength and modulus in one direction, low transverse strength and modulus, and relatively low shear and bond strength. Therefore, development of new or modified test methods was required. Final test specimen geometries are shown in Fig. 8. The test methods developed are discussed below.

TENSION

Transverse tension specimens require sufficient thickness to minimize damage in handling. Specimens fabricated with six plies of material were found adequate for this requirement and at the same time permitted material economies. Since it was desired to test both longitudinal and transverse specimens from the same panel six-ply longitudinal tension specimens were used also. Specimen thicknesses varied from 0.040 in. (1.02 mm) for T-50 to 0.062 in. (1.58 mm) for M-II. The longitudinal tension specimen was straight-sided and employed bonded, premolded fiber glass end tabs. The fiber glass tabs are required to prevent damage to the specimen when it is gripped by self-tightening, serrated-face jaws during test. A room-temperature curing adhesive was used (60 parts Epon 828[9] and 40 parts Versamid 125[10]) to avoid bond-line shear stresses due to thermal expansion differences between the tabs and the specimen. A centering jig as used in mounting the specimens within the jaws to provide alignment with the load train.

The transverse tension specimens were tested in a manner similar to the longitudinal tension specimens; however, pneumatically actuated rubber-faced jaws were

[9] Shell Chemical Co., Plastics and Resins Division, Downey, Calif.
[10] General Mills, Inc., Chemical Division, Kankakee, Ill.

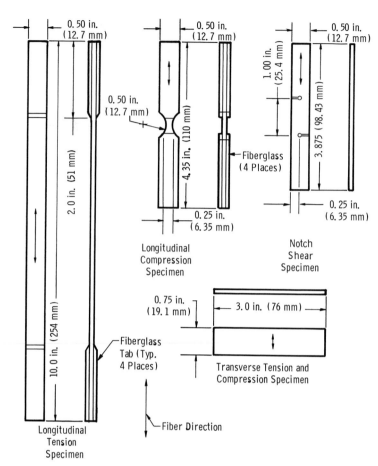

0.50 in.
(12.7 mm)

0.50 in.
(12.7 mm)

0.50 in.
(12.7 mm)

0.50 in.
(12.7 mm)

1.00 in.
(25.4 mm)

3.875 (98.43 mm)

2.0 in. (51 mm)

4.35 in. (110 mm)

Fiberglass
(4 Places)

0.25 in.
(6.35 mm)

Longitudinal
Compression
Specimen

0.25 in.
(6.35 mm)

Notch
Shear
Specimen

10.0 in. (254 mm)

0.75 in.
(19.1 mm)

3.0 in. (76 mm)

Fiberglass
Tab (Typ.
4 Places)

Transverse Tension and
Compression Specimen

Fiber Direction

Longitudinal
Tension
Specimen

Fig. 8—Test specimen geometries.

used. The lateral pressure exerted by these jaws is sufficient for the low loads encountered. Since the jaws do not damage the specimen no end tabs were required.

COMPRESSION

Compressive strength determinations of six-ply panels were complicated by three requirements: (1) specimens had to be supported to prevent instability failures, (2) specimen ends had to be restrained to prevent end brooming, and (3) the test fixture could not restrain the specimen in the applied load direction. These requirements applied to both longitudinal and transverse specimens.

The specimen configuration used for longitudinal compression is shown in Fig. 8. A single-ply of 181 style glass cloth was bonded to the specimen sides at each end leaving a 0.50 in. (12.7 mm) unsupported center area. The specimens were necked down in the gage length to a 0.25 in. (6.35 mm) width using a 0.50 in. (12.7 mm) radius on each side.

Compression tests were conducted using a specially designed fixture (Fig. 9). Each specimen end was clamped between recessed steel tabs. The fixture was designed with 0.10 in. (2.54 mm) gaps between tabs on both sides (when the ends of the specimen are flush with the tab ends) to allow for compressive deforma-

Fig. 9—Longitudinal compression test arrangement.

tion. The gaps on either side were offset by 1.00 in. (25.4 mm). In order to minimize bending and maintain alignment the specimen with tabs then was placed in a second (vise) fixture. With this arrangement (offset gaps) over 90 percent of the specimen failures occurred in the reduced-width gage length rather than in the gap areas. An exploded view of a compression specimen and the test fixture is shown in Fig. 10. The ball and blocks shown were used to ensure uniform load distribution across the specimens.

Transverse compression specimen configuration was identical to the transverse tension specimen (Fig. 8). Due to the relatively low loads encountered, the required test fixture was quite simple. Specimens were placed in the vise with short steel blocks placed on either side (Fig. 11) to prevent end brooming. The vise was tightened lightly, and the blocks were held in place by the small frictional forces between the specimen and blocks, and between the blocks and sides of the vise. With blocks slightly less than one half the specimen length a gage length of 0.125 in. (3.175 mm) was used.

SHEAR

Although interlaminar shear strength is useful in screening materials, in-plane shear strength is a more useful value for design of aircraft-type structural elements. The shear strengths determined, therefore, were based on a notch-shear specimen (Fig. 8) rather than the more widely used short-beam shear test. The notch-shear specimens were tested by placing them in a vise to maintain a vertical position and applying an axial compressive load. Failures generally occurred in shear between the holes. The shear strengths given in Fig. 1 were obtained by dividing the failure load by nominal shear area (specimen thickness times distances between holes).

FIG. 10—Longitudinal compression test fixture: exploded view.

It was realized that with the selected specimen configuration the shear stresses along the failure plane were not uniform. To assess shear stress concentration effects, a finite element analysis of the T-50 specimen was performed. Isostress curves are presented as a ratio of the local shear stress to the nominal shear stress in Fig. 12. The maximum calculated shear stress concentration, 1.57, was found to occur at the edges of the holes. Since the ultimate shear strength values given in Fig. 1 are for nominal strength, they represent somewhat conservative design allowables.

INSPECTION AND FABRICATION PROCEDURES

In the production of composite laminates for use in efficient and reliable structures, it is essential that the quality of the finished product be assessed adequately. To ensure maximum performance of the finished product, it is necessary to exercise control at three levels: (1) raw material, (2) in-process, and (3) finished product. While all of these areas must be controlled to the fullest possible extent, the last is perhaps most important. When a new material system is being evaluated, data scatter may be introduced by variations in test techniques or by variations in the material itself. As a means of evaluating inspection techniques and assessing the variability in experimental results attributable to differences in the specific material within each test specimen, selected panels were inspected dimensionally, visually, radiographically, and ultrasonically. Density, resin content, and void content were determined on specimens from the test panels.

For radiographic inspection a Seifert 150 kV beryllium window source was used. A three-minute exposure at 25 kV and 5 mA was found most appropriate for

FIG. 11—Transverse compression test arrangement.

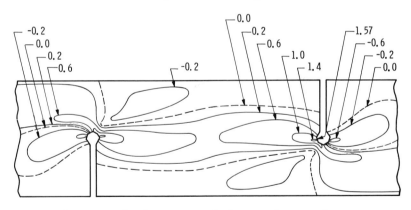

FIG. 12—Shear stress distribution in notch-shear specimen: $\tau_{xy}/\tau_{nominal}$.

the graphite composite panels. Ultrasonic inspection was conducted with an Automation Industries 721 Reflectoscope, research tank, and auxiliary equipment. Panels were interrogated with a 15 MHz focused transducer. The results of the interrogation were displayed on C-scan recording equipment. Both positive- and negative-gate techniques were investigated. Figure 13 shows a representative C-scan obtained with a negative-gate setting. The white areas indicate regions of signal attenuation caused by material nonuniformities such as voids, microcracks, variations in fiber content, etc.

Elkin et al on Graphite Composites

Fig. 13—Typical C-scan record: negative gate.

Preimpregnated (prepreg) materials were kept at 0 F (−18 C) in heat-sealed polyethylene bags containing dessicant. Material was allowed to stablize to room temperature in a vacuum chamber. Tool preparation during this time involved cleaning with chlorinated solvent, waxing all metal surfaces and baking them at 250 F (121 C), followed by application of Teflon release spray to surfaces which contact the layup. Pre-cut graphite tapes were handled carefully using white nylon gloves and laid on a ply of TX 1040[11] release cloth under which was a ply of Vac-Pak nylon film[12] on the laminating surface. MS-240 Quick-freeze[13] was used to facilitate removal of the prepreg tape from its release film, and a Teflon squeegee was used for collimating and debulking. Over the six-layer graphite tape layups, four plies of pellon bleeder cloth were applied, followed by six plies of 1B-301 yellow glass release fabric,[14] one ply of TX 1040 release cloth, and 0.020 in. (0.508 mm) stainless-steel sheeting. The laminate assemblies were than vac-bagged and cured in a heated platen press or in an autoclave. The cure schedules are given in Table 3.

Temperatures given in Table 3 were held to ±10 F (±5.5 C); time-hold tolerances were ±1 min for periods less than 1 h and ±5 min for periods greater than 1 h. Tolerance on pressure levels for both processes was ±5 percent. Autoclave pressurization rates were 20 psi/min (2300 N/m²/s) on pressurizing, and 10 psi/min (1150 N/m²/s) for reducing pressure. Autoclave temperature rate controls were 5 to 10 F/min (2.8 to 5.6 C/min) for both heat up and cool down. The platen press cure fabrication process (H in Table 3) utilized a male plug/female cavity die, whereas, due to the lower pressures, no constraints were required on lateral resin flow in the vac-bag/autoclave processes (A through G).

[11] Pallfax Corp., Putnam, Conn.
[12] Richmond Corp., Highland, Calif.
[13] Miller Stevenson Chemical Co., Danbury, Conn.
[14] Hexel Corp., Coast Manufacturing Division, Lancaster, Ohio.

CONCLUSIONS

A number of conclusions can be drawn from the results obtained:

1. The longitudinal and transverse compressive, transverse tensile, and in-plane shear strengths of unidirectional Morganite Type II graphite fiber laminates are significantly higher than those of laminates with Thornel-50 or HMG-50 fibers. Moreover, based on reported fiber tensile strengths, the Morganite II composites appear potentially stronger in longitudinal tension also.

2. For compressive applications involving low-loading intensities, where stability is the critical design factor, laminates fabricated from Thornel-50 fibers are more efficient than those using either HMG-50 or Morganite II fibers. For high compressive loadings, however, where strength becomes the critical design parameter, the Morganite II is considerably more efficient.

3. For Thornel-50 laminates a fiber content of between 60 and 65 percent by volume gives the highest longitudinal tensile and compressive strengths. In general, however, shear or transverse strengths are required also and a lower fiber content, approximately 55 percent, should result in a better overall balance of strength properties.

4. The coefficients of variation obtained for transverse and shear tests indicate the need for further refinement of the test methods. The test methods for longitudinal tensile and compressive strengths are adequate for currently available materials.

5. The correlations of ultrasonic (C-scan) records and composite strength are encouraging, and further efforts should yield a reliable NDT procedure.

ACKNOWLEDGMENTS

Work described in this paper was performed under Bell Aerosystems Company's Internal Research and Development funding. A portion of the fiber was supplied by the Air Force Materials Laboratory, Wright-Patterson Air Force Base, R. M. Neff, project engineer, as part of a U. S. Air Force/Industry Cooperative Program.

Acknowledgments are made to the following Bell personnel: F. M. Anthony, assistant chief engineer, Structural Systems Department, for overall guidance; R. Willard, manufacturing engineering, fabrication; D. Gullo, Nonmetallics Laboratory, testing; and R. Stauffis, quality control, NDT inspections.

References

[1] Ray, J. D., "Recent AFML Development on Graphite Fiber Reinforced Structural Plastics," meeting on Fiber Composites, *Development and Application of Graphite Fiber Composites and Stability and Multiaxial Stress Behavior of Composite Material Structures,* Case Western Reserve University, Cleveland, Ohio, 8–9 Oct. 1968.

[2] Barnet, F. R., Goan, J., and Simon, R., "Advanced Graphite Fiber Composite Studies at the Naval Ordinance Laboratory," meeting on Fiber Composites, *Development and Application of Graphite Fiber Composites and Stability and Multiaxial Stress Behavior of Composite Material Structures,* Case Western Reserve University, Cleveland, Ohio, 8–9 Oct. 1968.

[3] Petit, P. H., "Ultimate Strength of Laminated Composites," meeting on Fiber Composites, *Development and Application of Graphite Fiber Composites and Stability and Multiaxial Stress Behavior of Composite Material Structures,* Case Western Reserve University, Cleveland, Ohio, 8–9 Oct. 1968.

[4] "Structural Airframe Application of Advanced Composite Material," Ninth Quarterly Progress Report, Contract Number AF 33(615)-5257, Air Force Materials Laboratory, Wright-Patterson Air Force Base, Ohio, Sept. 1968.

[5] Blakslee, O. L. et al, "Fabrication, Testing and Design Studies with 'Thornel' Graphite-Fiber, Epoxy-Resin Composites," *12th National Symposium,* Society of Aerospace Materials and Process Engineers, Anaheim, Calif., 10–12 Oct. 1967.

[6] Private communications with E. Harmon, Northrup Norair Corp., Hawthorne, Calif.

Failure Criteria for Filamentary Composites

C. C. CHAMIS[1]

Reference:

Chamis, C. C., "Failure Criteria for Filamentary Composites," *Composite Materials: Testing and Design, ASTM STP 460*, American Society for Testing and Materials, 1969, pp. 336–351.

Abstract:

A two-level, linear, semiempirical theory for a failure criterion is described. The theory predicts the strength behavior of unidirectional filamentary composites under uniaxial and combined stress from basic constituent material properties and fabrication process considerations. Application of the theory to several filament-nonmetallic matrix composites are presented, and comparisons are made with experimental data. These results show good agreement between theory and experiment. Simple and combined strength envelopes are generated to illustrate the versatility of the theory and to point up problem areas in experimental work and design.

Key Words:

fiber composites, failure criteria, design, unidirectional, uniaxial load, combined stress, strength behavior, evaluation, tests

NOMENCLATURE

A Array arrangement, distribution
A Parameter, Eq 22
a_1, a_2 Constants, Eq 4
d Diameter
E Longitudinal modulus
F Failure function
G Shear modulus
K, K' Elastic and correlation coefficients, respectively, Eq 12
$k, \bar{k}$ Apparent and actual volume ratio, respectively
N Number of filaments or voids
N_x, N_y Applied loads
p Parameter, Eq 9
S Simple strength, failure or limit stress
t Thickness
β Theory-experiment correlation factor
ϵ Strain
θ Angle between load and filament directions
ν Poisson's ratio
σ Stress
φ_μ Strain-magnification factor

SUBSCRIPTS

B Interface bonding
C Compression
D Debonding

[1] Aerospace engineer, Lewis Research Center, National Aeronautics and Space Administration, Cleveland Ohio. 44106.

f   Filament property
l   Ply property
m Matrix property
p Limiting property
R Residual stress
S Shear
T Tension
v Void
x, y, z Load axes
1, 2, 3 Material axes (the 1-axis coincides with the filament direction)
α, β *T* or *C*, tension or compression

Experimental observations have shown that unidirectional fiber composites exhibit five primary failure stresses, also referred to as limit stresses and simple strengths. These stresses result when the composites are subjected to failure under uniaxial loading in their plane. The failure stresses are identified individually as follows: (1) longitudinal tensile (S_{l11T}), (2) longitudinal compressive (S_{l11C}), (3) transverse tensile (S_{l22T}), (4) transverse compressive (S_{l22C}), and (5) intralaminar (inplane) shear (S_{l12S}). These stresses are illustrated in Fig. 1 (see also Chapter 2[1][2]). Experimental observations also show that unidirectional fiber composites under combined loading fail at stress levels considerably different from those under simple loading (Chapter 3 [1]). Therefore, five simple strengths and a combined-stress strength criterion are needed to describe the strength (limit) behavior of a unidirectional filamentary composite (UFC).

Examination of the physical makeup of the UFC reveals that its strength depends on the properties of the constituents and the particular fabrication process. What is needed then is a mathematical formalism to relate the strength of the UFC to the properties of its constituents and to the particular fabrication process. This mathematical formalism can be constructed with the aid of a two-level theory. The first level is a theory to predict the simple strengths of the UFC from constituent properties and fabrication process considerations. The second level is a theory to predict the onset of failure (limiting condition) of the UFC from either the simple strengths (predicted in the first level) or from certain measured values. The failure criteria at both levels, to be effective and useful for design practice, must account directly or indirectly for the following desirable features: (1) they must have some theoretical basis; (2) they should reflect the particular fabrication process (void content, size, and distribution; filament spacing-nonuniformity and misalignment; differences in bulk and *in situ* constituent properties; imperfect interface

[2] The italic numbers in brackets refer to the list of references appended to this paper.

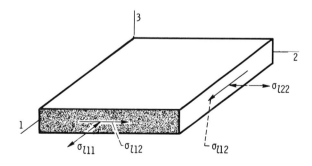

Fig. 1—Unidirectional filamentary composite (geometry and simple strength definitions).

NOTE: $S_{l11C} \leq \sigma_{l11} \leq S_{l11T}$; $\sigma_{l12} \leq S_{l12S}$; $S_{l22C} \leq \sigma_{l22} \leq S_{l22T}$.

bond; residual stress, etc.); (3) they must be applicable to both isotropic and anisotropic filaments and matrices; (4) the resulting equations should be relatively simple; and (5) the theories must be experimentally substantiated at both levels.

Various theories have been proposed in the literature [1,2]. Here, suffice it to say that all of these theories are deficient in either one or more of these desirable features. In this paper a two-level, semiempirical theory is described which meets (directly or indirectly) all of the desirable features. The theory at both levels is linear and developed primarily for nometallic composites. The basic hypothesis of this approach is that variables which cannot be accounted for directly are incorporated indirectly through the judicious introduction of theory-experiment correlation factors. The physical bases and justifications for these factors are discussed in the body of the report. A list of symbols is given in the Nomenclature.

UNIAXIAL SIMPLE STRENGTHS

The physical makeup of the UFC suggests that its simple strength will be related to its constituent properties and to the fabricaton processes as follows:

$$S_l = f[k,d,N,A)_{f,v},k_m,(E,\nu,G,S,\epsilon_p)_{f,m},S_B,S_R] \dots (1)$$

The variable S_l in Eq 1 denotes the UFC simple strength. The term $(k,d,N,A)_{f,v}$ denotes volume content, size, number, and distribution of filaments and voids. The variable k_m denotes the volume content of the matrix. The term $(E,\nu,G,S,\epsilon_p)_{f,m}$ represents the elastic and strength properties of the filaments and matrix. The variables S_B and S_R denote interface bond strength and residual stress, respectively. The void content, the bond strength, and the residual stresses are dependent on the filament surface treatment. They also depend on various matrix additives, hardeners, temperature, and pressure during fabrication and on the fabrication method of making the UFC. As can be seen the list of variables on which the UFC simple strengths depend is quite long.

Evaluation of the function in Eq 1 presents a formidable task which requires sophisticated statistical methods and a large number of experiments. This results in complex mathematical expressions not readily amenable to use in design. Variables such as void size and distribution, filament spacing and nonuniformity, interface bond strength, and residual stresses are influenced by the particular fabrication process. If it is assumed that the particular fabrication process remains approximately invariant, then it is reasonable to group all these variables into theory-experiment correlation factors. The concept just stated simplifies the derivations and resulting expressions of the simple strengths and yet retains all the essential parts of Eq 1. The details are described in the following sections.

LONGITUDINAL TENSILE FAILURE STRESS (S_{l11T})

The expression describing the longitudinal tensile strength is a modified rule of mixtures equation of the following form:

$$S_{l11T} = S_{fT} \left[\beta_{fT}\bar{k}_f + \beta_{mT}\bar{k}_m \left(\frac{E_{m11}}{E_{f11}}\right) \right] \dots (2)$$

The undefined variables in Eq 2 are as follows: S_{fT} is the filament bundle strength (or single filament strength for monofilament composites); β_{fT} and β_{mT} are the theory-experiment correlation factors which account for the particular fabrication process; $\bar{k}_f$ and $\bar{k}_m$ are actual filament and matrix volume contents are defined in the Appendix; and E_{m11} and E_{f11} represent the in situ longitudinal moduli of the matrix and filament, respectively.

Some important points should be noted: (1) Eq 2 is linear in $\bar{k}_f$ if β_{fT} and β_{mT} are independent of $\bar{k}_f$, (2) S_{l11T} is controlled by the filament bundle strength (single filament for monofilament composites), and (3) the proximity to unity of the coefficients β_{fT} and β_{mT} is a measure of the validity of the rule of mixtures and the relative insensitivity of S_{l11T} to the fabrication process.

Longitudinal Compressive Failure Stress (S_{l11C})

Using the rule of mixtures, the longitudinal compressive strength is related to constituent properties by the following equation:

$$S_{l11C} = S_{mC}\left[\beta_{mc}\bar{k}_m + \beta_{fc}\bar{k}_f \left(\frac{E_{f11}}{E_{m11}}\right)\right]\dots\dots\dots\dots\dots (3)$$

Under compressive loading, it is also possible that a UFC wil fail by a combination of debonding and intralaminar shear [3], microbuckling excluded. This condition is approximated by the following expression:

$$S_{l11D} = a_1 S_{l12S} + a_2 \dots\dots\dots\dots\dots\dots\dots (4)$$

S_{mC} is the matrix compressive strength; β_{mc} and β_{fc} are theory-experiment correlation factors analogous to those for S_{l11T}; a_1 and a_2 are empirical curve fit parameters; and S_{l12S} is the UFC intralaminar shear strength and will be defined subsequently. The remaining variables have been defined already. The curve-fit parameters a_1 and a_2 in Eq 4 are evaluated as is described in [1,3]. It is possible that they would remain the same for various systems with various filaments but one matrix. Though it is suspected that this might be the case, these coefficients should be evaluated for each filament-matrix system for reliable predictions. The important points here are: (1) S_{l11C} is very sensitive to β_{fc} since $(E_{f11}/E_{m11}) \gg 1$, Eq 3, and (2) the various fabrication process effects are introduced through S_{l12S}, Eq 4. It is suggested as a conservative measure that S_{l11C} be taken as the smaller of the two values computed from Eqs 3 and 4 or in equation form

$$S_{l11C} = \text{MIN} \left\{S_{mC}\left[\beta_{mc}\bar{k}_m + \beta_{fc}\bar{k}_f \left(\frac{E_{f11}}{E_{m11}}\right)\right], (a_1 S_{l12S} + a_2)\right\}\dots\dots\dots (5)$$

Transverse Tensile Failure Stress (S_{l22T})

The governing equation for the transverse tensile strength is based on the hypothesis that S_{l22T} is limited by the tensile strain in the matrix (Section 2.5[1]). In equation form this condition is expressed by

$$S_{l22T} = f(\epsilon_{mpT})E_{l22}\dots\dots\dots\dots\dots\dots (6)$$

where ϵ_{mpT} is the allowable matrix tensile strain defined in Fig. 2a (or any other suitable definition) and E_{l22} is the composite transverse modulus. For the linear case, the following relationship holds

$$S_{l22T} = \epsilon_{l22p}E_{l22}\dots\dots\dots\dots\dots\dots (7)$$

where ϵ_{l22p} is measured at the first knee (or point of linear deviation on the composite stress-strain curve as is illustrated in Fig. 2b. It is shown in Chapter 2 [1] that the strains ϵ_{mpT} and ϵ_{l22p} are related by the following equation

$$\epsilon_{l22p} = \beta_{22T} \frac{\epsilon_{mpT}}{\beta_v \varphi_{\mu22}}\dots\dots\dots\dots\dots\dots (8)$$

where β_{22T} is the theory-experiment correlation factor, β_v and $\varphi_{\mu22}$ are the void effect and the matrix-strain-magnification factor, respectively, and are in the Appendix. Substitution of Eq 8 into Eq 7 yields the following equation:

$$S_{l22T} = \beta_{22T} \frac{\epsilon_{mpT}}{\beta_v \varphi_{\mu22}} E_{l22}\dots\dots\dots\dots\dots\dots (9)$$

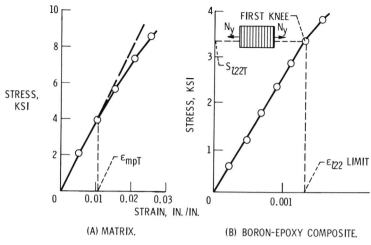

FIG. 2—Matrix and composite stress strain curves (Ref 4).

Equation 9 relates S_{l22T} to the limiting-matrix tensile strain, to the void effects, to the composite transverse modulus, and to the fabrication process through β_{22T}. It is interesting to note that both local effects (β_v and $\varphi_{\mu22}$) and average effects (E_{l22}) influence the transverse tensile strength. The coefficient β_{22T} is selected so that Eq 9 correlates with experimental data.

TRANSVERSE COMPRESSIVE FAILURE STRESS (S_{l22C})

The governing equation for the transverse compressive strength is derived in a fashion similar to that used for S_{l22T}. The result is

$$S_{l22C} = \beta_{22C} \frac{\epsilon_{mpC}}{\beta_v \varphi_{\mu22}} E_{l22} \dots \dots \dots (10)$$

where β_{22C} is the theory-experiment correlation factor, ϵ_{mpC} is the limiting-matrix compressive strain, and the remaining variables are the same as for S_{l22T}. Equations 9 and 10 differ only in the correlation coefficients and the allowable matrix strains.

INTRALAMINAR SHEAR FAILURE STRESS (S_{l12S})

The governing equation for the intralaminar shear strength is derived by a procedure similar to that used for S_{l22T}. The result is

$$S_{l12S} = \beta_{12S} \frac{\epsilon_{mpS}}{\beta_v \varphi_{\mu12}} G_{l12} \dots \dots \dots (11)$$

where β_{12S} is the theory-experiment correlation factor, ϵ_{mpS} is the allowable matrix shear strain, G_{l12} is the composite shear modulus, β_v and $\varphi_{\mu12}$ are the void and shear matrix-strain-magnification factors and are defined in the Appendix. The void effects are the same for S_{l22T}, S_{l22C}, and S_{l12S} (Chapter 2 [1]).

SELECTION OF THE CORRELATION COEFFICIENTS FOR SIMPLE STRENGTHS

The theory-experiment correlation coefficients in Eqs 2, 3, 9, 10, and 11 are selected from simple experimental setups. It is recommended that β_{12S} be evaluated from thin tubular test specimens rather than short beam specimens since the uniform shear strength of a ply is required. A numerical example in selecting β_{fT} is illustreated in the Appendix.

The important points to be noted in connection with Eqs 9, 10, and 11 are: (1) The failure stresses S_{l22T}, S_{l22C}, and S_{l12S} are very sensitive to the matrix properties, to the composite elastic properties, to the void effects and to the matrix-strain-magnification factor; therefore, it is important that the matrix-strain-magnification factors be determined with accuracy. (2) The nearness of the coefficients β_{22T}, β_{22C}, and β_{12S} to unity is a measure of the validity of the hypothesis and of the insensitivity of the failure stresses S_{l22T}, S_{l22C}, and S_{l12S} to the fabrication process. It should be clear from the discussion up to this point that all five UFC failure stresses, filament and matrix properties, filament and void content, matrix-strain-magnification factors, and the UFC elastic properties are needed to evaluate the correlation coefficients. Thus, the failure stress, as well as filament and matrix properties and filament and void content, should be made available by the material supplier. It cannot be overemphasized that these properties need to be known accurately for meaningful formulations of failure criteria and in particular for the selection of theory-experiment correlation factors. The matrix-strain-magnification factors and the UFC elastic properties can be computed when the constituent material properties are known (Appendix and Refs 1 and 2).

The simple strengths of UFC from several filament matrix systems are available in the literature. Some of these are listed in Table 1. Three points should be noted in this table: (1) the simple strengths are for one filament volume content (k_f), (2) the simple strengths of the last four composites are preliminary data and may be modified as more published data become available, and (3) the magnitudes of the tensile and compressive strengths are considerably different.

The correlation coefficients selected from the UFC simple strengths in Table 1 are listed in Table 2. As can be seen from Table 2, many of the correlation coefficients are near unity. Some exceptions are the coefficients β_{fC} for all the composites and β_{22T} for some Thornel composites. The point to be noted is that the rule of mixtures (matrix strength limited) is not a representative mechanistic model for S_{l11C}. The results of the Thornel composites indicate poor constituent bond, which behavior is similar to that of a composite with large void content. The important point to be kept in mind that the equations derived here describe the UFC simple strength behavior satisfactorily. An additional important point concerning the first-level semiempirical theory is that the correlation coefficients in Table 2 can be used to analyze, design, and develop for UFC with filament and void contents in the practical range $0.35 \lesssim k_f \lesssim 0.75$ and $0 \lesssim k_v \lesssim 0.20$. This should hold so long as the fabrication process variables noted in Eq 1 remain invariant for that particular

Table 1—*Unidirectional filamentary composite simple strengths.*

Material	Failure or Limit Stress, ksi					Filament Content, k_f	Reference and Notes
	S_{l11T}	S_{l11C}	S_{l22T}	S_{l22C}	S_{l12S}		
Boron....................	210[a]	195	8.1	26.4	12.1	0.50	6
E-glass....................	157	101	4.0[b]	20.0	6.0[b]	0.50	7
S-glass....................	268	207	3.3	21.0	5.5	0.60	8
Thornel 25................	92	67	1.0	21.0	4.0	0.50	9
Thornel 40................	140	91	1.0	19.0	3.7	0.67	9
Thornel 50................	115	60	3.6	17.0	2.6	0.60	10[c,d]
Modmor I treated	130	120	6.0	20.0	8.0	0.50	11[c,d]
Modmor II treated	150	130	8.0	20.0	11.0	0.50	11[c,d]
Beryllium................	67.4	60.0	5.0	22.0	11.7	0.50	12[d]

[a] Different k_f: (boron, 0.54).
[b] Estimates.
[c] Estimated from data reported in the reference.
[d] These values are preliminary and might change as more information becomes available.

Table 2—*Correlation coefficients for various composites (the β coefficients correspond to data in Table 1).*

Correlation Coefficient and Equation	Fiber-Matrix Composite								
	E-glass	S-glass	Boron	Thornel 25	Thornel 40	Thornel 50	Morganite I	Morganite II	Beryllium
β_{fT}, Eq 2	0.82	1.00	0.84	1.00	0.84	0.83	1.00	0.84	1.00
β_{fC}, Eq 3	0.33	0.55	0.12	0.12	0.08	0.05	0.09	0.16	0.08
β_{mT}, Eq 2	1.00	1.00	1.00	1.00	1.00	1.00	1.00	1.00	1.00
β_{mC}, Eq 3	1.00	1.00	1.00	1.00	1.00	1.00	1.00	1.00	1.00
β_{22T}, Eq 9	0.55	0.66	0.90	0.06	0.08	0.26	0.50	0.70	0.53
β_{22C}, Eq 10	1.10	1.70	1.22	0.49	0.50	0.50	0.65	0.70	0.90
β_{12S}, Eq 11	0.86	1.30	1.50	0.48	0.46	0.27	0.91	1.37	1.40
a_1,[a] Eq 4	13.30	13.30	13.3	13.3	13.3	13.3	13.3	13.3	13.3
a_2,[a] Eq 4	31900	31900	31900	31900	31900	31900	31900	31900	31900
$K_{l'12TT}$,[b] Eq 12	1.00	1.00	1.00	1.00	1.00	1.00	1.00	1.00	1.00
$K_{l'12CT}$,[b] Eq 12	1.00	1.00	1.00	1.00	1.00	1.00	1.00	1.00	1.00
$K_{l'12TC}$,[b] Eq 12	1.00	1.00	1.00	1.00	1.00	1.00	1.00	1.00	1.00
$K_{l'12CC}$,[b] Eq 12	1.00	1.00	1.00	1.00	1.00	1.00	1.00	1.00	1.00

[a] For glass-epoxy composite. Might need reevaluation for other composites.
[b] These values were assumed. Might need modification.

process. Of course if any of these fabrication process variables change, then the coefficients need to be reevaluated.

The coefficients in the last four lines of Table 2 are needed in the combined-stress strength criterion.

COMBINED-STRESS STRENGTH CRITERION

The governing equation for this criterion is derived from the following two postulates: (1) at the onset of failure, the distortion energy under simple and combined loading remains invariant, and (2) the tensile and compressive properties of UFC are the same up to the onset of failure. These two posulates are based on the von Mises criterion for isotropic materials and on the experimental observation that the distortion energy of UFC remains invariant under rotational transformation (Section 3.3 [1]). The formal derivations are described in detail in Section 3.3, Ref 1. The resulting equation (from Ref 1) is

$$F(\sigma_l, S_l, K_{l12}) = 1 - \left[\left(\frac{\sigma_{l11\alpha}}{S_{l11\alpha}} \right)^2 + \left(\frac{\sigma_{l22\beta}}{S_{l22\beta}} \right)^2 + \left(\frac{\sigma_{l12S}}{S_{l12S}} \right)^2 \right.$$
$$\left. - K'_{l12\alpha\beta} K_{l12} \frac{\sigma_{l11\alpha}\sigma_{l22\beta}}{|S_{l11\alpha}| \, |S_{l22\beta}|} \right] \quad ..(12)$$

where F denotes the combined-stress strength criterion as follows:

$F(\sigma_l, S_l, K_{l12}) > 0$ no failure
$F(\sigma_l, S_l, K_{l12}) = 0$ onset of failure
$F(\sigma_l, S_l, K_{l12}) < 0$ failure condition exceeded

The subscripts α and β denote T (tension) or C (compression); σ_l denotes the applied stress state determined from the stress analysis; S_l denotes the UFC simple strength either determined from the equations described previously or measured experimentally, $K'_{l12\alpha\beta}$ is the theory-experiment correlation coefficient and is determined as will be described subsequently. The coefficient K_{l12} is given by (from Chapter 3, [1])

$$K_{l12} = \frac{(1 + 4\nu_{l12} - \nu_{l13})E_{l22} + (1 - \nu_{l23})E_{l11}}{[E_{l11}E_{l22}(2 + \nu_{l12} + \nu_{l13})(2 + \nu_{l21} + \nu_{l23})]^{1/2}} \quad \cdots\cdots\cdots (13)$$

where E_l and ν_l denote UFC modulus and Poisson's ratio, respectively. The subscripts 11, 31, etc. refer to the corresponding axes in Fig. 1. For isotropic material, Eq 13 reduces to unity, as can be verified by direct substitution, and Eq 12 reduces to the well known von Mises criterion.

Several important points should be noted at this juncture: (1) UFC do not exhibit similar properties in tension and compression, as was stated in the second postulate made at the beginning of this section; therefore, $K'_{l12\alpha\beta}$ is introduced to compensate for this disparity. (2) Equation 12 describes failure at each quadrant by using, at most, four parameters. (3) The correlation coefficient $K'_{l12\alpha\beta}$ can have different values in different quadrants. (4) The product $K'_{l12\alpha\beta}K_{l12}$ can be 'defined as one constant and determined experimentally; however, this disguises the composite effects which are introduced into Eq 12 through K_{l12}. (5) The product $K'_{l12\alpha\beta}K_{l12}$ is not restricted to any range, that is, $-\infty < K'_{l12\alpha\beta}K_{l12} < \infty$ (Chapter 3, Ref 1). (6) Equation 12 is applicable to all materials exhibiting generally orthotropic elastic symmetry and is not restricted only to UFC.

Values of the variable K_{l12} versus k_f are plotted in Fig. 3 for several filament-resin systems. The graphs in Fig. 3 are applicable to composites with various void contents since K_{l12} is only slightly sensitive to the void content. A procedure to select the coefficients $K'_{l12\alpha\beta}$ is illustrated in Fig. 4. The experimental results in this figure are for JT-50 graphite at 10 percent porosity as reported in Ref 13. For this material, $K_{l12} = 0.85$. As can be seen from Fig. 4, the coefficients $K'_{l12\alpha\beta}$ can be selected so that a good theory-experiment correlation can be obtained. One important point to be noted in Fig. 4 is that the slope of $F(\sigma_l, S_l, K_{l12})$ is discontinuous across the quadrant junctures. This type of behavior is typical of orthotropic materials.

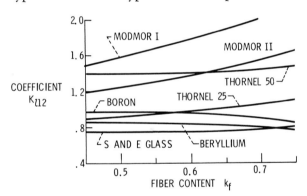

FIG. 3 — Combined-stress strength-criterion coefficient (Eq 13).

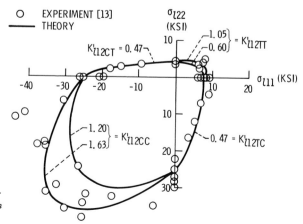

FIG. 4—Evaluation of correlation coefficient $K'_{l12\alpha\beta}$ (Eq 12).

Chamis on Filamentary Composites 343

Another important point is that the failure criterion is sensitive to $K_{l12\alpha\beta}$ in the tension-tension (TT) and compression-compression (CC) quadrants.

APPLICATIONS, RESULTS, AND DISCUSSION

The two-level semiempirical failure theory can be used in several ways: composite failure analysis, design, structural synthesis, generation of strength envelopes, and as an aid to experimental work. Here, the discussion is restricted to composite failure analysis (whch serves as a verification of the two level theory) and to the generation of strength envelopes (which point out problem areas in testing and design).

COMPOSITE FAILURE ANALYSIS AND THEORY VERIFICATION

The two-level, semiempirical failure theory has been applied to several multi-layered filament-matrix composites. The input data in these applications consisted of the constituent material properties, the filament and void contents, the correlation coefficients (Table 2), the composite geometry, and the failure or maximum load. The generation of other required properties (ply elastic constants, simple strengths, etc.), the composite stress analysis, and the failure test according to Eq 12 were carried out by a multilayered-filamentary-composite-analysis computer code. Typical results are presented in Table 3.

The results in Table 3 are for Thornel 40-epoxy composite specimens reported in Ref 9. The first four columns in this table contain the composite geometry, the fifth column the applied load, and the last two columns the value of the failure criterion (Eq 12) for the first and second plies. As can be seen in the last two columns of Table 3, the criterion predicts failure or nearly so in all composites except those noted by a. The large negative values of the criterion indicate primarily two possibilities: (1) the presence of load transfer difficulties from the outer to the inner plies and (2) the transverse plies failed at an early state of the loading process and the load was carried primarily by the longitudinal plies. The criterion values of the longitudinal plies are in accord with the second possibil-

Table 3—*Results of failure analyses of Thornel 40-epoxy multilayered composites based on data from Ref 9.*

Composite Ply Arrangement	Fiber Content, k_f	Ply Thickness, t_l, in.	Failure Load, lb/in.		Failure Criterion, Eq 12 $F(\sigma_l, S_l, K_{l12})$	
			N_x	N_y	First ply or first pair	Second ply or second pair
4(90,0,0,90)	0.64	0.0086	1960	0	−16.5[a]	0.63
	0.64	0.0086	0	2300	.49	−23.1[a]
	0.64	0.0086	−1520	0	.97	−.03
	0.64	0.0086	0	−1750	−.36	.95
4(90,10,−10,90)	0.57	0.010	1440	0	−7.80[a]	0.73
	0.57	0.010	0	2240	.57	−17.0[a]
	0.57	0.010	−1520	0	.96	.14
	0.57	0.010	0	−1960	.25	.88
4(10,−10,−10,10)	0.69	0.0079	2120	0	0.52[a]	0.52
	0.69	0.0079	0	31.6	.23	.23
	0.69	0.0079	−1710	0	−2.18	−2.18
	0.69	0.0079	0	−15.5	−.16	−.16
6(10,−10,45,−45,10, −10)	0.56	0.0102	3300	0	0.30	−0.67
	0.56	0.0102	0	428	−4.80	−.78
	0.56	0.0102	−2940	0	.65	−1.09

[a] Load transfer difficulties.

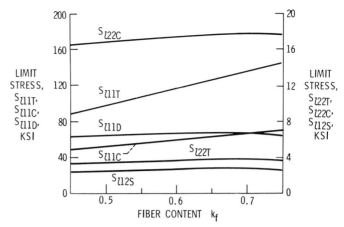

FIG. 5—Limit stresses for Thornel 50-epoxy composites (based on correlation coefficients in Table 2 and zero void content).

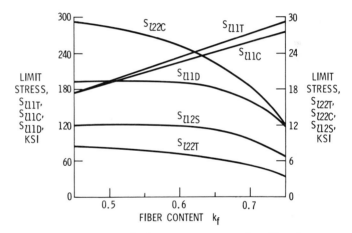

FIG. 6—Limit stresses for boron-epoxy composites (based on correlation coefficients in Table 2 and zero void content).

ity. This is another important use of the semiempirical theory; that is, it points out problem areas which need be either remedied or avoided. Results of several other systems are presented in Ref 2.

GENERATION OF STRENGTH ENVELOPES

Figures 5 and 6 contain graphs of the simple strengths S_{l11T}, S_{l11C}, S_{l11D} (Eq 4), S_{l22T}, S_{l22C}, and S_{l12S} versus filament content (k_f) with zero voids. These figures are for Thornel-50 and boron-filament UFC. The graphs in these figures were generated from the correlation coefficients in Table 1. The important points to be noted here: (1) The transverse and shear limit stresses decrease with increasing filament content for boron composites but remain rather invariant for Thornel 50 composites. (2) The decrease of these limit stresses (S_{l22T}, S_{l22C}, and S_{l12S}) is very rapid at high filament volume content values. (3) Test results for transverse and shear properties for isotropic filament composites should be reported with accurate volume content, particularly in the high range. Those for orthotropic filaments ($E_{f11}/E_{f22} \gg 1$) need not be very accurate. If longitudinal compressive failure

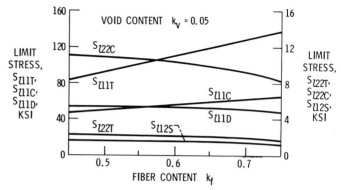

FIG. 7— Limit stresses for Thornel 50-epoxy composites (based on correlation coefficients in Table 2 and 5 percent void content).

of UFC is governed by S_{l11D} (constituent debonding and intralaminar shear) then this failure strength decreases with filament content (k_f), and the value of k_f should be measured fairly accurately.

Figures 7 and 8 contain simple strength envelopes for Thornel-50 and boron composites with five percent void content, respectively. Superposition of Figs. 5 with 7 and 6 with 8 reveals a considerable drop in the transverse and shear strengths of composites with voids. It is important, therefore, to report the void content accurately when presenting experimental results on transverse and shear strengths. Of course the void content should be reported with longitudinal compressive strength as well since this strength could be governed by S_{l11D}.

The combined-stress strength behavior for a Thornel 50-epoxy UFC is illustrated in Fig. 9. (Note that this figure has different horizontal and vertical scales.) The contours in this figure represent strength envelopes for various values of intralaminar shear expressed as a fraction of S_{l12S}. The three important points to be noted in Fig. 9 are: (1) Thornel 50 composites under the proper proportion of combined loading can resist considerably more load than their simple strengths would indicate, (2) the normal load capacity of this UFC is insensitive to small values of shear loads, and (3) the longitudinal compressive strength of UFC is very sensitive to transverse tensile loads while the longitudinal tensile strength is sensitive to trans-

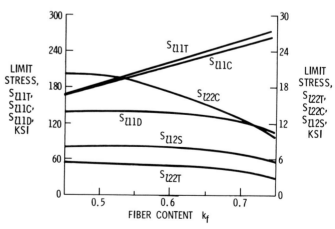

FIG. 8—Limit stresses for boron-epoxy composites (based on correlation coefficients in Table 2 and 5 percent void content).

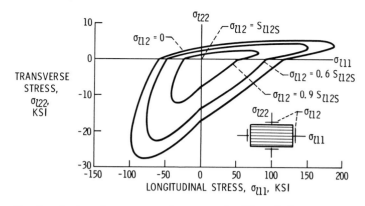

FIG. 9—Combined-stress-strength criterion for Thornel 50-epoxy composite (based on data in Table 1 and Eq 12).

verse compressive loads and very sensitive to small transverse tensile stresses. Therefore, it is important to bear in mind when testing for longitudinal strengths that the transverse stresses should be eliminated completely. A small amount of shear stress can be tolerated.

Figure 10 illustrates the strength envelope of UFC when loaded with normal loads at some angle to the filament direction. The upper part of the figure is for tensile load and the lower for compressive. These envelopes were obtained by expressing σ_l in Eq 12 in terms of σ_x and then solving the resulting expression for σ_x. As can be seen in this figure, the strength drops off very rapidly as the load direction deviates from the filament direction (for small angles). The composite has lost about 50 percent of its strength when the load is applied 5 deg to the filament direction. It is imperative, therefore, to have the load completely aligned with the filament direction for longitudinal tensile strength tests. In tests to determine strengths for $\theta > 40$ deg (also transverse strength ($\theta = 90$ deg)) the load alignment is not critical.

Figure 11 illustrates an analogous effect for the case of shear load. The upper part of the figure is for positive shear (tending to elongate the filaments) and the lower part for negative. One important effect brought out in the lower part of Fig. 11 (not widely recognized) is that UFC loaded with negative shear load 45 deg to the filament direction has a very low shear strength. This means that

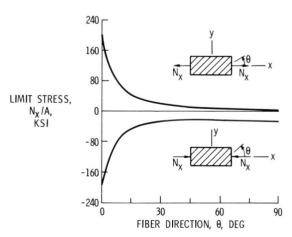

FIG. 10—Limit stress for off-axes normal load boron-epoxy composite. (simple strengths from Fig. 7 at 50 percent fiber content.)

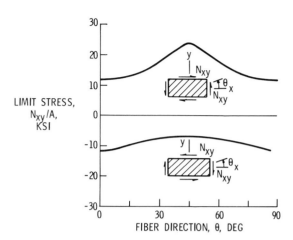

LIMIT STRESS,
N_{xy}/A,
KSI

FIBER DIRECTION, θ, DEG

Fig. 11— Limit stress for off-axis shear load boron-epoxy composite. (simple strengths from Fig. 7 at 50 percent fiber content.)

when ±45-deg plies are introduced to help carry shear loads, the −45-deg ply fails (perhaps not completely) at relative small loads and the +45-deg ply carries the load. The partially failed −45-deg ply causes the composite to exhibit a non-linear load response as the loading increases. Other important points to be kept in mind for shear strength tests and design are: (1) shear strength of UFC is insensitive to small angular deviations ($\theta \approx 0$ and $\theta \approx 90$ in Fig. 11), (2) the strength of UFC loaded with positive shear load at 45 deg is sensitive to small angular deviations while it is insensitive if loaded with negative shear load, and (3) the strength of a UFC loaded with positive shear load at 45 deg is approximately three times greater than the similar case with negative shear load. This ratio is approximately equal to S_{122C}/S_{122T}.

Several recommendations have been made already on how theoretical work can aid the experimental effort. On the other hand, controlled experimental work is a very important asset in formulating and verifying any theory. For this reason, it is imperative that the complete test record (type of specimen, constituents, voids, type of test and techniques, type of failure, strain rate, means for measuring strain and elongation, and any other factors with influence the result) be reported when presenting experimental data for filamentary composite properties.

CONCLUSIONS

The results of this investigation lead to the following conclusions:

1. The simple strengths are sensitive to the correlation-coefficients and thereby to the particular fabrication process.

2. The simple strengths are fairly sensitive to void content.

3. The transverse and shear strengths are decreasing functions of the filament content (particularly at the high range) for isotropic filament composites.

4. The combined-stress strength criterion is sensitive to its correlation coefficient in the tension-tension and compression-compression quandrants and relatively insensitive in the other two quadrants.

5. The longitudinal compressive strength is very sensitive to the presence of transverse tensile stress.

6. The longitudinal tensile strength is very sensitive to small transverse tensile stress, and it is sensitive to transverse compressive stress.

7. The normal load carrying capacity of a UFC decreases rapidly as the angle between filament and load direction increases. Consequently, the longitudinal tensile and compressive strengths are very sensitive to load misalignment.

8. The shear strength is insensitive to load misalignment.

9. The positive shear load carrying capacity of a UFC increases as the angle

between filament and load direction increases to a maximum at 45 deg, but it decreases for negative shear load to a minimum at 45 deg.

10. Plies introduced to resist shear forces may do so with nonlinear response.

11. The complete test record should be reported when presenting experimental data on composite materials.

ACKNOWLEDGMENTS

A portion of this work was supported by the Advanced Research Projects Agency, Department of Defense, through Grant no. AF 33(615)-3110, administered by the Air Force Materials Laboratory while the author was a member of the Engineering Design Center, Case Western Reserve University, Cleveland, Ohio.

Appendix

USEFUL RELATIONS

ACTUAL FILAMENT AND MATRIX VOLUME CONTENT

Let k_f and k_m denote the apparent filament and matrix volume content, respectively, and let $\bar{k}_v$, $\bar{k}_f$, and $\bar{k}_m$ denote the actual void, filament, and matrix volume content, respectively; then it can be shown (Appendix A, Ref 1) that

$$\bar{k}_f = (1 - \bar{k}_v)k_f \dots\dots\dots\dots\dots\dots\dots\dots(14)$$
$$\bar{k}_m = (1 - \bar{k}_v)(1 - k_f) \dots\dots\dots\dots\dots\dots(15)$$
$$k_m = 1 - k_f \dots\dots\dots\dots\dots\dots\dots\dots\dots(16)$$

MATRIX-STRAIN-MAGNIFICATION FACTORS AND VOID EFFECTS

The transverse and shear matrix-strain magnification factors are respectively given by (Chapter 2, Ref 1):

$$\varphi_{\mu 22} = \left[\frac{1}{1 + p(\bar{A} - 1)}\right]\left[1 + p(\nu_{f12} - \nu_{m12}\bar{A})\left(\frac{E_{l22}\sigma_{l11} - \nu_{l21}E_{l11}\sigma_{l22}}{E_{l11}\sigma_{l22} - \nu_{l12}E_{l22}\sigma_{l11}}\right)\right] \dots(17)$$

if $E_{l11}\sigma_{l22} - \nu_{l12}E_{l22}\sigma_{l11} = 0$, then

$$\varphi_{\mu 22} = 1 \dots\dots\dots\dots\dots\dots\dots\dots\dots(18)$$

$$\varphi_{\mu 12} = \frac{1}{1 - p\left(1 - \dfrac{G_{m12}}{G_{f12}}\right)} \dots\dots\dots\dots\dots\dots(19)$$

The maximum void effect factor is given by (Chapter 2, Ref 1):

$$\beta_v = \frac{1}{1 - \left(\dfrac{4k_v}{\pi k_m}\right)^{1/2}} \dots\dots\dots\dots\dots\dots(20)$$

The notation in Eqs 18 to 20 is as follows:

$$p = \left(\frac{4\bar{k}_f^{1/}}{\pi}\right)^2 \dots\dots\dots\dots\dots\dots\dots(21)$$

$$\bar{A} = \frac{1 - \nu_{f12}\nu_{f21}}{1 - \nu_{m12}\nu_{m21}}\left(\frac{E_{m22}}{E_{f22}}\right) \dots\dots\dots\dots\dots(22)$$

The variables E, G, ν, and σ denote longitudinal modulus, shear modulus, Poisson's ratio, and stress, respectively. The subscripts v, f, and m denote void, filament, and matrix, respectively, while the numerical subscripts correspond to the filament direction depicted in Fig. 1. The variable k_v is the void content and k_f and k_m are defined by Eqs 14 and 16, respectively.

EVALUATION OF CORRELATION COEFFICIENTS FOR SIMPLE STRENGTHS

Several ways can be used to evaluate the correlation coefficients. The simplest one is illustrated by the following example: (Thornel 25-epoxy):

$$S_{fT} = 180,000;\ S_{l11T} = 92,000\ \text{psi};\ \bar{k}_f = 0.50;\ \bar{k}_m = 0.5;$$
$$E_{f11} = 25 \times 10^6\ \text{psi};\ E_{m11} = 0.55 \times 10^6\ \text{psi}$$

Solving Eq 2 for β_{fT} yields

$$\beta_{fT} = \frac{1}{\bar{k}_f}\left(\frac{S_{l12T}}{S_{fT}} - \beta_{mT}\bar{k}_m\frac{E_{m11}}{E_{f11}}\right) \dots\dots\dots\dots\dots\dots\dots (23)$$

Assuming $\beta_{mT} = 1.0$ and substituting the values for all the variables in Eq 23 yields $\beta_{fT} \approx 1.0$. The remaining correlation coefficients can be evaluated in the same fashion.

References

[1] Chamis, C. C., "Design Oriented Analysis and Structural Synthesis of Multilayered Filamentary Composites," Ph.D. thesis, Case Western Reserve University, Cleveland, Ohio, 1967.

[2] Chamis, C. C., "Failure Criteria for Filamentary Composites," NASA TND-5367, National Aeronautics and Space Administration, 1969.

[3] Fried, N., "The Response of Orthogonal Filament-Wound Materials to Compressive Stress," 20th Annual Conference, Society of the Plastics Industry, New York, N. Y., 1965, pp. 1-1-C to 8-1-C.

[4] Rogers, C. W. et al, "Application of Advanced Fibrous Reinforced Composite Materials to Airframe Structures," (AFML-TR-66-313, Vol. 2, AD-803128L), General Dynamics/Fort Worth, Sept. 1966.

[5] Chamis, C. C., "Thermoelastic Properties of Unidirectional Filamentary Composites by a Semiempirical Micromechanics Theory," Science of Advanced Materials and Process Engineering, Vol. 14, Western Periodicals Co., North Hollywood, Calif., 1968, paper I-4-5.

[6] Waddoups, M. E., "Characterization and Design of Composite Materials," Composite Materials Workshop, Tsai, S. W., Halpin, J. C., and Pagano, N. J., eds., Technomic, Stamford, Conn., 1967, pp. 254–308.

[7] Hoffman, O., "The Brittle Strength of Orthotropic Materials," Journal of Composite Materials, Vol. 1, No. 2, April 1967, pp. 200–206.

[8] Noyes, J. V. and Jones, B. H., "Analytical Design Procedures for the Strength and Elastic Properties of Multilayer Fibrous Composites," Paper 68-336, American Institute of Aeronautics and Astronautics, April 1968.

[9] Blakslee, O. L. et al, "Fabrication, Testing, and Design Studies With "Thornel" Graphite-Fiber, Epoxy-Resin Composites," Science of Advanced Materials and Process Engineering, Vol. 12, Western Periodicals Co., North Hollywood, Calif., 1967, Section A-6.

[10] Hoggatt, J. T., Burnside, J. Y., and Bell, J. E., "Development of Processing Techniques for Carbon Composites in Missile Interstage Application," Report D2-125559-4 (AFML-TR-68-155, AD-839 857L), Boeing Company, June 1968.

[11] New Products Data Sheet, Morganite Research and Development Limited.

[12] Schwartz, H. S., Schwartz, R. T., and Mahieu, W., "Mechanical Behavior of Beryllium Wire Reinforced Plastic Composites," *Science of Advanced Materials and Process Engineering*, Vol. 10, Western Periodicals Co., North Hollywood, Calif., 1966, pp. A-41 to A-55.

[13] Weng, T. L., "Biaxial Fracture Strength and Mechanical Properties of Graphite-Base Refractory Composites," Paper 68-337, American Institute of Aeronautics and Astronautics, April 1968.

Shear Stability of Laminated Anisotropic Plates

J. E. ASHTON[1] AND
T. S. LOVE[1]

Reference:

Ashton, J. E. and Love, T. S., "Shear Stability of Laminated Anisotropic Plates," *Composite Materials: Testing and Design, ASTM STP 460*, American Society for Testing and Materials, 1969, pp. 352–361.

Abstract:

An analytical and experimental study of the shear stability of laminated anisotropic plates is presented. The plates considered are rectangular, have clamped edges, and are fabricated with laminated construction of boron-epoxy composite material. Such plates, when the laminates are midplane symmetric, are anisotropic in nature. The analytical solutions are obtained by means of the principle of stationary potential energy, using the Ritz method. The deflected shape is approximated with beam characteristic functions. Experimental results are presented for 14 boron-epoxy plates and for 2 aluminum plates, and the results compared with the theory. Good agreement is shown.

Key Words:

stability, anisotropy, plates, laminates, boron, epoxy laminates, fiber composites, evaluation, tests

The advent of the widespread use of fiber-reinforced composite materials has introduced a renewed interest in the analysis of plates composed of laminated orthotropic materials. Such plates, when the laminations possess midplane symmetry, are equivalent mathematically to homogeneous anisotropic plates. The stability analysis of such plates has not been considered in the literature except for a few special cases [1,2,3],[2] and experimental investigations of their behavior are also scarce [4,5]. The present paper presents an analytical and experimental study of the shear stability of rectangular laminated anisotropic plates with clamped edges. The analytical solutions are obtained by means of the principle of stationary potential energy, using the Ritz method. The deflected shape is approximated with beam characteristic functions. Experimental results are presented for 14 boron-epoxy plates and for 2 aluminum plates, and the results compared with the theory. Good agreement is shown.

ANALYSIS

The plate is shown in Fig. 1. The solution is obtained by equating the potential energy due to bending plus the potential energy of the inplane loads moving through the bending deflections to zero, and requiring this equivalence to be stationary. The potential energy due to bending can be written [6]

$$V = \tfrac{1}{2} \int_0^a \int_0^b D_{11}W^2_{,xx} + 2D_{12}W_{,xx}W_{,yy} + D_{22}W^2_{,yy}$$
$$+ 4W_{,xy}\{D_{16}W_{,xx} + D_{26}W_{,yy} + D_{66}W_{,xy}\}dxdy \ldots (1)$$

---

[1] Senior structures engineer and senior test engineer, respectively, General Dynamics Corp., Fort Worth, Tex. 76101.

[2] The italic numbers in brackets refer to the list of references appended to this paper.

352

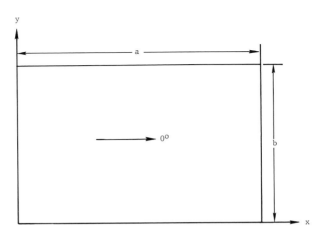

FIG. 1—Coordinates.

where:

W = plate deflection,

V = potential energy due to bending,

D_{ij} = plate bending stiffness coefficients which are defined in terms of the lamina orientations and stacking sequence as well as the lamina orthotropic elastic constants E_1, E_2, ν_{12}, and G_{12} [7], and

a comma denotes partial differentiation.

The potential energy of the inplane loads is

$$U = -\tfrac{1}{2} \int_0^a \int_0^b (N_x W_{,x}^2 + N_y W_{,y}^2 + 2N_{xy} W_{,x} W_{,y}) dx dy \dots \dots \dots (2)$$

where N_x, N_y, N_{xy} are the inplane stress resultants. For this analysis, only the shear stress resultant N_{xy} will be considered.

In order to find an approximately stationary energy balance between the bending energy and the potential energy of the inplane loads moving through the bending deflections, the deflection W has been assumed in the following form:

$$W(x,y) = \sum_{i=1}^{7} \sum_{j=1}^{7} a_{ij} X_i(x) X_j(y) \dots \dots \dots \dots \dots (3)$$

where:

a_{ij} = parameters to be determined, and

$X_i(x)$ = mode shape function for the ith mode of a clamped-clamped unit length beam.

The assumed series Eq 3 satisfies the boundary conditions of zero deflection and zero slope. The solutions are obtained by substituting the expression Eq 3 into Eqs 1 and 2 and equating the sum of these to zero. The requirement that this sum must be stationary is then approximately equivalent to the requirement that this energy balance must be a minimum. This minimum is found by differentiating the energy balance with respect to each a_{ij} which results then in 49 simultaneous equation for the parameters a_{ij}:

$$\sum_{m=1}^{7} \sum_{n=1}^{7} d_{ikmn} a_{mn} - \sum_{m=1}^{7} \sum_{n=1}^{7} e_{ikmn} a_{mn} = 0 \qquad i, k = 1 \cdots 7 \dots \dots \dots (4)$$

Ashton and Love on Laminated Anisotropic Plates   **353**

where:

$$d_{ikmn} = D_{11}\psi_{3im}\psi_{1kn}\frac{b}{a^3} + D_{12}(\psi_{5im}\psi_{5nk} + \psi_{5mi}\psi_{5kn})\frac{1}{ab}$$

$$+ D_{22}\psi_{1im}\psi_{3kn}\frac{a}{b^3} + 2D_{16}(\psi_{6mi}\psi_{4kn} + \psi_{6im}\psi_{4nk})\frac{1}{a^2}$$

$$+ 2D_{26}(\psi_{4im}\psi_{6nk} + \psi_{4mi}\psi_{6kn})\frac{1}{b^2} + 4D_{66}\psi_{2im}\psi_{2kn}\frac{1}{ab},$$

$$e_{ikmn} = N_{xy}(\psi_{4im}\psi_{4nk} + \psi_{4mi}\psi_{4kn}),$$

$$\psi_{1mn} = \int_0^1 X_m(\xi)X_n(\xi)d\xi,$$

$$\psi_{2mn} = \int_0^1 X_m(\xi),_\xi X_n(\xi),_\xi d\xi,$$

$$\psi_{3mn} = \int_0^1 X_m(\xi),_{\xi\xi} X_n(\xi),_{\xi\xi} d\xi,$$

$$\psi_{4mn} = \int_0^1 X_m(\xi),_\xi X_n(\xi)d\xi,$$

$$\psi_{5mn} = \int_0^1 X_m(\xi),_{\xi\xi} X_n(\xi)d\xi, \text{ and}$$

$$\psi_{6mn} = \int_0^1 X_m(\xi),_{\xi\xi} X_n(\xi),_\xi d\xi.$$

Equation 4 can be solved for the smallest value of N_{xy} which yields a zero determinant of the system of equations. For the solutions presented below, the critical value of this shear stress resultant, $N_{xy_{cr}}$, has been obtained by iterating the eigenvector [8].

The above analysis utilizing the Ritz method converges from above on the exact solution for the problem at hand as long as the series terms are complete and satisfy the geometrical boundary conditions. Since the assumed series Eq 3 satisfies these conditions, and since a large number of terms have been used in the approximating series, the solutions presented herein are quite accurate. In fact, all solutions presented herein are within one percent of the exact values for the properties and formulation presented herein.

Due to the anisotropic nature of the plate constitutive equations, the plate has a "preferred" direction of shear. That is, the value of $N_{xy_{cr}}$ is different in absolute value when applied in the positive shear direction than when applied in the negative shear direction. Thus one direction is more resistant to shear buckling than the opposite direction, and the lowest positive and negative eigenvalue (in absolute value) needs to be determined to completely define the shear buckling behavior. In the results presented below, the values of $N_{xy_{cr}}$ are those corresponding to the direction of shear enforced by the test fixture.

EXPERIMENTAL PROGRAM

The purpose of the experimental program was to determine the critical buckling load of 14 boron-epoxy laminated plates and 2 aluminum plates. The composite panels were fabricated of 16-plies of Narmco 5505 boron-epoxy preimpregnated tape (Narmco Materials Division of the Whittaker Corp.). The plates were tested under uniform shear on all edges in a "picture frame" fixture which clamped all edges. The panels were 21 by 9 in. with $1\frac{1}{4}$ in.^{2} notches cut in opposite corners. Each panel had 100 attachment holes precision drilled with a Cavitron (ultrasonic) to match the shear-loading frame. A typical panel is shown in Fig. 2.

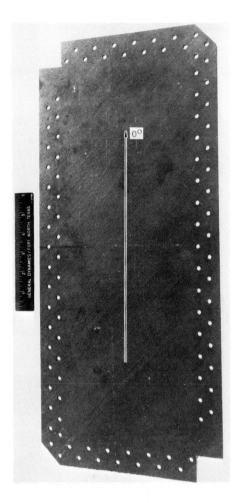

FIG. 2—Boron composite shear panel.

The principal variables in the panels tested were the orientations of the lamina. The panel lamination orientations with respect to the long axis of the panels are given in Table 1 along with the average panel thicknesses (the panel thicknesses were measured at 9 locations to obtain these average thicknesses; the variation in thickness over a panel was generally less than 1 percent).

The shear frame, Fig. 3, consisted of two sets of two pin-connected legs which were attached to the shear panel by 50 bolts each. The inside area of the shear panel after attachment of the frame was 6 by 18 in.

The shear tests were conducted in a Baldwin-Southwark 200,000-lb loading machine. The loading frame was installed in the test machine with two clevices. Five linear differential transformers, Sanborn Model 585 DT-100, were used to monitor deflections. These are shown attached to the panel in Fig. 4.

To proceed with a test, the external load was first increased to 500 lb at which time polyurethane tabs and the transformer rods were bonded to the panel. At this load the transformers were adjusted to indicate zero deflection.

The critical buckling load was determined in each test using the method recommended by Southwell [9] and described briefly as follows:

Southwell has shown that near the critical buckling load the relationship between

Table 1—*Shear panel configurations with respect to long axis.*

Panel	Configurations
Aluminum 1	$t_{av} = 0.121$
Aluminum 2	$t_{av} = 0.088$
1	0/90/0/90/0/90/0/90/90/0/90/0/90/0/90/0
	$t_{av} = 0.088$
2	16 @ 0 deg
	$t_{av} = 0.087$
3	16 @ 0 deg
	$t_{av} = 0.088$
4	same as (1)
	$t_{av} = 0.085$
5	90/−45/+45/0/90/−45/+45/0/0/+45/−45/90/0/+45/−45/90
	$t_{av} = 0.084$
6	same as (5)
	$t_{av} = 0.084$
7	same as (12)
	$t_{av} = 0.086$
8	+45/−45/+45/−45/+45/−45/+45/−45/−45/+45/−45/+45/−45/+45/−45/+45
	$t_{av} = 0.088$
9	same as (10)
	$t_{av} = 0.090$
10	0/+45/−45/90/0/+45/−45/90/90/−45/+45/0/90/−45/+45/0
	$t_{av} = 0.089$
11	same as (12)
	$t_{av} = 0.088$
12	−45/+45/−45/+45/−45/+45/−45/+45/+45/−45/+45/−45/+45/−45/+45/−45
	$t_{av} = 0.088$
13	16 @ −45 deg
	$t_{av} = 0.090$
14	16 @ −45 deg
	$t_{av} = 0.085$

Fig. 3—Shear frame with panel installed.

FIG. 4—Shear setup showing transformer rack and five linear transformers in position.

the deflections, externally applied load, and the critical buckling load can be expressed as:

$$\delta = \frac{a_1}{\dfrac{P_{cr}}{P} - 1}$$

where:
 δ = deflection,
 a_1 = constant,
 P = applied load, and
 P_{cr} = critical buckling load.
Rewriting Eq 5 in a more usable form:

$$\frac{\delta}{P} P_{cr} - \delta = a_1 \dotfill (6)$$

it is apparent that a plot of δ/P versus δ should yield a straight line (near the critical buckling load). Furthermore, the slope of this line is equal to $1/P_{cr}$.

Care must be used in selecting the region in which the Southwell hypothesis applies. Points on the δ/P versus δ plots corresponding to a maximum deflection greater than one half the plate's thickness can not be used (membrane stiffening action begins to become important), and points well below the buckling region also will not follow a straight line relationship [10]. In the present work, the

Ashton and Love on Laminated Anisotropic Plates 357

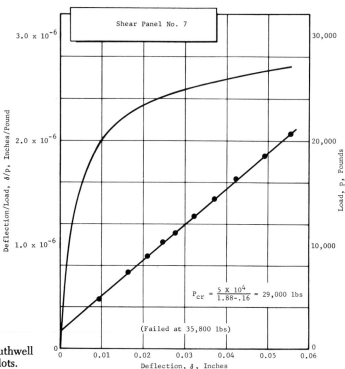

FIG. 5—Typical Southwell and load-deflection plots.

straight line region was determined graphically, emphasizing the larger deflections, but keeping the maximum deflection below half the plate thickness. The Southwell plot for a typical test (using the transformer showing the largest deflection), as well as the corresponding load deflection plot, are presented in Fig. 5.

RESULTS

The analysis procedure, as described above, was used to predict the buckling loads of the experimental plates. The input properties for the lamina making up the plates were obtained by independent tests on sandwich beam specimens. The properties used were:

$$E_1 = 3.1 \times 10^7$$
$$\nu_{12} = 0.28$$
$$E_2 = 2.7 \times 10^6$$
$$G = 0.75 \times 10^6$$

For the two aluminum plates, the assumed properties were:

$$E = 1.05 \times 10^7$$
$$\nu = 0.33$$

The results of the experimental program are tabulated in Table 2 and compared in Fig. 6. In the table and figure the experimentally determined K factor is given versus the analytically predicted K factor. The buckling load is defined as:

$$N_{xy_{critical}} = K \frac{D_{11}}{b^2} \dots\dots\dots\dots\dots\dots\dots\dots (7)$$

where D_{11} = calculated flexural rigidity in the x-direction.

Table 2—Shear panel buckling coefficients.

Panel	K (experimental)	K (theory)
Aluminum 1.........	91.1	95.2
Aluminum 2.........	88.5	95.2
1.................	53.2	54.4
2.................	13.5	13.4
3.................	13.8	13.4
4.................	49.4	54.4
5.................	137.8	137.8
6.................	143.0	137.8
7.................	109.5	118.5
8.................	131.9	149.3
9.................	63.0	65.3
10.................	70.0	65.3
11.................	111.2	118.5
12.................	125.5	118.5
13.................	48.8	37[a]
14.................	50.6	37[a]

[a] 44.4 with $E_2 = 3.4 \times 10^6$.

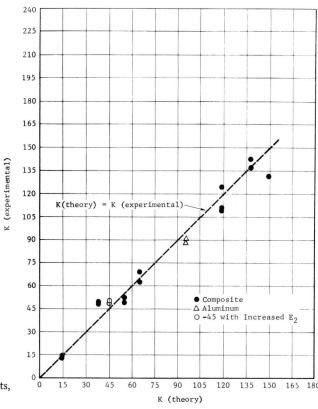

Fig. 6—Buckling results, shear plates.

Agreement between theory and experiment is good except for Panels 13 and 14, which were unidirectional panels with the principal axis of orthotropy at 45 deg to the plate edges. These panels are affected strongly by the transverse modulus E_2. Results for an increased E_2 of 3.4×10^6 are shown also in Fig. 6. An improvement between theory and experiment results. Such an increase in the transverse modulus has only a very small effect on the other plates tested; for example, the solution for a 16-ply ± 45-deg plate is only 2 percent greater with the increased transverse modulus, and thus this material property may be a source of the error.

One other source of error in the test setup used in this investigation would be important when testing Panels 13 and 14. Such panels are strongly anisotropic inplane, and consequently a uniform shear stress applied to the plate produces midplane strains ϵ_x and ϵ_y. Since the edges of these shear panels were bolted to the fixture, such inplane strain was restricted. This would cause, for Panels 13 and 14, induced tensile stresses that would increase the buckling resistance. The other plates tested were all orthotropic inplane, and thus no such error would be introduced.

An interesting point concerning the shear buckling behavior of anisotropic plates is that such plates are more resistant to edge shear in one direction than shear in the opposite direction. Panels 7, 11, and 12 were ± 45-deg plates with the fibers in the outside layer oriented in the tension direction (of the resolved shear stress). Panel 8 was a ± 45-deg plate with the fibers in the outside layer oriented in the compression direction (of the resolved shear stress). The predicted buckling resistance of Panel 8 was 26 percent higher than for Panels 7, 11, and 12. The average observed difference was 14 percent.

Panels 5, 6, 9, and 10 were "pseudoisotropic" inplane. The buckling resistance was directional, however, as the stiffer bending layup with respect to the short direction (Panels 5 and 6) was 17 percent more resistant to buckling.

CONCLUSIONS

An experimental and analytical determination of the critical buckling loads of boron-epoxy composite plates under shearing loads has been described. The analysis, based upon the basic lamina properties and an assumed mode energy solution, has proved to be an accurate means for predicting the shear buckling behavior of these plates.

ACKNOWLEDGMENT

This work was sponsored by the Air Force Materials Laboratory, Research and Technology Division, Air Force Systems Command, United States Air Force under Contract No. AF33(615)-5257.

References

[1] Thielemann, W., "Contribution to the Problem of the bulging of Orthotropic Plates," NACA Technical Memorandum 1263, National Advisory Committee for Aeronautics, Aug. 1950.

[2] Lekhnitski, S. G., Anisotropic Plates, 2nd ed., OGIZ, Moscow-Leningrad, 1947, English edition translated by Tsai, S. W. and Cheron, T., Gordon and Breach, New York, 1968.

[3] Ambartsumyan, S. A., Theory of Anisotropic Plates, Nauka, Moscow, 1967.

[4] Mandell, J. F. and Kicher, T. P., "An Experimental Study of the Buckling of Anisotropic Plates," Proceedings of the Meeting on Fiber Composites, Case Western Reserve University, Cleveland, Ohio, 8 and 9 Oct. 1968.

[5] Davis, J. G. and Zender, G. W., "Compressive Behavior of Plates Fabricated from Glass Filaments and Epoxy Resin," NASA TN-D-

3918, National Aeronautics and Space Administration, 1968.

[6] Chen, P. E., "Bending and Buckling of Anisotropic Plates," *Composite Materials Workshop*, Tsai, S. W., Halpin, J. C., and Pagano, N. J., eds., Technomic, Stamford, Conn., 1968.

[7] Azzi, V. D. and Tsai, S. W., "Elastic Moduli of Laminated Anisotropic Composites," *Experimental Mechanics*, June 1965.

[8] Thomson, W. T., *Mechanical Vibrations*, Prentice-Hall, Englewood Cliffs, N. J., 1962.

[9] Southwell, R. V., *Proceedings of the Royal Society*, London, Series A, Vol. 135, 1932, p. 601.

[10] Roorda, J., "Some Thoughts on the Southwell Plot," *Journal of the Engineering Mechanics Division*, American Society of Civil Engineers, Vol. 93, No. EM6, Dec. 1967.

Behavior of Fiber- Reinforced Plastic Laminates under Biaxial Loading

C. W. BERT, [1]
B. L. MAYBERRY, [2] **AND**
J. D. RAY [3]

Reference:
Bert, C. W., Mayberry, B. L., and Ray, J. D., "Behavior of Fiber-Reinforced Plastic Laminates under Biaxial Loading," *Composite Materials: Testing and Design, ASTM STP 460,* American Society for Testing and Materials, 1969, pp. 362–380.

Abstract:

This paper reports on 54 biaxial-loading tests on fiber glass-epoxy laminates of three lamination schemes: parallel ply, cross ply, and quasi-isotropic. The tests were conducted on a specially designed fixture using a cable-pulley system in conjunction with a universal testing machine. The loadings were biaxial tension with 1:1 and 1:2 principal-load ratios. The contours of the reduced-thickness test sections of the flat specimens were designed to achieve a uniform-stress field.

Key Words:

fiber composites, reinforced plastics, mechanical properties, evaluation, tests

The objectives were to develop equipment and procedures for determining the biaxial mechanical behavior of laminates, and to evaluate fiber glass woven cloth reinforced epoxy laminates under biaxial tension (1:2, 1:1, and 2:1 principal-load ratios).

Most previous evaluations of composite-material behavior were limited to simple loadings, that is, uniaxial tension and compression, flexure, and pure shear [1].[4] Effects of angular orientation of uniaxial loading with respect to the major material-symmetry axis have been studied [2]. However, in many practical structures, the material is subjected to biaxial loading, that is, loads in two directions.

Various kinds of specimens have been used to obtain biaxial-load test data for metallic alloys [3]. Some of these specimens have been used to obtain limited biaxial-load data for composites: especially closed-end, thin-walled cylinders subjected to various loadings [4–6].

Recently cruciform sandwich beams were used to determine biaxial yield strengths of boron epoxy [7]. The main disadvantages were the stress concentrations produced at the corner fillets and possible effects of the core-to-facing bond on laminate strength. In laminate tests reported after the present program began, flat specimens without a reduced-thickness test section were used [8].

The basic philosophy here was to evolve flat laminate specimens which were simple to fabricate and yet capable of developing high-stress levels and a uniform tension stress field of prescribed biaxial-load ratio in the test section.

The material was 181-style, E-glass cloth with Volan A finish and impregnated with Epon 828-Z epoxy. Table 1 lists the lamination arrangements and loadings used (approximately five tests for each combination).

[1] Project director and professor, School of Aerospace and Mechanical Engineering, University of Oklahoma, Norman, Okla. 73069.
[2] Research engineer, now 2nd Lieutenant, U.S. Air Force, Webb Air Force Base, Tex.
[3] Co-project director; now assistant professor, Memphis State University, Memphis, Tenn.
[4] The italic numbers in brackets refer to the list of references appended to this paper.

Table 1—*Lamination arrangements and loadings.*

Lamination Arrangement	Type of Loading	No. of Plies	Angular Orientation of Loading, deg
Parallel ply.............................	biaxial 1:1	4	0 and 45
	biaxial 1:2	4	0 and 90[a]
	shear	4	0 and 45
	uniaxial tension	4	0 and 90
Cross ply (balanced)....................	biaxial 1:1	4	0
	biaxial 1:2	4	0[b]
	shear	4	0 and 45
	uniaxial tension	4	0[b]
Quasi-isotropic (0,60,120 deg)............	biaxial 1:1	6	0 and 45
	biaxial 1:2	6	0 and 90[a]
	shear	6	0 and 45
	uniaxial tension	6	0 and 90

[a] A 1:2 biaxial loading at 90 deg is sometimes designated as a 2:1 biaxial loading at 0 deg.

[b] In a balanced cross-ply laminate, there is no difference between the effect of a loading at 90 deg and that of the same loading at 0 deg. Thus, only tests at 0 deg needed to be conducted.

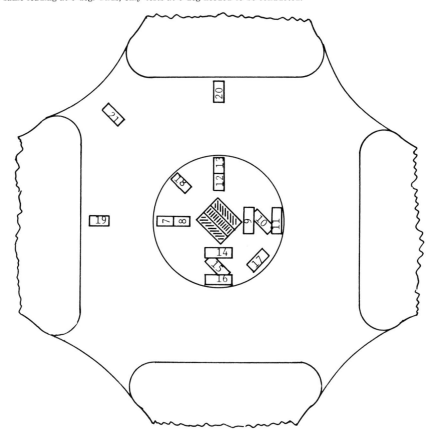

FIG. 1—Gage locations on preliminary 1:1 biaxial loading specimen.

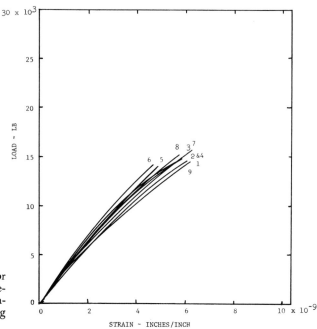

FIG. 2—Load-strain data for gages 1 through 9 on preliminary strain field evaluation of 1:1 biaxial loading specimen.

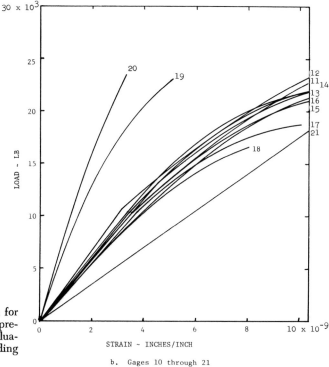

FIG. 3—Load-strain data for gages 10 through 21 on preliminary strain field evaluation of 1:1 biaxial loading specimen.

b. Gages 10 through 21

SPECIMEN DEVELOPMENT AND PRELIMINARY EVALUATION

To eliminate bending and twisting effects induced by in-plane loading, laminating arrangements which were symmetric about the midplane were used. The thicker section of the specimens was three times as thick as the reduced-thickness test section. With the use of tapered-thickness metal tabs to transfer load from the loading device to the specimen, a shear load-carrying capacity of 1500 psi was achieved.

In previous in-plane biaxial-load tests on metals, the reduced-thickness test-section shape had a square contour with rounded corners [9]. The test-section contours used here were synthesized to give a uniform biaxial-stress field in the test section (see the Appendix). The resulting contours were circular for the specimens subjected to a biaxial-load ratio of 1:1 and a $1:\sqrt{2}$ ellipse for the biaxial-load ratio of 1:2 (and 2:1).

In preliminary tests on 1:1 biaxial specimens, brittle lacquer was used to obtain a qualitative indication of the stress field. Over the reduced-thickness test section, the brittle-lacquer crack pattern was bidirectional like that shown by Durelli et al [10]; this indicated the stress field was biaxial throughout the test section.

Additional preliminary strain evaluations were carried out on a 1:1 specimen using metallic-foil, electric-resistance strain gages (Budd C6-141B single gages and C6-141R3V strain rosettes) located as shown in Fig. 1. The uniformity of the readings associated with the same orientation in the test section (Figs. 2 and 3) substantiates the validity of the specimen design. The slight difference due

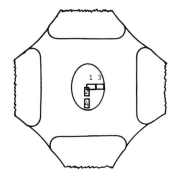

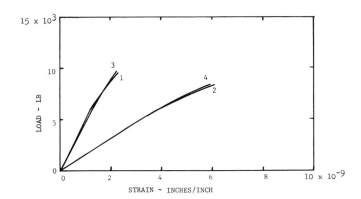

Fig. 4—Load-strain data from preliminary strain field evaluation of 1:2 biaxial load specimen.

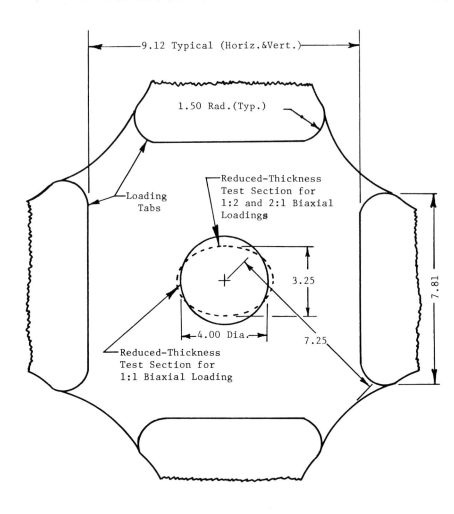

All dimensions Are in Inches

FIG. 5—Biaxial-loading specimen designs.

to orientation is attributed to the difference between the Young's moduli in the warp and weave directions. As a result of the strain uniformity in the test section, it was decided to use only two strain gages in the main data-gathering tests. Since the highest strains were recorded in the region between two adjacent loading tabs, the loading-tab external-corner radii in the final specimens were increased.

In view of the good strain distribution achieved in the 1:1 specimens, a more limited strain-field survey was conducted for the 1:2 specimens (Fig. 4).

Figure 5 shows the final specimen designs, with metallic-foil strain gages (Budd C6-141B) mounted as shown.

SPECIMEN FABRICATION

The material was used in the form of a two- or three-layer preimpregnated laminate (prepreg), made on a wide multilayer prepreg machine [11]. Prepreg 52 in. wide and in length up to 2800 in. was produced and placed in cold storage until needed. A dry layer of 181-style fiber glass cloth was used as a bleeder to improve uniformity of resin content and of venting. Teflon FEP film, 0.0005 in. thick and perforated with $\frac{3}{16}$-in.-diameter holes spaced $\frac{1}{2}$ in. apart in a rectangular array, was used as the bleeder/prepreg parting agent.

The B-stage time was approximately 7 h, from removal from cold storage to placement in the autoclave for curing. The cure was 160 F and a total pressure of 70 psi (10 psi vacuum plus 60 psi autoclave) for 100 min, with a zero-pressure precure of 23 min. Curing was conducted in a large autoclave [11], and the postcure was 350 F for 2 h.

To fabricate the specimens, the prepreg was arranged in the proper lamination arrangement and cut to conform to a sheet-metal template. For prepreg which was to become the top and bottom reinforcing material (that is, all of the specimen except the reduced-thickness test section), a central circular or elliptical hole (see Section on Specimen Development and Preliminary Evaluation) was cut out. For prepreg destined to become the test section and the central layers of the rest of the specimen, no hole was cut out. Slightly undersize sheet-metal inserts which fitted into the cut-out hole were used to prevent excess resin flow into the test section. Final curing resulted in a specimen having the desired reduced-thickness test section with naturally formed resin fillets. The fillets at the exposed edges between the loading tabs were machined, and the loading tabs were bonded to the specimen with 828 epoxy and V140 polyamide curing agent. The bonding cure served as the postcure for the specimens.

LOADING FIXTURE AND TEST PROCEDURE

In previous in-plane biaxial-loading fixtures, multiple hydraulic rams were used [9]; however, it was difficult to maintain the same biaxial-load ratio throughout the test of an anisotropic composite. To avoid this, an original cable-and-pulley system reacted by a rigid steel frame was used here to load the specimen biaxially (Fig. 6). The fixture can be converted from a 1:1 biaxial-load ratio to a 1:2 ratio by changing the cable path and pulley. Figure 7 shows the fixture in place in the 60,000-lb-capacity Riehle universal testing machine.

The cable size and type ($\frac{5}{8}$ in. 6 by 36) were chosen for static load capacity and flexibility to enable easy bending over the pulleys. Prior to testing specimens in a given configuration, the cables of the loading fixture were "set" by applying load to an aluminum plate of the same size as the specimens. To attach to the cable, the cable end was put into a tapered steel socket and flared until each strand was separated and bent back approximately 135 deg. Then a high-zinc No. 1 Babbitt was poured into the socket, resulting in a high load-carrying capacity as verified by preliminary tension tests.

Each 6-in.-diameter pulley contained a cylindrical roller bearing, which operated on a 1.5-in.-diameter hardened steel shaft pressed into the loading fixture.

To permit specimen self alignment, a load link was provided between each loading tab in the specimen and the loading fixture.

A total of eight loading tabs, by which the biaxial loads were transferred from the load links in the fixture to the specimen, were used for each specimen: one on each side (to minimize out-of-plane bending) at each of the four loading ears. To promote a more uniform load transfer from the hardened SAE 4140 steel loading tabs to the specimen, the tabs were tapered to a minimum of 0.0625 in.

The weight of the specimen plus the heavy loading tabs was balanced out to prevent loading the top part of the specimen in tension and the bottom in compres-

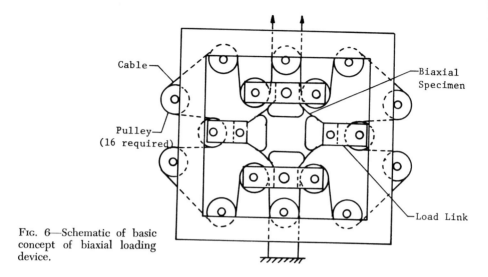

FIG. 6—Schematic of basic concept of biaxial loading device.

FIG. 7—Biaxial fixture installed in universal test machine.

Composite Materials

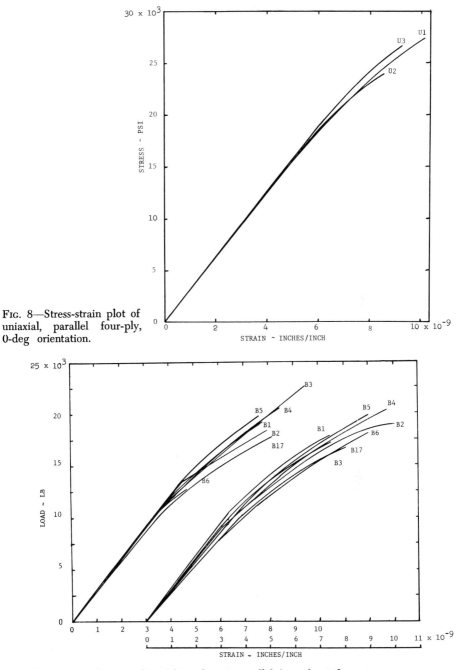

FIG. 8—Stress-strain plot of uniaxial, parallel four-ply, 0-deg orientation.

FIG. 9—Load-strain plot of biaxial, 1:1, parallel four-ply, 0-deg orientation.

Table 2—*Results of uniaxial tension tests.*

Specimen No.	E, psi	ν	S_L, psi	S_U, psi
FOUR-PLY PARALLEL LAMINATED, 0-DEG ORIENTATION				
U1............. 3.34×10^6		0.148	13 250	37 400
U2............. 3.24		0.144	11 750	41 150
U3............. 3.36		0.132	13 500	40 500
FOUR-PLY PARALLEL LAMINATED, 90-DEG ORIENTATION				
U4............. 3.14		0.193	10 000	41 400
U5............. 3.19		0.181	12 000	42 600
U6............. 3.11		0.192	12 750	42 200
FOUR-PLY CROSS LAMINATED, 0-DEG ORIENTATION				
U7............. 4.55		0.252	11 750	23 200
U8............. 3.10		0.152	11 250	25 500
U9............. 3.16		0.148	14 750	22 200
U10............ 3.33		0.095	10 500	25 600
SIX-PLY QUASI-ISOTROPIC LAMINATED, 0-DEG ORIENTATION				
U11............ 3.59		0.297	13 250	33 700
U12............ 3.42		0.324	12 500	35 000
U13............ 3.68		0.320	15 750	31 100
U14............ 3.79		0.323	19 000	34 000
U15............ 3.63		0.323	12 250	39 600
SIX-PLY QUASI-ISOTROPIC LAMINATED, 90-DEG ORIENTATION				
U16............ 3.63		0.286	10 500	36 400
U17............ 3.85		0.336	13 250	33 800
U18............ 3.62		0.312	13 500	33 700
U19............ 3.57		0.328	12 000	35 500
U20............ 3.61		0.291	11 250	34 500

sion. Then the slack was pulled out of the cable and the specimen loaded at an approximately constant rate.

EXPERIMENTAL RESULTS

The experimental stress-strain or load-strain data, from which the reduced data presented here were taken, are presented in Ref *12*. Figures 8 and 9 show typical data.

The stress values used in specifying the strengths are nominal *composite* stresses, that is, total load divided by original cross-sectional area, since these are the ones that a designer needs directly. In general, they are not the same as the stresses physically present in each individual ply of the laminate.

UNIAXIAL TENSION TESTS

The uniaxial tension specimens used in the main data-gathering series were designated U1-U20 and were 0.75 in. wide by 6.00 in. long (between grips). The strain values used are those measured by variable-electric-resistance strain gages.

The Young's moduli, E, and Poisson's ratios ν, were both determined in the initial straight-line elastic portion of the stress-strain curves.

In aircraft alloys, the criterion for limit strength is the 0.2 percent offset yield strength. However, many composite materials, including glass fiber reinforced plastics (FRP), do not reach this high of an offset strain. The limit-strength criterion used here is the stress corresponding to a 0.01 percent strain offset.

After strain data were taken, two small circular-arc notches were made to prevent failure in the grips and the test continued to failure. The ultimate tensile strength (UTS) is defined in the traditional manner as the stress corresponding to the maximum load reached in the test. For many composite materials, including FRP, the fracture strength coincides with the UTS.

Table 2 gives the reduced data for the uniaxial tests. The cross-ply specimens were not tested at a 90-deg orientation, since this is no different than 0-deg orientation for such a specimen (symmetrically laminated with an equal number of plies at 0 and 90 deg).

Table 3 lists mean values of the moduli (E and ν), used in determining biaxial stress values.

The minor Poisson's ratio for the parallel ply presented in Table 3 appeared to be high, so a calculated value was used for all data reduction.[5] Also the major and minor elastic moduli were not different within experimental error, so a mean value was used for each. This mean value was taken for the quasi-isotropic and was an average between the values. The minor and major moduli for the parallel- and cross-ply specimens were close and within experimental error; therefore, these data were not averaged for the data reduction.

TORSION-TUBE SHEAR TESTS

To obtain data on shear modulus for use in Eqs 4, below, 24 torsion-tube shear tests were conducted. The specimens were short to prevent torsional buckling; the original curves of shear stress versus shear strain were presented in Ref 12. Table 4 lists the reduced modulus and strength data.

BIAXIAL-LOADING TESTS

The individual biaxial-loading specimens used in the main data-gathering series (Fig. 5) were designated B1-B54, and the corresponding load-strain curves are presented in Ref 12.

[5] The calculated value was obtained from the reciprocal relationship: $0.1410 \times 3.15/3.304$.

Table 3—Summary of mean values of initial composite moduli.

Lamination Arrangement	Parallel Ply	Cross Ply	Quasi-Isotropic
Major elastic modulus, $\bar{E}_{11}$	3.304×10^6	3.196×10^6	3.62×10^{6b}
Minor elastic modulus, $\bar{E}_{22}$	3.15×10^6	3.196×10^6	3.66×10^{6b}
Major Poisson's ratio, $\bar{\nu}_{12}$	0.1410	0.1316	0.317^c
Minor Poisson's ratio, $\bar{\nu}_{21}$	0.188^a	0.1316	0.311^c
Shear modulus, $\bar{G}$	1.325×10^6	1.31×10^6	1.93×10^6

[a] This value appears to be erroneously high, so a value of 0.1342, calculated from the reciprocal relationship ($0.1410 \times 3.15/3.304$) was used in all subsequent data reduction.

[b] Since these values are not significantly different, within experimental error, a single value of 3.64×10^6 psi is used in all subsequent data reduction.

[c] Since these values are not significantly different, within experimental error, a single value of 0.314 is used in all subsequent data reduction.

Table 4—*Results of torsion-tube shear tests.*

Specimen No.	G, psi	S_{Ls}, psi	S_{Us}, psi
FOUR-PLY PARALLEL LAMINATED, 0-DEG ORIENTATION			
S1	1.47×10^6	7 000	30 100
S2	1.25	7 200	31 256
S3	1.53	4 000	79 862
S4	1.05	8 500	28 040
FOUR-PLY PARALLEL LAMINATED, 45-DEG ORIENTATION			
S5	3.68	6 800	57 400
S6	3.13	5 500	54 000
S7	7.7[a]	7 200	80 800[a]
S8	3.37	9 000	66 500
FOUR-PLY CROSS LAMINATED, 0-DEG ORIENTATION			
S9	1.3	4 800	28 125
S10	1.43	7 000	21 350
S11	0.96	10 000	24 766
S12	1.75	5 000	32 260
S13	1.11	7 000	25 620
S14	2.3[a]	10 000	32 238
FOUR-PLY CROSS LAMINATED, 45-DEG ORIENTATION			
S15	2.5	10 200	50 000
S16	3.34	10 500	46 875
S17	2.22	7 000	46 875
SIX-PLY QUASI-ISOTROPIC LAMINATED, 0-DEG ORIENTATION			
S18	2.0	9 800	55 510
S19	2.22	11 000	53 375
S20	1.82	10 500	55 510
SIX-PLY QUASI-ISOTROPIC LAMINATED, 45-DEG ORIENTATION			
S21	1.82	7 500	26 218
S22	1.62	14 800	49 105
S23	1.82	10 000	49 319
S24	2.22	16 400	50 386

[a] Values too high; not used in final data reduction.

The composite biaxial-stress values were obtained by multiplying the measured load value, P, by the following factors:

$$\sigma_1'/P = (\bar{Q}_{11}')(\epsilon_1'/P) + (\bar{Q}_{12}')(\epsilon_2'/P) \dotfill (1)$$
$$\sigma_2'/P = (\bar{Q}_{12}')(\epsilon_1'/P) + (\bar{Q}_{22}')(\epsilon_2'/P) \dotfill (2)$$

where directions 1 and 2 are the major and minor material-symmetry directions, ϵ_1'/P and ϵ_2'/P are taken from biaxial-load versus strain data presented in Ref 12, and $\bar{\nu}_{12}$ and $\bar{\nu}_{21}$ are composite Poisson's ratios from Table 3. For $\theta = 0$ deg, the primes can be removed and

$$\bar{Q}_{11}' = \bar{Q}_{11} = \bar{E}_{11}/\bar{\lambda}, \ \bar{Q}_{12}' = \bar{Q}_{12} = \bar{\nu}_{21}\bar{E}_{11}/\bar{\lambda}, \ \bar{Q}_{22}' = \bar{Q}_{22} = \bar{E}_{22}/\bar{\lambda} \dotfill (3)$$

where $\bar{\lambda} = 1 - \bar{\nu}_{12}\bar{\nu}_{21}$.

For $\theta = 45$ deg

$$\bar{Q}'_{11} = \bar{Q}'_{22} = (1/4\bar{\lambda})(\bar{E}_{11} + 2\bar{\nu}_{21}\bar{E}_{11} + \bar{E}_{22} + 4\overline{\lambda G})$$
$$\bar{Q}'_{12} = (1/4\bar{\lambda})(\bar{E}_{11} + 2\bar{\nu}_{21}\bar{E}_{11} + \bar{E}_{22} - 4\overline{\lambda G}).$$ \quad \right\} \quad \dots \dots \dots \dots (4)

Table 5 presents the results obtained by applying Eqs 1 and 2 to the load-strain data [12].

The limit and ultimate strength criteria are the same as used for the uniaxial specimens. Although there is theoretical basis for an "equivalent" offset strain for biaxial loading, design usage favors the same offset for all types of loading [3].

The method used here to determine stresses from measured strains is quite accurate in the linear elastic range. When the load-strain curve becomes nonlinear, the accuracy of the stress values is not known. When the material in the reduced test section "yields" (that is, undergoes higher strains than it would if it were elastic), its effective moduli are lower than the corresponding elastic moduli, and the actual stress is lowered. However, if material outside of the reduced-thickness test section yields first, most of the additional load is put into the test section causing the strain gages to read erroneously high, for a given value of load applied to the edge of the specimen. Thus, the calculated ultimate strength values presented are upper bounds.

Figure 10 shows photographs of typical fracture patterns for both specimen types (1:1 and 1:2 or 2:1). A number of the fractures of the 1:1 specimens

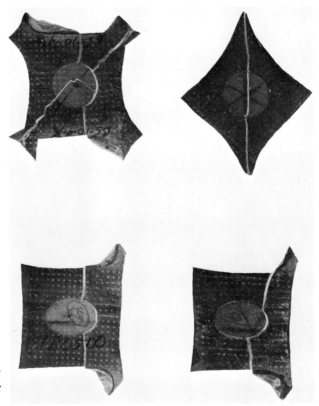

FIG. 10—Photographs of typical fracture patterns for biaxial-loading specimens.

Table 5—*Results of biaxial-loading tests.*

Specimen No.	S_{L1}', psi	S_{U1}', psi	S_{L2}', psi	S_{U2}', psi
FOUR-PLY PARALLEL LAMINATED, 0-DEG ORIENTATION, 1:2 NOMINAL LOADING				
B12	18 900	33 000	9 700	14 900
B13	19 650	26 500	9 150	12 420
B14	15 800	29 000	6 370	12 150
B15	17 900	35 100	7 560	15 550
B16	14 980	28 500	6 840	14 750
FOUR-PLY PARALLEL LAMINATED, 0-DEG ORIENTATION, 1:1 NOMINAL LOADING				
B1	13 200	29 500	13 280	28 400
B2[a]	18 000	28 000	17 200	30 800
B3[a]	16 820	30 850	14 650	29 650
B4	16 700	28 450	15 500	29 250
B5	20 500	28 200	15 850	29 000
B6	15 750	36 000	12 850	37 000
B17	12 950	33 700	15 200	35 900
FOUR-PLY PARALLEL LAMINATED, 0-DEG ORIENTATION, 2:1 NOMINAL LOADING				
B18	6 800	11 700	12 300	27 000
B19	7 650	13 000	25 700	36 300
B20	10 350	13 800	18 000	28 400
B21	8 320	14 650	19 300	35 100
B22	11 300	14 250	20 900	29 750
FOUR-PLY PARALLEL LAMINATED, 45-DEG ORIENTATION, 1:1 NOMINAL LOADING				
B7	25 200	38 800	23 200	38 800
B8	15 100	24 200	13 600	28 400
B9[a]	10 950	26 500	14 350	28 000
B10[a]	14 100	23 800	11 040	25 100
B11	10 250	20 850	16 100	29 300
FOUR-PLY CROSS LAMINATED, 0-DEG ORIENTATION, 1:1 NOMINAL LOADING				
B23[a]	14 200	23 000	14 350	26 960
B24	15 200	31 700	16 700	31 700
B25	19 100	26 500	19 100	26 500
B26	19 100	27 200	20 950	27 200
B27	18 400	26 400	15 400	26 400
B28	17 650	28 400	17 650	28 400
FOUR-PLY CROSS LAMINATED, 0-DEG ORIENTATION, 1:2 NOMINAL LOADING				
B29[a]	6 800	13 250	16 700	30 400
B30	9 520	16 350	21 000	38 200
B31	11 400	18 900	13 300	23 000
B32[a]	6 750	10 200	14 100	19 400
B33[a]	4 000	4 000	9 000	10 200
B34	9 370	13 350	16 000	25 300
B35[a]	6 790	11 430	15 250	26 700

Table 5—(Continued)

Specimen No.	S_{L1}', psi	S_{U1}', psi	S_{L2}', psi	S_{U2}', psi
Six-Ply Quasi-Isotropic Laminated, 0-deg Orientation, 1:1 Nominal Loading				
B36	12 600	26 850	13 580	28 000
B37[a]	13 500	39 100	9 450	32 500
B38	10 650	34 100	11 700	34 100
B39	13 000	38 600	16 200	41 600
B40	5 930	24 800	6 000	26 900
Six-Ply Quasi-Isotropic Laminated, 45-deg Orientation, 1:1 Nominal Loading				
B41[a]	14 200	27 700	14 950	27 700
B42	7 300	14 000	11 850	19 500
B43	15 600	28 300	17 350	24 700
B44	12 350	20 800	14 300	22 200
B45	15 750	18 950	13 200	20 900
Six-Ply Quasi-Isotropic Laminated, 0-deg Orientation, 2:1 Nominal Loading				
B50[a]	20 700	33 900	10 600	13 350
B51	25 300	36 800	13 050	13 900
B52[a]	22 000	41 200	12 200	16 600
B53[a]	21 100	34 600	11 350	12 400
B54[a]	22 700	41 000	14 600	15 300
Six-Ply Quasi-Isotropic Laminated, 0-deg Orientation, 1:2 Nominal Loading				
B46[a]	7 150	12 600	21 950	36 000
B47	7 360	9 700	18 600	23 600
B48[a]	6 600	11 350	14 500	33 000
B49[a]	9 600	13 420	20 600	36 000

[a] The fracture did not pass through the reduced-thickness test section.

were mutiple, thus giving additional testimony to the attainment of a uniform balanced-biaxial stress field. Of the 54 biaxial specimens in the data-gathering series, only 18 (footnotes[a] in Table 5) did not fracture through the test section. However, a study of apparent ultimate strength data in Table 5 shows that these specimens failed at values neither abnormally low nor abnormally high.

EVALUATION OF EXPERIMENTAL RESULTS

PARALLEL-PLY LAMINATES

Since the authors know of no composite micromechanics analysis applicable to fabric-reinforced composites, no quantitative prediction of the individual stiffnesses of parallel-ply laminates can be made on the basis of constituent-material properties. Since the reciprocal relationship is based on very fundamental thermodynamics principles, it should hold. The two parallel-ply moduli values in Table 3 appear to be reasonable for the resin content (34 to 36 percent by weight) and so does the value of $\bar{\nu}_{12}$. However, the $\bar{\nu}_{21}$ value appears to be erroneous, and, since the reciprocal relation was violated, a value of 0.1342 given by the reciprocal relation appears to be more valid. The $\bar{\nu}_{12}$ and the corrected $\bar{\nu}_{21}$, as well as the $\bar{E}_{11}$ and $\bar{E}_{22}$ values, are quite close to those reported by Refs 13, 14.

Bert et al on Fiber-Reinforced Plastic Laminates

The value of G in Table 3 (1,386,000 psi) is approximately 71 percent higher than the 810,000-psi value given in Ref *14*. Part of the difference may be due to the type of specimen, since the Ref *14* values were obtained by in-plane panel-shear tests. Sidorin [*15*] obtained 11 percent higher shear-modulus values by torsion-tube tests than by panel-shear tests for the same fabric orientation used here.

The value of $\bar{G}'$ corresponding to an orientation, θ, of 45 deg can be predicted from the values $\bar{G}$, $\bar{E}_{11}$, $\bar{E}_{22}$, and $\bar{\nu}_{12}$, by the following equation [*12*]:

$$1/\bar{G}' = (1/\bar{G}) + [(1 + 2\bar{\nu}_{12})\bar{E}_{11}^{-1} + (1/\bar{E}_{22}) - (1/\bar{G})] \sin^2 2\theta \ldots \ldots \ldots (5)$$

Equation 5 gives a value of 1,370,000 psi, which is considerably lower than the 3,390,000-psi value of $\bar{G}'$ determined experimentally for θ = 45 deg.[6]

Figure 11 presents the limit and apparent ultimate strength results from Table 5 as strength envelopes. The 2:1 and 1:0 data appear to be symmetric with the 1:2 and 1:0 data, respectively. No attempt was made to fit a strength theory to the data presented.

CROSS-PLY LAMINATES

Using laminate stiffness theory, it is predicted that the stiffness properties of a symmetrically laminated cross-ply laminate are the mean values of the individual-ply properties. Thus, the predicted values based on parallel-ply data are:

[6] Apparently the incompatibility between the torsion-tube properties and the properties of the uniaxial and biaxial specimens was due to a difference in resin content. The torsion-tube specimens did not bleed adequately, resulting in a resin content of approximately 40 percent, compared with 34 percent for the flat ones.

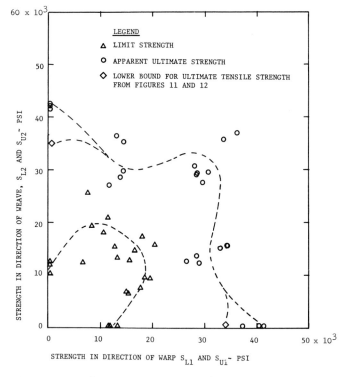

FIG. 11—Biaxial-tension strength envelope for four-parallel-ply laminate.

Composite Materials

$\bar{E}_{11} = \bar{E}_{22} = 3.23 \times 10^6$ psi, $\bar{\nu}_{12} = \bar{\nu}_{21} = 0.1376$. Both values are slightly higher than the corresponding ones listed in Table 3 for the cross-ply laminate.

Figure 12 shows the limit and apparent ultimate strength envelopes for the cross-ply laminate, using data from Table 5.

Quasi-Isotropic Laminates

The predicted quasi-isotropic stiffness properties obtained by using parallel-ply properties in equations due to Werren and Norris [16] are: $\bar{E}_{11} = \bar{E}_{22} = 3.04 = 10^6$ psi, $\bar{\nu}_{12} = \bar{\nu}_{21} = 0.160$, $\bar{G} = 1.31 \times 10^6$ psi. These values compare very poorly with the mean quasi-isotropic values listed in the last column of Table 3.

Figure 13 shows the limit and apparent ultimate strength envelopes for the quasi-isotropic laminate, using Table 5 values. The higher strength at a biaxial-load ratio of 2:1 as compared to 1:1 is unexplained.

Conclusions

1. Use of the specimens and loading fixture described here is believed to be an accurate, simple-to-fabricate method for obtaining strength data on flat-sheet composite materials subjected to biaxial-tension loads.

2. The E-glass, Volan A finish, 181-style fabric/828-Z epoxy laminates tested in this program generally exhibited considerably higher limit strengths (and slightly lower apparent ultimate strengths) under 1:1, 1:2, and 2:1 biaxial-load ratios than in uniaxial tension.

3. Of the three lamination arrangements evaluated (parallel-ply, cross-ply, and quasi-isotropic laminates), the parallel-ply laminate was slightly superior under bi-

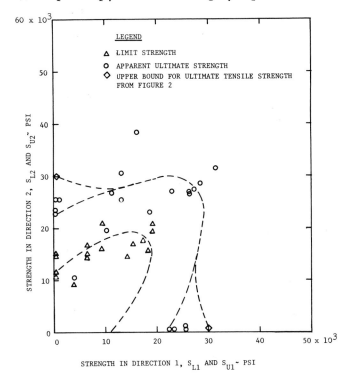

FIG. 12—Biaxial-tension strength envelope for four-ply, symmetrically cross laminated material.

Bert et al on Fiber-Reinforced Plastic Laminates

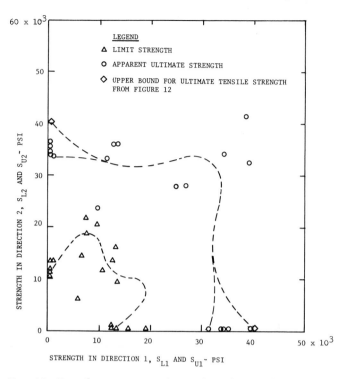

FIG. 13—Biaxial-tension strength envelope for six-ply quasi-isotropic laminate.

axial loading, although the differences were small. This suggests that for this particular weave, combination of materials, and type of loading (uniaxial or biaxial tension), parallel-ply lamination is advantageous structurally as well as being easier
to fabricate. (This same conclusion would not be expected to hold true for a
more highly unidirectional composite such as a tape or a filament-wound composite.)

ACKNOWLEDGMENT

The research reported was sponsored by the U. S. Army Aviation Materiel Laboratories, Fort Eustis, Virginia, with J. P. Waller as technical monitor.

Appendix

DESIGN OF TEST-SECTION CONTOUR TO ACHIEVE A UNIFORM STRESS STATE

Using two-dimensional theory of elasticity, the shape of reduced-thickness test
section necessary to achieve a uniform-biaxial-stress state is derived. To satisfy
the equations of equilibrium for generalized plane stresses, the Airy stress function,
Φ, is defined as follows:

$$N_x = \partial^2\Phi/\partial y^2; \quad N_y = \partial^2\Phi/\partial x^2; \quad N_{xy} = -\partial^2\Phi/\partial x\partial y \dots\dots\dots\dots(6)$$

where N_x, N_y, N_{xy} are the internal forces per unit length of run (normal to the x
and y axes, and shear, respectively).

A uniform stress state requires that N_x, N_y, and N_{xy} all be constants. Then integration of Eqs 6 yields the following expression for Φ:

$$\Phi = (\tfrac{1}{2})(N_x y^2 + N_y x^2) - N_{xy} xy + Ax + By + C \dots\dots\dots\dots (7)$$

where x and y are Cartesian position coordinates and A, B, and C are arbitrary constants of integration. The required contour shape (position coordinates x_c, y_c) of reduced-thickness test section is the one in which Φ is invariant with respect to position x_c, y_c; thus,

$$(\tfrac{1}{2})(N_x y_c^2 + N_y x_c^2) - N_{xy} x_c y_c + Ax_c + By_c + K = 0 \dots\dots\dots\dots (8)$$

where $K = \Phi(x_c, y_c) - C$. The constants A and B in Eq 8 can be omitted, since they determine only the origin of the coordinate system.

For any generalized-plane-stress state N_x, N_y (N_{xy}), by means of the Mohr's stress circle, an orientation can be found such that the shear force N_{xy} vanishes, and the associated loadings are the principal loadings N_p, N_q. By choosing the axes of the Cartesian coordinates used to define the test-section shape to coincide with the principal loading directions, N_{xy} can be omitted and N_x and N_y are replaced by N_p and N_q. Thus, Eq 8 can be simplified and rewritten in the standard form for an ellipse:

$$(x_c/a)^2 + (y_c/b)^2 = 1 \dots\dots\dots\dots\dots (9)$$

where a, b are the major and minor semiaxes, and the following relationship must hold:

$$b/a = (N_q/N_p)^{1/2} \dots\dots\dots\dots\dots (10)$$

Since only equilibrium considerations were used, the results are independent of the elastic behavior (anisotropic, orthotropic, or isotropic).

References

[1] "Testing Techniques for Filament Reinforced Plastics," AFML TR 66-274, Air Force Materials Laboratory, AD 801547, Sept. 1966.

[2] Tsai, S. W., "Strength Characteristics of Composite Materials," NASA CR-224, National Aeronautics and Space Administration, April 1965.

[3] Bert, C. W., Mills, E. J., and Hyler, W. S., "Mechanical Properties of Aerospace Structural Alloys under Biaxial-Stress Conditions," AFML TR 66-229, Air Force Materials Laboratory, AD 488304, Aug. 1966.

[4] Harrington, R. A. et al, "Design Information from Analytical and Experimental Studies of Filament Wound Structures Subjected to Combined Loading," B. F. Goodrich Aerospace and Defense Products, Final Report on Subcontract No. 89 to Hercules Powder Company, Allegany Ballistics Laboratory, AD 451216, 25 Feb. 1964.

[5] Ely, R. E., "Biaxial Fracture Stresses for Graphite, Ceramic, and Filled and Reinforced Epoxy Tube Specimens," RR-TR-65-10, U. S. Army Missile Command, Redstone Arsenal, AD 469036, June 1965.

[6] Weng, T.-L., "Biaxial Fracture Strength and Mechanical Properties of Graphite-Base Refractory Composites," AIAA Paper 68-337, 9th Structures, Structural Dynamics and Materials Conference, American Institute of Aeronautics and Astronautics/American Society of Mechanical Engineers, Palm Springs, Calif., April 1968.

[7] Waddoups, M. E., "Characterization and Design of Composite Materials," Composite Materials Workshop, Tsai, S. W., Halpin, J. C., and Pagano, N. J., eds., Technomic, Stamford, Conn., 1968, pp. 254–308.

[8] Grimes, G., Pape, B. J., and Ferguson, J. H., "Investigation of Structural Design Concepts for Fibrous Aircraft Structures, Vol. III, Technology Appraisal-

Experimental Data and Methodology," AFFDL TR 67–29, Vol. III, Air Force Flight Dynamics Laboratory, AD 824228, Nov. 1967.

[9] Terry, E. L. and McClaren, S. W., "Biaxial Stress and Strain Data on High Strength Alloys for Design of Pressurized Components," ASD TR 62-401, Aeronautical Systems Division, AD 283348, May 1962.

[10] Durelli, A. J., Phillips, E. A., and Tsao, C. H., *Introduction to the Theoretical and Experimental Analysis of Stress and Strain*, McGraw-Hill, New York, 1958, pp. 369–370.

[11] Nordby, G. M., Crisman, W. C., and Bert, C. W., "Fabrication and Full-Scale Structural Evaluation of Sandwich Shells of Revolution Composed of Fiber Glass Reinforced Plastic Facings and Honeycomb Cores," USAAVLABS TR 67-65, U. S. Army Aviation Materiel Laboratories, Fort Eustis, AD 666480, Nov. 1967.

[12] Bert, C. W., Mayberry, B. L., and Ray, J. D., "Behavior of Fiber-Reinforced Plastic Laminates Under Uniaxial, Biaxial, and Shear Loadings," USAAV-LABS TR 68-86, U. S. Army Materiel Laboratories, Fort Eustis, Jan. 1969.

[13] Youngs, R. L., "Poisson's Ratios for Glass-Fabric-Base Plastic Laminates," Report 1860, U. S. Forest Products Laboratory, Jan. 1957.

[14] "Plastics for Flight Vehicles, Part I: Reinforced Plastics," MIL-HDBK-17, Armed Forces Supply Support Center, Washington, D. C., Nov. 1959.

[15] Sidorin, Ya. S., "Experimental Investigation of the Anisotropy of Fiberglass-Reinforced Plastics During Shear," *Industrial Laboratory*, Vol. 32, 1967, pp. 723–726.

[16] Werren, F. and Norris, C. B., "Mechanical Properties of a Laminate Designed to be Isotropic," Report 1841, U. S. Forest Products Laboratory, May 1953.

Methods of Joining Advanced Fibrous Composites

R. N. DALLAS[1]

Reference:

Dallas, R. N., "Methods of Joining Advanced Fibrous Composites," *Composite Materials: Testing and Design, ASTM STP 460,* American Society for Testing and Materials, 1969, pp. 381–392.

Abstract:

Advanced fibrous composites have a high potential for aircraft structural applications; joint design limits this potential. How to transfer loads efficiently from composite components to supporting structures is a difficult problem. The approach taken in this study is to measure the strengths of adhesive bonded joints, riveted joints, and joints containing both rivets and bonds. The data generated have contributed to defining design allowables and have given a basis for the design of composite parts.

A significant variable has been found to be fiber orientation of the adherend. Three orientations of boron epoxy laminates have been bonded to titanium alloy Ti 6Al-4V: unidirectional, isotropic, and modified isotropic. By careful joint design it has been found possible to transfer over 90 percent of the tensile load carrying capability of the laminate through the joint.

Both ductile and semirigid adhesives have been tested, with the former giving higher shear stresses at ½ in. overlap distances. Shear stresses of 7000 psi have been measured, which equal metal-to-metal bonds with the same adhesive system. However, when the overlap is increased to 1½ in. the semirigid type will carry approximately 20 percent more load than the ductile type.

Adding rivets to bonded joints was found to have no effect on the static strength properties of composite-to-metal joints. This result is unexpected and may be because of nonoptimum design. It is anticipated that fatigue strengths would be improved by the combination joints.

Composite components have been designed, built, and tested. The joints did not fail when loaded up to over 4000 psi in shear.

Key Words:

joints, adhesives, adherends, mechanical properties, boron, fiber composites, epoxy laminates, fiber orientation, stress concentration, design, titanium, fatigue strength, flexible resin, L/t ratio, primer, overlap distance, mechanism of failure, faying surface, evaluation, tests

How can efficient joints be made in the new composites? To realize the full potential strength of high performance structural materials in a practical application is a universal problem. How to transfer loads efficiently from these materials to adjacent load bearing structures has not been solved completely; this is true even for the metals. Joints with the fibrous composite materials are even less developed. There is a welter of variables to contend with, and analytical methods of prediction are not developed highly. What data can be used to design joints in actual hardware so that all the benefits of the new material can be realized?

The objective of the present investigation is to generate data that designers could use to develop allowables for joints between advanced composites and metals. To accomplish this the strength of bonded joints was measured in single lap shear, and the ultimate bearing strengths of the laminates were determined. The adherends

[1] Lockheed-California Co., Division of Lockheed Aircraft Corp., Burbank, Calif. 95103

were boron/epoxy laminates jointed to titanium alloy (Ti 6Al-4V), a combination selected because of compatibility with respect to their coefficients of thermal expansion and the high specific strength and modulus of titanium which make it a promising candidate for structural applications in advanced aircraft designs.

Among the variables investigated were three fiber orientations, two adhesive systems, and the presence or absence of rivets. The values obtained from this study were used to design metal-to-boron laminate joints in the test section of a floor beam for the Navy P-3 Orion and a fire access door for the Air Force F-104 Starfighter. Both of these parts have been built, and the access door is now being flight tested in an Air Force airplane.

DISCUSSION OF JOINTS

Three general methods of joining structural materials are considered in this paper: adhesive bonding, riveting, and a combination of adhesive bonding and riveting. Bonding has many advantages. One, which provides major benefits, is the avoidance of bearing stress concentrations in the basic material. The principal disadvantage of bonded joints is the lack of a good nondestructive test procedure. Also bonded joints are susceptible to failures induced by peel stresses and stress concentrations at the ends of the joints. In the case of laminated composites, the outer ply in contact with the adhesives may fail by shearing away from the body of the laminate before the full strength of the bond is realized.

Mechanical fasteners obviate these defects of bonded joints. However, they have their weaknesses also. They are penalized by stress risers initiated at bolt or rivet holes and by reduction of the cross-section area that carries the load.

Combinations of mechanical fasteners with bonded joints have shown excellent fatigue characteristics with glass reinforced laminates. When riveted joints were supplemented with an adhesive bond, their cycle lives increased by a factor of 30 at their working stress level.

It has been reported by the McDonnell Douglas Corporation[2] that double lap shear joints between boron and titanium were improved substantially by the addition of bolts in the bond area. The apparent shear stress in static tests was raised from 3300 to 4300 psi in their investigation.

The work reported here does not confirm the advantages of the combination joint in static test strength. However, the designs of the combined joints tested were not optimized, and the investigation of this phase was exploratory only. The selection of fiber orientations in the laminates to be joined to metal structures was governed by the following consideration. The strength of laminate-to-metal joints is highly dependent upon the fiber orientation of the laminate. Therefore, compromises have to be made between the orientation that gives the best laminate properties for a particular application and the best joint properties. The laminate is of little value if loads cannot be transfered to it.

Unidirectional laminates give the best strengths of any in the direction of the fibers. Laminates with this orientation can be used advantageously in combination with metal structures. The composite contributes to resisting the primary load, and the metal resists all of the off-axis loads. This type of design utilizes the properties of the reinforcing fibers with maximum efficiency and introduces the minimum amount of risk in a new application. Furthermore, this fiber orientation gives the highest joint strength. Therefore, this unidirectional laminate was used as a basis for comparing joint strength properties.

While the most efficient uses for the advanced composites may be in combination with metal structures where unidirectional fiber orientations are satisfactory, they cannot be limited to such applications always. When multidirectional loads are

[2] "Investigation of Joints and Cutouts in Advanced Fibrous Composites for Aircraft Structures," DAC 68259, McDonnell Douglas Corp., St. Louis, Mo.

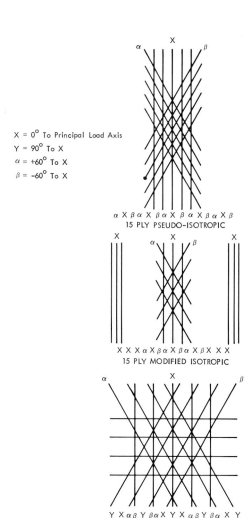

X = 0° To Principal Load Axis
Y = 90° To X
α = +60° To X
β = −60° To X

α X β α X β α X β α X β α X β
15 PLY PSEUDO-ISOTROPIC

X X X α X β α X β α X β X X X
15 PLY MODIFIED ISOTROPIC

Y X α β Y β α X Y X α β Y β α X Y
17 PLY BORON FIRE ACCESS DOOR

Fig. 1—Laminate fiber orientation.

encountered and metal backups are unavailable, the laminate must be designed to resist loads in all directions. To represent joints in such applications a 15-ply pseudo-isotropic laminate was made. The fiber orientation is shown in Fig. 1. Single lap shear tests were performed with the same adhesive systems and the same overlap distances as were used with the unidirectional laminate. Because the strengths of the joints with pseudo-isotropic laminates were well below those obtained on unidirectional laminates, a third orientation was investigated. The objective was to combine the high joint strength of unidirectional laminate with the multidirectional load carrying capability of the pseudo-isotropic layup. To achieve this result a modified-isotropic orientation was designed (Fig. 1). The rationale was the nine-ply isotropic center would give multidirectional capability, while the three outer plies could nest and therefore transfer higher loads into the body of the laminate. The test results show that, as expected, both the joint strength and laminate tensile strength of the modified-isotropic laminates are in between the unidirectional and the pseudo-isotropic orientations.

Two adhesive systems identified as Type A and Type B, were used in this study. Type A, the more flexible of the two, is an epoxy resin modified with

Table 1—_Loads carried by bonded joints._

Adhered Orientation	Adhesive Type	Joint Type[a]	½ in. Overlap; 6:1 L/t Ratio Load, lb	Stress, psi	1 in. Overlap; 12:1 L/t Ratio Load, lb	Stress, psi	1½ in. Overlap; 18:1 L/t Ratio Load, lb	Stress, psi
Titanium to titanium blank	A	SL	3500	7000	. . .	. . .	. . .	. . .
Unidirectional	A	SL	3500	7000	4700	4700	5700	3800
Titanium to titanium blank	B	SL	2500	5000	. . .	. . .	. . .	. . .
Unidirectional	B	SL	2500	5000	4600	4600	6900	4600
Unidirectional	B	DL	5400	5400	9000	4500	11000	3700
Pseudo-isotropic	A	SL	2200	4400	. . .	. . .	. . .	. . .
Pseudo-isotropic	B	SL	1400	2800	. . .	. . .	. . .	. . .
Modified isotropic	B	SL	2250	4500	3200	3200	3700	2700

[a] Joint type: SL = single lap shear; DL = double lap shear.

a polyamide. Type B is an epoxy resin modified with a nitrile. A third system, similar to Type B, was spot checked and found to have properties so similar to Type B that further work on this system was discontinued.

Combinations of rivets with adhesive bonds were investigated. Riveting was not useful for joining the unidirectional laminates because the host material split when the rivet was expanded in place. When coupons of multidirectional layups were riveted to metal (with no adhesive bond), the rivets induced tensile failures in the laminates at the rivet hole at modest stresses, implying a stress riser effect. When the rivet was combined with an adhesive bond, the joint supported the same load as similar bonded joints without rivets, and neither a penalty nor a useful advantage resulted.

PREPARATION OF TEST JOINTS

The joints made in this program were between Ti-6Al-4V titanium alloy and boron/epoxy laminates. The laminates were prepared from 3-in.-wide tape, SP 272, supplied by Minnesota Mining and Manufacturing Co. The laminates were laid up to the orientation specified, vacuum bagged, and cured in an autoclave under pressure at the cure temperature for 1 h. The rate of temperature rise was 5 F/min. The laminates were cooled under pressure to ambient and then cut into 1 by 4-in. coupons with a diamond saw. Before bonding, the surface glaze was broken by a light sandblast and the surface flushed clean with isopropanol. The 0.05-in.-thick titanium sheets were cut into 1 by 4-in. coupons which were cleaned and primed.

The primers and adhesives used in this study are proprietary designations of the Bloomingdale Department of the Plastics Division of American Cyanamide Co. FM1000 is referred to throughout this paper as Type A, and FM123-5 is referred to as Type B. BR1009-49 was used as a primer for the FM1000 adhesive and BR127 as a primer for FM 123-5 adhesive.

Single lap shear joints were made with a controlled bond line thickness of 0.005 in. and were held in alignment mechanically during cure. The rate of temperature rise to the cure temperature was 5 F/min. Cure was for 1 h at 350 F for Type A adhesive and 250 F for Type B adhesive.

Three coupons were used for each test. The results reported in Table 1 are the average of the three tests rounded off to two significant figures.

TEST OF BONDED JOINTS

Unidirectional

Joint specimens were prepared from unidirectional 15-ply laminate, 0.083-in. nominal thickness, to 0.050-in. titanium. To determine the joint strengths attainable, single lap shear joints were prepared at three overlap distances (three L/t ratios with constant thickness adherends). Both a flexible type adhesive and a semirigid adhesive system were evaluated. Metal-to-metal joints were fabricated and tested at a ½ in. overlap distance to serve as a basis for comparison with the laminate to metal joints.

Type A adhesive gives the highest average shear stress values at the ½ in. overlap, Fig. 2. The stress level is the same for titanium bonded to titanium or for titanium bonded to the unidirectional laminate. However, the average stress drops rapidly as the overlap distance is increased and drops to approximately half its value as the overlap distance increases from ½ to 1½ in.

Type B adhesive yields smaller average shear values at the ½ in. overlap, but the value does not fall off appreciably with increasing overlap distance. The load carrying capacity of a joint made with Type B adhesive at a 1½ in. overlap is approximately 20 percent greater than can be obtained with Type A adhesive. The joint is not as strong as the base laminate itself when tested in a single lap shear, but, when the joint is redesigned to test a specimen in double lap shear, the ultimate tensile strength of the laminate is approached.

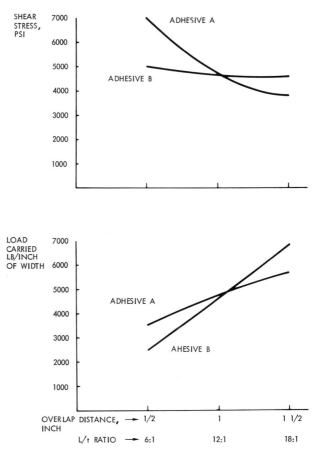

Fig. 2—Unidirectional laminates bonded with Adhesives A and B at three overlap distances.

The mechanism of failure has been discussed previously.[3] The first and second plies in contact with the faying surfaces appear to fail in tensile and then shear away from the underlaying plies. Figure 3 gives a schematic representation of the mode of failure. Figures 4, 5, and 6 show actual photographs of failed specimens. In some of these specimens damage extends into the third ply from the surface. To find out what fraction of the inherent tensile strength of the laminate could be transferred through a double lap shear joint, a group of joints of this type was fabricated. A comparison between these and single lap shear joints is given in Fig. 7. This series of tests included one laminate tensile failure at the 1½ in. overlap. Although the failure was below the expected strength of the laminate, it approached the measured tensile strength. Failure occurred at a tensile stress of 116,000 psi. The lowest pure tension test failure measured with rectangular coupons occurred at 126,000 psi, and failure in each test coupon was initiated at the grips where aluminum doublers were bonded to the coupons to aid in holding them. According to calculation, if a 0.004-in.-diameter boron filament with

[3] Dallas, R. N. and Stone, R. H., "A Starfighter Access Door from Boron Epoxy Laminate," Technical Paper EM 67-686, American Society of Tool and Manufacturing Engineers.

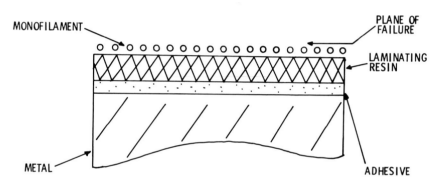

FIG. 3—Schematic of mode of failure.

FIG. 4—Bond failure at ½ in. overlap.

FIG. 5—Bond failure at 1 in. overlap.

FIG. 6—Bond failure at 1½ in. overlap.

an ultimate strength of 400,000 psi were laid into a tape containing 200 filaments per inch, then each ply of unidirectional tape should support approximately 1000 lb of load per inch of width. The unidirectional laminate on this basis would support 15,000 lb of load per inch of width, equivalent to a stress of 183,000 psi. So far it has been impossible to obtain this high value with flat tensile coupons or to approach it in joint strength, but the joint strength does yield over 90 percent of the laminate strength measured by tension tests on rectangular test coupons.

PSEUDO-ISOTROPIC

Joints were prepared from 15-ply laminates with the pseudo-isotropic fiber orientation shown in Fig. 1. A comparison of shear strength of two adhesives is shown in Fig. 8 of unidirectional laminate joints and pseudo-isotropic.

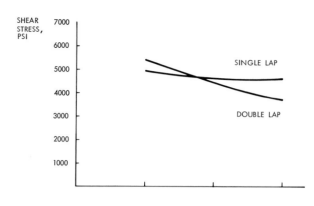

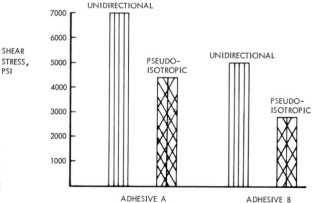

FIG. 7—Comparison of single and double lap shear joints of unidirectional laminate bonded with Adhesive B at three overlap distances.

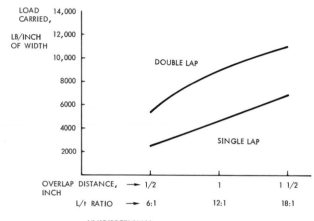

FIG. 8—Comparison of unidirectional and pseudo-isotropic orientations bonded with Adhesives A and B at ½ in. overlap.

In the 1 in. and the 1½ in. overlaps only, tensile failures of the laminate were encountered. The joint is stronger than the base laminate in these cases, but the strengths developed were rather disappointing and did not equal the strengths measured on tension tests of rectangular coupons machined from the same laminate. One significant finding was that the load supported by the ½ in. overlap is small in comparison with the load carried by the unidirectional laminate, only about 60 percent of the latter with either adhesive system. It appears that poor inter-laminar shear strength is the reason for this. When the joint fails, only the ply in contact with the adhesive fails leaving all the under layers intact and undamaged.

The outer ply strips away cleanly from the angled plies underneath it because plies at an angle to one another cannot nest together and the load is never transferred into the body of the laminate.

The term "nesting" as used here means that when adjacent plies have the same orientation the fibers intermesh and their identification with any particular ply is lost. However when adjacent plies are at an angle to one another they remain in distinct layers. In the latter case the stress is concentrated in the outer ply aggrevating the basic problem of a stress gradient through the adherends in bonded joints.

It was thought that rivets might aid in transferring loads into the isotropic orientations. Unbonded riveted joints either sheared rivets or, in some instances, failed the laminate in tensile through the rivet hole at low stresses. However, when 5/32-in rivets were placed in the center of a 1 in. overlap bonded with Type B adhesive, no significant difference in load carrying capability over similar un-riveted bonded joints could be detected.

As a consequence, it was decided to modify the ply orientation to combine the multidirectional strength of the isotropic layup with the ability to accept loads into the body of the laminate typical of unidirectional orientations.

MODIFIED ISOTROPIC

Joints were prepared from 15-ply laminates with the modified-isotropic fiber orientation shown in Fig. 1. The laminate had the ability to resist multidirectional loads because of the isotropic center section, while the plies at the faying surfaces

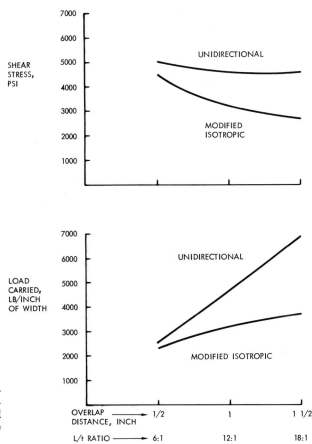

FIG. 9—Comparison of uni-directional and modified-iso-tropic laminates bonded with Adhesive B at three overlap distances.

Orientation	Tensile Strength, psi	Pin Bearing Strength, psi
Unidirectional..................	126 000	74 000
Pseudo-isotropic................	55 000	93 000
Modified isotropic..............	82 000	. . .
Fire access door laminate..........	51 000	113 000

had an opportunity to nest and thereby support a higher shear load in a bonded joint. Figure 9 shows a comparison of unidirectional laminate with the modified isotropic at three overlap distances. The unidirectional surface plies appear to improve load transfer, but the shear stress drops off rapidly with increasing overlap distance, perhaps due to a greater strain in the joint with this configuration.

A somewhat similar phenomenon was noted when an 11-ply isotropic laminate 0 deg, (60 deg, 0 deg, 120 deg)3, 0 deg was compared with a 15-ply (60 deg, 0 deg, 120 deg)5. The 11-ply laminate failed at a lower shear stress although it exhibited a higher tensile stress due to a larger proportion of 0-deg fibers. The strain for a given load would be greater with the thinner laminate.

Some tests were run with the 15-ply modified-isotropic orientation with rivets in addition to the adhesive bond. One 5⁄32-in. monel rivet was placed in the center of the 1⁄2-in. overlap joints, and the two rivets were placed in the 1 1⁄2 in. overlaps 1⁄2 in. from the end of the joint There was no significant difference in the load carried whether rivets were used or not. The result was similar to that reported for the pseudo-isotropic laminate.

The pin bearing and tensile strengths of the fiber orientations tested, as well as the orientation used for the fire access door, were measured and reported in Table 2. The values reported are ultimate bearing rather than the 4 percent offset typical of metal pin bearing strength.

Fɪɢ. 10—Metal bushings used to load beam.

HARDWARE COMPONENTS

A test section of an aircraft floor beam was redesigned in boron composite to replace an aluminum extrusion. The new design of the beam was a sandwich structure with cross-plied boron face sheets to carry shear loads and unidirectional laminate beam caps to resist bending loads. The loads were introduced into the sandwich by means of metal bushings bonded to the face sheets (Figs. 10 and 11). Since the part was a minimum weight design, weighing 25 percent less than the conventional structure, stresses were high and the joints critical.

The design required an allowable stress of 4000 psi in the metal-bushing to face-sheet joint; Type B adhesive was used for this purpose. When the part was static tested, it exceeded the design strength and ultimately failed the base laminate independent of the joint. Type B adhesive was used to bond the boron face sheets to the aluminum honeycomb core. Strengths exceeding 800 psi were obtained routinely in face sheet to core tension tests. Figure 12 shows a typical tag end specimen used for this test.

A second part designed in boron composite was a fire access door for the F-104 Starfighter, Fig. 13. The boron panel has the filament orientation shown in Fig. 1. It is joined to the titanium picture frame which attaches to the structure of the airframe. The composite-to-metal joint was a combination of a Type A adhesive bond and monel rivets used as a backup to increase reliability. Static tests of this design have given no indication of deterioration when subjected to ultimate design loads at 200 F. One of the boron composite fire access doors has been flight tested for several months on an Air Force F-104.

Fig. 11—Boron sandwich beam ready for test.

Fig. 12—Tag end section of beam sandwich for core to face sheet adhesion test.

Fig. 13—F-104 fire access door designed in boron.

CONCLUSIONS

Unidirectional laminates give the best adhesive bond strengths of any filament orientation. Double lap shear joints are substantially equal in strength to the tensile strength of the laminate up to 15 plies. Isotropic laminates exhibit low values for shear, about 60 percent of those obtainable with unidirectional laminates. The tensile strength of the isotropic laminate is so low, particularly when tested in a single lap shear joint, that the laminate fails in tensile before the joint is destroyed for overlap distances of one inch or more.

With modified-isotropic laminates, the joint strength is a compromise between that of the unidirectional and the isotropic layups. Since under a given load more strain occurs with the modified isotropic than with the unidirectional, joints made with the former will not tolerate high stresses at long overlaps.

It appears that a Type A adhesive would give the most efficient joint with a modified-isotropic laminate. Combinations of adhesive bonded and rivet joints have improved both the fatigue properties and static properties of riveted joints, according to previous work. Although these results would be expected, it has not been possible to confirm them in this study.

Composite Materials

Diffusion Bonded Scarf Joints in a Metal Matrix Composite

E. F. OLSTER[1] AND R. C. JONES[1]

Reference:

Olster, E. F. and Jones, R. C., "Diffusion Bonded Scarf Joints in a Metal Matrix Composite," *Composite Materials: Testing and Design, ASTM STP 460,* American Society for Testing and Materials, 1969, pp. 393–404.

Abstract:

Joints, approaching 100 percent efficiency, have been obtained in a 2024 aluminum alloy reinforced with type 355 stainless steel wires using a deformation diffusion bonding technique. A scarf joint, whose geometry is dependent upon the critical fiber length and the distance between layers of fibers, was developed for this study.

The effect of an aluminum oxide coating on the joining surfaces was investigated, and it was concluded that within the range of bonding parameters used in this study sound joints could be obtained only if the oxide coating was removed immediately before joining.

A wide range of joining parameters provided sound joints; however, the ultimate tensile strength of the joined specimen decreased as bonding times increased.

The high-strength scarf joint specimens failed with massive debonding along the diffusion bond planes resulting from the composite fabrication technique. These fracture surfaces examined with a scanning electron microscope and appeared nearly identical to fracture surfaces in unjoined, untreated material.

The lower-strength scarf joint specimens fractured with little or no debonding along these original bond planes, and when subjected to flexure tests could undergo massive deformation with no indication of macroscopic failure.

Key Words:

joining, diffusion bonding, fiber composites, metal matrix, evaluation, tests

The efficient joining of metal matrix composites is one of the major problems currently facing users of these materials. It is necessary to develop high efficiency joints and joining procedures which will provide reproducible strengths and failure modes. This study has been aimed at developing such joints in one metal matrix composite system, a stainless steel wire reinforced aluminum alloy composite. The composite used in this study is manufactured by Harvey Aluminum, Inc. of Torrance, Calif., by a diffusion bonding technique. The result is a unidirectional, continuous filament composite plate 0.25 in. thick having the mechanical properties described in Table 1.

A search through the literature reveals that virtually all of the industrial joining techniques have been utilized in attempts to join metal matrix composites. This includes adhesive joining, fusion welding, resistance spot welding, mechanical fastening, brazing, and diffusion bonding. Only rarely have these conventional methods proved satisfactory, however.

Diffusion bonding, although not thoroughly understood and not yet a standard commercial procedure, has been used as a fabrication technique and joining technique for several metal matrix composite systems. Preliminary investigations employ-

[1] Graduate student and associate professor, respectively, Department of Civil Engineering, Massachusetts Institute of Technology, Cambridge, Mass. 02139. Mr. Jones is a personal member ASTM.

Table 1—*Properties of a stainless steel wire, aluminum alloy matrix composite.*

UTS[a] of stainless steel wire (0.009 in. diameter)....	480 to 520 ksi
UTS of composite:	
RT[b]...	172 to 176 ksi
−320 F......................................	215 ksi
+700 F.....................................	100 ksi
Volume fraction..............................	25%
Modulus of elasticity of composite...............	15×10^3 ksi
Density of composite...........................	0.145 lb./in.^{3}
Strength to weight ratio........................	1.21×10^6 in.
Stiffness to weight ratio........................	1.03×10^8 in.
Stainless steel................................	Type 355
Aluminum alloy..............................	Type 2024

[a] UTS = ultimate tensile strength.
[b] RT = room temperature.

ing diffusion bonded joints in the present study provided encouraging results, and it was felt that further study of this technique should be pursued.

FUNDAMENTALS OF DIFFUSION BONDING

Diffusion bonding is a process whereby two pieces of material with closely fitting surfaces are placed in contact and eventually join, resulting macroscopically in a continuum. The process is implemented by the application of thermal and strain energy. Diffusion bonding can be separated into three categories: (*a*) solid state bonding, (*b*) deformation diffusion bonding, and (*c*) diffusion brazing. Table 2 describes the properties of each type of bonding.

The current literature [1,2,3][2] regarding the processes by which diffusion bonding occurs is imprecise and does little more than name two bonding theories: (1) the "film theory" and (2) the "energy barrier theory." Both of these propose that

[2] The italic numbers in brackets refer to the list of references appended to this paper.

Table 2—*Summary of diffusion bonding techniques.*

Type of Bonding	Characterization	Advantages	Disadvantages
Solid state bonding...	little or no plastic strain	good tolerance control	surface preparation critical and the oxide layer must be nearly eliminated before joining
Deformation diffusion bonding	massive plastic strain at the joint	surface preparation is only moderately important (the deformation breaks up the oxide layer)	oxide particles are found at the joint and may lead to crack formation; high pressures are required
Diffusion brazing.....	an intermediate layer either as a foil or as a plated surface is used to promote diffusion	promotes diffusion by providing a chemical gradient; high temperature —a eutectic liquid is formed providing intimate contact	possibility of forming brittle intermetallics exists; because of the Kirkendall effect voids may be formed at the joint

there is a driving force for bonding—the decrease in free energy of atoms originally at a surface, resulting from an increase in the coordination number as bonding occurs. The film theory implies that merely bringing two "clean" surfaces into contact results in a bond, whereas the energy barrier theory proposes that additional energy (an activation energy) must be supplied for bonding to occur.

The mating surfaces are brought into contact by external pressure. At the atomic level there are many gaps at the rugged, contacting surfaces. Some of these gaps are filled by plastic flow; the rest are filled by surface diffusion. Eventually, volume diffusion, relying upon a vacancy mechanism, completes the bond [4].

DIFFUSION BONDED SCARF JOINT

A basic knowledge of the behavior of composites and the manner in which they carry load is necessary for an understanding of the rationale behind joining schemes and joint geometry.

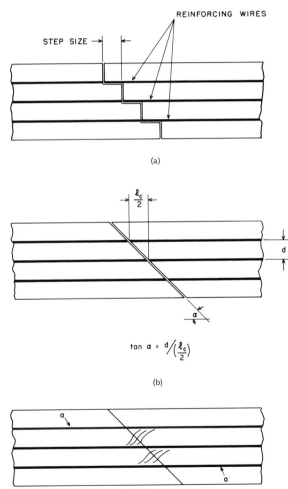

$$\tan \alpha = \frac{d}{\left(\frac{\ell_c}{2}\right)}$$

(b)

FIG. 1—(a) Idealized step type joint (step size $\geq$ $lc/2$), (b) scarf type joint (scarf angle $\alpha = \tan^{-1}$ $(2d/lc)$, and (c) shear transfer in scarf joint.

WAVY LINES INDICATE SHEAR TRANSFER.
WIRES DENOTED "a" HAVE INCOMPLETE TRANSFER.

(c)

Unidirectionally reinforced composites having discontinuous fibers and loaded in tension parallel to the reinforcement are assumed primarily to carry load axially in the fibers. The load is transmitted from fiber to fiber by shear stresses in the matrix. At a joint, a continuous filament composite can be analyzed as a discontinuously reinforced material. The critical fiber length (lc) can be defined as twice the minimum embedment length required to cause fiber fracture as opposed to fiber pullout. The value of lc for the aluminum-stainless steel composite used in this investigation was estimated to be approximately 0.12 in. This estimation was based on an assumed elastic-perfect plastic matrix.

If a technique were developed such that only a few wires were discontinuous at any cross section, a reasonably effective joint could result. One such technique is a step type joint shown in Fig. 1a. Full load transfer between fibers could be expected when the step size (overlap) is at least ½ lc. Practical composites have many closely spaced layers of reinforcement, and the step joint becomes impractical. The next logical transition is a scarf joint whose geometry is described in Fig. 1b. Notice that in this case, just as in the step joint, one layer of wires is left without a mate, as shown in Fig. 1c. To a first approximation[3] the rule of mixtures prediction for the estimated strength of the composite must be modified by a factor β.

$$S_c = \beta S_f V_f + S_m(1 - V_f)$$

where:
    S_c = strength of composite,
    S_f = strength of wires,
    S_m = strength of matrix,
    V_f = volume fraction of fibers,
    $\beta = (n - 1)/n$, and
    n = number of layers of reinforcing wires.

This slight reduction in estimated strength can be eliminated in low V_f composites if high enough temperatures and pressures are used, resulting in flow of the matrix. Figure 2a shows the necessary offset required to obtain $\beta = 1$. The final position of the layers of reinforcing wire is shown in Fig. 2b.

[3] It is assumed that the matrix material will bond ideally with the reinforcement and with adjoining material.

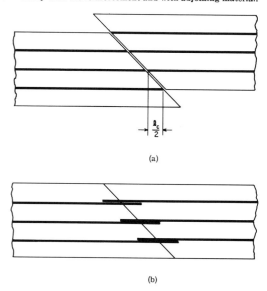

(a)

Fig. 2—Fabrication of a deformation scarf joint with full load transfer (a) before joining and (b) after deformation diffusion bonding.

(b)

Preliminary information regarding the range of the parameters used to diffusion bond aluminum was obtained from other investigators. A decision was made to restrict the joining parameters to the following: (a) time: 1 to 24 h, (b) temperature: 500 to 1000 F, and (c) pressure: 5000 to 18000 psi.

A mold was constructed to align the pieces to be joined and to prevent excessive macroscopic flow during the diffusion joining process. The mold was placed in a hydraulic press which was fitted with surface plates capable of temperatures from room temperature to 1000 F.

Specimens were prepared in the following way. Bars 6 by ¼ by ¼ in. were cut from a plate so that the reinforcing wires were parallel to the longitudinal axis of the bar. The bars were then cut into two pieces at the predetermined scarf angle. The faces of the scarf were polished with various grades of abrasive. This was a wet polishing ending with a No. 600 grit. The final prejoining technique is given in Table 3.

The specimens were placed in the mold in an overlapped position so that $\beta = 1$. Series B and C specimens were allowed to heat up to bonding temperature before pressure was applied, while Series A specimens were brought into intimate contact immediately. A concise description of these procedures is given in Table 3.

TESTING PROGRAMS

MECHANICAL TESTS

Mechanical tests were performed on diffusion bonded specimens which were machined to the geometry shown in Fig. 3. Tests to determine the ultimate tensile strength were performed on an Instron testing machine with a crosshead movement of 0.05 in./min.

Two specimens, which failed repeatedly in the grips of the Instron, were equipped with an electromechanical extensometer and subjected to a moderate tensile load so that the modulus could be determined. These specimens were then subjected to three-point flexure tests in an attempt to evaluate the effectiveness of the diffusion bond plane in shear.

The results of both the modulus tests and the flexure tests were compared to similar tests on unjoined, untreated material which is denoted by the words "as-

Table 3—Summary of joining procedures.

Specimen Type	Initial Preparation	Prejoining Procedure	Final Cleaning	Apply Heat	Apply Pressure
Series A:	wet, hand grinding, with Nos. 180, 240, 400, and 600 grit emery	wet, hand grinding with No. 400 grit emery	wash in acetone		
Time at start (min)	t_0	$t_0 + 20$	$t_0 + 25$	$t_0 + 26 = T_1{}^a$	$t_0 + 26 = T_1$
Time at finish (min)	$t_0 + 20$	$t_0 + 25$	$t_0 + 26$		
Series B:	wet, hand grinding, with Nos. 180, 240, 400, and 600 grit emery	none	wash in acetone		
Time at start (min)	t_0		$t_0 + 1440$	$t_0 + 1441$	$t_0 + 1456 = T_1$
Time at finish (min)	$t_0 + 20$		$t_0 + 1441$		
Series C:	wet, hand grinding with Nos. 180, 240, 400, and 600 grit emery	electropolish the joining surfaces	wash in acetone		
Time at start (min)	t_0	$t_0 + 20$	$t_0 + 1495$	$t_0 + 1496$	$t_0 + 1521 = T_1$
Time at finish (min)	$t_0 + 20$	$t_0 + 35$	$t_0 + 1496$		

^{a} T_1 = bonding begins.

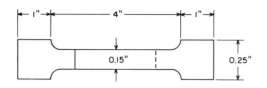

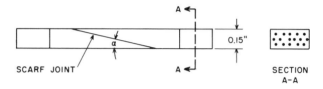

Fig. 3—Geometry of tension specimens.

α = SCARF ANGLE = $6\frac{1}{2}°$

received material." A qualitative estimate of the effectiveness of the composite material in shear was obtained by comparing values of τ_m

where:

$\tau_m = (3/2)(V/A)$,
V = maximum shear force, and
A = cross-sectional area.

It is not suggested that this value of τ_m is quantitatively the value for the maximum horizontal shear stress, but is felt that this value has some merit for comparative purposes.

METALLOGRAPHIC TESTS

A scanning electron microscope was used to observe the fracture and failure[4] surfaces resulting from the tension tests. An attempt was made to disclose reasons for failure and to correlate these with the cleaning procedure and the bonding history. The fracture surface of these joined composites was compared also to a similar surface on as-received material.

Metallographic examination of a cross section cut through the scarf joint of a typical diffusion bonded specimen was performed to observe the bond and to note any difference in the diffusion bond at the joint as compared to the diffusion bond planes resulting from the original manufacturing process.

DISCUSSION OF EXPERIMENTAL DATA

MECHANICAL TESTS

The experimental data obtained from Series A specimens, where the abrasive removal of the oxide layer immediately before bonding coupled with the application of bonding pressure immediately minimized contact of the bonding surface with air is given in Table 4. This treatment was undoubtedly responsible for the high performance of Series A specimens as compared to Series B or C, which are described in Fig. 4. The data reveal that with the proper choice of bonding parameters an efficient, high-strength joint can be made by diffusion bonding. Specimens A4 and A5 failed at an ultimate tensile strength (UTS) of 170 ksi which is within 2 percent of the unjoined material value.[5]

[4] On a macroscopic scale debonding essentially along the original scarf joint occurred in these regions.
[5] Refer to Table 1.

Table 4—*Experimental tensile data from Series A Specimens.*

	Bonding Parameters				
Specimen Designation	Time from T_1,[a] h	Temperature, deg F	Pressure, ksi	Failure Stress, ksi	Failure Mode[c]
A3.............	1	1000	12.2	128	2
A4.............	1	1000	14.6	170	4
A5.............	1	1000	14.6	170	4
A8.............	16	1000	12.2	154	4
A9.............	22.5	1000	9.8	148	4
A10............	4	1000	12.2	135	4
A11............	7	850	14.6	110[b]	4
A12............	4	500	17.8	0	1
A13............	6	900	13.3	107	2
A14............	5	950	13.3	148	4
A15............	4.5	950	13.3	118[b]	4
A16............	5	950	8.9	122	2
A17............	1.5	1000	17.8	142	3

[a] See Table 3.
[b] Failed in grips.
[c] See Fig. 4.

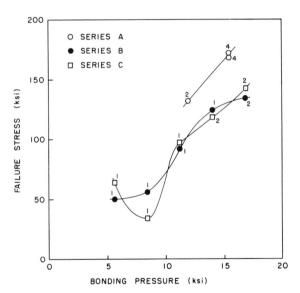

Fig. 4—Failure stress versus bonding pressure (superscripts refer to failure mode given directly below the graph).

Specimens A11 and A15, both of which failed in the grips of the Instron testing machine during tensile loading, were subjected to modulus and flexure tests. The moduli compare favorably to values obtained from tests on as-received material [5]. These specimens deformed continuously in flexure, and finally a "plastic hinge" was formed. A flexure test on a specimen of as-received material of comparable size resulted in debonding and buckling of the uppermost compression layer for a distance of ¼ in. along the length of the specimen, which resulted in a sharp decrease in load. At continuing center line displacement, the load increased slightly until gross shear failure of a plane 30 percent of the depth down from the top propagated rapidly from one end of the beam to approximately the middle of the specimen. The flexural behavior is shown in Fig. 5.

A graphical display of the data obtained from uniaxial tension tests on Series B and C specimens (Fig. 4) implies a nearly linear relation between failure stress and bonding pressure which suggests that if higher pressure are acceptable perhaps the failure stress will reach the desired value.

With both Series B and C the cleaning (except for an acetone rinse) took place approximately 24 h before bonding, and the specimens were preheated before pressure was applied. These procedures allowed a relatively thick oxide coating to form on the mating surfaces which hopefully would be broken up by the deformation diffusion bonding process. The major difference between these groups of specimens was the surface roughness which was several orders of magnitude greater for Series B specimens. It was expected that the rougher surface on Series B specimens would deform readily and break up the oxide coating resulting in higher failure strengths. This was not observed and is possibly due to the fact that more surface area was exposed in Series B specimens resulting in an increase in the total amount of oxide in the vicinity of the joint.

METALLOGRAPHIC OBSERVATIONS

Metallographic observations of a typical diffusion bonded joint indicate that the bond along the scarf joint is similar to, but not as complete as, diffusion bonds

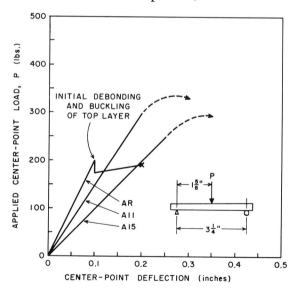

SPECIMEN	HEIGHT (inches)	WIDTH (inches)	τ_m (psi)
A11	0.214	0.154	8350
A15	0.206	0.137	7600
AR (as received)	0.214	0.132	4800

Fig. 5—Load-deflection behavior of flexure specimens.

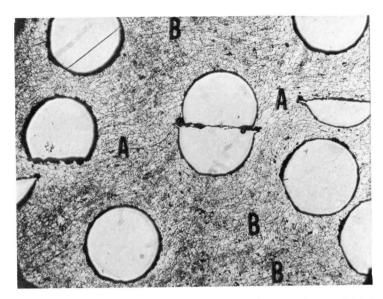

Fig. 6—Cross section through Specimen A8 showing the scarf joint bond line (A) and the fabrication bond lines (B).

formed during manufacture of the composite and subjected to the additional temperature and pressure history incurred during the joining process. A typical cross section through a joint is shown in Fig. 6. Note the fiber-fiber contact and massive flow of material near the joint.

A micrograph typical of the fracture surfaces of high-strength joints is shown in Fig. 7a. The high-strength specimens resulted when the bonding periods were sufficient to allow full shear transfer at the joint but short enough to cause no substantial fiber degradation or change in matrix properties. Figure 7a reveals that the fibers fail in a cup and cone mode and that debonding occurred at both the fiber-matrix interface and the matrix-matrix interface. This failure mode is somewhat typical of as-received material [6]; however, note that the failure in the matrix-matrix interface is quite severe in the as-received material shown in Fig. 7b.

Increased bonding periods result in decreased tensile strength, perhaps due to fiber degradation in the form of a partial annealing. Figure 7c shows the fracture surface of such a low-strength specimen. The major difference observed in the fracture surface of the low-strength specimens as compared to the high-strength and as-received specimens is the continuity at the diffusion bonds between adjacent layers of the matrix. This increased matrix continuity is attributed to the additional hot pressing of the original fabrication diffusion bond joints. It resulted in the high flexural performance of specimens A11 and A15 compared to as-received material.

The failure surfaces[6] of specimens which exhibited slight or partial joint failure were examined in an effort to detect flaw sites; however, positive identification of such a flaw has not been obtained as yet. In general, the failure surfaces were rugged, as shown in Fig. 8, with portions of the matrix ripped from the reinforcing wires.

Examination of Series B and C specimens revealed fracture and failure surfaces similar to those described above. The roughness of the failure surface increased

[6] Failure near or along the scarf joint.

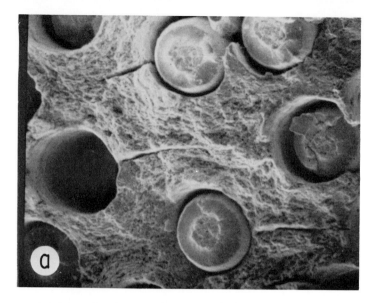

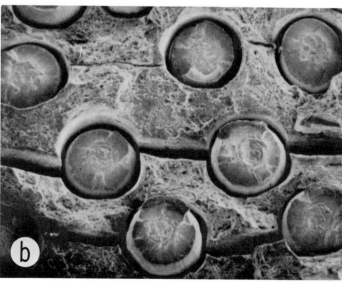

FIG. 7—(a) Micrograph of Specimen A4 showing cup and cone fiber fracture, and failure at both the fiber-matrix interface and at the matrix-matrix interface; (b) micrograph of as-received material, note the more severe matrix-matrix interface failure; and (c) micrograph of Specimen A9, note the absence of failure at the matrix-matrix interface.

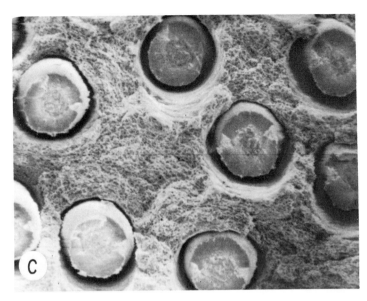

Fig. 7—Continued.

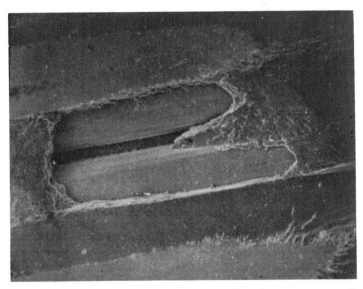

Fig. 8—Micrograph of a typical failure surface showing regions of satisfactory bond (where matrix was torn from fibers), and regions of poor bond (where surface appears flat).

with increased bonding pressure and correlates qualitatively with the observed increases in failure stress.

CONCLUSIONS

It has been shown that this aluminum-stainless steel composite can be joined satisfactorily in air if the oxide layer is removed by abrasion immediately before joining.

Olster and Jones on Bonded Scarf Joints

To date, joints of nearly 100 percent efficiency have been obtained in the aluminum-stainless steel composite used for this research. Sound joints have been obtained over a wide variation in bonding parameters. The average efficiency for sound joints is 90 percent, based on an expected value of 172 ksi. This efficiency undoubtedly can be improved as the technique is perfected.

The strongest specimens had an ultimate tensile strength of 170 ksi. These specimens were joined at 1000 F for 1 h with an applied pressure of 14,600 psi. The fracture surface of these specimens closely resembled the fracture surface of as-received material which exhibited massive debonding along original diffusion bond planes.

Less strong specimens having an average UTS of 155 ksi resulted when the bonding time was significantly greater than 1 h. The fracture surface of these specimens showed little or no debonding along the diffusion bond planes. Flexure tests of these specimens revealed that they could undergo massive deformation without debonding, whereas similar tests on as-received material resulted in failure along a diffusion bond plane.

ACKNOWLEDGMENT

Financial support for this research has been provided by the Dow Chemical Company of Midland, Mich., through the Dow-Massachusetts Institute of Technology (MIT) Program in Structural Materials, and through the MIT Inter-American Program in Civil Engineering, funded by the Ford Foundation. The composite material employed in this study was provided by the Harvey Engineering Laboratories of Torrance, Calif.

References

[1] Cline, C. L., "An Analytical and Experimental Study of Diffusion Bonding," Welding Research, Supplement to the Welding Journal, Nov. 1966, pp. 481s–489s.

[2] Tylecote, R. F., "Diffusion Bonding," Welding and Metal Fabrication, Vol. 35, No. 12, Dec. 1967, pp. 483–489.

[3] Tylecote, R. F., "Diffusion Bonding," Welding and Metal Fabrication, Vol. 36, No. 1, Jan. 1968, pp. 31–35.

[4] Hansen, D., Kammer, P., and Martin, D., "Fundamentals of Solid State Welding and their Applica-

tion to Beryllium Aluminum and Stainless Steel," Contract No. DA 01-021-AMC-11706(Z), Battelle Memorial Institute, pp. 12–20.

[5] "Composite Materials for Turbine Compressors," Solar Division of International Harvester Company, Contract No. AF 33(615)-5189, pp. 38–39.

[6] Jones, R. C., "Deformation of Wire Reinforced Metal Matrix Composites," Metal Matrix Composites, ASTM STP 438, American Society for Testing and Materials, 1968, pp. 183–217.

Tungsten Alloy Fiber Reinforced Nickel Base Alloy Composites for High-Temperature Turbojet Engine Applications

**D. W. PETRASEK[1] AND
R. A. SIGNORELLI[1]**

Reference:

Petrasek, D. W. and Signorelli, R. A., "Tungsten Alloy Fiber Reinforced Nickel Base Alloy Composites for High-Temperature Turbojet Engine Applications," *Composite Materials: Testing and Design, ASTM STP 460,* American Society for Testing and Materials, 1969, pp. 405–416.

Abstract:

The potential of tungsten-2 percent thoria, 218 tungsten, and tungsten-1 percent thoria wire reinforced nickel-base superalloy composites for turbine bucket applications was evaluated based on stress-rupture strength, oxidation, and impact resistance. The results indicate that refractory metal alloy fiber-superalloy composites have potential for turbine bucket use based on the properties measured. Composites were produced having stress-rupture properties superior to conventional cast superalloys at use temperatures of 2000 and 2200 F. The 100 and 1000-h stress-rupture strength of the composite at 2000 F was 49,000 and 37,000 psi, respectively. A few thousands of an inch of matrix or cladding material was found to be sufficient to protect the fibers from oxidation at 2000 F for times up to 300 h. At 300 F and above the impact resistance of the composite compares favorably with that of superalloys.

Key Words:

fiber reinforcement, fiber composites, superalloy, mechanical properties, turbojet engine, turbine bucket alloy, oxidation, high-temperature alloy, refractory fiber, tungsten wire, evaluation, tests

A need exists for materials to be used as high-temperature components in advanced turbojet engines. Materials currently available for use in these engines either have limitations of strength or of oxidation resistance. The strength of superalloys, for example, is inadequate, and the refractory metals, which have adequate strength at high temperatures, have poor oxidation resistance. One of the most promising types of materials for applications in the 2000 to 2400 F temperature range is refractory fiber reinforced superalloy composites, if advantage can be taken of the high strength of the refractory metal fiber and the relatively good oxidation resistance of the superalloy matrix.

The principal property requirement for bucket materials is high creep rupture strength combined with adequate oxidation resistance. The effects of alternate or contributory failure mechanisms must be considered also; these include impact, thermal and mechanical fatigue, thermal shock, and hot corrosion damage. The evaluation of the potential of new materials for turbine bucket use is performed normally using laboratory tests to determine these properties.

Previous work conducted at the NASA Lewis Research Center demonstrated

---

[1] Materials engineer and head, respectively, Fiber Metallurgy Section, NASA Lewis Research Center, Cleveland, Ohio. 44135.

that 70 volume percent refractory metal fiber superalloy composites could be produced having a 100-h rupture strength at 2000 F of 35,000 psi as compared to 12,000 psi for conventional cast superalloys [1].[2] The results of this work demonstrated the necessity to minimize reaction between the fiber and matrix material by proper fabrication procedure and proper fiber diameter selection in design of a composite to obtain high-strength properties. The lamp filament wire used in this composite was the strongest commercially available wire at the time. Improved high-strength wire, however, has been made available under contract by Lewis Research Center as part of a continuing effort to obtain higher strength fiber materials.

The object of this investigation was to determine the stress-rupture strength and conduct exploratory studies of impact and oxidation resistance of tungsten alloy fiber-superalloy composites at 2000 F. A determination of these properties would then permit discussion of the potential of such materials for use in turbojet engine components. In this program additional work was conducted using improved high-strength fibers of tungsten-2 percent thoria as compared to conventional lamp filament wire used previously. Composites of a nickel base alloy reinforced with tungsten-2 percent thoria wire were fabricated containing up to 70 volume percent wire. The composite specimens were then evaluated in stress-rupture at 2000 and 2200 F. Exploratory studies also were made of the oxidation resistance and impact resistance of refractory metal fiber superalloy composites. Composites were made using tungsten-1 percent thoria and 218 tungsten wire as the reinforcement material. Continuous weight-gain oxidation tests were conducted in static air at 2000 F. Izod or Charpy impact tests were conducted on the composite specimens at room temperature to 2000 F.

EXPERIMENTAL PROCEDURE

WIRE AND MATRIX MATERIAL

The wire materials selected for use in this investigation were tungsten-2 percent thoria and commercial lamp filament wire of 218 tungsten and tungsten-1 percent thoria. The wires were in the as-drawn, cleaned, and straightened condition. The wire diameters were 0.015 in. for the tungsten-2 percent thoria and 218 tungsten material and 0.020 in. for the tungsten-1 percent thoria material.

The composition of the nickel base alloy matrix material was selected based upon its compatibility with the fiber, as determined in a prior investigation [1]. The nominal composition of the nickel alloy was 56 percent nickel, 25 percent tungsten, 15 percent chromium, 2 percent aluminum, and 2 percent titanium. The nickel alloy powder was vacuum cast and atomized into fine powder ranging in size from −325 to 500 mesh. Vacuum cast stress-rupture specimens for the alloy were obtained from the same master melt used for making the powder.

COMPOSITE SPECIMEN FABRICATION

Composites containing the tungsten alloy wire and the nickel alloy were made by a slip casting process described in detail in Ref 1. The metal powder slip consisted of the nickel alloy powder and a solution of ammonium salt of alginic acid in water. Composite specimens were made by inserting continuous length tungsten alloy wires into an inconel tube and infiltrating the wire bundle with the metal slip. After slip casting, each specimen was removed from its tube, dried in air, and then sintered at 1500 F for 1 h in hydrogen. The specimens were then reinserted into the inconel tubes, sealed at both ends, and isostatically hot pressed at 1500 F for 1 h, and then at 2000 F for 1 h using helium pressurized

[2] The italic numbers in brackets refer to the list of references appended to this paper.

to 20,000 psi. Fully densified specimens of over 99 percent theoretical density were produced and machined into specimens to be tested in stress rupture.

TESTING PROCEDURE

Stress-rupture tests on the wire material were conducted at 2000 and 2200 F in a measured vacuum of 5×10^{-5} torr using a stress-rupture apparatus specifically designed for the testing of small diameter fibers. A detailed description of this apparatus may be found in Ref 2.

Stress-rupture tests at 2000 and 2200 F were conducted on vacuum cast nickel alloy matrix specimens and on composite test specimens using conventional creep machines and a helium atmosphere.

METALLOGRAPHIC STUDIES

Stress-rupture specimens were examined metallographically after fracture to determine the depth of reaction between the matrix material and the fiber as a function of time at temperature, and to determine the volume percent of fiber content of the specimens. The microstructure of the fiber-matrix interface was examined metallographically and the depth of reaction measured optically on transverse sections of composite specimens using a filar eyepiece at a magnification of $\times 500$. The depth of the reaction zone is defined as the distance from the fiber-matrix interface to the interface in the fiber where a microstructural change is observed. The volume percent fiber content for all composite specimens were obtained as follows: each specimen was sectioned transversely in an area immediately adjacent to the fracture, the sections were mounted, polished, and photographed at a magnification of $\times 25$, and a wire count was obtained from the photographs, from which the volume percent fiber contents were calculated.

OXIDATION STUDY

An exploratory study was made of the oxidation resistance of refractory metal fiber superalloy composites at 2000 F. Conventional lamp filament wire, 218CS and tungsten-1 percent thoria, was used in the composites. The matrix material had the same composition as that used for the stress-rupture study. The oxidation specimens were approximately ¼ in. diameter by ½ in. in length and contained fiber contents ranging from 40 to 70 percent. The composite specimens were fabricated using the same procedure as that used to make the stress-rupture specimens. The as-pressed composite specimen thus was encased completely with inconel. Three different type oxidation specimens were machined from the as-pressed specimens. The three types of machine specimens were as follows: (1) specimens were machined such that a 0.006 in. layer of the inconel container remained on all exterior surfaces of the specimens, (2) specimens in which the cross-sectional ends of all wires were exposed to the oxidizing environment, and (3) specimens machined to expose both the cross-sectional ends of all wires and a longitudinal section through those wires intersecting the outer periphery of the cylindrical specimen.

Continuous weight-gain tests were conducted on oxidation specimens. Specimens were suspended by a platinum wire from an analytical balance positioned above a resistance wound furnace. Static air oxidation tests were conducted at 2000 F.

IMPACT RESISTANCE STUDIES

Impact tests were made on composite specimens of 218CS and tungsten-1 percent thoria wire nickel alloy specimens having fiber contents ranging from 50 to 60 percent. A standard Charpy tester and a low capacity Izod impact tester were used. The low capacity Izod impact tester is described in more detail in Ref 3. Charpy specimens were machined to ASTM Specification for Notched Bar Impact Testing of Metallic Materials (E 23-66), while Izod specimens were machined to one half size of ASTM specifications. The Charpy specimens were 0.394 by 0.394 by 2.165-in. bars while the Izod specimens were ³⁄₁₆ by ³⁄₁₆ by 1.5-in. bars. The entire testing procedure was conducted in accordance with ASTM Specification E 23-66. Standard unnotched bar type specimens and specimens in the V-notched condition were studied. Charpy impact tests were conducted at 2000 F, while miniature Izod impact specimens were tested at room temperature to 300 F.

The small specimens and low capacity of the miniature Izod tests were used to better differentiate between relatively small impact values. The very large capacity Charpy was preferred for the high values obtained above the brittle transition temperature of the tungsten alloy wire.

RESULTS AND DISCUSSION

STRESS RUPTURE PROPERTIES

The stress-rupture results obtained on the nickel base alloy matrix material and the tungsten-2 percent thoria wire are shown in Tables 1 and 2. Conventional stress versus rupture life plots of these data indicated a 100-h stress rupture strength at 2000 F for the nickel alloy of 3200 psi and for the wire of 95,000 psi. An extrapolated 1000-h stress-rupture strength at 2000 F of 2000 psi was obtained for the nickel alloy matrix material, while that for the wire was approximately 93,000 psi.

The stress-rupture properties obtained for the composites are shown in Table 3. The composite specimens contained fiber contents from 25 to 70 volume percent and were tested at stress levels from 35,000 to 55,000 psi. In order to determine the composite stress-rupture strength for a specific life and at a specific fiber content, it was necessary to determine the fiber strength contribution in the composite. The stress on the fiber was calculated by neglecting the stress on the matrix and dividing the composite specimen load by the area of fiber contained in the composite. It was assumed that the fiber carries the major portion of the load during stress-rupture and that the matrix contribution was negligible which is in agreement with the analysis of the stress-rupture properties of composites as reported in Ref 4. The calculated stress on the fiber is shown in Table 3. Figure 1 is a plot of the calculated stress on the tungsten-2 percent thoria wire at 2000 F based on composite data as a function of rupture time. Also shown in the plot is the stress-rupture strength of tungsten-2 percent thoria wire which was tested in vacuum at the same temperature as a function of rupture life. A least square fit was made

Table 1—*Stress-rupture properties of nickel base alloy material.*

Stress, psi	Life, h	Reduction in Area, %
	2000 F	
3000	108.6	29.0
3000	125.0	35.9
3500	63.6	33.4
4000	32.2	37.3
5000	8.3	42.4
	2200 F	
1000	116.7	12.5
1500	17.8	11.7

Table 2—*Stress-rupture data for tungsten —2 percent thoria wire (0.015-in.-diameter).*

Stress, psi	Life, h	Reduction in Area, %
2000 F		
100 000	6.5	45.0
97 000	43.3	26.2
95 000	228.1	42.2
93 000	218.5	38.2
2200 F		
80 000	17.8	15.4
75 000	26.1	14.2
73 000	51.7	...
70 000	89.6	29.3
70 000	120.6	49.1
65 000	116.0	20.4
60 000	146.6	17.8

of the composite wire data. The calculated stress-rupture strength contribution of the wire contained in the composite is shown to be lower than the wire tested in vacuum. The lower strength properties of the wire contained in the composite was shown in a previous investigation [1] to result from reaction with the matrix material. The properties of the composite may be related to the degree of reactivity between the fiber and matrix material as well as to the initial properties of these components. Generally the smaller the depth of penetration into the fiber the higher the composite properties. The results of the depth of penetration measurements for the composites are shown in Table 3. It can be seen from Table 3 that as the time of exposure at temperature increases the depth of penetration into the fiber also increases. The strength of the fibers in the composite also continuously

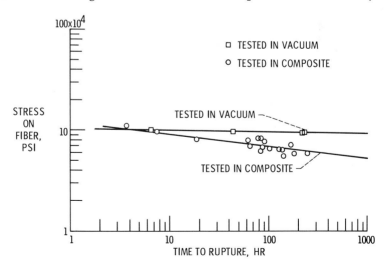

FIG. 1—Stress rupture strength of tungsten-2 percent thoria wire in composite and of wire tested in vacuum at 2000 F.

Table 3—*Stress-rupture data for nickel alloy-tungsten—2 percent thoria (0.015-in.-diameter) wire composites.*

Stress, psi	Life, h	Volume % Fiber	Calculated Stress on Fiber, psi	Penetration, in.
		2000 F		
30 000	7.5	31.1	96 500	0.00053
	86.3	44.9	66 800	0.00156
	83.7	48.5	62 000	0.00117
35 000	140.2	63.8	54 900	...
	180.6	60.8	57 700	0.00374
	244.0	59.0	59 300	0.00208
40 000	0.1	24.7	161 800	0.00041
	3.7	36.0	111 200	0.00027
	88.9	52.3	76 400	0.00132
	101.5	61.3	65 300	...
	126.4	61.5	65 100	0.00213
45 000	65.4	64.4	70 000	0.00208
	134.1	70.8	63 600	0.00132
	167.2	62.0	72 500	0.00213
50 000	0.2	40.6	123 200	0.00026
	77.4	60.9	82 100	0.00125
	81.0	60.0	83 300	0.00125
55 000	0.3	48.5	113 500	0.00026
	0.8	38.3	143 500	0.00052
	18.7	67.8	81 200	0.00130
	61.1	69.1	79 700	0.00117
		2200 F		
15 000	14.8	53.5	28 100	0.00416
	18.4	58.6	25 600	0.00500
	23.8	59.3	25 300	0.00390

decreases with time at temperature as a result of the increased penetration as a function of time. The stress-rupture strength of the fiber in the composite thus shows a continuous decrease in strength as compared to that for a fiber which is not reacted and which was tested in vacuum as shown in Fig. 1. Even after 100-h exposure, however, the reacted fiber retains 72 percent of the properties of a fiber which has not been reacted. Extrapolation of the data would suggest that after exposure for 1000 h the reacted fiber retains 57 percent of the properties of an unreacted fiber.

The plot shown in Fig. 1 can be used to determine the stress-rupture properties of composites containing varying volume fractions of fibers. The stress on the fibers to cause rupture in a specific time is multiplied by the volume fraction of fiber contained in the composite. From the data shown in Fig. 1, for example, it would be expected that a composite containing the tungsten-2 percent thoria wire and having a fiber content of 70 volume percent would have a 100-h stress-rupture strength of 49,000 psi (0.70 × 70,000 psi) at 2000 F. The validity of this calculation can be supported by actual composite data. It can be seen from Table 3 that a specimen stressed a 45,000 psi containing 70.8 volume percent fiber had a rupture life of 134 h and that a specimen stressed at 55,000 psi and having a fiber content of 69.1 ruptured in 61 h. The calculated 100-h rupture strength value of 49,000 psi for a composite containing 70 volume percent thus appears to be valid. Using the method indicated above, the 100 and 1000-h rupture strength at 2000 F was calculated for composites containing 70 volume percent fiber contents.

Composite Materials

Figure 2 shows a plot of the stress for rupture in 100 and 1000 h for the nickel alloy matrix material, the composite containing 70 volume percent tungsten-2 percent thoria wire and the unreacted tungsten-2 percent thoria wire all tested at 2000 F. The 100-h rupture strength of the composite at 2000 F was found to be 49,000 psi, while the 1000-h rupture strength obtainable from extrapolation is 37,000 psi.

Figure 3 compares the 100 and 1000-h rupture strength of conventional cast nickel alloys and tungsten-thoria wire reinforced nickel alloy composites tested

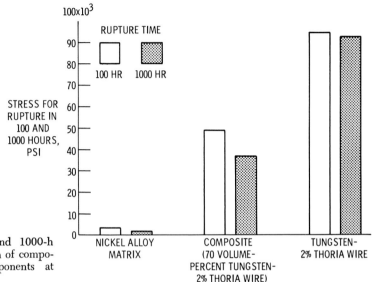

FIG. 2—100 and 1000-h rupture strength of composite and components at 2000 F.

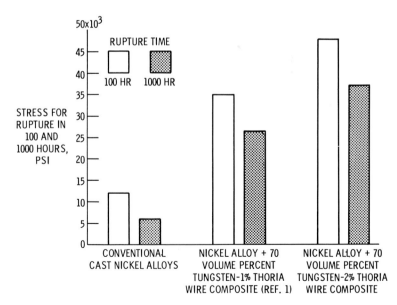

FIG. 3—100 and 1000-h rupture strength of cast nickel alloys and tungsten-thoria wire reinforced nickel alloy composites tested at 2000 F.

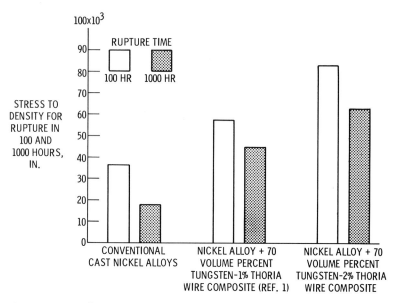

FIG. 4—100 and 1000-h specific rupture strength of cast nickel alloys and tungsten-thoria wire reinforced nickel alloy composites tested at 2000 F.

at 2000 F. The rupture strength of the composite containing 70 volume percent tungsten-1 percent thoria wire was obtained in previous work reported in Ref 1. A 40 percent improvement in 100 h rupture strength was obtained using tungsten-2 percent thoria as the reinforcement material as compared to tungsten-1 percent thoria wire as a reinforcement. Both composite systems were made by the same fabrication process and contained a matrix material having the exact same composition. A similar improvement was obtained for a 1000-h rupture strength by use of tungsten-2 percent thoria wire rather than tungsten-1 percent thoria wire. The tungsten-2 percent thoria wire composite is seen to have a 100-h rupture strength four times that of the strongest conventional cast superalloys at 2000 F and a 1000-h rupture strength which is six times that of the superalloys.

The density of the composite material is much greater than that of cast nickel alloys and must be taken into consideration. The tensile stresses in turbine blades, for example, are a result of centrifugal loading; therefore, the density of the material is important. Tungsten has a density about 2.3 times that of most nickel base alloys, and thus a composite containing 70 volume percent tungsten fibers has a density approximately 1.9 times that of most nickel base alloys. A comparison of the specific strength properties of the composite and conventional superalloys is therefore significant. Figure 4 is a plot of the 100 and 1000-h specific rupture strength comparison of cast nickel alloys and tungsten-thoria wire reinforced nickel alloy composites at 2000 F. Even when density is taken into consideration the composite containing tungsten-2 percent thoria wire is much stronger than the conventional cast superalloys. The tungsten-2 percent thoria wire composite has a 100-h specific rupture strength over twice that of the superalloys at 2000 F, and a 1000-h specific rupture strength over three times that of the superalloys. Comparison with the stress/density properties of superalloys may be used as a basis for indicating promise of composite materials for turbine bucket use. Standard nickel alloys used as turbine buckets have a stress/density for 1000-h rupture at 1800 F of about 60,000 in. The fiber composite has similar stress/density values at 2000 F. Based on this strength comparison, the fiber composite has a 200 F use temperature advantage over conventional superalloys.

OXIDATION AND IMPACT PROPERTIES

Stress-rupture strength alone does not satisfy the requirements of a material for turbine bucket applications. The material must be able to resist the oxidation and impact imposed by engine operating conditions as well as other failure mechanisms. Studies of oxidation resistance and impact failure are in progress at Lewis Research Center, and encouraging preliminary results have been obtained.

OXIDATION STUDIES

Specimens in which the ends of the as-pressed specimens were sectioned to expose all of the refractory metal wires to the oxidizing environment exhibited excessive oxidation in air at 2000 F. The tungsten alloy wires oxidized at their normal high rate of oxidation. Specimens in which the container material was removed completely to expose all of the refractory metal wires in the longitudinal direction and surface wires in the transverse direction also exhibited excessive oxidation at 2000 F. The majority of the oxidation, however, occurred in the longitudinal direction. In the transverse direction only the wires exposed to the atmosphere were oxidized, as shown in Fig. 5a. The matrix material protected the internal fibers from oxidation in the transverse direction. Specimens completely encased with a 0.006-in. layer of inconel exhibited good oxidation resistance at 2000 F for exposure times over 300 h. Figure 5b shows a cross-sectional view of such a specimen tested in air at 2000 F for 50 h. The inconel cladding was oxidized and a coherent oxide scale formed with the inconel. Oxidation had not progressed to the composite, and the surface wires were not affected by the oxidation of the cladding. The reaction zone formed with the wire after exposure for 50 h at 2000 F, as seen in Fig. 5b, was due to alloying of the matrix material with the wire. The results of the preliminary oxidation study indicated that advantage can be taken of the better oxidation properties of the matrix material or of a cladding material on the outer surface of the composite specimen. The tungsten alloy fiber oxidizes at its normal high rate when exposed to the oxidizing environment. A few thousands of an inch of matrix or cladding material, however, protected the fiber from oxidation for exposure times over 300 h at 2000 F.

The best superalloys may require additional oxidation protection for some turbine applications. Excellent progress has been made in developing coatings to provide this protection, and some of the promising coatings currently emerging may be an ideal cladding material for fiber composites. These coatings would be expected to be even more effective than the inconel cladding used on the oxidation test specimens.

IMPACT STUDIES

An exploratory study was also made to determine the impact resistance of refractory metal fiber-superalloy composites. The notched Izod test specimens yielded the following results: room temperature–4.7 in·lb, 250 F–4.7 in·lb, and 300 F–26.7 in·lb. The unnotched Izod specimens results were as follows: room temperature–3.25 in·lb, 185 F–7.1 in·lb, and 300 F–greater than 26.7 in·lb. Two Charpy tests were conducted at 2000 F. An unnotched specimen had a measured impact strength of 29 ft·lb, while a notched specimen had an impact strength of 27.5 ft·lb. The low impact strength of the composites below 300 F may be attributed to the brittleness of the tungsten wires below their transition temperature as indicated in previous work [5]. Above 300 F the impact resistance of the composite compares favorably with that of the superalloys.

The impact resistance necessary for turbine buckets is not related clearly to laboratory impact test data. Cast nickel alloys, however, which have been run successfully in turbojet engine tests, have been reported to have low capacity Izod impact strengths in the unnotched condition of less than 15 in·lb at room tempera-

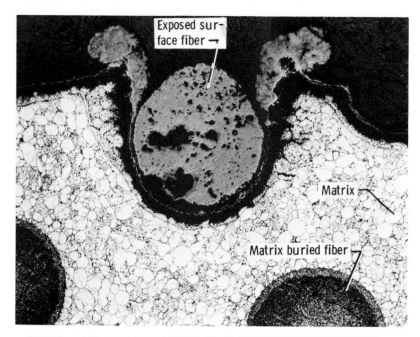

(*a*) Fiber, 0.015-in.-diameter 218 tungsten. Test condition, 5 h in air at 2000 F (×100).

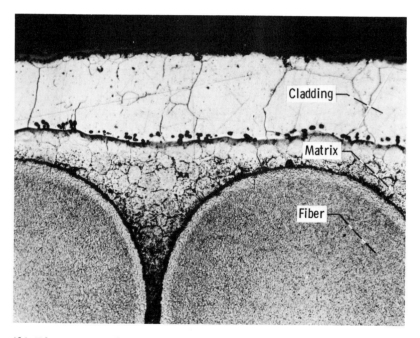

(*b*) Fiber, 0.020-in.-diameter tungsten. 1 percent thoria. Test condition, 50 h in air at 2000 F (×150).

Fɪɢ. 5—Transverse section of oxidized tungsten alloy fiber-nickel alloy composite.

ture and 1650 F [6,7]. The impact strength of the composites compare favorably with superalloys at test temperatures of 300 F and above. However, the low impact strength values obtained at 250 F and below may be troublesome. Impact resistance at room temperature may be increased by decreasing fiber content and by improving the ductility of the matrix. Volume fiber content along the length of a bucket could be varied because of the normal variation of stress along the length of turbine buckets. The improvement of impact strength possible by variation of fiber content should be studied. No effort has been made to study increased matrix impact strength. The matrix composition of this study sacrificed impact strength to maximize compatibility. Optimization of both impact and compatibility is an area for future work. Improvement of room-temperature impact strength thus should be investigated.

CONCLUSIONS AND SUMMARY OF RESULTS

The potential of tungsten-2 percent thoria, 218 tungsten, and tungsten-1 percent thoria wire reinforced nickel base superalloy composites for turbine bucket applications has been evaluated. The evaluation was based on the following laboratory test data: stress-rupture properties at 2000 and 2200 F, oxidation resistance at 2000 F, and impact resistance from room temperature to 2000 F. The results obtained and the conclusions based on them are as follows:

1. Composites were produced having stress-rupture properties superior to conventional cast superalloys at 2000 F. The 100 h stress-rupture strength obtained for the composites at 2000 F was 49,000 psi as compared to 11,500 psi for the best cast nickel alloys. The 1000-h rupture strength obtainable (by extrapolation) for the composite was 37,000 psi as compared to 6000 psi for cast superalloys at 2000 F.

2. The composite containing tungsten-2 percent thoria wire was also much stronger than the conventional cast superalloys when density was taken into consideration. The composite had a 100-h specific rupture strength at 2000 F over twice that of superalloys and an extrapolated 1000-h specific rupture strength over three times that of the superalloys.

3. The composite has a 200 F turbine bucket use temperature advantage over conventional superalloys based on comparison of laboratory stress-rupture test data.

4. Advantage can be taken of the good oxidation resistance of the superalloy matrix material by surrounding the fibers with the matrix material or a cladding material so that the fibers are not exposed to an oxidizing environment. The tungsten alloy fiber oxidizes at its normal high rate when exposed to the oxidizing environment. A few thousands of an inch of oxidation resistant material, however, was found to be sufficient to protect the fibers from oxidation at 2000 F for times up to 300 h.

5. Above 300 F the impact resistance of the composite compares favorably with that of superalloys. Considerably lower impact strength values, attributable to the brittleness of the tungsten wires below their transition temperature, were obtained below 300 F.

References

[1] Petrasek, D. W., Signorelli, R. A., and Weeton, J. W. "Refractory-Metal-Fiber-Nickel-Base-Alloy Composites for Use at High Temperatures," NASA TN D-4787, National Aeronautics and Space Administration, 1968.

[2] McDanels, D. L. and Signorelli, R. A., "Stress-Rupture Properties of Tungsten Wire from 1200 to 2500 F," NASA TN D-3467, National Aeronautics and Space Administration, 1966.

[3] Probst, H. B. and McHenry, H. T., "A Study of the Toss Factor in in the Impact Testing of Cermets by the Izod Pendulum Test," NACA TN 3931, National Ad-

visory Committee on Aeronautics, 1957.

[4] McDanels, D. L., Signorelli, R. A., and Weeton, J. W., "Analysis of Stress-Rupture and Creep Properties of Tungsten-Fiber Reinforced Copper Composites," NASA TN D-4173, National Aeronautics and Space Administration, 1967.

[5] Petrasek, D. W., "Elevated-Temperature Tensile Properties of Alloyed Tungsten Fiber Composites," NASA TN D-3073, National Aeronautics and Space Administration, 1965.

[6] Signorelli, R. A., Johnston, J. R., and Weeton, J. W., "Preliminary Investigation of Guy Alloy as a Turbojet Engine Bucket Material for Use at 1650 F," NACA RM E56I19, National Advisory Committee on Aeronautics, 1956.

[7] Waters, W. J., Signorelli, R. A., and Johnston, J. R., "Performance of Two Boron-Modified S-816 Alloys in a Turbojet Engine Operated at 1650 F," NASA Memo 3-3-59E, National Aeronautics and Space Administration, 1959.

Composite Materials for High-Temperature Magnetic Applications

D. M. PAVLOVIC[1] AND R. J. TOWNER[1]

Reference:

Pavlovic, D. M. and Towner, R. J., "Composite Materials for High-Temperature Magnetic Applications," *Composite Materials: Testing and Design, ASTM STP 460*, American Society for Testing and Materials, 1969, pp. 417–429.

Abstract:

Different types of composite materials containing dispersoids, fibers, or aligned-eutectic structure in a matrix of either cobalt or iron or their alloys were investigated for potential use as solid-rotor core materials operating at 1200 to 1600 F under stress. Composite-strengthening methods deteriorated the soft-magnetic quality of the matrix material. Magnetic saturation induction (B_s) decreased in direct proportion to the volume fraction of nonmagnetic phase present. In regard to structure-sensitive magnetic properties, such as coercive force (H_c) and d-c magnetization curve, the effects of matrix composition, amount and distribution of secondary phase, internal stress, demagnetization effects, and temperature were studied experimentally and interpreted by reference to theory. The overall consideration of both magnetic and mechanical performance, as well as ease of processing and thermal stability, favored dispersion-strengthened iron + 27 weight percent cobalt-base material for further development.

Key Words:

dispersion strengthening, fiber composites, eutectic strengthening, iron, cobalt, iron-cobalt alloys, cobalt-columbium eutectic, magnetic properties, coercive force, high temperatures, demagnetization effects, evaluation, tests

Efficient operation of space electric power systems at high temperatures imposes severe requirements on the magnetic and mechanical quality, and environmental stability of core materials used in solid-rotor alternators. Solid-rotor core materials are expected to display a combination of high mechanical strength and good soft-magnetic quality at elevated temperatures. This requirement is impossible to meet with conventional magnetic steels hardened by conventional processes. Compared with high-strength steels, the synthetically strengthened materials offer considerable flexibility with respect to both matrix composition and morphology of the strengthening agent in developing optimum combinations of the mechanical and magnetic properties.

The feasibility of employing fiber reinforcement to strengthen soft-magnetic materials for high-temperature applications was investigated recently by Collins [1].[2] Kossowsky and Colling [2] investigated the mechanical and magnetic behavior of the cobalt-columbium eutectic.

The purpose of this paper is to present the hysteresis behavior of cobalt and cobalt-iron composites prepared by either dispersion strengthening or fiber reinforcement, and to compare the overall capabilities of these systems as well as of eutectic-

---

[1] Advisory scientist and fellow scientist, respectively, Westinghouse Electric Corp., Aerospace Electrical Division, Lima, Ohio 45802. Present address for both authors is Bettis Atomic Power Laboratory, West Mifflin, Pa. 15122.

[2] The italic numbers in brackets refer to the list of references appended to this paper.

strengthened cobalt-columbium system. Emphasis is placed on the development of experimental and theoretical information rather than on the optimization of properties.

PROCEDURE

Effects of structural parameters on the magnetic behavior of composites were characterized. The specific saturation (from which saturation magnetization, B_s, was calculated) was measured on specimens by means of a magnetic balance in a field gradient of 975 oersteds per centimeter with a mean applied field of 11,500 oersteds [3]. The coercive force (H_c)—the reversed field required to reduce the induction to zero—was measured with the longitudinal direction of the specimen parallel to the magnetic field using a precision coercive force meter [4]. D-C magnetization curve (which describes the magnetization induced in a ferromagnetic material by an external magnetic field) was determined on rods or single strips using the Fahy permeameter [4].

MATERIALS

Through the cooperation of the New England Materials Laboratory (now Teledyne Materials Research), specimens of iron-base extrusions containing oxide dispersions were obtained [5]. These materials contained from 2.5 to 10 volume percent (v/o) of thorium oxide (ThO_2), aluminum oxide (Al_2O_3), or magnesium oxide (MgO) particles with an effective average diameter in the range from 0.2 to 1.0 μm. The extrusions were made with a billet temperature of 1500 to 1550 F. Also obtained from the same source [6] through the cooperation of the Cobalt Information Center were extrusions of pure cobalt and cobalt + 10 v/o Al_2O_3, the latter having an oxide particle diameter of 0.2 μm. The extrusion temperature was 1850 F. A cobalt + 2 v/o ThO_2 composition was provided by DuPont [7]. In the iron + cobalt alloy series, iron + 27 weight percent (w/o) cobalt-base extrusions containing up to 21 v/o dispersoid and an iron + 48 w/o cobalt + 2 w/o vanadium + 2 w/o zirconium extrusion containing 3 v/o zirconium oxide (ZrO_2) particles of 0.3 μm average size were made with a billet temperature of 2000 F.

The continuous-fiber-reinforced specimens were processed by diffusion bonding at Battelle Memorial Institute [8]. The matrix material were sheets, either iron + 27 w/o cobalt or iron + 49 w/o cobalt + 2 w/o vanadium, both 0.010 in. thick, or cobalt, 0.020 in. thick. The boron and silicon carbide fibers were 0.004 in. diameter. The boron fiber was coated with a layer (less than 0.0001 in. thick) of vapor desposited chromium. The metal filaments used were tungsten wire 0.005 in. in diameter and a high-strength stainless maraging steel wire (AFC-77, National Standard Co.) containing 13.5 w/o chromium, 5.2 w/o molybdenum, 13.6 w/o cobalt, and 0.14 w/o carbon. The filament was machine wound on one sheet and then sandwiched between two additional sheets. The specimens were encapsulated in carbon steel sheets for densification; fiberfax sheets were used to prevent bonding between the specimens and the bonding containers. Diffusion-bonding parameters were 1 h at 2050 F and 15,000 psi.

The cobalt + 18 w/o columbium + 15 w/o iron material was made available by R. Kossowsky, Westinghouse Research Laboratories. Centerless-ground and annealed rods 10 in. long were used for magnetic tests. The microstructure of this material was lamellar except for several coarse dendrites which occupied 10 to 20 percent of the structure. Microhardness testing showed the dendrites to be cobalt rich. The continuous phase was cubic cobalt which contained approximately 5 w/o columbium in solution, and interspersed in this matrix were thin lamellae of the intermetallic columbium cobalt phase $CbCo_3$. The average lamella thickness was 2 μm. The interlamellar spacing was on the average 3 to 4 μm.

RESULTS AND DISCUSSION

In a discussion of magnetic properties, it is important to distinguish between structure-insensitive and structure-sensitive properties. Saturation induction, B_s, and Curie temperature, T_c, as well as anisotropy constants (magnetocrystalline anisotropy and magnetostriction) are structure-insensitive properties. Permeability, coercive force, and remanence are structure-sensitive properties. These are sensitive to the microstructure as determined by such variables as grain size and internal strain, as well as size, geometry, and distribution of nonmagnetic inclusions.

The response of the structure-sensitive properties to the presence of second phase and induced strain may be characterized by the reactions falling into the following three major categories:

1. Increase in magnetic hardness characterized by an increase in coercive force and a decrease in permeability.

2. Demagnetization effects which arise from the interaction of magnetic flux with the second phase resulting in a decrease in the values of permeability and remanent induction.

3. Directionality in magnetic properties brought about by either the geometry of the strengthening agent or the interaction between the latter and the basic characteristics of the matrix or both.

In this paper, the factors contributing to the decrease in B_s and permeability, and an increase in H_c are discussed. The latter includes the effects of dispersion parameters, dislocations, and internal strain. The directionality in the magnetic properties of composites will be discussed in a separate paper.

SATURATION MAGNETIZATION

The presence of a nonmagnetic phase (regardless of its size and distribution) in a ferromagnetic matrix leads to a reduction in B_s in direct proportion to the volume percent introduced. This effect on B_s at 1200 and 1600 F is indicated in Fig. 1 for cobalt, iron, and iron + 27 w/o cobalt-base matrices. The iron + 27 w/o cobalt and cobalt-base materials are favored for magnetic applications at 1200 to 1600 F because of their high Curie temperatures and high values of B_s. The iron + 27 w/o cobalt-base material has saturation magnetization values in the 1200 to 1600 F range which are approximately 25 percent higher than those of cobalt. Also it has a B_s value at 1600 F which is approximately the same as cobalt at 1200 F. Iron + 50 w/o cobalt has B_s values very similar to iron + 27 w/o cobalt.

Dispersion-strengthened materials, which may contain only 2 to 10 volume percent (0.02 to 0.10 volume fraction) of nonmagnetic dispersoid, have an inherent advantage in saturation magnetization over the fiber-reinforced and aligned-eutectic-strengthened types of metal-matrix composites, which contain 20 volume percent or more of nonmagnetic constituent.

COERCIVE FORCE OF DISPERSION-STRENGTHENED COMPOSITES

Dispersoid Particle Size Effect

The coercive force of cobalt (cubic form), iron, and iron + 27 w/o cobalt-base materials containing spherical, nonmagnetic dispersoids having an effective average particle diameter in the range from approximately 0.1 to 1 μm was found to increase linearly with the volume fraction (V) of dispersoid [5,9], and with the inverse of the particle diameter (d). The linear dependence of coercive force (H_c) on the dispersoid parameter V/d is illustrated for iron-base materials containing from 0.025 to 0.10 volume fraction of ThO_2, Al_2O_3, or MgO in Fig. 2. This behavior

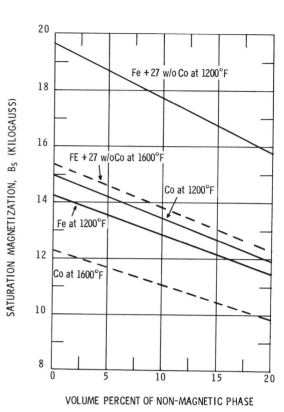

FIG. 1—Saturation magnetization of cobalt, iron, and iron + 27 weight percent cobalt-base materials at 1200 and 1600 F decreasing linearly with volume percent of nonmagnetic phase introduced.

may be expressed by the following relationship for cubic cobalt and iron + 27 w/o cobalt, as well as for iron-base dispersion-strengthened materials:

$$H_c = H_{c_o} + C\left(\frac{V}{d}\right)$$

where:

H_{c_o} = coercive force of the base material without dispersoid,
C = coefficient dependent on matrix composition and test temperature,
V = volume fraction, and
d = average effective particle diameter.

The increase in coercive force:

$$\Delta H_c = H_c - H_{c_o} = C\left(\frac{V}{d}\right)$$

due to the presence of dispersoid was analyzed in terms of dispersoid and basic magnetic parameters for the matrix.

Mager [10] reviewed several theories of coercive force and their applicability to nonmagnetic inclusions of various sizes and shapes. His domain wall energy (γ) model led to the following proportionalities for H_c:

$$H_c \propto \left(\frac{\gamma}{I_s}\right) \quad \left(\frac{V}{d}\right) \text{ for } d \gg \delta$$

$$H_c \propto \left(\frac{\gamma}{I_s}\right) \quad \left(\frac{V}{\delta}\right) \text{ for } d \ll \delta$$

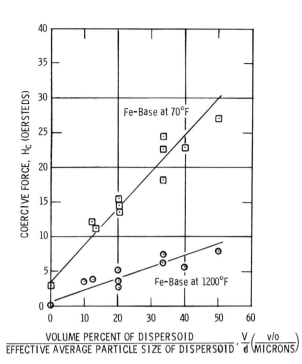

FIG. 2—Coercive force at 70 and 1200 F of iron-base extruded rod increasing linearly with dispersoid parameter $\dot{V}/d$. Materials contained from 2.5 to 10 volume percent of 0.2 to 1.0 μm particles.

$$\frac{\text{VOLUME PERCENT OF DISPERSOID}}{\text{EFFECTIVE AVERAGE PARTICLE SIZE OF DISPERSOID}}, \frac{V}{d}\left(\frac{v/o}{\text{MICRONS}}\right)$$

where:

γ = domain wall energy (ergs/cm²),
I_s = saturation intensity of magnetization (emu/cm³),
V = volume fraction of inclusions,
d = inclusion diameter (cm), and
δ = domain wall thickness (cm).

A strong dependence on particle size was evident as predicted by the preceding proportionalities when the ΔH_c curves were "normalized" [9] in regard to the composition and temperature-dependent parameters by multiplying ΔH_c by $(1/V)$ (I_s/γ). Curves for normalized ΔH_c of iron, cobalt, and iron + 27 w/o cobalt were linear, parallel, and almost superimposed, as illustrated for iron and cobalt in Fig. 3. This figure shows data points for the iron-base material at room temperature and for the DuPont Co + 2 v/o ThO₂ and nickel + 2 v/o ThO₂ [11] with d = 0.02 μm. For the cases where $d \ll \delta$ in Fig. 3, the ΔH_c term was plotted versus $1/\delta$ instead of $1/d$.

Relative Effects of Dispersoid and Internal Stress

Three equations and the prevailing conditions for which they were developed by Néel on the basis of free magnetic pole formation were employed [11] in order to interpret the temperature dependence of coercive force. This permits interpretation in terms of relative effects of the interactions between magnetocrystalline anisotropy constant K and dispersoids (H_{c3}) on one hand, and magnetostriction constant λ_S and internal stress (H_{c1} and H_{c2}) on the other. These equations do not contain a term for particle size (d), and are most applicable when $d \leqq \delta$.

In applying the Néel equations, the volume fraction (f) of the material subject to a large disturbing stress was taken to be 0.10, the internal stress (σ) was 1×10^9 dynes/cm² (14,500 psi or 9.8 kg/mm²), and V was the actual volume fraction

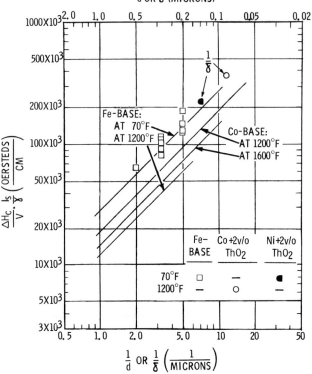

FIG. 3—Increase in measured coercive force (ΔH_c) at 70, 1200, and 1600 F from dispersoids of volume fraction (V) and diameter (d) in iron, cobalt, and nickel extrusions.

of dispersoid. The values of f and σ were estimates of stress distribution in the dispersion-strenghtened materials, and in line with values used by earlier investigators [12]. When compared at equivalent ratios of test temperature to Curie temperature, cubic cobalt had high values of K [13] which were approximately the same as those for iron, but the λ_s values for the cubic cobalt were much higher than those for iron.

The coercive force measured from room temperature to 1112 F (600 C) of the iron + 2.5 v/o ThO$_2$ composition was in rather good agreement with that calculated from the H_{c3} equation, Fig. 4, indicating that the K-inclusion interaction did predominate over the λ_s-stress interaction. Furthermore, the measured coercive force was plotted in Fig. 4 for the secondary worked condition, 12 cycles each consisting of 15 percent cold reduction by swaging at room temperature followed by a 15-min heating period at 1400 F (760 C) applied to the original hot-extruded rod. However, there was essentially no difference in the measured H_c versus T curves for the extruded and secondary worked conditions, the latter expected to contain a somewhat higher level of internal stress.

For cobalt + 2 v/o ThO$_2$, the measured values of coercive force were obtained on the cubic cobalt structure at low as well as at high temperatures. The coercive force for cubic cobalt, calculated on the basis that $\Sigma H_c = H_{c3} + H_{c1}$, agreed very well with that measured at all temperatures, Fig. 5. The H_{c1} contribution had to be considered at the lower temperatures, however, in order to obtain this agreement. Although the H_{c3} contribution (K-dispersoid) was substantial below approximately 1200 F (649 C), it decreased with increasing temperature at a much greater rate than that from the λ_s-stress interaction. This reflected the more rapid decrease in K than in λ_s with temperature.

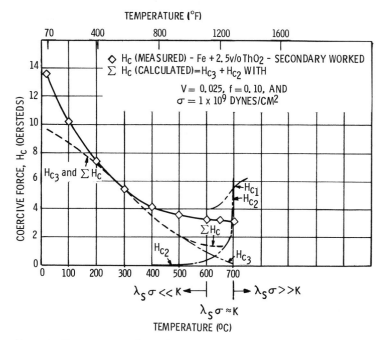

Fig. 4—Temperature dependence of coercive force of dispersion-strengthened iron as calculated from Néel equations and as measured.

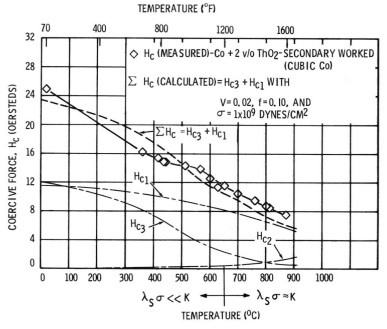

Fig. 5—Temperature dependence of coercive force of dispersion-strengthened cobalt (cubic) as calculated from Néel equations and as measured.

Dislocation Effect

Relationships have been developed from theoretical models indicating that the contribution to coercive force from dislocations in the matrix is proportional to the square root of the dislocation density. The relationship presented by Huzimura [14] for nickel and which is very similar to that given in Kneller's review [15], is as follows:

$$H_c = \left[\frac{\frac{3}{2}\lambda_s}{I_s} \right] \left[\frac{(Gb\delta)}{L} \quad \left(\ln \frac{L}{2\delta} \right)^{1/2} (N)^{1/2} \right]$$

where:
λ_s = magnetostriction constant,
I_s = saturation intensity of magnetization,
G = shear modulus,
b = Burger's vector,
2δ = thickness of the domain wall,
L = edge length of the domain wall, and
N = dislocation density per unit area.

This equation was applied to the DuPont Co + 2 v/o ThO$_2$ material in the secondary worked condition in order to determine its applicability to cubic cobalt by calculating the coercive force values at 1200 F (649 C) and 1600 F (871 C) for a composite in which cubic cobalt phase was predominate, and using a dislocation density of 1×10^{11} lines per cm^2. The results presented in Table 1 indicate a contribution to coercive force from dislocations of 8.5 oersteds at 1200 F and 7.7 oersteds at 1600 F.

These values correlated well with the values of 8.3 and 5.7 oersteds, respectively, calculated from the Néel equation for the internal stress (H_{c1}) contribution to coercive force. For the latter, the λ_s-stress interaction was considered using 0.10 volume fraction (10 volume percent) of disturbed material subject to peak internal stresses of the order of 1×10^9 dynes/cm^2 (14,500 psi).

By way of explaining the basis for these calculations, the average dislocation density of 1×10^{11} lines per cm^2 was an approximate value calculated from the cell-like structure (substructure) of the cobalt + 2 v/o thoria (ThO$_2$) extruded and swaged bar. This dislocation density value was consistent with a cell size of about 1 μm (corresponding to a 1 μm interparticle spacing of thoria), a cell wall thickness of high dislocation density of 0.02 to 0.05 μm (roughly comparable to the 0.02 μm effective average particle size of the thoria), the cell walls occupying

Table 1—*Calculation of coercive force* (H$_c$) *for cobalt-base alloy at elevated temperatures from dislocation equation of Huzimura and internal stress* (H$_{c1}$) *equation of Néel.*

Property	Cobalt Base	
	At 1200 F	At 1600 F
Magnetostriction constant, λ_s(cm/cm)...................	22.6×10^{-6}	12.5×10^{-6}
Saturation intensity of magnetization, I_s(emu/cm^3)........	1193	980
Shear modulus, G(dynes/cm^2)...........................	5.52×10^{11}	4.21×10^{11}
Burger's vector, b(cm)................................	2.52×10^{-8}	2.52×10^{-8}
Thickness of the domain wall, 2δ(cm)................	870×10^{-8}	1830×10^{-8}
Edge length of the domain wall, L(cm).................	1×10^{-4}	1×10^{-4}
Dislocation density per unit area, N(lines/cm^2)...........	1×10^{11}	1×10^{11}
Calculated coercive force, H_c(oersted):		
From dislocation equation of Huzimura................	8.5	7.7
From internal stress(H_{c1}) equation of Néel.............	8.3	5.7

0.10 volume fraction of the matrix, and a lattice disorientation between adjacent cells in the range of 2 to 5 deg.

The preceding structural values were consistent with information on dispersion-strengthened cobalt [7] and other base materials [16,17]. It has been reported that dispersion-strengthened nickel sheet (TD Nickel) has a well-defined structure consisting of cells with dimensions approximately 0.5 to 1.0 μm, and disorientation between cells of not more than 2 to 3 deg [18]. A primary shear of 20 percent was reported to produce an angular spread of approximately 5 deg between plane normals in the matrix adjacent to a particle and the matrix further away [19, p. 74]. The dislocation density of 1×10^{11} lines/cm^2 also was consistent with values reported for worked crystals [20,21,22].

COERCIVE FORCE OF FIBER-REINFORCED COMPOSITES

The mechanism of magnetic hardening through fiber reinforcement is basically not different from that applying to dispersion strengthening. Discontinuous nonmagnetic fiber, such as in eutectic strengthened cobalt + iron + columbium alloy, requires some special considerations because of the shape anisotropy associated with the geometry of the aligned eutectic platelet, but similar cases have been discussed before [3].

The continuous-fiber reinforced composites are expected to show only a slight increase in coercive force, since subsidiary magnetic domains form at large inclusions [23]. Furthermore, according to the inclusion theory [3], no increase in the coercive force value should be expected as the number of large rod-shaped inclusions increases. Experimental data obtained in this study substantiate some of these predictions. For instance, an addition of 20 v/o tungsten wire increased the coercive force at 70 F of the iron matrix to only 3.5 oersteds as compared to 13.5 oersteds measured on a dispersion-strengthened composite of iron + 2.5 v/o ThO$_2$. A smaller increase in H_c to about 16 oersteds was observed in a matrix sensitive to the stress (or strain) magnetostriction interaction, iron + 49 w/o cobalt + 2 w/o vanadium reinforced with 15 v/o steel wire (Fig. 6), as compared with 44.5 oersteds for the same matrix dispersion-strengthened with 3 v/o ZrO$_2$ (Fig. 7).

A special case arises when structural changes at high temperatures take place either in the matrix or in the filament. Internal stresses (strains) are generated

FIG. 6—Effect of cyclic heating and cooling on coercive force of iron + 49 weight percent cobalt + 2 weight percent vanadium sheet reinforced with 15 volume percent nickel maraging steel wire.

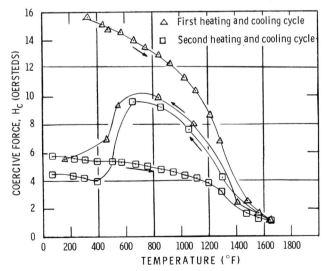

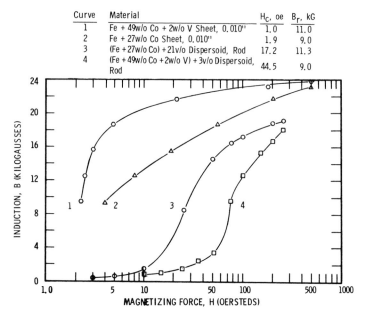

Curve	Material	H_c, oe	B_r, kG
1	Fe + 49w/o Co + 2w/o V Sheet, 0.010"	1.0	11.0
2	Fe + 27w/o Co Sheet, 0.010"	1.9	9.0
3	(Fe + 27w/o Co) + 21v/o Dispersoid, Rod	17.2	11.3
4	(Fe + 49w/o Co + 2w/o V) + 3v/o Dispersoid, Rod	44.5	9.0

FIG. 7—D-C magnetization curves of iron-cobalt base alloys with and without dispersoid.

which may increase the coercive force in certain temperature regions in a matrix with a high value of magnetostriction. This pattern of coercive force change was observed in a composite incorporating maraging steel wire in a matrix of iron + 49 w/o cobalt + 2 w/o vanadium. Both the matrix and the filament undergo structural changes, the matrix through an atomic ordering reaction which commences at about 1300 F and the filament through austenite reversion which begins at about 800 F. Although both the matrix and the filament are ferromagnetic, the magnetic behavior of the matrix determines the H_c versus temperature pattern of the composite.

The coercive force of the matrix has been observed [24] to increase in the temperature range 500 to 1100 F. In terms of basic magnetic parameters, the high value of magnetostriction (λ) and a low value of magnetocrystalline anisotrophy (K) are major contributors to this behavior. The very low value of K of this alloy at elevated temperatures permits the $\lambda\sigma$ product to dominate H_c, and as a result an increase in stress (σ) at elevated temperatures increases H_c value to a peak. A further increase in temperature decreases both λ and σ; hence, H_c decreases toward low values at high temperatures. On cooling, K increases with decreasing temperature, thus diminishing the influence of the $\lambda\sigma$ term on H_c. Consequently H_c decreases from the peak (which coincides with zero-K value) as the temperature is lowered.

The same pattern, Fig. 6, applies to the above mentioned composite, but the H_c values are higher. The reason for the lack of a H_c peak on heating appears to be associated with the fact that the material was already in the ordered condition at room temperature; hence, no substantial increase in stress (or strain) took place as the temperature was raised. The material disorders at temperatures above 1300 F; hence, strain sets in upon cooling through the ordering region. A stress-relief effect is responsible for a decrease in the 70 F H_c value after each heating cycle.

Figure 7 displays the d-c magnetization curves of iron + 27 w/o cobalt and iron + 49 w/o cobalt + 2 w/o vanadium sheets, both with and without [25] dispersion strengthening. Dispersion strengthening made each material magnetically harder essentially by displacing the magnetization curve toward higher magnetizing fields and, as a result, lowering permeability. The curve for dispersion strengthened iron + 49 w/o cobalt + 2 w/o vanadium rod also shows a considerable strain effect; a high value of magnetostriction of this alloy and internal strain generated by the effects of atomic ordering and processing appear to be primarily responsible for high degree of magnetic deterioration. When annealed at temperatures above 1250 F, followed by rapid cooling, the coercive force of this composite decreased from 44.5 to 16 oersteds.

Figure 8 displays room-temperature magnetization curves of dispersion-strengthened cobalt and iron + 27 w/o cobalt rods, H-11 tool steel forged ring, and a eutectic-strengthened cobalt + 18 w/o columbium + 15 w/o iron rod. The dispersion-strengthened iron + 27% cobalt rod reaches an induction of 17 kg at 100 oersteds; it has a higher permeability at all induction levels than H-11 tool steel which is considered a potential core material for solid rotors in space power alternators. The high value of K in cobalt and its considerably lower B_s value than that of iron + 27 w/o cobalt, result in a lower permeability of the cobalt composite at higher fields than of the iron + cobalt matrix composites.

The eutectic-strengthened cobalt + 18 w/o columbium + 15 w/o iron rod displays a pronounced demagnetization effect which flattens out the magnetization curve and lowers considerably the remanent magnetization.

The demagnetization factors assume different values depending on the position

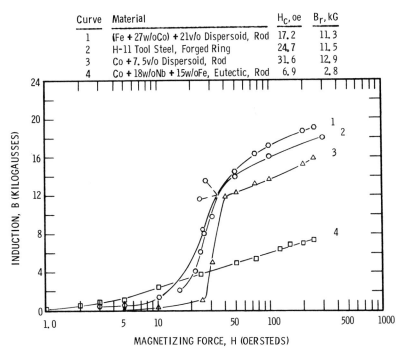

Curve	Material	H_c, oe	B_r, kG
1	(Fe + 27w/oCo) + 21v/o Dispersoid, Rod	17.2	11.3
2	H-11 Tool Steel, Forged Ring	24.7	11.5
3	Co + 7.5v/o Dispersoid, Rod	31.6	12.9
4	Co + 18w/oNb + 15w/oFe, Eutectic, Rod	6.9	2.8

FIG. 8—D-C magnetization curves of iron-base and cobalt-base composites. H-11 tool steel included for comparison.

of the flux with respect to the geometry of the strengthening agent [3]. A rough sequence of increasing demagnetization is as follows: long rod or needle in the longitudinal direction, sphere, and spheroid. When translated in terms of composites (for the same volume of nonmagnetic secondary phase) this means: continuous-fiber reinforcement in the longitudinal direction, dispersion strengthening, and discontinuous-fiber (eutectic) strengthening.

CONCLUSIONS

1. Composite strengthening deteriorated the magnetic quality of the matrix material. The amount of nonmagnetic secondary phase decreased magnetic saturation induction (B_s) in accordance with the rule of mixtures. A matrix material for high-temperature magnetic applications, therefore, must have a high Curie temperature and high saturation induction.

2. An iron + 27 w/o cobalt-base matrix is favored for further development of composite-strengthened soft-magnetic materials for applications involving stress in the 1200 to 1600 F range. This matrix provides the highest magnetic induction for applied fields of 100 oersteds and above at temperatures up to approximately 1725 F (end point of the gradual decrease in B_s of this material with increasing temperature).

3. The structure sensitive properties, such as coercive force (H_c) and permeability were shown to be influenced by the amount and distribution of secondary phase and stress through their interaction with the basic magnetic parameters of the matrix material, that is, magnetocrystalline anisotropy and magnetostriction constants. For high-temperature magnetic applications involving stress, the matrix should have low values of magnetocrystalline anisotropy and magnetostriction in order to minimize the magnetic hardening effects of the presence of the second phase and stress.

4. Internal demagnetization effects led to a marked flattening of the magnetization curve when large amounts of a nonmagnetic discontinuous phase were present in the matrix, such as platelets in eutectic-strengthened cobalt-iron-columbium alloy.

5. Considering the present state of technology, the dispersion-strengthening mechanism is superior to continuous fiber reinforcement and aligned-eutectic strengthening because overall it has a less deteriorating effect on the magnetic properties of the matrix. This results primarily from the fact that a much smaller amount of nonmagnetic constituent has to be present in order to achieve mechanical strength at high temperatures.

6. The potential of fiber reinforcement to provide a soft-magnetic material for high-temperature applications would be much enhanced if continuous fibers of higher strength became available in production quantity, and if the fibers themselves were both magnetic and unreactive with the matrix.

References

[1] Collings, D. A., "Fiber-Reinforced Fe-5ONi and Fe-1.35Si Alloys," Journal of Applied Physics, Vol. 39, No. 2 (Part I), 1 Feb. 1968, pp. 606–607.

[2] Kossowsky, R. and Colling, D. A., "Unidirectionally Solidified Cobalt-Based Alloys for Elevated Temperature Magnetic Application," Advanced Techniques for Material Investigation and Fabrication, Vol. 15, Society of Aero-

space Materials and Process Engineers, Los Angeles, Calif., April 1969.

[3] Hoselitz, K. Ferromagnetic Properties of Metals and Alloys, Clarendon, Oxford, 1952, pp. 37, 70, 78, and 131.

[4] Pavlovic, D. M. et al., "High-Temperature Magnetic Test Techniques and Analyses," Advanced Techniques for Material Investigation and Fabrication, Vol. 14, So-

ciety of Aerospace Material and Process Engineers, Cocoa Beach, Fla., 5–7 Nov. 1968.

[5] Bufferd, A. S. and Towner, R. J., "Magnetic and Mechanical Properties of Dispersion-Strengthened Iron," Second European Symposium on Powder Metallurgy, Stuttgart, Vol. 2, No. 6.23, May 1968.

[6] Bufferd, A. S. and Grant, N. J., "Oxide Dispersion—Strengthening of Cobalt-Base Alloys," Journees Internationales des Applications du Cobalt, 9–11 June 1964.

[7] Mincher, A. L., "Dispersion-Strengthened Cobalt Alloys," Cobalt, No. 32, Sept. 1966, pp. 119–123.

[8] Battelle Memorial Institute, Columbus, Ohio report (authored by Fleck, J. N., to Westinghouse Aerospace Electrical Division, 1968.

[9] Towner, R. J. and Pavlovic, D. M., "Effect of Dispersoid Parameters on High-Temperature Behavior of Cobalt and Iron + 27 Percent Cobalt," Journal of Applied Physics, March 1969.

[10] Mager, A., "Über die Wirkung von Fremdstoffen in weichmagnetischen Metallen und Legierungen," Zeitschrift für Angewandte Physik, Vol. 14, 1962, pp. 198–204.

[11] Towner, R. J. et al., "Effect of Dispersoids on Magnetic Properties of Iron, Cobalt, and Nickel," Journal of Applied Physics, Vol. 39, No. 2 (Part I), 1 Feb. 1968, pp. 601–603.

[12] Kneller, E., Beiträge Zur Theorie des Ferromagnetismus and der Magnetisierungskurve, Köster, W., ed., Springer-Verlag, Berlin, 1956, p. 122.

[13] Rodbell, D. S., Journal of the Phyical Society of Japan, Vol. 17, Supplement B-I, 1962, p. 313.

[14] Huzimura, T., "Dislocations in Ferromagnetic Materials," Transactions, Japan Institute of Metals, Vol. 2, 1961, pp. 182–186.

[15] Kneller, E., Ferromagnetismus, Springer-Verlag, Berlin, 1962, pp. 293, 517.

[16] Ashby, M., "The Hardening of Metals by Non-Deforming Particles," Zeitschrift für Metallkunde, Vol. 55, No. 1, 1964, pp. 5–17.

[17] Klein, M. J. and Huggins, R. A., "The Structure of Cold Worked Silver and Silver-Magnesium Oxide Alloys," Acta Metallurgica, Vol. 10, Jan. 1962, pp. 55–62.

[18] Inman, M. C. and Smith, P. J., "An Electron Transmission Study of Oxide Particle Statistics In TD-Nickel Alloy," The Pennsylvania State University, June 1966, Grant No. AF-AFOSR 669-65, Air Force Office of Scientific Research, Washington, D. C., AD 636487.

[19] Ashby, M. F., "The Theory of the Critical Shear Stress and Work Hardening in Dispersion-Hardened Crystals," Harvard University, Dec. 1966, Office of Naval Research Contract Nonr-1866(27), NR-031-503, AD 647991.

[20] Biorci, G., Ferro, A., and Montalenti, G., "Magnetic After-Effect in Iron due to Motion of Dislocations," Physical Review, Vol. 119, No. 2, 15 July 1960, pp. 653–657.

[21] Cottrell, A. H., Dislocations and Plastic Flow in Crystals, Oxford, London, 1956, pp. 102, 154.

[22] Humphreys, F. J. and Martin, J. W., "The Effect of Dispersed Silica Particles on the Recovery and Recrystallization of Deformed Copper Crystals," Acta Metallurgica, Vol. 14, June 1966, pp. 775–781.

[23] Nix, W. K. and Huggins, R. A., "Coercive Force of Iron Resulting from the Interaction of Domain Boundaries with Large Non-magnetic Inclusions," Physical Review, Vol. 135, No. 2A, 20 July 1964, pp. A401–407.

[24] Bozorth, R. M., Ferromagnetism, Van Nostrand, Princeton, N. J., 1951.

[25] Kueser, P. E. et al., "Properties of Magnetic Materials for Use in High-Temperature Space Power Systems," NASA SP-3043, National Aeronautics and Space Administration, 1967, pp. 246, 135.

Cryogenic Boron-Filament-Wound Pressure Vessels

E. E. MORRIS[1] AND
R. J. ALFRING[1]

Reference:

Morris, E. E. and Alfring, R. J., "Cryogenic Boron-Filament-Wound Pressure Vessels," *Composite Materials: Testing and Design*, ASTM STP 460, American Society for Testing and Materials, 1969, pp. 430–443.

Abstract:

Filament tensile strengths and pressure-strain characteristics for a high-modulus, boron-filament-wound/resin composite, pressure vessel were obtained. Five 8-in.-diameter by 13-in.-long boron-filament-wound, metal-lined pressure vessels were fabricated and burst tested (three at 75 F and two at −320 F) with the following results:

1. The average ultimate boron-filament strength at room temperature was 236,600 psi. The values obtained were highly consistent, varying only 1.9 percent from the average.

2. At −320 F, the boron-filament strength increased about 16 percent to 273,800 psi.

3. Naval Ordnance Laboratory (NOL) rings fabricated from the same boron-filament tape used in the test vessels had an average ultimate filament strength of 309,000 psi at room temperature. The vessels achieved 77 percent of this value.

4. Good correspondence was obtained between the expected and the measured pressure strain characteristics at 75 and −320 F. At the vessel burst points, the strains were about 0.30 to 0.40 percent, roughly one tenth of the strain encountered in glass-filament-wound vessels.

5. The resin content in the boron filament-wound composite ranged from 37 to 47 volume percent.

The vessels displayed very low biaxial strains and high strength-to-weight ratios. Such low strains permit strong metal liners (for example, titanium) to work elastically up to the ultimate stress of boron filament. The present strength and potential strength of boron are high enough to ensure that boron-filament-wound, metal-lined vessels will be lighter than competing homogeneous metal tanks for cryogenic service.

Key Words:

boron, fiber composites, filament wound materials, pressure vessels, cryogenic properties, pressure-vessel testing, mechanical properties, evaluation, tests

Composite materials are replacing metals in many aerospace structures. The use of composite structures for cryogenic pressure vessels can yield considerable weight savings because the winding material has a much higher strength for its weight than do metal-tank materials. Recent test results for 50 metal-lined glass filament-wound tanks (under Contracts NAS 3-6287, 3-6292, and 3-6297 with the Lewis Research Center of the National Aeronautics and Space Administration) demonstrated the great potential for weight savings in cryogenic use. Certain limitations may exist, however, in the application of glass-filament-wound vessels with thin metal liners for service involving many cycles of pressure loading.

Glass-filament-wound vessels strain approximately 2 percent at the design operating pressure and cryogenic temperatures, because of the relatively low modulus

---

[1] Engineering supervisor and engineering specialist, respectively, Structural Products Department, Aerojet-General Corp., Azusa, Calif. 91702.

and high strength of the glass filaments; when they are pressurized to the burst point at cryogenic temperatures, the ultimate strain is from 3 to 4 percent. For cryogenic service, metal liners are being used inside the tanks because elastomers and polymers cannot now provide the necessary leaktightness at low temperatures. Large glass-filament strains at the operating stress, however, cause the thin, light-weight liner to exceed its yield point and plastically deform, and repeated applications of such strains during simulated service have presented difficult problems in material selection and the design of compatible liners that can sustain fatigue cycling.

Boron filaments have a high strength-to-weight ratio, and 4 to 5 times the tensile modulus of glass filaments and provide a composite material uniquely suited for use with metal liners in pressure vessels. The use of high-strength, high-modulus boron filaments would greatly reduce the large biaxial strains imposed on liners by glass filaments and result in a high-performance structure. Efficient stress levels can be developed in the boron filaments at low strains below the elastic limit of titanium (and of other high-strength metals).

Aerojet's program was undertaken to develop otherwise unavailable preliminary cryogenic-temperature mechanical-property data for a high-modulus, boron-filament-wound/resin composite, pressure-vessel structure. A complete report on the program was issued.[2]

PRESSURE-VESSEL DESIGN

The metal-lined, filament-wound, test vessel was an 8-in.-diameter by 13-in.-long, closed-end cylinder. It was designed to achieve a longitudinal-to-circumferential

[2] Morris, E. E. and Alfring, R. J., "Cryogenic Boron-Filament-Wound Containment Vessels," NASA CR-72330, Aerojet-General Report 3475 under Contract NAS 3-10282, Nov. 1967.

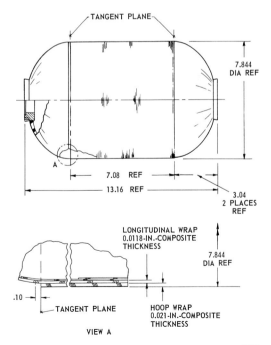

FIG. 1—Pressure vessel.

VIEW A

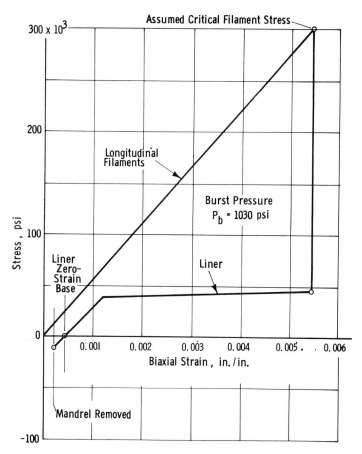

Fig. 2—Pressure-vessel stress-strain relationships, longitudinal direction of cylinder.

strain ratio of approximately 1 : 1 at as low an internal pressure as possible with state-of-the-art materials and winding procedures. This type of vessel, fabricated from longitudinally and circumferentially oriented filaments wound over a 0.006-in.-thick Type 304 stainless steel foil liner, was selected on the basis of experience acquired in previous development efforts.[3,4] Dimensional coordinates of the heads and other vessel characteristics were defined with the aid of a computer program that analyzed the tanks. Details of the vessel design may be found in footnote 2.

The analysis indicated that the burst pressure would be 1030 psig at 75 F, with ultimate boron-filament stresses of 330,000 psi in the hoop-wound filaments and 300,000 psi in the longitudinally would filaments, and with a hoop-wound boron composite thickness of 0.021 in. and a longitudinally wound boron composite thickness of 0.013 in. in the cylindrical section.

[3] Soffer, L. M. and Molho, R., "Cryogenic Resins for Glass-Filament-Wound Composites," NASA CR-72114, Aerojet-General Report 3343 under Contract NAS 3-6287, Jan. 1967.
[4] Lewis, A. and Bush, G. E., "Improved Cryogenic Resin/Glass-Filament-Wound Composites," NASA CR-72163, Aerojet-General Report 3392 under Contract NAS 3-6297, April 1967.

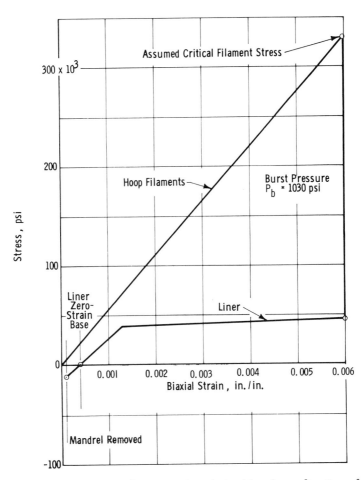

FIG. 3—Pressure-vessel stress-strain relationships, hoop direction of cylinder.

The pressure-vessel design is shown in Fig. 1. The liner is first overwrapped longitudinally to a composite thickness of 0.0118 in. with two layers of side-by-side, unidirectional, resin-preimpregnated, boron-filament tape of 0.125-in. width and containing 29 boron monofilaments. After longitudinal winding, four layers of the tape are applied in a side-by-side pattern along the cylinder to produce a hoop-wound composite thickness of 0.021 in. The filament-winding pattern used is shown in Fig. 1.

The pressure-vessel design was analyzed to determine stresses and strains in the filaments and liner under various loading conditions. The 75 F stress-strain relationships for the longitudinal direction of the cylinder in the filaments and the liner are shown in Fig. 2, and for the hoop direction in Fig. 3. Vessel strains in the longitudinal direction at a filament stress of 300,000 psi are 0.52 percent, and in the hoop direction at 330,000 psi are 0.59 percent.

The computer output was used to construct pressure-strain curves for 75 F test conditions. The predicted curve for 75 F is presented in Fig. 4. The initial steep portion is due to the load-carrying capability of the liner. As the liner undergoes

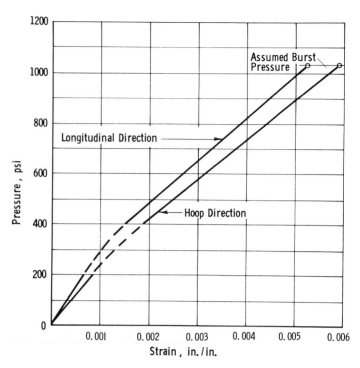

Fig. 4—Pressure-strain relationships.

plastic deformation above 350 psi, the increasing load is dumped into the filament-wound composite.

FABRICATION

The 0.006-in.-thick liners were made from AISI Type 304 stainless steel by pressure forming the end domes, machining the polar bosses, rolling a cylindrical section, and roll-resistance seam welding the segments.

Two leak checks were made on each tank liner—the first a soap-solution leak test under an internal pressure of 7 psi, and the second a helium-leak check with a mass spectrometer at a pressure differential of 20 psi. All liners successfully passed the tests.

Boron-filament tape of 0.125-in. width, consisting of 29 single 0.004-in.-diameter filaments collimated in a side-by-side orientation and preimpregnated with epoxy resin, was procured from the Narmco Materials Division of the Whittaker Corp. The tape was made from monofilaments having a strength range from 420,000 to 480,000 psi and an elastic modulus of 55×10^6 psi. The resin system was Shell 58-68R, which consists of Epon 828, Epon 1031, nadic methyl anhydride (NMA), and benzyldimethylamine (BDMA) in the ratio of 50/50/90/0.3 parts by weight. The boron-preimpregnated (prepreg) tape has a resin content of 33 weight percent.

The tanks were filament wound according to the design requirements. One revolution (two layers) of boron-filament tape was wound longitudinally over the liner assembly (Fig. 5), followed by four layers of hoop windings in the cylindrical sections, running between the tangency planes of the two heads (Fig. 6). After

Fig. 5—Longitudinal winding of pressure vessel.

winding, the vessels were vacuum-bag cured at 200 F for 2 h, 250 F for 2 h, and 300 F for 4 h.

TEST PROGRAM

Test Plan, Facility, and Instrumentation

Vessels were subjected to single-cycle burst tests at a rate of pressurization that produced a strain of approximately 0.25 percent/min in the longitudinal and circumferential directions. Data were recorded continuously on the internal pressure, exterior-surface temperature, and deflection versus pressure relationships at three points distributed to provide hoop and longitudinal strains. One set of hoop-strain measurements was made at the cylindrical-section center, and two sets of axial-strain measurements were made along the cylinder.

Inhibited water was used for pressurization in the room-temperature tests.

Fig. 6—Hoop winding of pressure vessel.

Liquid nitrogen (LN_2) was used to pressurize the vessels for the -320 F tests. The cryogenic-test facility (Fig. 7) consisted of a vacuum chamber with provisions for instrument leads and vacuum-jacketed pressurization lines. The vacuum chamber was pumped down to 4×10^{-4} mm Hg to assure the required temperatures. Thermal equilibrium was obtained before testing was initiated. Thermal equilibrium was defined as a vessel flange or skin temperature of -300 F or less, with -310 F or less at the vessel-outlet vent line.

Figure 8 shows the locations of instruments used to monitor temperatures, longitudinal and circumferential strains, and internal pressures throughout testing. Platinum resistance thermometers and copper-constantan thermocouples were used in the -320 F tests. Two measurements were made on the exterior surface (90 deg apart circumferentially) near the tangency at each end of the tank. In addition, the temperatures of the cryogenic fluids inside and outside the tank were recorded. Iron-constantan thermocouples were used to monitor tank-surface temperatures during ambient burst tests.

An Aerojet-developed extensometer was used to make strain measurements.

Table 1—*Pressure-vessel performance data.*

Tank	Test Sequence	Resin Content		Test Temperature, deg F	Burst Pressure, psia	Burst-Pressure Stresses, psi		Strain at Burst, %		Apparent Failure Origin
		Weight %	Volume %			Hoop Filaments	Longitudinal Filaments	Hoop Filaments	Longitudinal Filaments	
A-1....... 3		30.31	47.03	75	752	236 200	227 800	0.30	0.37	hoop filaments in cylinder
A-2....... 1		25.48	41.09	75	765	241 000	232 100	...[a]	~0.37	hoop filaments in cylinder
A-3....... 5		22.71	37.48	75	742	232 500	224 400	...[b]	...[b]	longitudinal filaments
A-4....... 4		26.37	42.22	-320	915	291 200	281 600	0.38	0.44	undetermined
A-5....... 2		22.12	36.69	-320	815	256 500	251 800	0.29	0.35	hoop filaments in cylinder

[a] Not obtained; extensometer-clamp malfunction.
[b] Not measured.

FIG. 7—Liquid nitrogen burst-test setup, boron filament-wound pressure vessel.

The test results for the vessels at 75 and −320 F, are summarized in Table 1.

Tank A-1 was pressurized at room temperature to its burst point of 752 psig. Failure appeared to originate in the hoop filaments of the cylindrical section and then propagate into a circumferential fracture in the head area of the composite. The failure, shown in Fig. 9, is representative of the other failures obtained in the program. Excellent strain data were obtained (Fig. 10). At failure, the stresses were 236,200 psi in the hoop filaments and 227,800 psi in the longitudinal. The strains at burst were 0.37 percent in the longitudinal direction and 0.30 percent in the hoop direction. The resin content of the cured composite was 30.31 weight percent or 47.03 volume percent.

Tank A-2 was hydrostatically pressurized at room temperature to its burst point of 765 psig. At the burst, the stresses were 241,000 psi in the hoop filaments and 232,100 psi in the longitudinal. The longitudinal strain at burst was about 0.37 percent. The resin content of the cured composite was 25.48 weight percent or 41.09 volume percent. Tank A-3 had a room-temperature burst point of 742 psig. The initial fracture in Tank A-3 appeared to be in the longitudinal filaments, followed by propagation into the hoop filaments. At failure, the stresses were 232,500 psi in the hoop filaments and 224,400 psi in the longitudinal. The resin content of the cured composite was 22.71 weight percent or 37.48 volume percent.

Tank A-5 was installed in the vacuum chamber and filled with liquid nitrogen; after steady-state conditions were attained, it was pressurized with LN_2 to its burst

Symbol	Measurement
P_s	Supply Pressure
P_c	Specimen Pressure
T_s	Supply Temperature
T_0	Specimen Temperature
SG_1	Specimen Deflection, Hoop
$SG_{2,3}$	Specimen Deflections, Longitudinal
$TSG_{1,2,3}$	Deflection Beam Temperatures
$TC_{1,2}$	Specimen (Skin) Temperatures

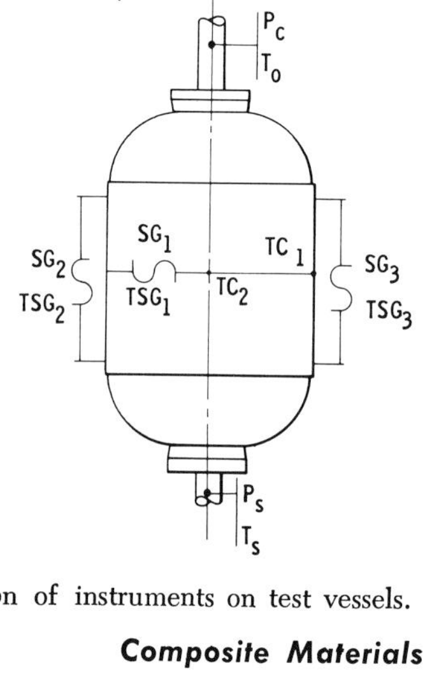

Fig. 8—Location of instruments on test vessels.

point of 815 psi. The failure appeared to originate in the hoop filaments and then propagate into a circumferential fracture in the head area of the composite similar to the failure modes obtained in ambient-temperature tests of Tanks A-1 and A-2, and the liner bulged without fracture. Satisfactory data were obtained on longitudinal and hoop strains. At failure, the stresses were 256,500 psi in the hoop filaments and 251,800 psi in the longitudinal. The strains at burst were 0.35 percent in the longitudinal direction and 0.29 percent in the hoop. The tank-surface temperatures at the burst point were about −308 F, as shown in Fig. 11. The resin content of the cured composite was 22.12 weight percent or 36.69 volume percent.

Tank A-4 was cooled down and pressurized with LN$_2$ to its burst point of 915 psi. Excellent data were obtained on longitudinal and hoop strains. The stresses at failure were 291,200 psi in the hoop filaments and 281,600 psi in the longitudinal. The strains at burst were 0.44 percent in the longitudinal direction and 0.38 percent in the hoop. The tank-surface temperatures at the burst point were about −308 F. The resin content of the cured composite was 26.37 weight percent or 42.22 volume percent.

Fig. 9—Boron filament-wound pressure vessel following water burst test.

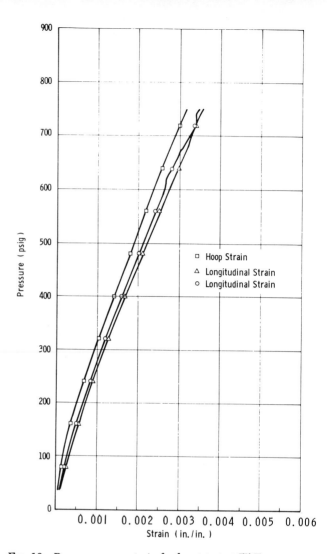

FIG. 10—Pressure versus strain for burst test at 75 F.

EVALUATION OF TEST RESULTS AND STRENGTH LEVELS

Figure 12 shows the ultimate boron-filament strength levels attained in the 75 and −320 F burst tests, the average values, and the data range. The metal-liner load was considered in filament-stress computations. The average levels were 236,000 psi at 75 F and 273,800 psi at −320 F, or a 16 percent improvement over the room-temperature value.

To determine filament strengths and the composite strengths of the boron-filament tape from which the vessels were fabricated, nine tests were performed on Naval Ordnance Laboratory (NOL) rings made with prepreg tape from each of the three spools used in tank fabrication. In these nine tests, the ultimate strength of the filament averaged 147,800 psi in the composite and 309,000 psi in the filament. The average ambient-temperature hoop-filament strength of 236,600 psi obtained in the pressure-vessel tests was 77 percent of the filament strength attained in the NOL ring tested during this program.

440   *Composite Materials*

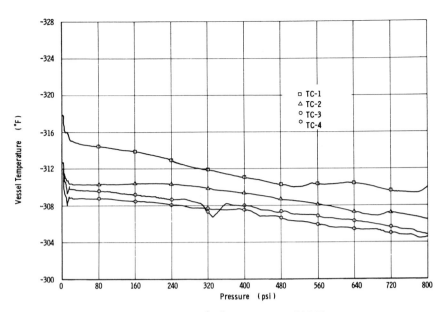

Fig. 11—Temperature versus pressure for burst test at −320 F.

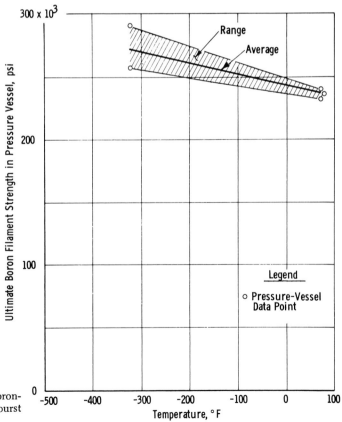

Fig. 12—Ultimate boron-filament strength at burst pressure.

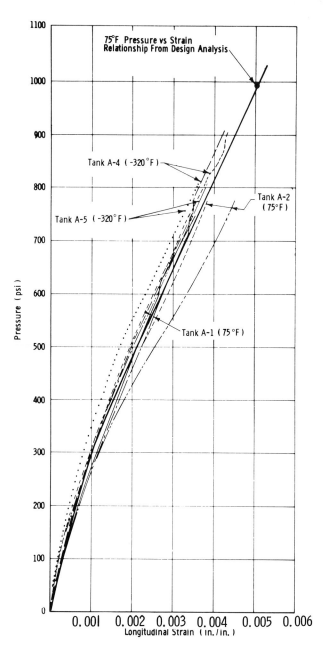

Fig. 13—Predicted and measured pressure versus longitudinal strain characteristics in burst tests.

Pressure versus strain data were obtained for all vessels except A-3. Figure 13 provides predicted (computer-design analysis as shown in Fig. 4) and measured pressure versus longitudinal-strain curves for 75 and −320 F burst tests. Excellent correspondence was obtained. Figure 14 compares the measured pressure versus hoop-strain characteristics with the predicted values (Fig. 4). The correspondence

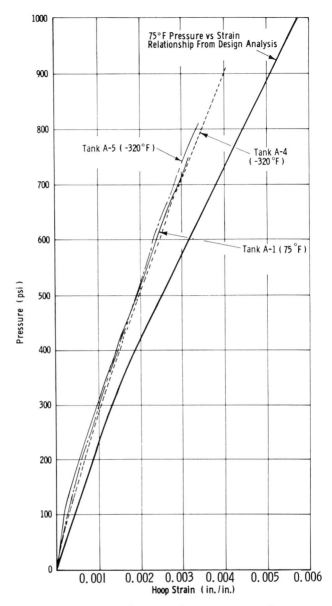

FIG. 14—Predicted and measured pressure versus hoop-strain characteristics in burst tests.

was not as close as for the longitudinal-strain data. The measured hoop strains were less than predicted by the design analysis.

ACKNOWLEDGMENT

Guidance and many helpful suggestions were provided throughout the program by the NASA project manager, R. F. Lark of the Liquid Rocket Technology Branch, Lewis Research Center.

Morris and Alfring on Pressure Vessels

Structural Synthesis of Composite Materials for Ablative Nozzle Extensions

N. N. AU,[1]
E. R. SCHEYHING,[2] AND
G. D. SUMMERS[2]

Reference:

Au, N. N., Scheyhing, E. R., and Summers, G. D., "Structural Synthesis of Composite Materials for Ablative Nozzle Extensions," *Composite Materials: Testing and Design*, ASTM STP 460, American Society for Testing and Materials, 1969, pp. 444–459.

Abstract:

Complexities associated with the application of composite materials for the thermal protection of aerospace structures are discussed in the light of a detailed investigation of the Titan III Stage II ablative nozzle extension. This paper describes the methods used in a structural evaluation of the nozzle extension that led to an understanding of several structural failures. Results of an experimental investigation of the composite material properties and analyses of the nozzle extension's response to mechanical and thermal loads are presented. Particular attention is focused on the challenges encountered in testing to obtain the properties of the liner and shell materials over their operating temperature ranges. The importance of individual material properties and fabrication variables is explored. The effect of moisture content on thermal expansion and stress levels is shown to be of primary importance. Also, the ability to vary stress levels by changing the tape wrap angle is demonstrated. The experience gained from this investigation is translated into a structural synthesis approach tailored to the specific requirements of current aerospace programs. The importance of integrating the various interacting technical disciplines in terms of the design configuration, environmental conditions, material thermal and mechanical properties, thermal and structural analyses, fabrication techniques, and process controls is discussed. Finally, conclusions drawn from the nozzle extension study are extended to other applications of composite materials.

Key Words:

ablative materials, rocket nozzles, composite materials, structural design, mechanical properties, stress analysis, fabrication, structural compatibility, evaluation, tests

The ability of certain composite materials to sustain extremely high temperatures has led to their extensive use for the thermal proection of aerospace structures. When composite insulators are integral with a load carrying substructure, complexities in design, test, and analysis result. A structural synthesis in this case requires coordination of materials research, experimental determination of material properties, and results of thermal and structural analyses. Such a synthesis is shown here for an actual composite structure, the Titan III Stage II ablative nozzle extension,

[1] Associate head, Solid Mechanics Department, Applied Mechanics Division, Aerospace Corp. 90045.
[2] Member of Technical Staff, Launch Vehicle Structures Section, Solid Mechanics Department, Applied Mechanics Division, Aerospace Corp. 90045.

in order to identify structural failure modes, predict structural failures experienced, and guide a successful modification program.

TITAN III STAGE II NOZZLE EXTENSION

GEOMETRY

The Stage II engine and nozzle extension (skirt) assembly are shown in Fig. 1. The primary structural element of the skirt is a glass phenolic honeycomb sandwich shell which is protected from the combustion gases by an asbestos-phenolic, char-forming ablative liner. The plume from a roll control nozzle impinges on one side of the skirt and requires silica batt insulation over nearly one third of the outer laminate. The axial length of the nozzle extension is 52.95 in. Other dimensions are shown in the longitudinal section of Fig. 2.

FABRICATION

Skirt fabrication begins when preimpregnated asbestos-phenolic tape is wound around a mandrel at room temperature. The warp of the tape is in the hoop direction, while the tape width initially forms an angle of 45 deg with the mandrel. The liner is then vacuum bagged, heat is applied, and excess resin is bled out. In this way, the liner density is increased, and the angle of plies with respect to the mandrel is reduced by 10 to 15 deg. The outside surface of the liner is then machined to its final dimensions.

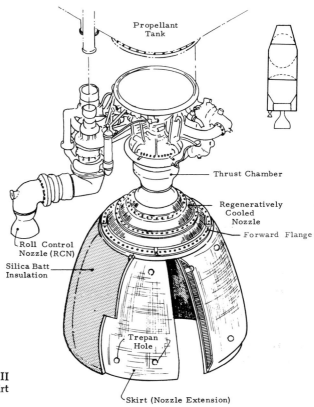

FIG. 1—Titan III Stage II engine and ablative skirt assembly.

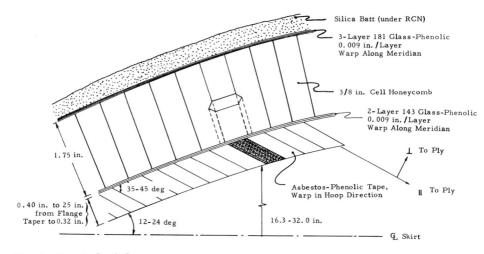

Silica Batt (under RCN)

3-Layer 181 Glass-Phenolic
0.009 in./Layer
Warp Along Meridian

3/8 in. Cell Honeycomb

2-Layer 143 Glass-Phenolic
0.009 in./Layer
Warp Along Meridian

1.75 in.

⊥ To Ply

35-45 deg

Asbestos-Phenolic Tape,
Warp in Hoop Direction

‖ To Ply

0.40 in. to 25 in.
from Flange
Taper to 0.32 in.

12-24 deg

16.3 -32.0 in.

₵ Skirt

FIG. 2—Longitudinal skirt section.

Two plies of 143 glass-phenolic, ⅜-in. cell honeycomb, and three plies of 181 glass-phenolic laminate are bonded to the liner in succession, with a vacuum bag cure cycle required for each additional component. The warp directions of both laminates are aligned with the meridian of the skirt. At each end, the honeycomb is contoured such that the inner and outer laminates meet and are wrapped with glass roving.

FLIGHT FAILURE HISTORY

Prior to the structural modification described in this paper, the flight history of Titan III and II skirt was marred by four structural failures. Failure locations were from 12 to 32 in. from the forward flange; the average time of the failures was 180 s after ignition.

Skirt failures can be catastrophic since a decrease of thrust at a critical point in the firing may prevent the achievement of desired orbit. Also, an asymmetric thrust may be sufficient to cause loss of vehicle control. To prevent such events, a structural evaluation of all aspects of the Stage II skirt was undertaken [1].[3]

MATERIALS INVESTIGATION

The gathering of sufficient data for a complete stress analysis of a structure becomes a more extensive and complex task when composite materials are used instead of homogeneous, isotropic materials. In general, difficulties in testing are posed by: the orthotropic nature of most composites, the influence of fabrication procedures on material properties and specimen size, the response of the material to storage and flight environments, and the effects of structural constraints on possible failure modes. These problem areas are discussed with reference to the Stage II skirt.

The material's response to applied stress, that is, whether it is elastic, elastic plastic, or viscoelastic, determines the type of stress analysis needed. In the present case, preliminary tests indicated that the asbestos-phenolic and glass laminates had essentially linear stress-strain relations until the onset of failure. Consequently, an elastic analysis is adequate for the prediction of stresses. The required properties for an axisymmetric analysis of the skirt are the Young's moduli and coefficients of thermal expansion in the three principal material directions, three independent

[3] The italic numbers in brackets refer to the list of references appended to this paper.

Poisson's ratios, and the shear modulus in the meridional plane. To evaluate margins of safety, it is also necessary to know the tensile and compressive strengths in the principal directions, the interlaminar shear strength, and the shear and block tensile strengths of the various bondlines. Tests to determine the rate of release of volatiles from the skirt also were conducted to predict the pressure in the unvented honeycomb cells.

The determination of liner properties proved to be most challenging, since block specimens of asbestos phenolic did not yield data comparable to that obtained from specimens wound under initial tension. It thus became necessary to take specimens from existing liners whose 0.400 in. thickness required nonstandard specimens in directions normal and parallel to the plies.

The addition of heat to the liner initiates a series of irreversible chemical processes which drastically alter its macrostructural characteristics. At temperatures above 600 F, gases are evolved from the phenolic resin. In so doing, they cool the residual char which is formed and also retard the penetration of the high surface temperature to the cooler portions of the liner. The asbestos phenolic is essentially fully charred at 1000 F, and postfiring measurements indicate that negligible dimensional changes take place in the char. The typical temperature distribution through the liner shown in Fig. 3 was derived on the basis of data from material tests and test firings.

The liner's coefficients of thermal expansion were found to be the most significant parameters. Although it is standard practice to oven-dry expansion specimens prior to testing in order to have reproducible results, a comparative study was made of two specimens conditioned for 40 h at 300 F and at 100 percent relative humidity, respectively. The rather dramatic differences in expansion characteristics are illustrated in Fig. 4, where it can be seen that, even for a relatively low heating rate, the wet specimen contracts substantially more than the dry one. A specimen with no preconditioning then was tested and found to have essentially the same contraction as the wet specimen. All further testing then was performed on specimens preconditioned for 8 h at 100 percent relative humidity to simulate the liner's moisture content prior to flight.

Reducing the specimen diameter to 0.150 in. from the standard 0.325 in. was found to have no effect on the measured expansion. The smaller diameter then

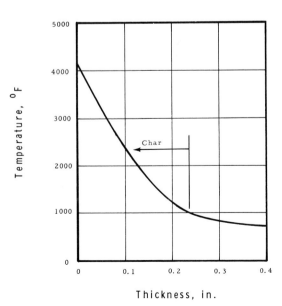

FIG. 3—Typical radial temperature distribution through the liner.

Thickness, in.

Au et al on Ablative Nozzle Extensions

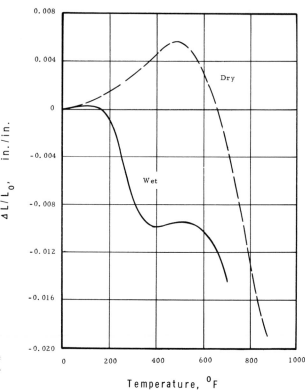

FIG. 4—Effect of moisture
on thermal expansion of asbestos phenolic.

Temperature, °F

was adopted, along with a preheated fixture, in order to increase the heating rate
from 12 to 180 F/min. The test method used to determine the thermal expansion
characteristics was ASTM Test for Coefficient of Linear Thermal Expansion of
Plastics (D 696-44) and the temperature range extended to 1500 F. The changes
in length per inch in the hoop, parallel, and perpendicular-to-ply directions are
recorded on Fig. 5 where it is shown that the tendency to contract is most pronounced in the latter direction.

The tensile strength and modulus of the liner were found from room temperature
to 800 F in the hoop and meridional directions by ASTM Test for Tensile Properties
of Plastics (D 638-58). The required tensile properties in the perpendicular and
parallel-to-ply directions could not be found with standard test specimens; so, the
following procedure was adopted. Compression tests were run on small prismatic
specimens in the hoop, meridional, perpendicular, and parallel-to-ply directions.
The values of compressive modulus obtained for the latter two directions were
then multiplied by the ratio of the meridional tensile modulus to compressive modulus to determine the tensile moduli in these directions. Strain gages mounted on
the sides of the compression specimens were used to find Poisson's ratios. Perpendicular-to-ply tensile strength specimens were made by bonding cylindrical liner
specimens to tapered aluminum plugs.

The measurement of tensile properties at temperature above 1000 F was simplified considerably by knowledge of the liner's contraction characteristics, preliminary
stress analyses, and observation of test-fired liners. Asbestos-phenolic liners in the
postfired state typically contain a uniform distribution of hoop cracks to the depth
of the charred material. Preliminary analysis predicted that these cracks occur

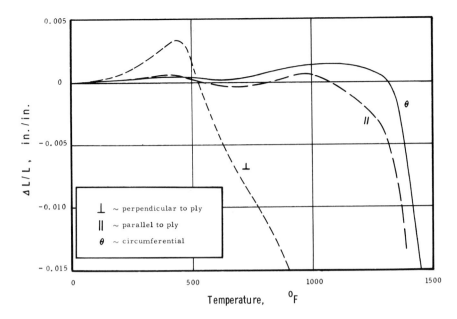

FIG. 5—Thermal expansion characteristics of asbestos phenolic.

during firing as a result of interply tensile stresses. Additional parametric analyses indicated that the tensile modulus and strength in the hoop direction were of importance at temperatures over 1000 F.

A test method recently developed involves an electric plasma arc to obtain high temperatures and heating rates of tension, compression, and shear specimens [2,3]. It was found from meridional tension specimens that asbestos phenolic has negligible interply strength at temperatures above 1000 F, thereby substantiating the prediction that cracking occurs during firing. Further tests determined tensile strengths and moduli in the hoop direction at temperatures up to 3000 F.

The shear modulus in the meridional plane can be computed from the transformation equations of an elastic, orthotropic material [4] with a knowledge of the ply angle and strains of the tension specimen of Fig. 6.

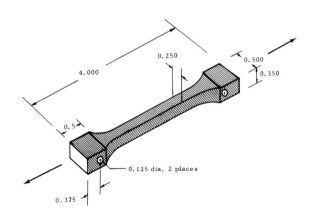

FIG. 6—Meridional plane shear modulus test specimen.

Au et al on Ablative Nozzle Extensions

Testing of the laminates also posed interesting experimental problems, basically because the modes of failure are dependent upon the amount of lateral restraint offered by adjacent parts of the structure. Because the inner laminate is supported fully by the liner, it can be expected to fail in compressive shear. When tests were conducted in accordance with Method 1021 of FED-STD 406, however, it was found that the short unsupported section of the specimen failed by splitting between plies. A new fixture which supports the specimen over its entire length, therefore, was devised [5] in order to induce a compressive shear failure. Substantially higher strength values were obtained with this technique. The outer laminate is supported by the honeycomb cell walls and is dimpled into the cells. Its effective modulus is reduced by the dimples, and it also has a tendency to fail by intercell buckling. To evaluate these conditions, a biaxial test fixture was developed to support a square specimen, including liner and honeycomb, taken from a skirt. Uniaxial as well as biaxial properties were determined in this manner by loading the outer laminate directly.

It is clear from this brief review of the test program that the determination of composite material properties requires versatility beyond the scope of existing standardized tests. Additionally, the present investigation demonstrates the importance of conducting stress analyses and the test program concurrently for maximum effectiveness.

Stress Analysis

The preceding description of the nozzle extension, its environment, and the problems associated with the determination of material properties suggest that a rigorous stress analysis is a formidable task. The computer program described below, however, is believed to be adequate when its results are correlated with the postfired condition of tested skirts.

Stresses due to differential thermal expansion and pressure loads were predicted by a finite element computer program [6] using the structural model shown in Fig. 7. The material is assumed to be linearly elastic and to have temperature dependent, orthotropic properties. One principal elastic axis lies in the hoop direction, while the orientation of the other two axis in the meridional plane is arbitrary within each element. Loads, temperature distributions, and material properties are taken to be axisymmetric. Pore pressures, residual stresses, and aerodynamic shears are not accounted for, although liner cracking can be simulated by adjustment of the elastic constants. Although numerous loading conditions and times were

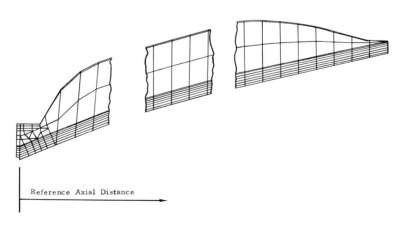

Reference Axial Distance

Fig. 7—Structural model of Titan III Stage II skirt.

Composite Materials

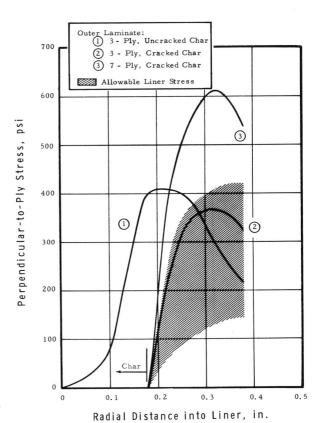

FIG. 8—Perpendicular-to-ply
liner stresses.

Radial Distance into Liner, in.

examined in the course of the analysis, only a few selected results will be discussed below.

Calculated stress distributions and allowable ranges under the roll control nozzle (RCN) are plotted in Figs. 8, 9, and 10 at 200 s from ignition. These results are believed to be representative of the local areas under the RCN, although the asymmetry of the temperature distribution and material properties are not accounted for.

The first analysis of the existing design considers the liner to be uncracked; values of Young's modulus for temperatures above 1000 F were extrapolated. The stresses computed on this basis exceed their respective maximum allowable values in the liner and both laminates. To continue, it was recalled that examination of numerous test-fired skirts revealed uniformly distributed hoop cracks in the char. This observation, together with the very low perpendicular-to-ply tensile strength of the char revealed by the test program, led to a modified structural model in which hoop cracks were simulated to the char depth.

The remarkable changes in stress distribution and magnitude throughout the skirt caused by hoop cracking of the char are illustrated. In this case, both laminates are marginal, that is, there is a possibility that they may fail. This prediction agrees with firing history; also, the outer laminate is shown to be critical at an axial location typical of flight failures.

Pressure induced stresses are computed to be much smaller than thermal stresses. The large difference in expansion properties between the liner and the glass laminates is the primary cause of stress in the skirt.

Au et al on Ablative Nozzle Extensions

Axial Distance, in.

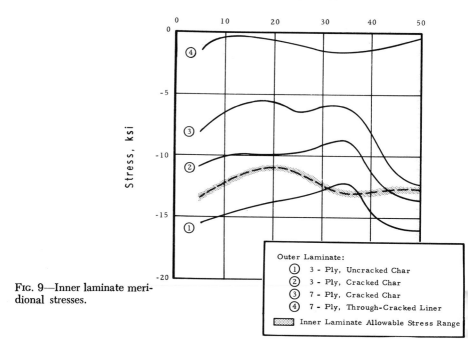

FIG. 9—Inner laminate meridional stresses.

Outer Laminate:
① 3 - Ply, Uncracked Char
② 3 - Ply, Cracked Char
③ 7 - Ply, Cracked Char
④ 7 - Ply, Through-Cracked Liner
▨ Inner Laminate Allowable Stress Range

Axial Distance, in.

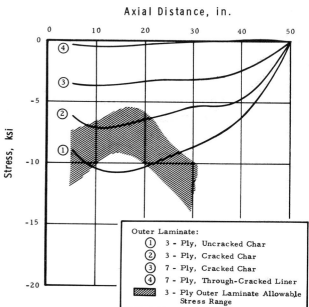

FIG. 10—Outer laminate meridional stresses.

Outer Laminate:
① 3 - Ply, Uncracked Char
② 3 - Ply, Cracked Char
③ 7 - Ply, Cracked Char
④ 7 - Ply, Through-Cracked Liner
▨ 3 - Ply Outer Laminate Allowable Stress Range

Trapped gas pressure in the honeycomb cells can add as much as 25 psi to the honeycomb-to-laminate bondline tension. Although this stress is low when compared to the strength of the basic structure, it can cause local defects near unbonded or crushed honeycomb to propagate. For this reason perforated cell honeycomb and positive external venting are desirable.

As a result of the foregoing analyses, the following possible local failure modes have been postulated:

(a) Intercell buckling of the outer face sheets under the RCN.

(b) Compressive shear failure of the inner laminate.

(c) Loss of honeycomb to laminate bond by cell pressure in areas of undiscovered defects.

SKIRT MODIFICATION PROGRAM

To capitalize on the beneficial effects of liner cracking, layers were added to the outer laminates of the existing nozzle extensions. The new configuration was analyzed, and the results are shown in Figs. 8, 9, and 10 for the condition of hoop cracks to the depth of the char. The interply stresses in this case exceed their maximum allowable values throughout the uncharred liner material. The analysis thus indicates that the stiffer seven-ply outer laminate tends to prevent the liner from shrinking and thus causes hoop cracking through to the inner laminate. The stresses for the fully cracked liner are very low, indicating that the modification leads to acceptable margins of safety. A small hole also was drilled through the outer laminate into each honeycomb cell to relieve trapped pressure.

These procedures were implemented on two skirts that subsequently were subjected to static test firings until burn-through occurred at almost 300 s without prior structural damage. To date, 16 successive Titan vehicles have flown successfully with these modified Stage II skirts.

STRUCTURAL SYNTHESIS

A successful structural synthesis of composite materials for ablative nozzle extensions must integrate effectively the various factors affecting the design configuration, environmental conditions, material thermal and mechanical properties, thermal and structural analyses, fabrication techniques, and process control. Current aerospace programs require this cooperative effort of the interfacing disciplines if a satisfactory total effect is to be achieved. The participating technical disciplines essential to the successful design of an ablative nozzle extension are examined briefly below.

DESIGN

Anisotropic materials such as fiber reinforced composites have enabled the designer to tailor the materials to his design. However, this increased design versatility has created a host of new problems. Design intuitions gained with isotropic materials are often not applicable. For example, it has been demonstrated by Au [7] that even with a simple cylindrical configuration exposed to an axisymmetric linear temperature gradient through the thickness, the thermoelastic behavior of the anisotropic design cannot be predicted from a knowledge of the thermoelastic behavior of the corresponding isotropic design. Consequently, the designer must be reoriented to the behavior of these new materials by working closely with the structural engineers and the material scientists.

ENVIRONMENT

For structures designed to withstand thermally induced loads, it has been traditional to stipulate a unit factor of safety for thermal stresses because the strength of materials varies vastly with increasing temperatures. For this reason, the temperature distribution must be predicted very carefully. In any event, the thermal properties such as thermal conductivity, density, specific heat, thermal gravimetric constants, heat of decomposition, heat transfer coefficient, rate of surface loss or ablation, etc. must be known fairly accurately if a realistic hyperthermal environment is to be predicted. A close working association among the material scientists, heat transfer analysts, and those engaged in the testing of material is an absolute necessity so that their mutual needs and requirements may be established and appreciated.

MATERIALS

A current difficulty encountered by structural analysts in predicting the structural performance of advanced composite structures in hyperthermal environments is the inadequacy of the mechanical data as they are being accumulated generally. In many instances, only room-temperature data are available. In other instances, precharred specimens of fiber reinforced plastics are used to obtain elastic properties and strengths which are not representative of the material as used in an actual structure. At this writing, there are no ASTM standard tests established for advanced composites and no standardized tables of allowable strengths. Consequently, while the features are useful and reproducible with good quality control, their use in controlled design is limited. To popularize advanced composites, the structural engineers must inform the materials researchers of their needs and requirements. Those engaged in the testing of materials to obtain engineering data must work closely with the structural analysts so that the information accumulated will suit the particular objective. The material researchers must consider also the macroscopic behavior of materials in real environments in order to develop a useful material and must interact closely with the thermal, structural, and design engineers to seek their assistance to overcome any inadequacy of the materials. Indeed, the final product of any advanced development program on materials should be a handbook which both the materials and the structural design engineers can use with confidence. Until this is accomplished, only a qualitative rather than a quantitative assessment of the design integrity can be made at best.

PARAMETRIC STRUCTURAL ANALYSES

One of the most valuable services that can be rendered by the structural analyst to the material researchers and those engaged in the testing of materials is to rank the various material properties in the order of their stress-inducing effects on the particular design. The usefulness of this parametric stress analysis will be illustrated by considering the original configuration of the Titan III Stage II ablative nozzle extension. By using high, nominal, and low values of Young's moduli in accordance with Fig. 11, the stresses on the inner and outer laminates vary, as shown in Fig. 12. It is seen that the liner's Young's moduli have a pronounced effect on the outer and inner laminate stresses. Similar analyses are performed in which the coefficient of thermal expansion of the liner in the perpendicular-to-ply direction (a) is varied and (b) corresponds to the liner in the wet and the dry conditions. The results of Figs. 13 and 14 show that the coefficient of thermal expansion of the asbestos-phenolic liner has a primary influence on nozzle extension stresses. In general, the stresses are found to increase as the moduli increase and the expansion coefficients become less positive or more negative.

The intrinsic value of such a parametric stress analysis is self evident. Not only will the results greatly enlighten the material researchers and the test engineers, but also the designer of these complex structures similarly will be benefited by a deeper comprehension of the load carrying characteristics of his creation. The structural behavior of the ablative nozzle extension is so complex that the functions of the structural analyst are certainly indispensable if a balanced design based upon both chemical and mechanical performance is to be achieved.

454

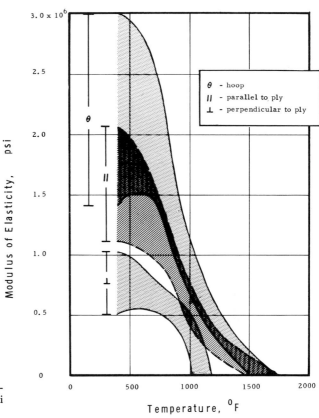

FIG. 11—Variation and dispersion of Young's moduli with temperature.

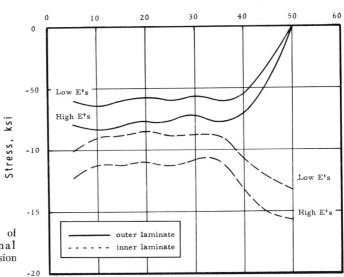

FIG. 12—Variation of laminate meridional stresses with dispersion in Young's moduli.

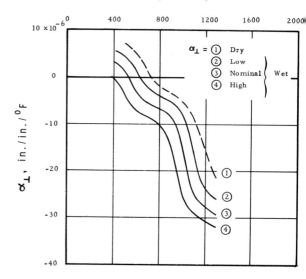

FIG. 13—Variation of liner perpendicular-to-ply coefficient of thermal expansion with liner condition and temperature.

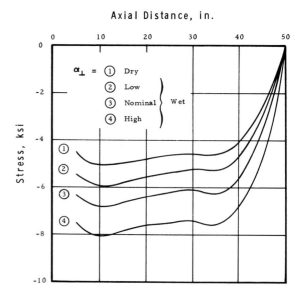

FIG. 14—Variation of outer laminate meridional stresses with liner perpendicular-to-ply coefficient of thermal expansion.

FABRICATION AND PROCESS CONTROL

The importance of fabrication techniques and procedures on the quality of the end product cannot be overemphasized. In some instances, the process of skirt fabrication is largely responsible for difficulties encountered in material testing, stress analysis, and the ultimate structural integrity of the design. For example, the dimples in the outer laminate lower its compressive strength. This lower strength is one of the causative factors for the initial failure of the outer laminate.

Another very significant fabrication parameter is the tolerance to which the liner wrap angles can be controlled. Figures 15 and 16 show the effect of the liner wrap angle on the stresses in the glass laminates. The profound influence of the liner wrap angle readily is seen.

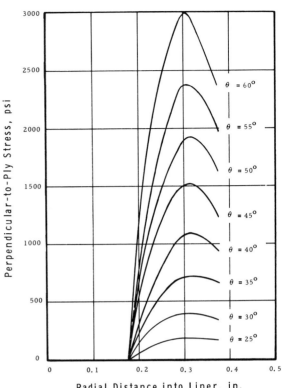

Perpendicular-to-Ply Stress, psi

$\theta = 60°$

$\theta = 55°$

$\theta = 50°$

$\theta = 45°$

$\theta = 40°$

$\theta = 35°$

$\theta = 30°$

$\theta = 25°$

Radial Distance into Liner, in.

FIG. 15—Effect of liner wrap angle on liner stress.

Axial Distance, in.

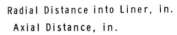

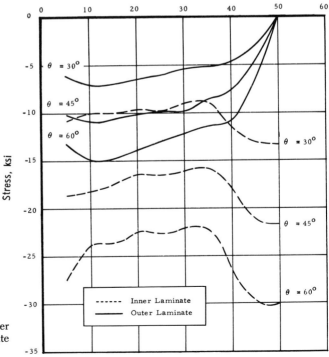

$\theta = 30°$

$\theta = 45°$

$\theta = 60°$

Stress, ksi

$\theta = 30°$

$\theta = 45°$

$\theta = 60°$

------ Inner Laminate
——— Outer Laminate

FIG. 16—Effect of liner wrap angle on laminate stresses.

The effects of variance in material properties have been investigated previously. Parameters such as coefficients of thermal expansion, Young's moduli, char strength, etc. of the liner have been shown to have a significant influence on the structural integrity of the nozzle extension. Until fabrication technique and process control can be advanced to the point of controlling these variations to predictable limits, a balanced design based upon both chemical and mechanical performance may not be possible.

CONCLUSION

The experience gained during the Stage II skirt investigation can be extended to other structures in which ablative materials are used. Most significantly, the mechanical and thermal properties of the ablator must be known over its entire range of operating temperatures if the structural performance is to be evaluated meaningfully. A search for candidate liner materials definitely should not be limited to consideration of thermal characteristics alone, since in many cases the success of a design will hinge on its ability to withstand the thermal stresses induced by differential expansion. Several materials whose thermal properties are comparable may exhibit vastly different structural behavior; the stress analyst must decide among them on the basis of their mechanical properties.

Structural compatibility of a liner and its supporting shell depends on the skirt geometry and the Young's moduli and coefficients of expansion of all the materials in the composite construction. An ideal way of minimizing thermal stresses is to select a liner material with very low Young's moduli. This property may be a natural one, or it may be induced. If the latter form of stress relief is desired, as when a tape-wrapped liner cracks under tensile stresses, a very stiff outer shell should be used, for example, a metal case. Uniformity of cracking and the effect on temperature gradients then become possible problem areas that must be resolved by testing. Paradoxically, materials that offer much greater char strength may be detrimental to the nozzle extension due to the prevention of stress relief by cracking and accordingly may be unattractive as liners. If liner cracking is deemed undesirable, a small positive coefficient of expansion is preferred for the liner, so that compression in the liner and tension in the shell result. The wrap angle of tape-wrapped liners can be optimized to cause cracking or minimize stress levels, depending on the design philosophy. Finally, the effects of moisture and heating rate on the behavior of liner materials must be investigated, and, if they are important, service conditions must be simulated during all element tests. The elimination of honeycomb in the outer shell structure is desirable from a quality control viewpoint. If honeycomb is deemed necessary to prevent skirt buckling during ignition, it must be vented between cells as well as to the exterior.

Progress currently is being made with several types of nondestructive testing methods which promise to improve the quality control of reinforced plastics. For example, microwave and sonic techniques have been developed and are employed for the inspection of ablative skirts.

The present investigation has contributed to the understanding of ablative skirt behavior. It also has identified many areas of endeavor where additional effort will be required in the future. Although it can be expected that these complex structures will continue to pose challenges, successful designs will be achieved whenever materials are selected on the basis of their contribution to the mechanical and thermal behavior of the structure.

ACKNOWLEDGMENT

The results reported herein are based upon a cooperative investigation of the Stage II nozzle extension by the Aerojet-General and Aerospace Corps. In a comprehensive study of this type, success is attained when the efforts of many individuals are unified by a well conceived and systematic program plan. The authors wish, therefore, to acknowledge the numerous contributions of the materials sciences, heat transfer, and structures personnel of both organizations, as well as the Titan Thrust Chamber Engineering and Liquid Propulsion Development Offices of Aerojet-General and Aerospace Corp., respectively.

References

[1] Welsh, W. E., Jr., and Ching, A., "A New Technique for Mechanical Strength Testing of Rapidly Charred Ablative Materials," Aerospace Corporation Report TR-1001-(2240-10)-7, Feb. 1967.

[2] Ching, A. and Welsh, W. E., Jr., "Strength and Stress-Strain Properties of Rapidly Heated Laminated Ablative Materials," Aerospace Corporation Report TR-0158(3240-30)-2, July 1967.

[3] Wilson, E. L., "Structural Analysis of Axisymmetric Solids," *Journal*, American Institute of Aeronautics and Astronautics, Vol. 3, No. 12, Dec. 1965, pp. 2269–2274.

[4] Au, N. N., "Stresses and Strains in Multi-Layer Anisotropic Hollow Cylinders," Aerospace Corporation Report TDR-469(5560-30)-4, 30 June 1965.

[5] Scheyhing, E. R. "Stress Analyses of Titan Family Stage II Ablative Skirts," Aerospace Corporation Report TOR-0158(3116-40)-1, 9 Feb. 1968.

Structural Evaluation of Long Boron Composite Columns

J. W. GOODMAN[1] AND
J. A. GLIKSMAN[2]

Reference:

Goodman, J. W. and Gliksman, J. A., "Structural Evaluation of Long Boron Composite Columns," *Composite Materials: Testing and Design, ASTM STP 460*, American Society for Testing and Materials, 1969, pp. 460–469.

Abstract:

Boron filament-epoxy composite tubes were evaluated for use as long column members in a spacecraft. Struts, 8 ft long and 3.2 in. in diameter, were designed, fabricated, and tested. After evaluation of test results, three-ply and four-ply laminates were selected to satisfy design load requirements of 5000 and 7000 lb., respectively. These designs were 33 and 45 percent lighter than 6061-T6 aluminum tubes with the same column design load. Effects of eccentricities were considered for column design as well as theoretical Euler buckling and local crippling failure loads.

Key Words:

boron, column, epoxy laminates, composite materials, compression, design, evaluation, tests

Truss structures are being utilized as the primary load path between satellite payloads and their boost vehicles. These trusses must be designed with the capability to withstand the quasi-static loads imposed on the payload during the booster staging operations. Stiffness constraints also are imposed in addition to the strength requirements. Individual strut members are loaded in both tension and compression during the various staging conditions. For long slender truss members, the design concept is that of a pin-ended long column. The optimum structural cross section for a long compression member is a thin-walled circular tube.

An evaluation, based on achieving minimum weight, was made during the design phase of a spacecraft truss structure. The bolt-to-bolt length of an individual strut was 8 ft. This strut was designed using the conventional material for such an aerospace application, aluminum, and a thin-walled circular cross section. Strength and stiffness requirements resulted in a strut weight of 3.6 lb. As a potential weight savings, various other materials were evaluated.[3] Beryllium would have been a candidate material because of its outstanding stiffness-to-density ratio, except that production techniques for long thin-walled beryllium tubes had not been developed. The most promising material appeared to be a boron composite with an estimated weight of 1.9 lb. In order to complete the evaluation, certain weight, stiffness, and local strength properties had to be determined. A test program was conducted to determine the required information. A change in design loads during the test program resulted in development of two composite struts.

[1] Member of Technical Staff, Applied Materials, TRW Systems Group, Redondo Beach, Calif. 90278. Personal member ASTM.

[2] Member of Technical Staff, Structural Analysis, TRW Systems Group, Redondo Beach, Calif. 90278.

[3] A comprehensive comparison of aluminum, titanium, steel, glass composite, and boron composite tubular columns is given in Ref 7.

DESIGN CRITERIA

The basic design criteria were identical for both struts except for the maximum design compression loads. The original compression load was 5000 lb, while the maximum load applied to the later strut was 7000 lb. Buckling is one type of failure for long thin-walled columns. For the ideally manufactured column member, two types of buckling can occur: (1) general instability (Euler buckling) or (2) local compressive buckling. For a perfect column with no initial eccentricity, failure will occur when the applied load, P, equals the Euler load, P_{cr}, or the local compression buckling load, P_{ccr}, whichever is lower. Every real column, however, has at least a small degree of curvature and eccentricity of applied load. These effects produce bending stresses and lateral deflections which increase as the applied load approaches the load for Euler buckling [1].[4] The basic tubular section straightness was specified as 0.010 in. for each 10 in. of length [2]. Since the column was comprised of an 86-in.-long tubular section plus two 5-in. end fittings, the straightness eccentricity could be a maximum of 0.086 in. The allowable end fitting eccentricity was placed at 0.06 in. at each end or a worst case sum of 0.120 in if the eccentricities were in opposite directions. Based on these potential eccentricities, 0.200 in. was chosen as an appropriate value for design purposes.

COLUMN ANALYSIS

The relationship between the axial compression load and the bending moment [3] at failure is:

$$R_c + R_b = 1.0 \dots\dots\dots\dots\dots\dots\dots\dots\dots\dots (1)$$

where R_c is the stress ratio for the axial compression load and R_b the stress ratio in bending. When the sum of R_c and R_b equals 1.0, failure occurs.

$$R_c = \frac{P}{P_{ccr}} \dots\dots\dots\dots\dots\dots\dots\dots\dots\dots (2)$$

$$R_b = \frac{M}{M_{cr}} \dots\dots\dots\dots\dots\dots\dots\dots\dots\dots (3)$$

The bending moment, M, on the column as a function of the initial eccentricity (e) is:

$$M = (\delta + e)P \dots\dots\dots\dots\dots\dots\dots\dots\dots\dots (4)$$

where $(\delta + e)$ is the total lateral deflection which varies with the applied load, P [1].

$$(\delta + e) = e \sec\left(\frac{L}{2}\sqrt{\frac{P}{EI}}\right) \dots\dots\dots\dots\dots\dots\dots\dots (5)$$

The allowable bending moment, M_{cr}, is a function of the material properties and the D/t ratio of the tubular cross section [3].

Utilizing the above equations for a 3.20-in.-diameter, 96-in.-long 6061-T6 aluminum tube with a 0.20-in. initial eccentricity, the thickness required to support a 5000-lb compressive load was 0.041 in. This resulted in a weight of 3.6 lb for the 86-in. tubular section. The thickness required to support a 7000-lb load was 0.057 in. The weight of this column was 4.9 lb.

The stiffness of the entire truss structure is dependent on the axial stiffness

---

[4] The italic numbers in brackets refer to the list of references appended to this paper.

(*AE*) of the individual column members. Utilizing a thickness of 0.041 in. with the 3.20-in.-diameter aluminum tube, the axial stiffness was 4.14×10^6 lb. This stiffness was sufficient to satisfy the dynamic constraints on the satellite. The use of a material other than aluminum was therefore dependent on its capability to provide adequate strength to support 5000 lb and on its stiffness being a minimum of 4.14×10^6 lb.

Test results on the boron composite specimens provided data to verify the capability to provide both adequate strength and stiffness. The design equations are the same as those for an aluminum column. The local buckling allowable, P_{ccr}, was determined from tests, as was the allowable bending moment, M_{cr}. Initial design of a boron column was based on a modulus of elasticity for a unidirectional boron laminate of 36×10^6 psi. Using a 3.20-in.-diameter tube with three plies of 0.0053-in. boron filament tape, the Euler buckling allowable, $P_{cr} = \pi^2 EI/L^2$, was calculated to be 7800 lb. This three-ply composite was chosen as the initial test column configuration. Therefore, based on the equations which govern the load carrying capability of the boron composite column, the following information was required from the testing phase of the investigation:

1. Axial stiffness, AE.
2. Local buckling allowable, P_{ccr}.
3. Bending moment allowable, M_{cr}.

MATERIALS AND FABRICATION

The materials, tooling, and processing are defined in this section. The various composite configurations are given, and the test specimens are described. The boron filament tape employed and the typical properties of laminates are given in Ref 5. The preimpregnated tape was made 10 in. wide and 100 in. long.

Long boron composite tubes were fabricated on a production mandrel containing spacer blocks in a hollow core; some short tubes were made on a laboratory mandrel with a solid core. These two tools resulted in laminates which were apparently resin lean and resin rich, respectively. Resin may have flowed from the laminate to the hollow core of the former while it was confined to the laminate with the latter. Laminates were laid up by hand, taking extreme care to prevent wrinkles. Each lamination was compacted by shrink tape or vacuum bag before adding the next. The final lay up was compacted before adding the bleeder and release cloths. Despite these precautions, one tube contained a full length wrinkle in the outer ply of glass cloth. The finished lay up was then placed in a vacuum bag and cured in an autoclave at 50-psi-min pressure. A step cure was used, 1 h at 200 F followed by 2 h at 300 F.

A minimum composite construction of three longitudinal plies of boron filaments was required to prevent long column (Euler) buckling, according to the preceding discussion. The magnitude of circumferential strength and stiffness required was not known. Three approaches were considered: (1) addition of a hoop oriented boron filament tape, (2) placement of the longitudinal plies of boron filaments at an angle to the tube axis, or (3) addition of glass cloth. Scheme 1 was rejected because of anticipated difficulty in wrapping the boron tape around a 1.6 in. radius. Scheme 2 was not employed because a very large ply angle would have been required for significant increase in hoop properties while the axial stiffness would have been greatly degraded [6]. Method 3 was finally adopted, based on the assumption that moderate additions of glass cloth reinforcement would be sufficient. As a starting point for experimentation, inner and outer plies of Style 112, 0.003-in.-thick, glass cloth were used to support the longitudinally oriented boron filaments. The interleaving plies of 104 cloth, 0.001 in. thick, employed as a carrier for the boron filaments were believed to be too light to significantly affect the structural

Table 1—*Boron composite designs.*

Design 1:
  Inner layer: one ply of 112[a] glass cloth.
  Center: three plies of boron filament tape[b] cross plied at ±5 deg to the tube axis, inter-
      laminated with plies of 104 cloth scrim employed as carrier for the boron filaments.
  Outer layer: one ply of 112 glass cloth.
  Nominal composite thickness 0.021 in. Composite density = 0.076 lb/in.³
Design 2:
  Same as Design 1 with the subsequent addition of one ply of hoop oriented 143 directional
  glass cloth. Nominal composite thickness 0.034 in. Density = 0.074 lb/in.³.
Design 3:
  Same as Design 1 with the subsequent addition of a fourth ply of boron tape at −5 deg
  and an overwrap of 143 glass cloth, hoop oriented. Nominal thickness 0.037 in.
Design 4:
  Inner layer: one ply of 112 glass cloth.
  Center: four plies of boron filament tape cross plied at ±5 deg, interlaminated with
      plies of 104 cloth scrim.
  Outer layer: one ply of 112 glass cloth.
  Nominal composite thickness 0.029 in.
Design 5:
  Same as Design 4 except that the boron tape was laid ±15 deg to tube axis.
Design 6:
  Same as Design 4, except outer ply was hoop oriented 143 glass cloth instead of 112.
  Nominal composite thickness 0.031 in. Density = 0.071 lb/in.³

---

[a] Glass cloth minimum thicknesses, per MIL-G-9084, for Styles 104, 112, and 143 are 0.001, 0.003, 0.008 in.,
respectively.
[b] Epoxy resin for all laminates was NARMCO 5505.

behavior. The boron filaments were cross plied at ±5 deg to the tube axis to ensure uniform stacking and spacing of the filaments.

Composites of three and four plies of boron filaments were evaluated. Table 1 lists the constructions. The initial configuration received the most extensive testing. Three tubes about 90 in. long were fabricated with the Design 1 construction. One of these was sectioned into shorter lengths for short compression specimens and a flexure specimen. The other two were used, unsectioned, for testing as long columns. When test results indicated a need for improved hoop strength, one tube (after failure as a long column) was sectioned to provide more short compression specimens and another flexure specimen. These were wrapped with a hoop oriented ply of Style 143 directional glass cloth, 0.008 in. thick; the configuration was designated as Design 2. Four-ply composites were prepared and evaluated as a result of the requirement for a higher load capability (7000 lb). Short column specimens of Designs 3, 4, and 5 were fabricated as part of that evaluation. Finally, a 90-in. tube was produced, Design 6, incorporating four plies of boron filaments and a reinforcing ply of Style 143 glass cloth. The composites tested were not symmetric about the middle plane, the boron filaments being alternately cross plied.

The short compression specimens consisted of a tube segment 12 in. long, bonded and riveted to aluminum end fittings. The same mode of end fitting attachment was employed for the flexure and long column specimens. Each end was reinforced with several plies of Style 181 glass cloth, 0.080 in. total thickness and 2.25 in. wide. The 7075-T6 aluminum alloy fittings were bonded into the composite tube, then holes were drilled and 24 blind rivets (³⁄₁₆ in. diameter) were installed at each end. The use of both bonding and riveting for attachment of the end fittings was a conservative approach, resulting in a trouble-free test configuration. The development of a reliable bonded attachment was not part of the scope of this test program.

Table 2—*Boron composite test results.*

Test	Design	Failure Load, lb	Axial Stiffness, lb	Boron,[c] Modulus, psi
Short column compression.........	1	7 790	...	...
	1	7 990	5.82×10^6	36.2×10^6
	1	8 000	...	...
	1	6 700	...	...
	2	13 120	...	...
	2	14 390	...	...
	3	19 720	7.90	36.9
	4	17 840	8.40	39.2
	5	17 960	7.09	33.1
Flexure.......................	1	7 350[a]	5.93	36.8
	2	12 180[a]	...	...
Long column[b].................	1	5 900	5.69	35.3
	1	7 050	5.57	34.6
	6	11 750	7.62	35.6

[a] Units are in · lb.
[b] Bolt-to-bolt length, 99 in. for Design 1, 93.5 in. for Design 6.
[c] Based on the boron thickness of the composite, 0.0053 in. per ply.

TEST METHODS AND RESULTS

This section presents the experimental methods employed and the results obtained in tests of short compression, flexure, and long column specimens.

Short column tests were performed to determine axial stiffness, AE, and local crippling failure loads, P_{ccr}. Strain gages were installed on some specimens, and the average strain was used to calculate axial stiffness. Results are summarized in Table 2. Specimens of the original three-ply laminate, Design 1, failed at 7790, 7990, 7995, and 6700 lb. The last value is anomalous; fracture initiated at a scrap of polytetrafluorethylene sheet unintentionally laminated between the second and third plies of boron filament tape. The fracture modes were characterized by large axial cracks indicating a hoop strength deficiency rather than buckling of the longitudinal fibers. The second group of three-ply specimens, Design 2, incorporated an additional layer of hoop oriented directional glass cloth. The compressive load for local failure was increased 70 percent, and the failure mode was an axisymmetric or circumferential wrinkle type. Short column specimens of four-ply composites (Designs 3, 4, and 5) were tested to provide preliminary data for selection of a four-ply long column configuration. All three designs exhibited the axisymmetric type of failure. Design 5 was fabricated using a ±15-deg cross ply and tested to provide a direct check on classical composite theory [4] in comparing the strength and stiffness of cross-plied composites. The composite cross plied at ±15 deg exhibited stiffness of 16 percent less than that of the ±5-deg composite, Design 4. The failure mode included cracking aligned with the filaments in the outer layer of boron tape as well as formation of a circumferential wrinkle.

Flexure specimens were tested as cantilever beams to determine the critical bending moment, M_{cr}, and the bending stiffness, EI. Three-ply composite specimens of Design 1 and reinforced Design 2 were tested. The failure mode was axial splitting for Design 1 and classical diamond pattern buckling for the glass cloth reinforced tube. Six strain gages were installed near the fixed end of one specimen from which data were obtained on the uniformity of flexural strain around the circumference. The stiffness obtained from the strain gage readings, as each of the six gages were oriented at the point of maximum strain, showed a variation of 14 percent.

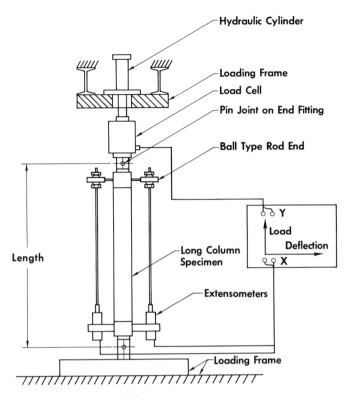

Fig. 1—Long column failure test setup.

Long column tests were run to verify the structural adequacy of the design. The prime verification was the determination of the failure load. Instrumentation also was added to estimate the axial stiffness, AE. These long column specimens were tested as shown in Fig. 1. The first three-ply composite tube, Design 1, failed at 5900 lb, compared to a short column crippling load, P_{ccr}, of 7900 lb and an estimated Euler buckling load, P_{cr}, of 7240 lb. The failure region, shown in Fig. 2, was 8 in. long in the center of the column. This failure mode resembled the short compression and flexure specimens in that there were a number of axial cracks and tears. Considerable lateral deflection was evident. Axial displacement was proportional to load until failure.

The second three-ply composite long column, also Design 1, failed at 7050 lb. Unlike the first, this column did not bow, as indicated by measurements at midlength. The failed region occupied 8 in. at 6 in. from the edge of the lower end fitting. The mode of failure appeared to be formation of a circumferential wrinkle and lateral translation of the two ends of the failed region. Axial cracks were present, but they were less prominent than in the fracture of the first column. The load displacement curve was again linear until fracture.

The third long column, Design 6, was fabricated using four plies of boron filaments. Failure occurred at 11,750 lb compared to an estimated Euler buckling load of 11,100 lb and an estimated short column crippling load of 19,720 lb (Design 3 test data). The axial displacement was proportional to load until failure. The composite fracture occurred approximately at midlength. Circumferential wrinkles were prominent. Slight lateral bowing of the tube was seen as the load approached 11,000 lb.

Goodman and Gliksman on Boron Columns 465

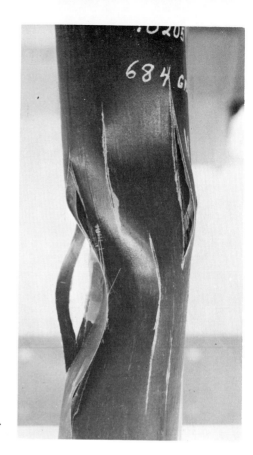

FIG. 2—Long column fracture, Design 1.

DISCUSSION

The objectives of this investigation were to select a composite strut design having a significant weight saving compared to an aluminum strut, and to demonstrate the load capability in tests of full scale columns. Figure 3 shows the composite strut design. The following observations primarily relate to these objectives.

Composite Design 2 was selected for a design compression load of 5000 lb and Design 6 chosen for the 7000-lb requirement. An end fitting for the aluminum tube would be employed directly on the boron tube; therefore, no weight penalty would be incurred.

Design Load, lb	86-in. Tubular Section		Savings, %
	Aluminum, lb	Boron, lb	
5000	3.6	2.4	33
7000	4.9	2.7	45

A portion of the total boron configuration weight for each design was 0.3 lb of glass cloth end reinforcement necessary to distribute the local rivet loads. Other important design criteria influence the potential weight savings attainable by advanced composites on spacecraft. Addition of a 0.004-in. aluminum coating to satisfy

Composite Materials

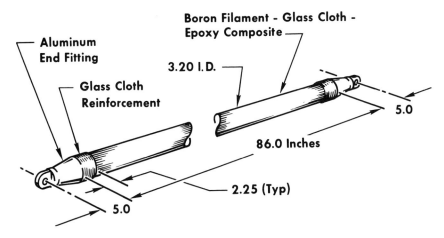

Aluminum
End Fitting

Boron Filament - Glass Cloth -
Epoxy Composite

3.20 I.D.

Glass Cloth
Reinforcement

5.0

86.0 Inches

2.25 (Typ)

5.0

Fig. 3—Long column configuration.

electrical and thermal conductivity requirements was found to reduce the weight savings by 0.35 lb.

The boron modulus, Table 2, was derived from the various composite designs by neglecting the effect of stiffness contribution of the glass cloth. The difference in failure loads of the two identical three-ply long columns, 5900 and 7050 lb, apparently was caused by eccentric loading of the composite tube either by eccentricity of the end fittings, curvature of the column, or asymmetry of composite stiffness of the cross section. Some evidence of asymmetric stiffness around the circumference of the tube was found in the flexure test. Initial column curvature was measured and found to be less than 0.020 in. on the four-ply column and on the stronger of the three-ply columns, but unfortunately was not measured on the column which failed at 5900 lb. The interpretation of eccentric loading most amenable to analysis, and the one employed here, assumes an eccentricity of placement of the end load on a straight homogeneous column. Test results in conjunction with Eqs 1 through 5 were used to predict column failure loads as a function of initial eccentricity, Fig. 4. The failure load for a perfect column $(e = o)$ was the lower value of the Euler and local compression crippling loads. The conservatism of the chosen boron composite designs was shown by the permissible initial eccentricity of 0.4 in. (Fig. 4) for each which exceeds the design requirement of 0.200 in. Optimization of the hoop reinforcement would result in an additional weight savings.

CONCLUSIONS

The results of the investigation have illustrated the relative ease with which boron composite columns can be employed to achieve substantial weight savings. The empirical approach and conservative design practices are acknowledged; however, they were appropriate to the short development time and limited scope of the investigation.

The following conclusions are made:

1. Eight-foot long tubular boron-epoxy composites have been fabricated, tested, and proven satisfactory for design compressive loads of 5000 and 7000 lb.

2. The conservatively designed boron composite tubes were approximately 30 percent lighter than aluminum tubes of conventional design. Additional weight savings can be achieved with boron composite tubular compression members by determination of minimum hoop reinforcement required and by optimizing the tube ends.

Goodman and Gliksman on Boron Columns   **467**

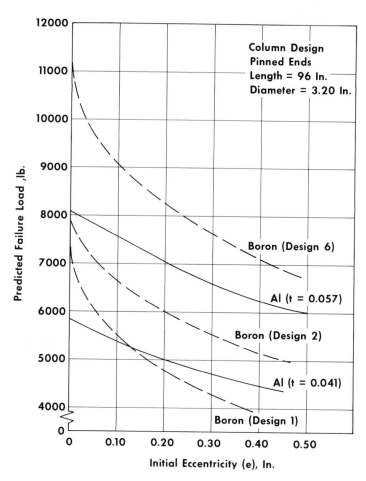

FIG. 4—Predicted column failure load versus initial eccentricity.

3. Column eccentricities, including curvature, alignment of end loading, and asymmetry of stiffness on a cross section must be considered in the selection of composite designs for long columns as well as the theoretical Euler buckling load and local crippling load.

4. A resin-lean three-ply boron composite tube cross plied at ±5 deg required hoop reinforcement to preclude axial cracking failures in compression. Hoop reinforced three-ply composites and resin-rich four-ply composites at ±5 and ±15 deg failed at higher loads in a circumferential wrinkle mode.

5. Hand lay-up techniques were adequate for fabrication of three- and four-ply long boron composite tubes, although care was necessary to avoid wrinkles.

ACKNOWLEDGMENT

The Air Force Materials Laboratory provided the boron filaments for this investigation.

References

[1] Timoshenko, S. P., *Strength of Materials,* Part I, Van Nostrand, New York, 1955, pp. 266 and 275.

[2] "Standards for Aluminum Mill Products," The Aluminum Association, 1966, p. 96.

[3] "Metallic Materials and Elements for Aerospace Vehicle Structures," MIL-HDBK-5A, Change Notice 2, 24 July 1967.

[4] "Structural Design Guide for Advanced Composite Application," prepared under contract No. AF 33(615)-5142 by Southwest Research Institute, San Antonio, Tex., for the Advanced Composites Division, Air Force Materials Laboratory, Sept. 1967, pp. VII–26 and IX–32.

[5] "Rigidite 5505 Epoxy—Boron Filament Prepreg Tape Advanced Composite Materials," Product Information data from Narmco Materials Division, Whittaker Corp., Costa Mesa, Calif., Aug. 1967.

[6] "Plastics for Flight Vehicles, Part I Reinforced Plastics," MIL-HDBK-17, Military Handbook, May 1964.

[7] Cole, B. W. and Cervelli, R. V., "Comparison of Composite and Noncomposite Tubes," *9th Structures, Structural Dynamics, and Materials Conference*, American Institute of Aeronautics and Astronautics/American Society of Mechanical Engineers, Palm Springs, April 1969, Paper 68–340.

Fracture Mechanics

Fracture Processes in Fiber Composite Materials

A. S. TETELMAN[1]

Reference:

Tetelman, A. S., "Fracture Processes in Fiber Composite Materials," *Composite Materials: Testing and Design, ASTM STP 460*, American Society for Testing and Materials, 1969, pp. 473–502.

Abstract:

The process of fracture in homogeneous and anisotropic composite materials involves three steps: (1) the initiation of a microcrack, (2) the stable growth of this microcrack under increasing load, out to macrocrack size, and (3) the unstable propagation of this crack at a critical stress level. The conditions for each of these processes are reviewed for homogeneous materials and then discussed in detail for fiber composite materials. The conditions for fracture initiation by fiber cracking depend primarily on the length and diameter of the fiber, their perfection, the fiber modulus, the test temperature, and the degree of interaction of the fiber with the surrounding matrix. Matrix and interface strength depend primarily on temperature, strain rate, and interfacial void content (for resin matrices). Unless the matrix is extremely strain-rate sensitive, premature fiber fracture does not lead to instability, and the composite strength is determined primarily by the average fiber strength. Several mechanisms for increasing the matrix toughness are discussed. The toughness associated with unstable, longitudinal crack propagation in composites then is considered in terms of fiber content, matrix toughness, interfacial bond strength, and matrix shear strength. It is shown that in certain composites the toughness decreases with increasing fiber content, and consequently maximum load carrying capacity is achieved at a particular fiber content V_f°, that increases with decreasing crack length. In other systems, particularly those containing cold-drawn metal fibers, both toughness and ultimate strength increase with fiber content.

Key Words:

fractures (materials), composite materials, fiber composites, mechanical properties, crack propagation, fracture mechanics, evaluation, tests

Composite materials, containing long filaments or short fibers surrounded by a matrix, are attractive for structural applications where high stiffness-to-weight and ultimate strength-to-weight ratios are required. A considerable amount of effort has been conducted to develop stronger reinforcements, means of fabricating structural shapes, and means of computing failure modes for composites. Most of the latter effort has been devoted to elastic failure (for example, buckling, gross yielding) of composite structures. Comparatively little attention has been given to the inelastic behavior of composites, particularly the modes of and criteria for crack propagation and fracture.

An understanding of the fracture behavior of composites is important for several reasons. Certain composites can be designed and fabricated to have unusually high resistance to crack propagation, and a composite of elements A and B can actually be tougher than either A or B individually (for example, fiber glass). This behavior offers exciting possibilities for applications where high structural reliability is re-

[1] Professor, Materials Division, College of Engineering, University of California, Los Angeles, Calif. 90025.

473

quired. However, the toughness of composites containing high strength, brittle filaments in a ductile resin or metal matrix often decreases as the volume fraction of the filament is increased. Consequently, there is a possibility that pre-existing flaws may be able to propagate and cause complete fracture at design stress levels that are lower than predicted by elastic considerations alone (for example, for buckling, gross yielding). The safe working stress then is determined by the stress at which unstable crack propagation will occur. This behavior is particularly important when the composite is to be utilized in an aggressive environment, at extremes of temperature, or under alternating or dynamic loading. Static tests on flaw-free structures may then be of little use in predicting composite failure modes.

While these cosiderations also apply to design with high-strength metallic structures, there are certain considerations with respect to composites that hinder our understanding of fracture behavior. First, the cost of fabricating composite material test specimens is presently one or two orders of magnitude greater than the cost of fabricating metallic test specimens. Consequently, a given number of research and development dollars simply does not produce as many test results that can be correlated and interpreted adequately to develop a rational picture for fracture-safe design. Since only certain aspects of the fracture behavior of metallic structures even now are being understood adequately [1],[2] it is apparent that understanding of composite behavior will not be inexpensive and that test programs for composites need to be planned carefully to obtain the maximum amount of relevant information per test. Secondly, the wide variety of fabrication procedures used for making components of the same "nominal" composite material, and the fact that composites are most attractive for individual "tailor-made" applications, make the need for characterization more important, and at the same time more difficult, than for high-strength metals. Finally, the anisotropy inherent in fiber composites complicates an exact analytic description of composite behavior, in terms of the behavior of the individual constituents.

The various fracture modes in a composite depend on the relative strengths and ductilities of the filament, the matrix, and the interface between them. These parameters, in turn, depend on the chemical composition of the constituents, processing history, and external conditions (temperature, strain rate, environment) under which the composite is loaded. Although our understanding of interfacial behavior is somewhat limited, we are in a position to synthesize the large body of knowledge on the behavior of reinforcements and matrices to make some meaningful predictions of the fracture behavior of composites. A description of those aspects of fracture initiation and crack propagation that are documented fairly well is contained in the body of this paper. In the following section, the criteria for the initiation, stable growth, and unstable propagation of cracks in homogeneous materials are reviewed. Following that section is one which deals with criteria for microrock initiation and stable crack growth in fiber composites. After that is a section in which the principles of elastic fracture mechanics are applied to describe the macroscopic conditions for longtudinal crack propagation in fiber composites, in terms of microscopic variables (for example, fiber size, matrix ductility, etc.). Due to space limitations, the important questions of fatigue, creep, and environmental effects on composite behavior will not be treated.

FRACTURE CRITERIA FOR HOMOGENEOUS MATERIALS

Fracture in homogeneous materials generally involves three processes [1]: (1) the initiation and initial growth of a microcrack, out to a size c_0 characteristic of a microscopic feature (for example, a grain diameter), a process called *initiation;* (2) the growth of this microcrack, or the linking of this microcrack with other microcracks of similar size, under increasing applied stress or strain to form a

---

[2] The italic numbers in brackets refer to the list of references appended to this paper.

macrocrack, a process called *stable crack growth;* and finally (3) the unstable propagation of this crack, under decreasing applied stress or strain, until fracture is complete, a process called *rapid, or unstable,' crack propagation.* Each of these processes is characterized by a particular "failure criterion." For example, when iron is tested at very low temperature, the number of cleavage microcracks n is proportional to σ^9, where σ is the applied tensile stress [2]. Consequently, the total crack length nc_0 increases sharply with applied stress (Fig. 1a). Alternatively, at higher temperatures where inclusions break and voids are formed, and these voids in turn coalesce by plastic strain concentration to form regions of rupture, the density of rupture regions approximately one grain diameter in length increases as a function of both the applied tensile stress level and the plastic strain level [3]. The criterion for each of these processes depends on metallurgical variables (for example, grain size, inclusion content) and external conditions (temperature), and that process which occurs at a lower level of applied stress (or strain) will be observed first in a tension test.

During stable crack growth under increasing stress or strain, the total cracked area nc_0 increases, and a certain number $n' < n$ of cracks link together to form a macrocrack of length c_F which can propagate unstably at a stress σ_F (or plastic strain ϵ_F) and cause complete failure (Fig. 1b). The details of this linking process, even in materials containing a fairly homogeneous distribution of second phase particles, are understood only partially at present [4,5]. Stages 1, 2, and 3 are observed only in plain tension tests conducted with fairly ductile materials. When very brittle materials are tested in tension, for example, the initial crack that is formed by plastic deformation is able to propagate unstably, and hence the initiation

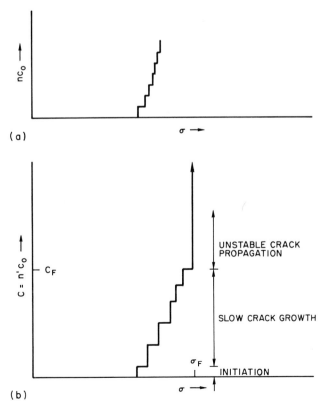

Fig. 1—(*a*) Total crack length (nc_0) as a function of applied stress σ. (*b*) initiation, stable growth, and unstable propagation of critical crack as a function of applied stress σ.

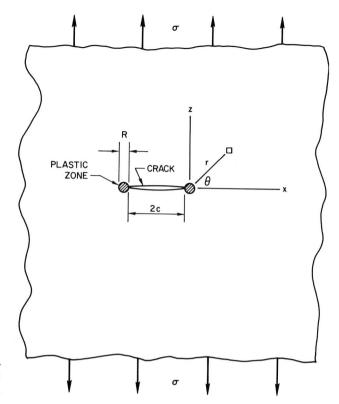

FIG. 2—Plastic zone formation near crack tip under tensile loading.

stress and fracture stress are the same; Stages 2 and 3 are not observed. When precracked $(c \gg c_o)$ specimens are tested, Stage 1 obviously is not required. Slow crack growth prior to instability occurs with materials having a certain degree of toughness; more brittle materials fracture without showing any evidence of slow crack growth prior to instability and then only Stage 3 is observed [1].

The principles of linear elastic fracture mechanics [6] can be used to predict the conditions which cause the unstable fracture that occurs before general yielding (that is, when the fracture stress σ_F is less than the net section yield stress σ_Y). Consider (Fig. 2) a specimen containing a sharp crack of length c that is much greater than the microstructural element size c_o but is still relatively small compared with the specimen width W. It is convenient to define a parameter K

$$K = \sigma \sqrt{\alpha \pi c} \quad \dots \dots \dots \dots \dots \dots \dots \dots \dots \dots \dots \dots \dots (1)$$

known as the stress intensity factor to describe the degree of stress (and strain) intensification that occurs near the border of the crack when a gross tensile stress σ is applied at the ends of the specimen. α is a parameter that depends on specimen geometry and ratio of σ to yield stress σ_Y, and approaches unity as c/W and σ/σ_Y approach 0. (For simplicity we will assign α a value of unity in subsequent discussion.) Initially, the local elastic stresses σ_{zz} near the crack tip vary with distance r from the tip as $K/\sqrt{2\pi r}$. Close to the crack tip $(r \to 0)$ $\sigma_{zz} \geqq \sigma_Y$, and consequently local plastic yielding can occur in a plastic zone of diameter R near the crack tip, where

$$R = \frac{1}{\pi} \frac{K^2}{\sigma_Y^2} \quad \dots \dots \dots \dots \dots \dots \dots \dots \dots \dots \dots \dots \dots (2)$$

476

Once yielding occurs, the tensile stress level in the plastic zone approaches a value somewhere between σ_Y and $3\sigma_Y$, depending on the degree of triaxial constraint that exists in the plastic zone.

The plastic extension that occurs in the plastic zone ahead of the crack tip causes the crack tip to be opened by an amount (COD) that is approximately [7]

$$COD \cong \frac{8}{\pi} \frac{\sigma_Y}{E} R \dots\dots\dots\dots\dots\dots\dots\dots\dots (3)$$

where E is the elastic modulus.

Consequently, from Eqs 1 and 2, we have

$$(COD) \cong \frac{K^2}{E\sigma_Y} \cong \frac{\sigma^2 \pi c}{E\sigma_Y} \dots\dots\dots\dots\dots\dots\dots\dots (4)$$

for $8/\pi^2$ and α set equal to unity. Since the tensile stress level in the volume element right at the crack tip is of the order of σ_Y, and since this element is being opened by an amount equal to the (COD), we can define the *work* done at the crack tip G as the product of this force per unit area times the displacement [8]

$$G \cong \sigma_Y(COD)$$

Unstable fracture occurs when the crack opening displacement reaches a critical value [8,9], COD^*, at which point G reaches a critical value G_c

$$G_c = \sigma_Y(COD)^* \dots\dots\dots\dots\dots\dots\dots\dots (5)$$

and hence K reaches a critical value K_c

$$K_c = \sqrt{EG_c} \dots\dots\dots\dots\dots\dots\dots\dots (6)$$

so that the fracture stress σ_F is given by

$$\sigma_F = \sqrt{\frac{EG_c}{\pi c_F}} \qquad (\sigma_F \ll \sigma_Y) \dots\dots\dots\dots\dots\dots (7)$$

where c_F is known as the critical crack size. It is noteworthy that as crack extension proceeds and c becomes greater than c_F, the tensile stress required for unstable propagation decreases; consequently, the propagation process is unstable.

Equation 7 can be used to describe the conditions for any form of unstable crack propagation, micro or macro, provided that the size of the propagating crack is an order of magnitude greater than the size of the region ("process zone") near the crack tip which is physically separating, and provided that $\sigma_{zz} < \sigma_Y$ in the vicinity of the crack tip *outside* the plastic zone [10]. Thus the equation is applicable for microcrack propagation caused by atomic separation (or dislocation coalescence) at the microcrack tip, provided that the crack size is greater than 10 atomic (or dislocation slip band) spacings. Most generally, the equation is used to describe macroscopic crack propagation where the crack size is greater than about 5 to 10 grain diameters.

The physical significance of the critical crack opening displacement depends on the nature of the separation that is occurring in the process zone near the crack tip. When microcracks propagate through grains, or macrocracks propagate in large size brittle crystals, $(COD)^*$ is of the order of the interatomic spacing. In metals, localized plastic strain ϵ_f is required for separation at the macrocrack tip, and, if the crack has a root radius ρ, then this separation is localized over a distance 2ρ such that [8]

$$(COD)^* \cong 2\rho\epsilon_f$$

and hence

$$G_c \cong \sigma_Y 2\rho\epsilon_f \dots\dots\dots\dots\dots\dots\dots\dots\dots (8)$$

The separating region ahead of the crack behaves as a miniature test specimen, superimposed in a plastically deforming matrix, having ductility ϵ_f. In general, the "local tensile ductility" ϵ_f decreases rapidly as σ_Y increases [1] so that G_c, (and hence K_c) decrease with increasing yield strength. It is noteworthy, particularly when discussing the fracture behavior of composites, that G_c increases as the root radius of the crack increases. Also, when ρ becomes of the order of the crack size c, the parameter α in Eq 1 becomes very small as the stress concentration factor of the crack is decreased drastically. Increasing toughness, therefore, can be achieved by crack tip blunting and by ensuring that plastic deformation be spread homogeneously over a large region near the crack tip.

This short discussion on fracture mechanics should be sufficient to provide understanding of particular aspects of fracture in composites that are discussed below. For more detailed treatment, the reader is referred to Refs *1, 6, 11*.

FRACTURE INITIATION AND INITIAL CRACK GROWTH IN FIBER COMPOSITES

STRESS DISTRIBUTIONS IN FILTER COMPOSITES

For purposes of discussion it is most convenient to consider fiber composites as composed of either continuous or discontinuous fibers surrounded by a matrix. The former are essentially "infinitely" long filaments that transmit load directly (Fig. 3a); the latter have finite lengths (Fig. 3b) and axial load is transferred to the fibers through shear stresses set up at the fiber-matrix interface when the matrix attempts to flow past the fibers. When a composite containing uniaxially aligned fibers in a residual stress-free matrix is stressed along the fiber axis, the deformation behavior is typically *isostrain* (that is, the strains in the matrix and fiber are equal, and hence equal to the composite strain ϵ). The modulus of the reinforcing fibers E_f is generally much greater than that of the surrounding matrix E_m so that most of the load is carried by the fibers.

The theory of the load-carrying capacity of uniaxially loaded fiber composites has been treated previously in detail [*12–15*], and only the results that are pertinent to subsequent discussion of fracture behavior will be presented here. For *continuous*

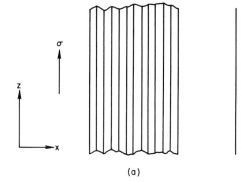

FIG. 3—Continuous (*a*) and discontinuous (*b*) fibers in a matrix under longitudinal stress σ.

(a)   (b)

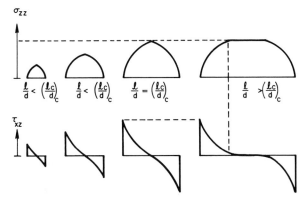

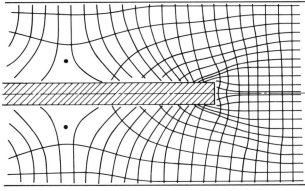

FIG. 4—Tensile stress distributions σ_{zz} in fiber, shear stress distribution in matrix, and tensile stress distributions in matrix, when discontinuous fibers are subjected to longitudinal load [16, 17].

fibers the nominal stress σ carried by the composite when both fibers and matrix are elastic obeys the rule of mixtures

$$\sigma = [E_f V_f + E_m(1 - V_f)]\epsilon \dots\dots\dots\dots\dots\dots(9)$$

where V_f and $(1 - V_f)$ are the volume fraction of fiber and matrix, respectively. When strains that are of the order of the yield strain of the matrix are achieved, the matrix deforms plastically, but the fibers, having a higher yield strength, remain elastic. Then

$$\sigma = EV_f\epsilon + (1 - V_f)\sigma_m(\epsilon) \dots\dots\dots\dots\dots(10)$$

where $\sigma_m(\epsilon)$ is the flow strength of the matrix at a strain ϵ.

Many composite materials contain discontinuous fibers prior to final fracture, either because: (1) continuous fibers were not used in fabricating the composite, (2) continuous filaments fractured during fabrication, or (3) continuous filaments fractured during loading prior to final fracture of the composite. The stress distribution in the matrix and discontinuous fibers have been studied experimentally [16,17] and theoretically [13,18,20] by several investigators. Briefly, the shear stress at the fiber-matrix interface τ_{xz} is a maximum at or close to the end of the fiber and then decreases to zero at the center of the fiber (Fig. 4). The tensile stress in the matrix parallel to the fiber axis $\sigma_{zz(m)}$ is very high close to the ends of the fibers and decreases to zero near the center of and close to the fiber; also, the tensile stresses in the matrix decrease rapidly as the perpendicular distance from the fiber increases. To a first approximation, the tensile stresses *in the fiber,* parallel to the fiber axis σ_{zz}, build up linearly from the ends of the fibers. For a fiber of radius r,

$$\sigma_{zz} = \frac{2\tau z}{r} \quad\dots\dots\dots\dots\dots\dots\dots\dots\dots\dots\dots\dots (11)$$

where τ is the maximum shear stress that can be supported in either the matrix or the interface, whichever is smaller (that is, the flow strength of the matrix at the composite strain ϵ, or the interface shear strength), and z is the distance measured along the fiber axis from one of its ends.

Discontinuous filaments of nonmetallic materials (for example, graphite, boron, silicon carbide) fracture when a *critical tensile stress* σ_f is produced in the fiber, parallel to the fiber axis.

From Eq 11 this stress will achieve a distance

$$z = \frac{r\sigma_f}{2\tau} \quad\dots\dots\dots\dots\dots\dots\dots\dots\dots\dots\dots\dots (12)$$

from the end of the fiber, and consequently the fiber must have a length

$$l_c = 2\left(\frac{r\sigma_f}{2\tau}\right) = \frac{d\sigma_f}{2\tau} \quad\dots\dots\dots\dots\dots\dots\dots\dots\dots (13)$$

before it can be broken. d is the fiber diameter, and

$$l_c/d = \frac{\sigma_f}{2\tau} \quad\dots\dots\dots\dots\dots\dots\dots\dots\dots\dots\dots\dots (14)$$

is defined as the *critical aspect ratio.* When $l = l_c$, the *average* stress carried by the fiber $\bar{\sigma}_f = \frac{1}{2}\sigma_f$ (Fig. 5a). When $l \gg l_c$, $\bar{\sigma}_f$ approaches σ_f (Fig. 5b), so that more efficient strengthening can be achieved. For example, when $l = 10l_c$, $\bar{\sigma}_f = 0.95\sigma_f$. The tensile stress supported by the composite is then

$$\sigma = \bar{\sigma}_f V_f + \sigma_m(\epsilon)(1 - V_f) \quad\dots\dots\dots\dots\dots\dots\dots (15)$$

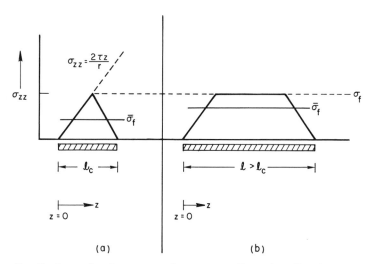

FIG. 5—Stress distribution in a discontinuous fiber when fiber fractures.

The load carrying capacity of composites containing metallic reinforcements (for example, high-strength wire) can be estimated in a similar manner, except that the ultimate tensile strength of the wire is taken usually as a measure of the maximum fiber stress when the wires are ductile and neck down before they fracture.

FRACTURE INITIATION IN FIBER COMPOSITES

When a fiber composite is loaded continuously parallel to the fiber axis, fracture generally *initiates* either by:

(a) fiber fracture, $\sigma_{zz} = \sigma_f$;
(b) interface shear fracture (debonding), $\tau_{xz} = \tau_i(f)$;
(c) matrix shear fracture, $\tau_{xz} = \tau_m(f)$; and
(d) matrix tensile fracture, $\sigma_{xx} = \sigma_m(u)$.

depending on which of these processes occurs first (that is, at a lower value of applied strain (composite stress)).

In practice, an exact analysis of the initiating failure mode is complicated because of: (1) the presence of residual stresses and strains introduced during fabrication resulting from differences in thermal expansion coefficients of matrix and filament; (2) the fact that the criterion for interfacial fracture (b) is not well understood, since $\tau_i(f)$ is difficult to either measure or calculate and the shear stress concentration $\tau_{xz}/(\sigma/2)$ at the *end* of the fiber is not known particularly as a function of fiber shape; and (3) the matrix shear and tensile fracture criteria (c) and (d) are not well understood for the same reasons. Consequently, only certain generalizations can be made at the present time. For metal matrices that are well bonded to the fibers, the high shear stresses at the fiber ends will promote plastic deformation prior to either matrix or interface shear fracture. The maximum shear stress carried by the matrix is then assumed [13] for the purposes of calculation to be one half the ultimate tensile strength of the matrix $\sigma_m(u)$. Consequently, the critical aspect ratio required for the fiber fracture is from Eq 14,

$$l_c/d \doteq \frac{\sigma_f}{\sigma_m(u)} \dots\dots\dots\dots\dots\dots\dots\dots(16)$$

and hence increasing fiber fracture strength or decreasing matrix (or interface) shear strength favor debonding rather than fiber fracture as the primary mode of fracture initiation. For metal matrix composites containing weak (or embrittled) interfaces, or resin composites in which shear fracture often occurs along the interface rather than in the matrix, the Naval Ordnance Laboratory (NOL) ring or horizontal beam tests [21,38] are used to determine $\tau_{i(f)}$, and the critical aspect ratio required for fiber fracture is

$$l_c/d = \frac{\sigma_f}{2\tau_f} \dots\dots\dots\dots\dots\dots\dots\dots\dots(17)$$

Many factors influence the fracture strength σ_f of reinforcing filaments, and it is impossible to discuss all the most important of these here. The length and diameter of filaments and whiskers, particularly those of nonmetallic materials, have a large effect on fiber strength. Figure 6 indicates that the average strength

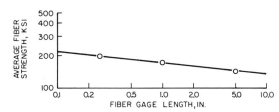

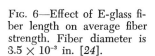

FIG. 6—Effect of E-glass fiber length on average fiber strength. Fiber diameter is 3.5×10^{-3} in. [24].

of glass fibers decreases as the fiber length l increases [22,23]. It has been shown [23] that this type of behavior is consistent with fibers whose strength distribution is of the Weibull type

$$g(\sigma) = l\alpha\sigma^{\beta-1} \exp{(-l\alpha\sigma^\beta)} \dots\dots\dots\dots\dots\dots\dots (18)$$

where α and β are constants for a particular material. Physically, the results of Fig. 6 can be interpreted to indicate that flaws (that is, small microcracks) are present on the surface or in the interior of the fibers and that the probability of encountering a more dangerous flaw increases with filament length. Some fibers contain populations of both internal and surface flaws, each of which is characterized by a different distribution. Figure 7 illustrates this type of behavior for boron filaments, where the severe A-type flaws, several inches apart, appear to be associated with the surface, whereas the B-type flaws appear to lie within the filament and may be associated with the formation of crystalline tungsten boride formed during the deposition of the boron onto the tungsten core [25]. Practically, the results of Figs. 6 and 7 indicate that composite structures, initially containing continuous filaments, undoubtedly contain broken filaments after relatively (to the average) low stresses have been applied, and consequently composites must possess sufficient tolerance for these flaws to allow them to be loaded fully. For example, if a microcrack formed in one filament of the "as-received boron" could propagate unstably and cause brittle fracture of the entire composite, the upper bound limit of fiber strengthening would be less than 100 ksi, as compared with 300 ksi if 30 percent of the fibers could break without causing ultimate failure of the structure.

The fracture strength of reinforcing fibers, particularly fine monocrystalline whiskers, increases as the fiber diameter d decreases. When the data [26,27] for several whiskers are compared, (Fig. 8). it is observed that

$$\sigma_f = \frac{\lambda\kappa}{d} \cong \frac{\lambda E}{10} \frac{d^*}{d} \dots\dots\dots\dots\dots\dots\dots\dots (19)$$

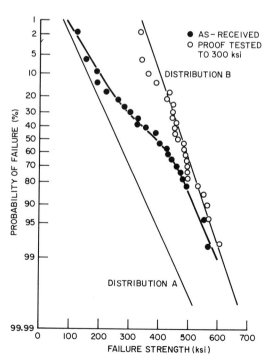

FIG. 7—Failure strength distributions of as-received and proof-tested boron fibers, 1 in. long [25].

Composite Materials

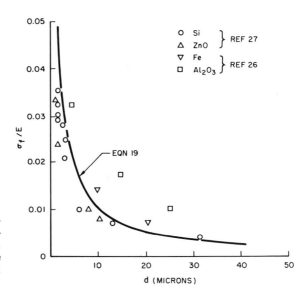

EQN 19

σ_f / E

d (MICRONS)

FIG. 8—Effect of fiber diameter on ratio of the (maximum) fracture strength to modulus ratios that have been observed for several whiskers.

where $3 \leq 1$ is a "degradation factor" that accounts for the presence of surface or internal flaws, and κ is a parameter that appears to be directly proportional to the theoretical cohesive strength $E/10$ and a characteristic diameter $d° \approx 1$ μm at which the theoretical strength can be achieved in the absence of flaws. This relationship illustrates the importance of high modulus for strengthening as well as stiffening purposes. The ultimate strength of cold-drawn wires also increases with decreasing fiber diameter [28,29] (Fig. 9); most of this increase probably results from increasing amounts of strain hardening as the cross-sectional areas of the wire are reduced.

In addition to geometrical variables, the fracture strength of reinforcements dispersed in a matrix will be influenced also by metallurgical variables. The strength of high-strength, cold-worked metal wires decreases with increasing temperature [29] (Fig. 10a) particularly above one half the absolute melting temperature, due to processes of recovery and thermally activated dislocation motion. Thermally activated microcrack formation, by localized dislocation nucleation, may be responsible for the decrease in strength of alumina whiskers with increasing temperature that is shown in Fig. 10b [26].

Probably the most important factor that influences the fracture strength of fibers that are incorporated in a matrix is the degree of reaction that can occur at the interface between fiber and matrix. In some cases (for example, beryllium in aluminum [30], tungsten in stainless steel [31]) a brittle intermetallic compound forms

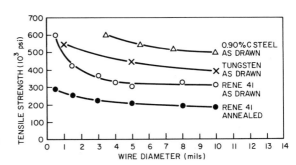

FIG. 9—Effect of wire diameter on the strength of cold-drawn wires [28].

TENSILE STRENGTH (10³ psi)

WIRE DIAMETER (mils)

0.90%C STEEL AS DRAWN

TUNGSTEN AS DRAWN

RENE 41 AS DRAWN

RENE 41 ANNEALED

Tetelman on Fiber Composite Materials

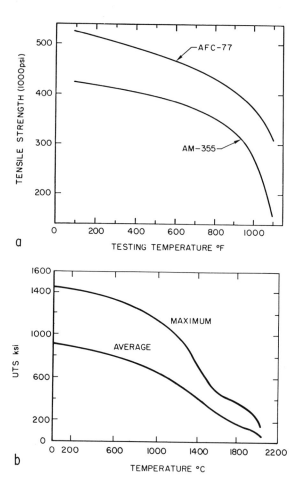

FIG. 10—(a) Effect of test temperature on the ultimate strength of cold-drawn steel wires [29], (b) effect of temperature on the ultimate strength of Al₂O₃ whiskers [26].

during fabrication which causes the composite to have a lower strength than the matrix material alone at ambient temperature. Similarly, TiB_2 and $TiSi_2$ are formed during the diffusion bonding of boron and silicon carbide filaments to α titanium alloys [25]. Figure 11 illustrates the effect boride thickness (proportional to the square root of the reaction time) on the fracture strength of boron filaments in a titanium matrix. The boride fractures at a strain at which the stress level in the filaments is 150 ksi, introducing surface cracks in the filaments whose length are equal to the boride thickness. The stress required to propagate these cracks through the filaments, causing filament fracture, varies inversely with the square root of the crack length (boride thickness), as shown in Fig. 11. At thickness below about 1600 Å the stress concentration factor of the boride crack is less than that due to intrinsic defects in the boron itself, and boride thickness has no effect on fracture strength.

In addition to promoting strength reductions by the formation of brittle interface compounds, the matrix can also lower fiber strength by promoting surface recrystallization of the fiber. When the degree of recrystallization is small, the strength of the fiber itself obeys a "rule of mixtures" governed by the strengths and volume fractions of the recrystallized and unrecrystallized portions [32]. However, when a relatively large amount of the fiber is recrystallized, and the recrystallized material is brittle (for example, in refractory metal fibers), the sharp cracks formed in

484   *Composite Materials*

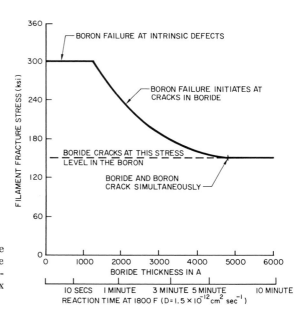

FIG. 11—Effect of boride thickness and reaction time on the strength of boron filaments in a titanium matrix [25].

this zone can propagate through the unrecrystallized portions of the fiber at stress levels below the ultimate strength normally observed for unrecrystallized material; the fiber strength and composite strength then decrease sharply (Fig. 12). Aluminum, nickel, and cobalt appear to have a large effect of this sort on tungsten [32], and nickel is able to promote recrystallization and weakening of graphite fibers [33].

Several factors influence the shear strength $\tau_m(f)$ of the metal or resin matrix and the interface strength. For metal matrices, the shear strength depends primarily on the strain hardening capacity, the degree of particle dispersion hardening, and the grain size. The strength of resin matrices depends primarily on the degree of cross linking between polymer chains, [51], as well as on the density of voids formed during processing [34]. Nonlinear deformation is a thermally activated process, and consequently the shear strength of metal and resin matrices [35] decreases with increasing temperature and decreasing strain rate (Fig. 13).

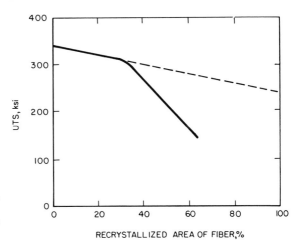

FIG. 12—Effect of recrystallization of tungsten fiber on the strength of a tungsten-copper-cobalt composite [32].

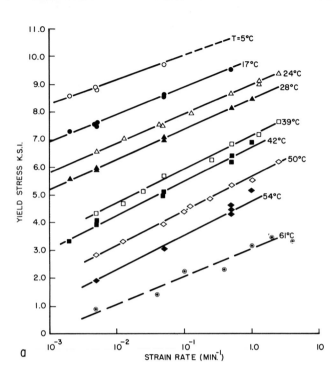

a

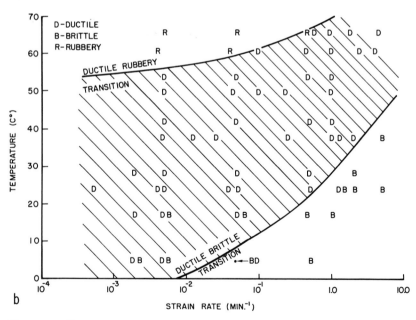

b

FIG. 13—Effect of strain rate and temperature on the yield strength (a) and fracture behavior (b) of epoxy-versamid resin [35].

Composite Materials

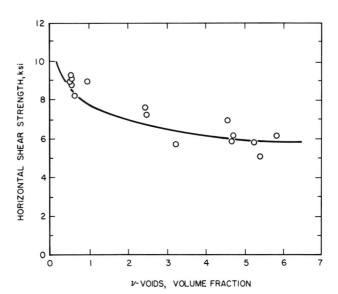

Fig. 14—Effect of void content on horizontal shear strength of resin-fiber composites [37].

ν-VOIDS, VOLUME FRACTION

Relatively little is known about the atomistic factors that affect the interface shear strength $\tau_i(f)$ of composites. Good bonding can be achieved with metal matrices provided a small degree of chemical reaction occurs [52], and interface fracture generally does not pose any serious problems. However, the situation is quite different for resin matrix composites whose service behavior often is limited by low values of interfacial shear strength, since cracking parallel to the fiber can lead to delamination in multiply composites as well as to fracture of an individual ply [36]. In addition to any atomistic factors, it appears that the content of voids (entrapped air bubbles) introduced during fabrication can have a significant effect on interfacial shear strength [36] (Fig. 14), as well as on transverse strength [34].

The preceding discussion has indicated that the mode of fracture initiation depends on the ratio of the tensile strength of the fibers to the shear strength of the matrix or interface, and that these parameters in turn are influenced by several factors. For example, if the fiber strength decreases slightly with temperature while the matrix strength decreases sharpy with temperature, then there will be some temperature T^* below which fibers of a given l/d will break (Eq 14 satisfied) and above which they will pull out (Eq 14 cannot be satisfied), before matrix or interfacial failure occurs. Likewise, if the fiber strength varies inversely with diameter according to a relation of the form $\sigma_f = \kappa'/d$, then fiber diameter will not affect the mode of fracture initiation, and fiber fracture will occur only if

$$l_c > \frac{\kappa'}{\tau}$$

The orientation of the applied stress axis relative to the fiber axis θ also will influence the mode of fracture initiation at a given temperature. Let

$\sigma' = \sigma \cos^2 \theta =$ tensile stress parallel to fiber axis ....(20a)
$\tau'_m = \sigma \cos \theta \sin \theta =$ shear stress in matrix or along interface ....(20b)
$\sigma'_m = \sigma \sin^2 \theta =$ tensile stress in matrix or perpendicular to interface ....(20c)

If shear fracture in the matrix occurs when $\tau' = \tau_m(f)$, or in the interface when $\tau' = \tau_i(f)$, then by setting Eq 20a and b equal we find that

$$\theta^* \cong \tan^{-1}\left(\frac{\tau_m(f)}{V_f \sigma_f}\right) \quad \text{or} \quad \theta^* \cong \tan^{-1}\frac{(\tau_i(f))}{V_f \sigma_f} \dots\dots\dots\dots (21)$$

is the transition angle, below which fracture initiates by fiber fracture and above which fracture initiates by matrix or interfacial shear. Likewise for

$$\theta > \theta^{**} = \tan^{-1}\left(\frac{\sigma_m(u)}{\tau_m(f)}\right). \dots\dots\dots\dots\dots\dots (22)$$

where $\sigma_m(u)$ is the ultimate strength of the matrix or the interfacial fracture strength, tensile fracture will initiate at the interface or in the matrix. These failure criteria appear to hold for metal matrix systems. For resin matrices, it appears that for $\theta > \theta^*$ a failure criteria based on a combination of resolved tensile and shear stress components appears to apply [38]:

$$(\tau'_m)^2 + (\sigma'_m)^2 = \sigma^2 \sin^2 \theta = \text{constant at fracture} \dots\dots\dots\dots (23)$$

Initial Crack Growth

Consider a composite structure containing either continuous filaments or discontinuous filaments of length $l > l_c$ such that fracture initiation occurs by fiber fracture. Several events can then occur, depending on the properties of the fiber, matrix, and interface.

1. All Filaments Have the Same Strength σ_f

If all filaments have the same σ_f, and if V_f is greater than a minimum value of about 0.10 [17], then the failure of all the filaments at the same stress level immediately leads to composite failure, as the bridges of matrix between the fibers cannot support the load, previously carried by the fibers, that is suddenly applied to them. This type of behavior is observed when metal matrix-metal fiber composites are broken in uniaxial tension (for example, tungsten-copper, molybdenum-copper) [13,34]. (The degree of matrix deformation during this type of fracture is discussed later in connection with crack propagation.)

2. Filaments Have a Distribution of Strengths, Matrix Is Brittle or Semibrittle

This type of behavior occurs for nonmetallic filaments imbedded in certain epoxy resins. Many resins are strain rate sensitive and fracture in a relatively ductile manner at low strain rates and moderate temperatures, (that is, large amounts of necking precede final fracture) and in a brittle manner at high strain rates and at low temperature (small reductions of area and absence of necking precede fast fracture [35]). The sudden fracture of a filament imbedded in a matrix of this type will cause the matrix adjacent to the filament break to be loaded rapidly, and the sudden release of stored energy in the filament will be available to allow the filament crack to propagate into the matrix. In a carefully controlled set of experiments [40], it has been shown that if the composite is being loaded at a fairly rapid rate, such that the nominal strain rate in the matrix is high, crack propagation can continue until either the specimen fails (single filament specimen) or the crack reaches the next adjacent filament, causing it to fracture, and so forth until the specimen breaks completely. However, if the composite is being strained at a slow rate, the initially propagating crack begins to slow down and eventually stops in the matrix. Undoubtedly, the transition strain rate at which

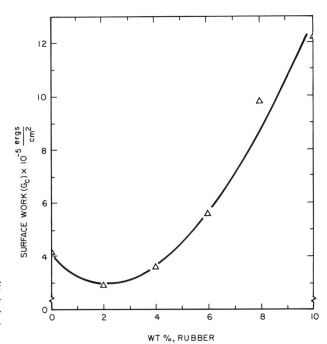

FIG. 15—Toughness of 828-epoxy resin as a function of weight percent Hycar 1312 rubber formulations [41].

a propagating crack will stop will increase with increasing temperature, but this remains to be shown.

As mentioned earlier, the fracture strength of a composite will be determined by the strength of the weakest filament if a crack that formed in the filament can propagate unstably and cause complete failure. As the filament spacing decreases (V_f increases at constant d or d decreases at constant V_f) the likelihood of unstable crack propagation to an adjacent fiber and beyond increases, particularly under impact loading and if the matrix is brittle. The probability of unstable crack propagation can be reduced by several means. First, the matrix can be chosen to have a higher resistance to crack propagation. This can be accomplished through the addition of "plasticizers" such as versamid [35] or hycar rubber [41] to a standard epoxy such as Epon 828, Fig. 15. A second possible means of preventing unstable, initial crack growth would be through the introduction of ductile filaments at various intervals between the brittle filaments, to act as crack stoppers by a splitting mechanism described later [40]. Third, the strain rate experiments on B-epoxy described earlier indicate that if a composite were loaded slowly to a certain stress level, unloaded, and then reloaded at a rapid rate, the composite strength was considerably higher than if the specimen had been loaded at the high rate with no preload. It was proposed that the stopped cracks formed in weak filaments at the slow rate would be "blunted out" and would not initiate fracture during the subsequent loading at higher rates. Fracture could occur only when new breaks occurred at high strain rates, and this required a higher load than for a virgin composite. This experiment suggests that it might be possible to increase the strength of similar systems by prestressing them at a higher temperature than they will be operated at in service, to break any weak filaments, and blunt out the resulting cracks under controlled conditions where crack formation is not harmful. Similar procedures have been applied to improving crack toughness of metallic materials [42,43].

3. Filaments Have a Distribution of Strengths, Matrix Is Relatively Tough

This type of behavior occurs for nonmetallic filaments imbedded in metal or tough resin matrices. Under increasing applied stress, the density of cracked filaments increases [23,44] (Fig. 16). If the interface strength is relatively weak, the high shear stresses set up around the fiber ends can cause debonding, which relieves the tensile stress concentration set up at the fiber break. It has been proposed [40] that the good agreement between either statistical or rule of mixture type models of glass-epoxy composite strength results from the fact that debonding prevents brittle crack propagation through the matrix and allows the stresses in the fibers to build up to their average fracture strength, at which point the composite fails as described in (1) earlier. This result suggests that greater load carrying capacity might be achieved with resin matrices if the bond strength were high enough to allow the average fiber strength to be realized, and give adequate transverse strength, but low enough such that debonding could occur after premature, low stress fiber fracture. (Similar considerations apply for improved crack propagation resistance, as shown later.) Since our understanding of the factors that determine bond strength are still incomplete, it is not possible to predict at this time how the bond strength could be controlled in such a manner.

Most metal matrices that are being considered for structural components (for example, nickel, aluminum, titanium alloys) are sufficiently tough and nonstrain rate sensitive that fracture in a filament does not lead to matrix crack propagation. If filament failure is not followed by debonding, then the stress concentration

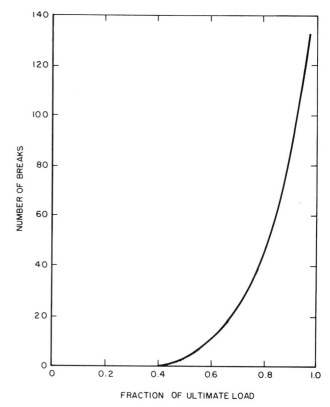

Fig. 16—The number of breaks in glass fibers, embedded in epoxy resin, at indicated fractions of the ultimate load of the composite [24].

FRACTION OF ULTIMATE LOAD

factor set up in the broken filament will cause large localized plastic strains which in turn load the adjacent fibers above the average of the composite. These fibers or their interface may then crack; alternatively, the next fracture initiation site could occur elsewhere in the composite, depending on the ratio of the strength of the filaments adjacent to the break relative to the strength of other fibers, and the magnitude of the plastic strain concentration factor at the break. Both a reduction of the statistical spread of the fiber strengths and a decreased fiber spacing tend to favor cracking in fibers adjacent to the first break, and the development of a crack several fiber spacings long. This may account for the fact that the strength of composites falls below the predicted rule of mixtures at large volume fractions of fiber.

LONGITUDINAL CRACK PROPAGATION IN FIBER COMPOSITE MATERIALS

Macrocracks, several fiber spacings in length, can be formed in fiber composites by a variety of means, such as fatigue, particle impact, poor joining procedures, or under increasing load as described in (3) previously. As in the case of fracture initiation, the relative strengths of fiber, matrix, and interface will determine the mode of crack propagation. Consider the crack shown in Fig. 17. In the vicinity of the tip the crack opening displacement (COD) will be accommodated by tensile strains in both the fiber and matrix that decrease as r^{-n} where $(\frac{1}{2} < n < 1)$ probably varies with fiber content (if the fiber fraction were 1.0, and the fibers were elastic, $n = \frac{1}{2}$, whereas n would equal 1.0 if $V_f = 0$ and the matrix were perfectly plastic). In addition to the tensile stresses and strains set up perpendicular to the fracture plane, transverse tensile stresses σ_{xx} and shear stresses τ_{xz} also are produced, along with transverse tensile strains and shear strains. Several modes

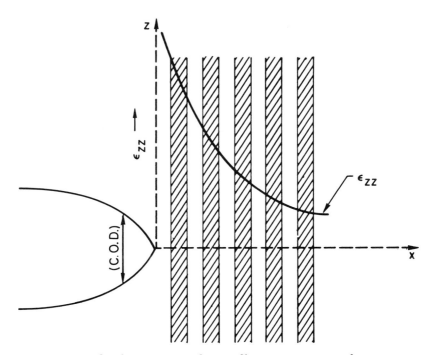

Fig. 17—Strain distribution near crack tip in fiber composite material.

of longitudinal crack propagation behavior can occur[3] depending on constituent properties.

SYSTEMS CONTAINING BRITTLE FIBERS, STRONGLY BONDED TO A TOUGH MATRIX

This type of behavior has been noted in composites that contain silicon or tungsten fibers imbedded in a copper matrix [47]. Figure 18 indicates that fibers ahead of the advancing crack front are broken and that the remaining bridges of matrix then neck down and fracture in a completely ductile manner. Since the fibers are extremely brittle, any toughness in the composite results from the work done

[3] Only a small amount of work has been done on the important problem of determining transverse toughness and mechanism of crack propagation [45,46]. Consequently, this subject will not be discussed here.

Courtesy of G. Cooper [47].

Fig. 18—Crack propagation in vacuum cast copper-tungsten fiber composite.

Composite Materials

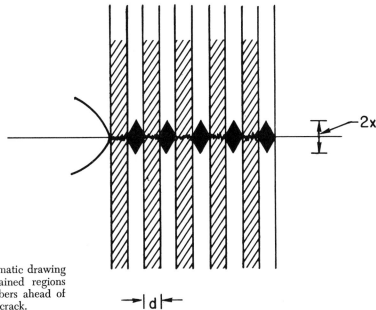

FIG. 19—Schematic drawing of heavily strained regions and cracked fibers ahead of an advancing crack.

in drawing down the matrix material. Consider for simplicity a two-dimensional composite (Fig. 19) containing fibers of diameter d separated by a distance X such that

$$V_f = \frac{d}{x + d}$$

Since Fig. 18 indicates that heavy deformation occurs in a triangular region extending a distance x above and below the fracture plane, the volume of material undergoing heavy plastic strain is approximately x^2, and consequently the volume of material being deformed per unit area of crack propagation is

$$\frac{x^2}{x + d} = \frac{x^2 V_f}{d} = \frac{d(1 - V_f)^2}{V_f} \dots\dots\dots\dots\dots\dots (24)$$

The work per unit volume required to fracture the matrix at a strain ϵ_f is

$$U = \int_0^{\epsilon_f} \sigma_m d\epsilon \dots\dots\dots\dots\dots\dots\dots\dots (25)$$

so that the work per unit length of crack propagation, G_c, is

$$G_c = \frac{d(1 - V_f)^2}{V_f} \int_0^{\epsilon_f} \sigma_m d\epsilon \cong \frac{d(1 - V_f)^2}{V_f} \sigma_m(u)\epsilon_m(u) \dots\dots\dots\dots (26)$$

To a first approximation U is equal to the product of the ultimate strength of the matrix $\sigma_m(u)$ and the uniform elongation $\epsilon_m(u)$. Figure 20 illustrates experimental data for the effect of fiber diameter on the work of fracture (taken as the area under the load-elongation curve for notched bend specimens) for tungsten-copper composites, with $V_f = 0.53$ and $U = 10^4$ in. No./in.^{2} The improvement in toughness associated with increased fiber diameter is clearly apparent. The agreement between the experimental data and Eq 26 is fairly good at fiber diameters larger than 0.02 in.,

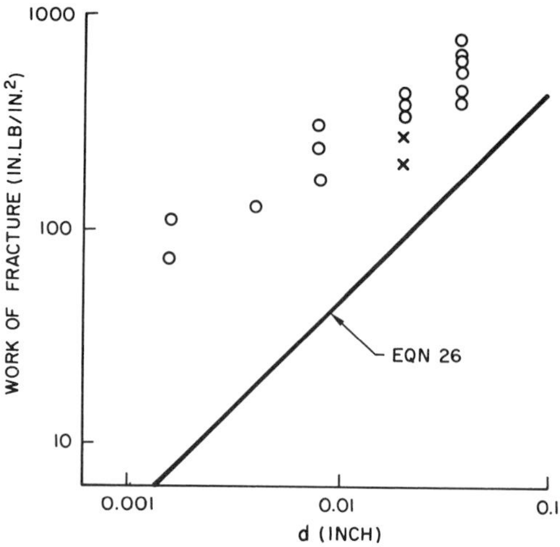

FIG. 20—Effect of fiber diameter on the work fracture for copper-tungsten composites. o-calculated from area on load-elongation curve; x-measured by fracture mechanics approach [47]. The theoretical curve predicted by Eq 26 also is shown.

particularly for the two data points obtained by fracture mechanics type tests as compared with measurements of area under the load-elongation curve. The latter measurements will include the work due to hysteresis prior to unstable fracture [47] and could account for the discrepancy between the theoretical model and experiment. There are several points of practical interest with respect to Eq 26. First, since G_c decreases with increasing V_f (given matrix, constant value of d), the notched fracture strength σ_F will also decrease with V_f according to Eq 7

$$\sigma_F = \sqrt{\frac{E_c G_c}{\pi c}} = \sqrt{\frac{E_f d (1 - V_f)^2 U}{\pi c}} \dotfill (27)$$

where $E_c = V_f E_f$ is the composite modulus and c is the size of pre-existing flaws. Alternatively, the strength of the uncracked composite obeys the rule of mixtures and increases with V_f.

$$\sigma_c = \sigma_m(\epsilon)(1 - V_f) + \bar{\sigma}_f V_f \approx \sigma_f V_f \dotfill (28)$$

for fiber strengths much greater than matrix strength and $l > 10 l_c$. Figure 21 illustrates the effect of fiber content on the cracked (Eq 27) and uncracked (Eq 28) strength of a boron-6061 aluminum composite ($\sigma_f = 330$ ksi, $E_f = 55 \times 10^6$ psi, $d = 4 \times 10^{-3}$ in., $U = 1.2 \times 10^4$ in. No./in.^{3}). Since σ_F decreases with increasing V_f at a given crack size c, while σ_c increases with V_f, for a given c there will exist some critical volume fraction V_f^* below which the composite will fail homogeneously with all the fibers loaded to their average strength ($\sigma_c < \sigma_F$) and above which the composite fails by unstable crack propagation ($\sigma_F < \sigma_c$) before the average strength of a fiber σ_f is achieved homogeneously. Maximum loadcarrying capacity is attained when $V_f = V_f^*$. As the crack size increases, V_f^* decreases, as shown in Fig. 22. Note that even if this method of estimating G_c is too low by an order of magnitude, the predicted value of optimum critical crack size at fiber volume fractions of practical interest (that is, $V_f \cong 0.5$) is still extremely small (0.01 to 0.10 in.). If relatively long cracks exist in composite structures, maximum load carrying capacity will be achieved at relatively low volume fractions of brittle fibers dispersed in a ductile matrix. Also, since conservative design practice requires that one assume that all flaws that cannot be observed by nondestructive inspection are in fact present, improvements in load-carrying capacity

Composite Materials

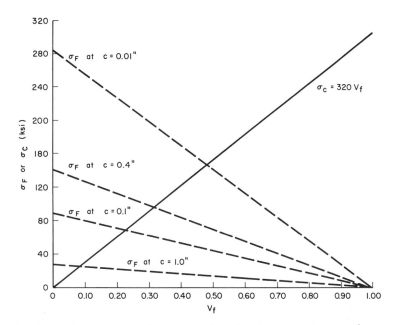

FIG. 21—Effect of volume fraction on the ultimate composite strength σ_c, and predicted fracture strength σ_F for cracks of indicated length in boron-6061 aluminum composite. For given crack size c, maximum load carrying capacity occurs where $\sigma_F = \sigma_c$; at which point $V_f = V_f{}^*$.

(that is, higher V_f^*) are tied closely to improvements in nondestructive inspection which can reduce the observable flaw size.

Second, Eq 26 predicts that G_c (and hence σ_F and V_f^*) decrease with decreasing fiber diameter, and consequently the toughness of composites containing whisker reinforcements may be too low to have adequate fracture toughness for reliable design [47]. However, as d is reduced, σ_f increases (Figs. 8 and 9) and if σ_f is then sufficiently large relative to the shear strength of the matrix or interface, crack propagation may be arrested by splitting as discussed below. In this case, real benefits could occur from refinements in fiber size.

Finally, Eq 26 indicates that composite toughness is directly proportional to matrix toughness. In general, the matrix toughness will be low when the matrix yield strength is very high and when the volume fraction of voids and inclusions is high [1] (low $\epsilon_m(u)$) and when the operating temperature is close to the melting temperature (low $\sigma_m(u)$). Consequently, while unnotched composite strength σ_c can be increased if high yield strength matrices are utilized, (Eq 28) this gain can be more than offset by the decrease in σ_F and V_f^* resulting from decreases in G_c, particularly when large cracks are (or may be) present. Similarly, processing techniques that introduce a large amount of porosity in the matrix can have particularly deleterious effects on toughness and load carrying capacity if interfacial splitting does not occur.

SYSTEMS CONTAINING WEAK INTERFACES OR WEAK MATRICES

When the fiber-matrix interface or the matrix is weak compared with the fiber strength σ_f (for example, in fiberglass), the transverse tensile stresses and shear stresses at the advancing crack tip may be able to cause longitudinal splitting through the formation and propagation of a longitudinal crack, as shown schemati-

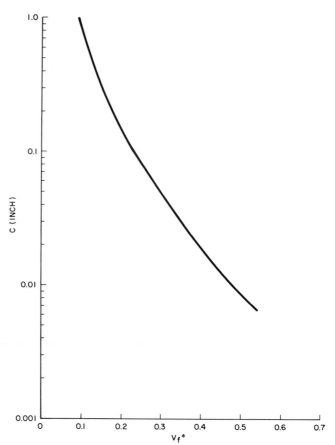

Fig. 22—Effect of crack length on V_f^*, for boron-6061 aluminum composite.

cally in Fig. 23, and in practice [47] for thin sheets of copper-silica (Fig. 24). The splitting action can blunt out completely the advancing crack, when the split extends over large distances, and this renders the material essentially insensitive to the propagation of cracks in planes that are perpendicular to the fibers. The criteria for splitting are not well understood, and are complicated by our lack of quantitative understanding of the triaxial stress and strain distribution at the

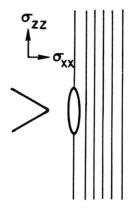

Fig. 23—Schematic drawing of splitting ahead of an advancing crack.

Courtesy of G. Cooper [47].

Fig. 24—Experimental evidence for splitting in a copper-silica composite.

tip of the propagating crack in a composite material, and the effect of triaxiality on interface, matrix, and fiber failure. For completely elastic solids, it has been shown [48] that the transverse tensile stresses σ_{xx} are about one fifth the longitudinal tensile stresses at the crack tip, so that if the fiber is less than five times stronger (in tension) than the matrix or interface, splitting should occur. In real (elastic-plastic) solids, this ratio appears to be too low. From measurements of the effect of thickness on transverse (interface) strength $\sigma_i(f)$ of relatively thin sheets, Cooper and Kelly [47] showed that $\sigma_f > 13.2 \ \sigma_i(f)$ for splitting to occur. Even this estimate may be too low for thick plates where full triaxiality probably would favor fiber fracture rather than splitting. In any case, it is apparent that increases in fiber strength (for example, by reducing fiber diameter or flaw distribution) could lead to large increases in toughness by promoting splitting at the crack tip. Alternatively, splitting is favored if the interfacial strength is reduced. This is acceptable for composites that are loaded uniaxially. However, low off-axis loads could produce shear or tensile interface failure (see Eq 21) if $\tau_m(f)$ or $\tau_i(f)$ were too low, and nullify any advantages gained from improved longitudinal crack propagation resistance in many structural applications.

In composites containing discontinuous fibers of length $l \leq l_c$, fiber fracture cannot occur near the tip of an advancing crack, as even the large plastic strains extending perpendicular to the crack plane will not be able to transfer sufficient load to break the fibers. Crack propagation then requires that the fibers be pulled out of the matrix [10,17] to a distance of the order of $l_c/2$, at which point the remaining bridges of matrix cannot support the load that is transferred to them and fail by local instability.

If the shear stress τ is maintained during pullout (that is, stable process), then the work W' done in pulling N fibers of radius r out to an average distance l' perpendicular to the fracture plane is

$$W' = N\pi r^2 \int_0^{l'} \sigma_{zz} \, dz = N\pi r^2 \int_0^{l'} \frac{2\tau z \, dz}{r} = \pi r \tau (l')^2 \ldots \ldots \ldots \ldots (29)$$

Since l' decreases from $l_c/2$ to zero as the fiber is being pulled out, the average work W done in pulling out N fibers to a (maximum) distance $(l_c/2)$ is

$$W = N \frac{\int_0^{l_c/2} \pi r \tau (l')^2 \, dl'}{\int_0^{l_c/2} dl'} = N \frac{\pi r \tau}{3} (l_c/2)^2 \dots\dots\dots\dots\dots (30)$$

and hence the toughness G_c associated with the pullout of a volume fraction of fibers $V_f = N\pi r^2$ is [10]

$$G_c = \frac{V_f \tau}{3r} (l_c/2)^2 = \frac{V_f}{12} \sigma_f l_c = \frac{V_f}{24} \frac{d}{\tau} \sigma_f^2 \dots\dots\dots\dots\dots (31)$$

Physically, Eq 31 indicates that in contrast to the case of crack propagation due to fiber cracking the fracture toughness (and hence σ_F) increases with increasing fiber content. Also, G_c increases with increasing fiber diameter if σ_f is independent of d. However, if $\sigma_f = (\kappa/d)^n$, where $n > 0.5$, then G_c will increase as the fiber diameter is decreased. Since σ_f will, in general, decrease more slowly with increasing temperature than $\tau \leq \sigma_m(u)/2$, fracture toughness should increase with increasing temperature as the fracture mode changes from fiber fracture to fiber pullout. A comparison of Eqs 26 and 31 indicates that at $\epsilon_m(u) = 0.5$, and $V_f = 0.5$, the toughness will be increased by a factor of $(\sigma_f/\sigma_m(u))^2$ which can be considerable (that is, a factor of 30 to 50). Similar relations apply for fiber pullout in resin systems, except that all shear forces are transmitted by friction and then $\tau = \mu p$, where μ is the coefficient of friction between fiber and matrix and p is the normal pressure of the resin on the fibers resulting from shrinkage during curing [17].

SYSTEMS CONTAINING DUCTILE METAL FIBERS

Although not as attractive from the point of view of stiffness/weight ratio, composites containing cold-drawn metal wires imbedded in a metal or resin matrix (for example, stainless steel in aluminum) are of interest where high strengths are of prime interest. Stainless steel wires having fracture strengths between 450 and 600 ksi have been prepared [28,29]. Furthermore, these wires neck down considerably before they fracture, and consequently metal fiber composites can have extremely high values of toughness, even at high unnotched strength levels. In fact, as shown in Fig. 25, the displacement associated with the fracture of the stainless steel wire filament exceeds that required to fracture the 2024 aluminum matrix [49]; since the load carrying capacity of the wires at fracture is approximately 6 times that of the matrix, it is not surprising to note that the fracture toughness increases with increasing fiber content in this system (Fig. 26), and consequently the ratio of K_c/σ_Y is considerably higher than that of conventional alloy steels at comparable yield strength to density ratios.

The high degree of toughness and yield strength that can be obtained in these systems results from the fact that the wires are able to deform to practically 100 percent reduction of area, even though their flow stress is of the order of 500 ksi. Unfortunately, this behavior generally is not obtained in bulk sections of similar material. (If it were possible, there would be no advantage to be gained from fabricating metal filament composites.) The high degree of local ductility and toughness that can be achieved in the wires as compared with the bulk material probably results from two effects. First, the wires are able to split away from the surrounding matrix (Fig. 25) and deform under essentially plane stress conditions. The reduction of tri-axial constraint considerably hinders void formation and coalescence and greatly increases ductility [5]. Secondly, the wires themselves exhibit extremely anisotropic properties by virtue of the conditions under which they were processed. The large drawing strains tend to refine and align harmful inclusions parallel to the fiber axis, reducing their strain concentration factor perpendicular to the fiber

498

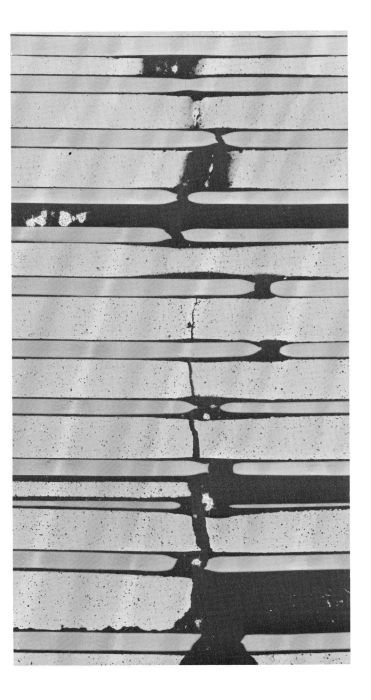

Courtesy of W. Gerberich [49].

Fig. 25—Cracking of 2024 aluminum matrix and extensive drawing down of high strength stainless steel ahead of an advancing crack.

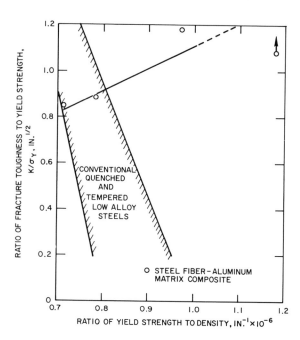

FIG. 26—Effect of yield strength to density ratio on the toughness to yield strength ratio of conventional quenched and tempered low-alloy steels and stainless steel-2024 aluminum composites [49].

axis. It generally has not been practical to process bulk material in a similar manner to achieve the required degree of anistropy. (The high strength and anisotropic toughness of ausformed steel results primarily from large scale splitting along prior austenite grain boundaries, similar to that described in Refs 2 and 50.)

It has been observed that the wires undergo heavy deformation over a distance x on either side of the fracture plane that is of the order of the fiber diameter d. Consequently, the volume of material undergoing heavy deformation per unit area of crack propagation is $(2\ dV_f)$, and hence the work done in breaking the fibers is

$$G_f = 2\ dV_f \int_0^{\epsilon_f} \sigma_f\ d\epsilon \cong 2\ dV_f \sigma_f(u)\epsilon_f(u) \dots\dots\dots\dots (32)$$

where $\sigma_f(u)$ and $\epsilon_f(u)$ are the ultimate strength and uniform elongation of the fibers. If the matrix makes a substantial contribution to the work of fracture (which is not the case with high yield strength matrices), then G_c will be the sum of G_f and the G_c value given by Eq 26.

References

[1] Tetelman, A. S. and McEvily, A. J., *Fracture of Structural Materials*, Wiley, New York, 1967.

[2] Kaechele, L. and Tetelman, A. S., "A Statistical Investigation of Microcrack Formation," *Acta Metallurgica*, Vol. 17, 1969.

[3] Bluhm, J. I. and Morrissey, R. J., "Fracture in a Tensile Specimen," *International Conference on Fracture*, Sendai, Japan, D-11, 73, 1965.

[4] McClintock, F. A., *Transactions*, American Society of Mechanical

Engineers, 1958, p. 582.

[5] McClintock, F. A., *Journal of Applied Mechanics*, Paper No. APN-14, 1968.

[6] McClintock, F. A. and Irwin, G. R., "Plasticity Aspects of Fracture Mechanics," *Fracture Toughness Testing and Its Application*, ASTM STP 381, American Society for Testing and Materials, 1965, p. 84.

[7] Goodier, J. N. and Field, F. A., *Fracture of Solids*, Interscience, New York, 1965, p. 103.

[8] Cottrell, A. H., "Strong Solids," *Proceedings of the Royal Society*, Vol. A282, No. 2, 1964.

[9] Wells, A. A., *British Welding Journal*, Vol. 855, 1963.

[10] Cottrell, A. H., "Mechanics of Fracture in Large Structures," *Proceedings of the Royal Society*, Vol. A285, No. 10, 1965.

[11] Paris, P. C. and Sih, G. C., "Stress Analysis of Cracks," *Fracture Toughness Testing and Its Applications*, ASTM STP 381, American Society for Testing and Materials, p. 30.

[12] Kelly, A. and Davies, G. J., "The Principles of Fibre Reinforcement of Metals," *Metallurgical Review*, Vol. 10, No. 2, 1965.

[13] Kelly, A. and Tyson, W. R., *High Strength Materials*, Wiley, New York, 1965, p. 578.

[14] Broutman, L. J. and Krock, R. H., *Modern Composite Materials*, Addison-Wesley, Reading, Mass., 1967.

[15] McDanels, D. L., Jeck, R. W., and Weeton, J. W., "Analysis of Stress-Strain Behavior of Tungsten-Fiber-Reinforced Copper Composites," *Transactions*, American Institute of Mining, Metallurgical, and Petroleum Engineers, Vol. 233, 1965, p. 636.

[16] Schuster, D. M. and Scala, E., *Fundamental Aspects of Fiber Reinforced Plastic Composites*, Interscience, New York, 1968, p. 45.

[17] Tyson, W. R. and Davies, G. J., "A Photoelastic Study of Shear Stresses Associated with Transfer of Stress During Fibre Reinforcement," *British Journal of Applied Physics*, Vol. 16, No. 199, 1965.

[18] Cox, H. L., "Elasticity and Strength of Paper and Other Fibrous Materials," *British Journal of Applied Physics*, Vol. 3, No. 72, 1952.

[19] Dow, N. F., "Study of Stress near a Discontinuity in a Filament Reinforced Composite Metal," General Electric Report TISR63SD61, 1963.

[20] Hedgepeth, J. M. and Van Dyke, P., "Local Stress Concentrations in Imperfect Filamentary Composite Materials," *Journal of Composite Materials*, Vol. 7, No. 294, 1967.

[21] Prosen, S. P., *Fiber Composite Materials*, American Society for Metals, Cleveland, 1965, p. 157.

[22] Metcalfe, A. G. and Schmitz, G. K., *Proceedings*, American Society for Testing and Materials. Vol. 64, 1964, p. 1073.

[23] Rosen, B. W., "Tensile Failure of Fibrous Composites," *Journal* American Institute of Aeronautics and Astronautics, Vol. 2, 1964, p. 1985.

[24] Rosen, B. W., *Fiber Composite Materials*, American Society for Metals, Cleveland, 1965, p. 37.

[25] Metcalfe, A. G., "Influence of Interaction on Fracture of Metal Matrix-Filament Composites," presented at Monsanto *Symposium on High Performance Composites*, Oct. 1967.

[26] Brenner, S. S., *Fiber Composite Materials*, American Society for Metals, Cleveland, 1965, p. 11.

[27] Evans, C. C., Gordon, J. E., and Marsh D. M., "A Mechanism for the Control of Crack Propagation in All-Brittle Systems," *Proceedings of the Royal Society*, Vol. A282, No. 518, 1964.

[28] Roberts, D. A., "Physical and Mechanical Properties of Some High-Strength Fine Wires," DMIC Memo 80, Defense Metals Information Center, 1961.

[29] Chandhok, V. K., Kasak, A., and Wochtmeister, G. C., *Proceedings of the Royal Society*, Vol. A282, No. 518, 1964.

[30] Farrell, K. and Parikh, N. M., IIT TR (R1-B241), Illinois Institute of Technology Research Institute, Chicago, 1963.

[31] Baskey, R. H., "Fiber Reinforcement of Metallic and Nonmetallic Composites: Phase I—State of Art and Bibliography of Fiber Metallurgy," Final Report Contract AF33(457)-7139.

[32] Petrasek, D. W., and Weeton, J. W., "Effects of Alloying on Room-Temperature Tensile Properties of Tungsten-Fibre-Reinforced-Copper-Alloy Composite," *Transactions American Institute of Mining, Metallurgical, and Petroleum Engineers*, Vol. 230, No. 977, 1964.

[33] Jackson, P. W. and Marjoram, J. R. "Recrystallization of Nickel Coated Carbon Fibres," *Nature*, Vol. 218, No. 83, 1968.

[34] Jones, B. H. and Noyes, J. V., Douglas Aircraft Co. Paper 4706 presented at Monsanto *Symposium on High Performance Composites,* Oct. 1967.

[35] Ishai, O., "The Effect of Temperature on the Delayed Yield and Failure of 'Plasticized' Epoxy Resin," Monsanto Research Corporation Technical Report, HPC-68-59, Sept. 1968.

[36] Corten, H. T., *Fundamental Aspects of Fiber Reinforced Plastic Composites,* Interscience, New York, 1968, p. 89.

[37] Fried, N., "Response of Orthogonal Filament Wound Materials to Compressive Stress," *Proceedings of the 20th Conference,* Reinforced Plastics Division, Society of the Plastics Industry, 1965, Section 1C.

[38] Ishai, O., Anderson, R. M., and Lavengood, R. E., "Failure-Time Characteristics of Continuous Unidirectional Glass-Epoxy in Flexure Composites" Section 3 of Monsanto Research Corporation Technical Report HPC-67-54, Dec. 1967.

[39] Tyson, W. R. and Kelley, A., "Tensile Properties of Fibre-Reinforced Metals: Copper/Tungsten and Copper/Molybdenum," *Journal of the Mechanics and Physics of Solids,* Vol. 13, No. 329, 1965.

[40] Mullin, J., Berry, J. M., and Gatti, A., "Some Fundamental Fracture Mechanisms Applicable to Advanced Filament Reinforced Composites," *Journal of Composite Materials,* Vol. 2, No. 82, 1968.

[41] McGarry, F. J., *Fundamental Aspects of Fiber Reinforced Plastic Composites,* Interscience, New York, 1968, p. 63.

[42] Tetelman, A. S., "The Effect of Plastic Strain and Temperature on Crack Propagation in Iron-3 % Silicon," *Acta Metallurgica,* Vol. 12, No. 993, 1964.

[43] Steigerwald, E. A., "Influence of Warm Prestressing on Notch Properties of Several High-Strength Alloys," *Transactions,* American Society for Metals, Vol. 54, No. 445, 1964.

[44] Darwish, F., to be published.

[45] Salkind, M. J. and George, F. D., "Investigation of the Impact Resistance of Al_3Ni Whisker Reinforced Aluminum Prepared by Unidirectional Solidification," AFML TR-66-234, Air Force Materials Laboratory, 1966.

[46] Wu, E. M., TAM Report No. 248, NRL Contract No. NONR 2947 (02)x, University of Illinois, Urbana, Ill., 1963.

[47] Cooper, G. A. and Kelly, A., "Tensile Properties of Fibre-Reinforced Metals: Fracture Mechanics," *Journal of the Mechanics and Physics of Solids,* Vol. 15, No. 279, 1967.

[48] Cook, J. and Gordon, J. E., "A Mechanism for the Control of Crack Propagation in All-Brittle Systems," *Proceedings of the Royal Society,* Vol. A282, No. 508, 1964.

[49] Gerberich, W. W., presented at University of California at Berkeley, Course on Composite Materials, June 1968.

[50] McEvily, A. J. and Bush, R. H., *Transactions,* American Society of Mining, Metallurgical, and Petroleum Engineers, Vol. 55, No. 654, 1962.

[51] Nielson, L. E., "Crosslinking-Effect on Physical Properties of Polymers," Monsanto Research Corporation TR-HPC-68-57, May 1968.

[52] Sutton, W. H. and Feingold, E., *Materials Science Research,* Vol. 3, 1966, Chapter 31.

Crack Phenomena in Cross-Linked Glassy Polymers

F. J. McGARRY[1] AND
J. N. SULTAN[1]

Reference:
McGarry F. J. and Sultan, J. N., "Crack Phenomena in Cross-Linked Glassy Polymers," *Composite Materials: Testing and Design, ASTM STP 460,* American Society for Testing and Materials 1969, pp. 503–511.

Abstract:
Previous work established that tensile stress cycling of fibrous glass reinforced resin composites degrades their macroscopic mechanical properties by inducing microcracks in the cross-linked glassy polymer matrix. While the greatest damage occurs in the first stress cycle, it continues to accumulate during subsequent loadings, and composite modulus reductions of 20 to 25 percent after 100 to 200 cycles are common. Detailed studies of unreinforced matrices indicate that much of the work required to drive a crack is consumed by molecular cold drawing and partial orientation in the glassy material at the crack tip. Further, this work can be increased substantially by suspending small particles of certain elastomers throughout the cross-linked matrix material; increases in the fracture surface work term of an order of magnitude, or more, are possible. When such toughened epoxies or polyesters are reinforced with fibrous glass and then stress cycled, the internal microcracking previously observed in the composite material is eliminated.

The degree of toughening conferred by the elastomer particles is related to their average size, the distribution of sizes, their composition, the composition of the matrix, the volume fraction of elastomer present, and the adhesion between particles and matrix. Recent work has revealed that other, additional factors are important. These latter include the initial molecular weight of the elastomer, the nature of elastomeric copolymers which can be used, the conditions of plane stress or plane strain under which toughness measurements are made, the detailed nature of the catalyst used to harden the glassy matrix, and the morphological features of the particles. All of these factors will be described and their significance discussed.

Key Words:
composites, cracking (fracturing), glass fibers, fiber composites, mechanical properties, microcracks, reinforced plastics

When a glassy cross-linked polymer matrix is reinforced with stiff fibers, stress concentrations in it are unavoidable. These arise because the physical and mechanical properties of the matrix and the reinforcement differ. A measure of the severity of such concentrations has been given by Kies [1][2] who analyzed several composite models; one of his findings showed that the local matrix strain between two adjacent fibers is inversely proportional to their spacing and directly proportional to the ratio of fiber modulus to matrix modulus. This means that local matrix strains can be 15 to 30 times greater than the ones experienced macroscopically by the composite material. Unless the matrix possesses at least microductility, it will fail cohesively when the composite has been loaded to only a small fraction of its ultimate capacity.

---

[1] Professor and research assistant, respectively, Department of Civil Engineering, Massachusetts Institute of Technology, Cambridge, Mass. 02139. Mr. McGarry is a personal member ASTM.

[2] The italic numbers in brackets refer to the list of references appended to this paper.

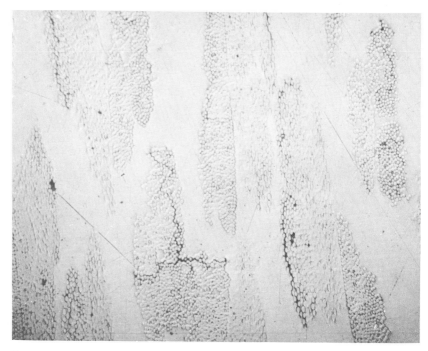

Fig. 1—Micrograph of fiber glass cloth (Style 181 fabric) reinforced epoxy resin composite. Taken from specimen which had undergone 100 cycles of pull-release tensile loading to 75 percent of ultimate tensile strength. Loading parallel to plane of figure. Note microcracks in regions of high fiber density.

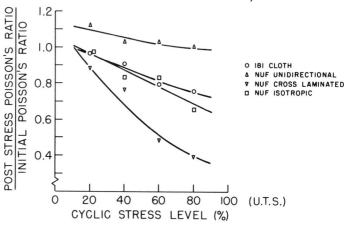

Fig. 2—Reduction in Poisson's ratio after one cycle of tensile loading to various levels of ultimate tensile strength. Epoxy resin based composites reinforced with woven (181) or nonwoven unidirectional fiber (NUF) glass fabrics.

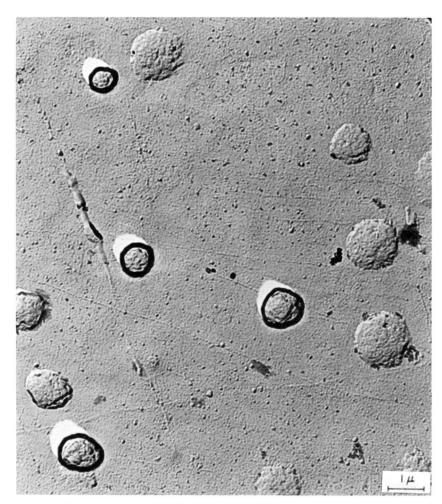

FIG. 3—Electron micrograph of cured epoxy-rubber system showing precipitated particles which induce crack toughening. Initial elastomer concentration: 10 pph. System cured at 100 C.

In fibrous glass reinforced epoxies and polyesters, such is the reason for their poor resistance to mechanical and thermal fatigue: microcracks form in high fiber density regions, propagate along fiber axis directions, and partially decouple the components in the composite (Fig. 1). Stiffness, strength, and chemical resistance all decline [2,3] (Fig. 2). It is also the reason why chemically resistant piping made of these materials cannot be pressurized for extended service at more than a small fraction of its shorttime burst strength; the microcracks which will otherwise form permit the corrosive contents to leak (weeping) and to directly attack the reinforcing glass fibers. If the latter occurs, gross failure of the pipe is likely.

THE RESIN MATRIX

Referring now only to the matrix material containing no fibers, when it undergoes brittle fracture a complicated series of actions takes place. Despite the interconnected or three-dimensional nature of its molecular network, apparently a large amount of microplastic flow occurs locally at the crack tip. This drastically rearranges the molecular structure there before finally disrupting it completely. The evidence is quite conclusive: the energy required to propagate a brittle crack through such systems is 1 to 3 orders of magnitude greater than that needed simply to pull apart the primary bonds in the chains [4]. Birefringence studies of fracture surfaces support this orientation inference also [5].

The same phenomena occur when glassy thermoplastics break, and these have been studied in detail by several investigators [6,7,8]. Here the discrepancy between theroretical and experimental values of fracture surface energy is even greater: 3 to 4 orders of magnitude are quite common. The oriented layers on the fracture surfaces are thicker (3000 to 5000 Å), and the voids associated with craze structure on the surfaces can be observed. Apparently the absence of chemical cross-links creates the potential for greater local flow in the thermoplastics, when they are in the glassy state.

Thermoplastics containing rubber particles, the impact resistant glasses, take great practical advantage of this potential. The particles induce flow, orientation, and crazing throughout a large fraction of all the glassy polymer phase which is present, in contrast to the crack orienting just the fracture surfaces when the particles are absent [9]. The macroscopic result is an enormous increase in resistance to

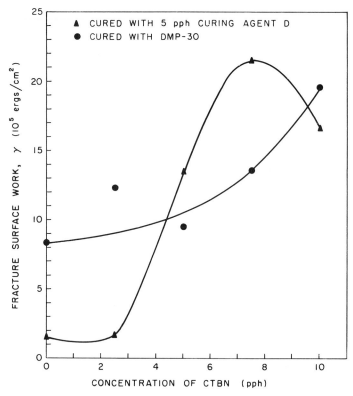

FIG. 4—Variation of fracture surface work of Epon 828 epoxy with initial concentration of added elastomer (CTBN). Toughening effect is sensitive to curing agent used.

Composite Materials

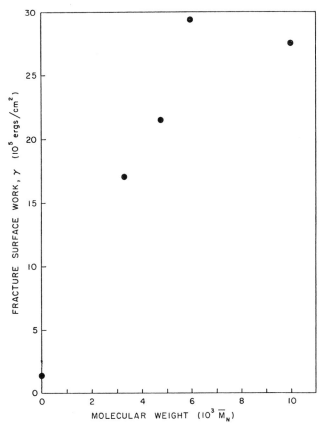

FIG. 5—Variation of fracture surface work of Epon 828 epoxy, cured with Agent D, with initial molecular weight of butadiene/acrylonitrile copolymer. Initial rubber concentration: 10 pph.

mechanical abuse, or capacity to absorb mechanical energy prior to gross cohesive fracture. They are very tough materials.

TOUGHENED CROSS-LINKED GLASSY POLYMERS

Recent research suggests the same toughening modification can be used with cross-linked glassy polymers to reduce or eliminate the microcracking they experience when reinforced with stiff fibers. To date, most of the work has used low molecular weight, butadiene-acrylonitrile copolymers for the rubbery phase[3] [10]. These liquid polymers are initially soluble in many liquid epoxies and polyesters but then precipitate out as a second phase when the surrounding matrix is cured in the normal fashion; the morphology of a typical product is shown in Fig. 3. As the content of added elastomer is increased, the work required to propagate a crack through the cured epoxy or polyester increases also: one order of magnitude is obtained easily, and, in some systems, two orders have been achieved. Typical data are shown in Fig. 4.

As might be expected, the precise degree of toughening depends upon a number of factors. Among these are the cross-link density in the glassy matrix, the size of the second phase particles, their composition, the concentration of them, and the degree of adhesion between them and the matrix. Some of the parameters

[3] Hycar CTBN, B. F. Goodrich Chemical Co., Cleveland, Ohio.

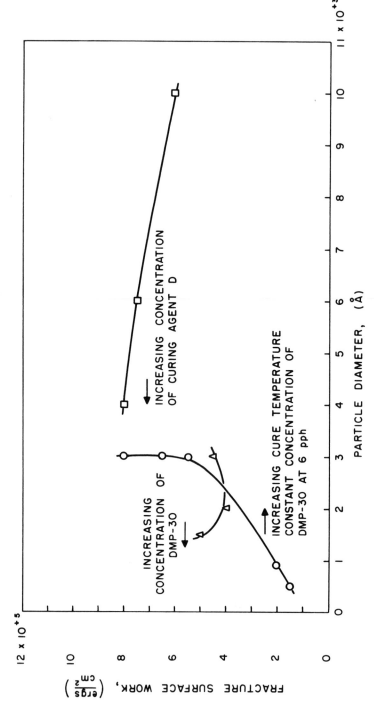

FIG. 6—Parameters influencing particle size and fracture toughness of CTBN/Epon 828 system. Initial rubber concentration 10 pph throughout.

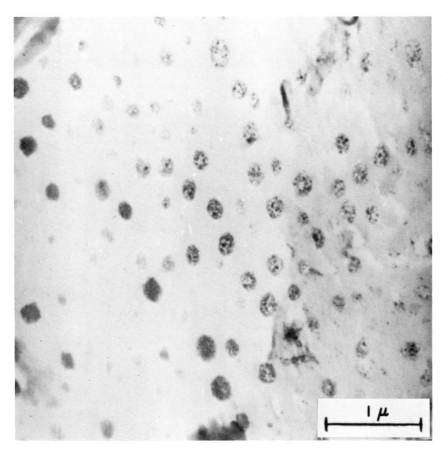

FIG. 7—Transmission electron micrograph of OsO₄ stained ultramicrotomed specimen: 10-pph CTBN elastomer added to Epon 828 epoxy and cured with 4-pph DMP-30 agent.

influencing these factors, in turn, are the curing temperature, the curing agent identity and concentration, the initial molecular weight of the rubber copolymer, its concentration, and its initial composition. The variations produced by some of these are shown in Figs. 5 and 6.

There is quite good evidence that the precipitated particles are not pure elastomer. Figure 7 is a transmission electron micrograph through a thin section of toughened epoxy which had been exposed to osmium tetroxide vapor for an extended period of time. The osmium preferentially deposits in the rubber-rich regions of the particles where it impedes transmission of the electrons; the variations in the micrograph in these regions are distinct, leading to the conclusion that a finite amount of epoxy is present in the particles also. Reaction rate studies and transition temperature measurements also lead to the same conclusion.

TOUGHENED MATRIX COMPOSITES

The initial goal, to reduce microcracking in stress-cycled fibrous glass composites, can be realized with such toughened matrices, Fig. 8. No sacrifice of composite tensile strength is incurred, and, in fact, often a modest improvement can be mea-

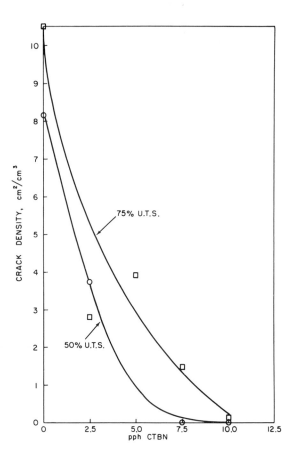

FIG. 8—Internal fracture surface area density in elastomer modified polyester matrix fibrous glass reinforced composites. Specimens tensile cycled 100 times to either 50 or 75 percent of ultimate tensile strength. When polyester matrix toughened with 7.5 to 10 percent added CTBN elastomer, microcracking in composites virtually eliminated.

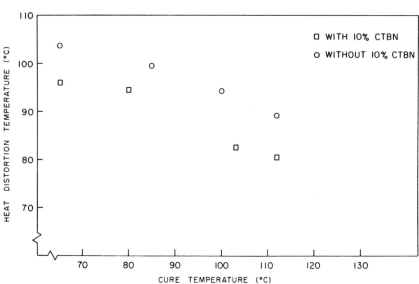

FIG. 9—Heat distortion temperature versus cure temperature, Epon 828 + 6 percent DMP30, with and without 10 percent CTBN.

sured. The elevated temperature resistance of the matrix is reduced somewhat, Fig. 9, though the presence of the glass fiber reinforcement frequently offsets this. No studies of changes in chemical resistance with toughening have been done as yet, but probably some losses of this property could result from the presence of the rubber particles.

Several additional points should be mentioned. Enough work has been done with various epoxies, polyesters, phenolics, and other cross-linked matrices to suggest that the principle of elastomer particle toughening is a general one, applicable to all thermosets to varying degrees. Work with other types of rubbers which also produce toughening effects leads to the same conclusion.

The idea can be used to improve the peel strength of metal-to-metal thermoset adhesives; very substantial gains in this property are realized from modest additions of rubber, if the proper microstructure is produced. Finally, it should be possible to formulate thermoset analogs to the impact resistant thermplastics by this same technique, if enough additional research and development effort is expended. Such plastics would be useful in many applications.

References

[1] Kies, J. A., "Maximum Strains in the Resin of Fiberglass Composites," NRL Report 5752, U. S. Naval Research Laboratory, 26 March 1962.

[2] McGarry, F. J. and Willner, A. M., "Microcracking in Fibrous Glass Reinforced Resin Composites," *Polymer Preprints*, American Chemical Society, Vol. 8, No. 2, Chicago, Sept. 1967, pp. 1056–1063].

[3] Broutman, L. J. and Sahu, S., "Progressive Damage of a Glass Reinforced Plastic During Fatigue," *Proceedings*, 24th Annual Meeting, Reinforced Plastics Division, Society of the Plastics Industry, 1969.

[4] McGarry, F. J. and Selfridge, Jr., "Fracture Surface Work of Modified Polystyrenes and Certain Crosslinked Polymers," *Proceedings*, 145th National Meeting of the American Chemical Society, New York, Sept. 1963.

[5] Kambour, R. P., "Refractive Index and Composition of Poly (methyl Methacrylate) Fracture Surface Layers," *Journal of Polymer Science*, A, Vol. 2, 1964, pp. 4165–4168.

[6] Berry, J. P., "Fracture In Glassy Polymers," *Nature*, Vol. 185, No. 91, 1960.

[7] Broutman, L. J. and McGarry, F. J., "Fracture Surface Work Measurement on Glassy Polymers by a Cleavage Technique, I and II," *Journal of Applied Polymer Science*, Vol. 9, No. 2, Feb. 1965, pp. 589–626.

[8] Kambour, R. P., "Mechanism of Fracture in Glassy Polymers. I. Fracture Surfaces in Polymethyl Methacrylate," *Journal of Polymer Science*, A. Vol. 3, 1965, pp. 1713–1724.

[9] Bucknall, C. B. and Street, D. G., "The Fracture Behavior of Polyblends Over a Range of Temperatures," *Advances in Polymer Science*, Technical Monograph 26, 1967.

[10] McGarry, F. J., "Toughening Crosslinked Glassy Polymers," *Proceedings*, International Symposium on Macromolecular Chemistry, International Union of Pure and Applied Chemistry, Toronto, Ont., 6 Sept. 1968.

Fractography of Aluminum-Boron Composites

R. C. JONES[1]

Reference:

Jones, R. C., "Fractography of Aluminum-Boron Composites," *Composite Materials: Testing and Design, ASTM STP 460*, American Society for Testing and Materials, 1969, pp. 512–527.

Abstract:

The basic objective of this study has been to relate the observed behavior of metal matrix composites to the micromechanical behavior of each component, as surveyed at the fracture surface. The principal method of investigation has been scanning electron microscopy.

Scanning electron microscope examination of fracture surfaces of metal matrix composites has been found to be an extremely useful tool in interpreting micromechanical behavior. Application of the scanning microscope to the boron filament-aluminum alloy composite system has allowed detailed examination of the components and their interaction *in situ*. The boron filaments are observed to undergo substantial break-up in the compositing and loading sequence. Incomplete bonding at diffusion bond planes in the matrix metal is found to be a major problem in cross-ply aluminum-boron composites. Debonding of diffusion bond planes, observed at fracture surfaces, has been observed on a large scale in both cross-ply and unidirectionally reinforced composites. It is not yet clear whether such debonding cracks are a contributory cause or an effect of the final fracture process. Potential fracture initiation sites are discussed, and the effect of a triaxial stress state on failure of the matrix material between fibers is presented graphically.

Observations made of fracture surfaces containing large amounts of incomplete bonding or debonding at diffusion bond planes lead to the conclusion that further optimization of fabrication parameters is needed to improve current state-of-the-art aluminum-boron materials.

Key Words:

fractography, fractures (materials), boron, aluminum, reinforced metals, mechanical properties, composite materials, metal matrix, evaluation, tests

It is often possible to determine mechanisms of deformation and fracture through detailed study of the fracture surfaces of materials which have been loaded to failure. The objective of the study reported here has been to relate observed macroscopic behavior of the metal matrix composite system to the micromechanical behavior of each component. The principal source of information on the micromechanics of the components of the aluminum-boron composite *in situ* has been detailed fracture surface examination.

Due to the great depth of field possible, scanning electron microscopy has been employed profitably for examination of the fracture surfaces of composites and of their individual components [1–7].[2] The components of the composites examined in this study were sufficiently electrically conductive that no preparation of fracture surface specimens was needed prior to examination with the scanning electron microscope. Secondary emission electron images were employed to accentuate topographic features. Since relatively low magnifications have been used in this study,

---

[1] Associate professor of civil engineering, Massachusetts Institute of Technology, Cambridge, Mass. 02139. Personal member ASTM.

[2] The italic numbers in brackets refer to the list of references appended to this paper.

the limited resolution of the scanning electron microscope in comparison with current transmission electron microscopes has not been a problem.

Metal matrix composite material studied is state-of-the-art commercially available aluminum boron. Boron filaments used in the fabrication of these composites were 0.004 in. in diameter, containing a tungsten boride core.[3] Prior to compositing their specified strength level was 400,000 psi or greater. Composites were fabricated by a diffusion bonding process, where alternate layers of boron filaments and aluminum alloy foil sheets were hot pressed together. The matrix metal in all cases in the materials studied here was 6061 aluminum alloy.

Since composite materials examined in this study were commercially produced by proprietary processes, exact fabrication parameters cannot be ascertained. The intent of this paper is to report on the state of the art and, through micromechanics observations, to indicate areas of fabrication technology where improvements should be sought. The composite materials examined in this study are thought to be representative of aluminum-boron composites fabricated by the diffusion bonding process described above.

Three different arrangements of boron filaments in the diffusion bonded composites have been studied: unidirectional reinforcement, ±30-deg cross ply, and 0 and 90-deg cross ply. All test specimens examined had been subjected to tension testing to failure at room temperature. The ±30-deg cross-ply material was loaded at 0 deg. Longitudinal and tranverse tension tests are included in the unidirectional and 0/90-deg materials.

In order to introduce the capabilities of scanning electron microscopy for fractography of metal matrix composites and to demonstrate the kind of observations made in this study, a variety of micrographs of fracture surface will be discussed in terms of micromechanical behavior. A more general discussion of common features and their bearing on macroscopic behavior will then be based upon these micrographs.

UNIDIRECTIONALLY REINFORCED MATERIAL

Previous studies have shown that in some metal matrix composite systems diffusion bond joints are prone to debonding as the material is loaded to failure [5,6]. The scanning electron micrographs of fracture surface of an aluminum-boron composite, shown in Fig. 1, indicate that such debonding also may occur in this system.

The composite material shown in Fig. 1 is a 6061 aluminum alloy reinforced with 25 volume percent of boron filaments in a unidirectionally reinforced panel 0.050 in. thick. In this and subsequent micrographs, scale can be determined by noting that the boron filaments are 0.004 in. in diameter. The fracture surface shown resulted from a longitudinal tension test, that is, load applied parallel to the fiber direction, in which the test specimen failed at 100 ksi. Massive debonding is observed, in Fig. 1a, at all diffusion bond planes through the thickness of the specimen. Detailed inspection of the interaction of filaments with the debonding process is made in Fig. 1b. Most filaments appear to have separated from the matrix as the diffusion bond planes have opened up, but one filament shown in Fig. 1b has been split at the apparent continuation of the debonding crack.

Well bonded aluminum-boron composites can be fabricated, as shown in Figs. 2 and 3. This material is 6061 aluminum alloy reinforced by 50 volume percent of boron filaments, fabricated in a 0.020-in.-thick unidirectionally reinforced panel. The room-temperature longitudinal tension test specimen shown here failed at 123 and 187 ksi, respectively. The very low magnification view of the fracture surface in Fig. 2a shows that the fracture path was not nearly planer, but rather left a fracture surface both generally and locally very irregular. A closer view of the

[3] During the fiber production process the original 0.0005 in. tungsten wire becomes a mixture of tungsten borides, primarily W_2B_5 and WB_4.

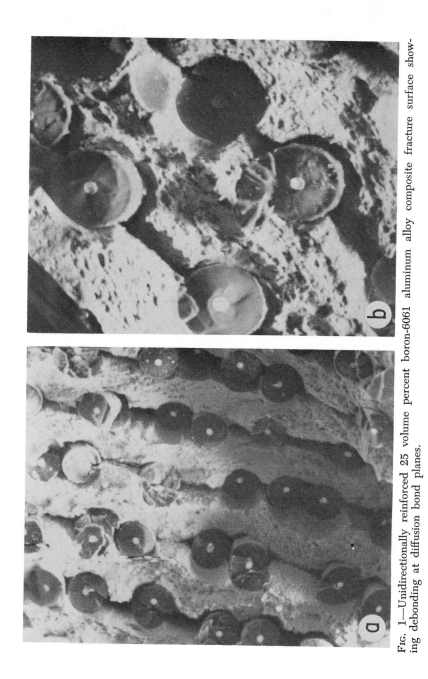

FIG. 1—Unidirectionally reinforced 25 volume percent boron-6061 aluminum alloy composite fracture surface showing debonding at diffusion bond planes.

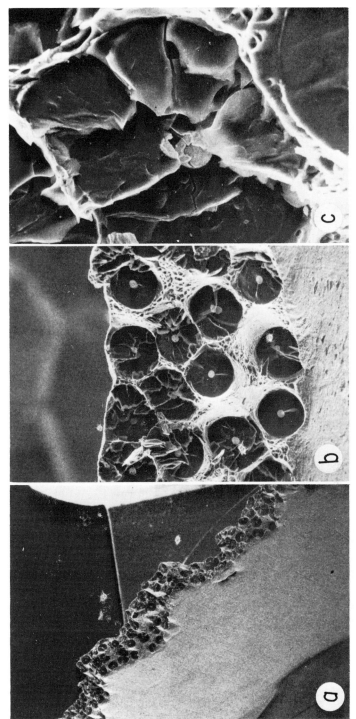

FIG. 2—Fracture surface of 50 volume percent boron-6061 aluminum alloy unidirectionally reinforced composite showing filament damage but no evidence of debonding.

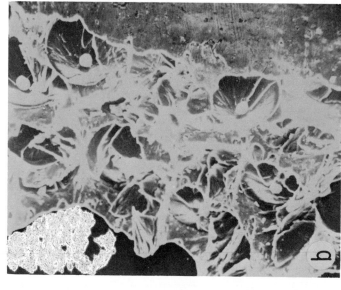

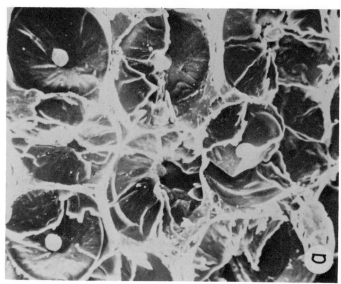

FIG. 3—Unidirectionally reinforced composite tested in tension to 187 ksi. Shear lip at outer edge of composite in inclined view indicates different stress state than that in material between filaments.

516

fracture surface is shown in Fig. 2b. No evidence of debonding at diffusion bond planes is observed. Fibers have fractured near the plane of the matrix fracture surface, indicating that substantial continuity existed between matrix and filament so that the fatal fracture crack was not deflected much between these two components. Compare this behavior with Fig. 5b, where fiber fracture surfaces are well above or below the matrix fracture surface indicating separation of the fracture processes in the aluminum and boron components.

The higher strength 50 volume percent specimen shown in Fig. 3 similarly shows no evidence of debonding at diffusion bond planes and fracture of filaments near the plane of the matrix fracture surface. An interesting feature of the fracture surface shown in Fig. 3b is the presence of shear lips at the outer surfaces of the composite sheet. This phenomenon suggests that the matrix metal in the interior of the composite sheet is more highly constrained against plastic deformation than the material at the surface.

Filament alignment and placement may not be perfect, even in unidirectionally reinforced composites. In Fig. 1a, filaments are seen to be virtually in contact at some points along bonding planes and widely spaced at other points. Similar spacing irregularities can be seen in the higher volume fraction composite shown in Fig. 2b. In this case overly close placement of fibers in local areas has led to substantial filament damage, probably during the hot pressing operation. A detailed view of one of the damaged fibers shown in Fig. 2b is presented in Fig. 2c. Both radial and transverse cracking of the filament are noted. Other filaments seen in Fig. 2b, where they have not been bunched together as tightly, show clean fracture surfaces and little evidence of damage due to fabrication. Filament damage in the 25 volume percent material, Fig. 1, is generally much less than that seen in the 50 volume percent composite shown in Fig. 2.

Additional indication of the high quality of bonding in the 50 volume percent unidirectionally reinforced material can be obtained from examination of the fracture surfaces of a tension specimen loaded transverse to the reinforcement direction. The specimen whose fracture surface is shown in Fig. 4 failed at 14 ksi in a room-temperature test. Many of the fibers can be seen in Fig. 4 to have split near their center, through the tungsten boride core. The fact that these fibers have split rather than pulled out may indicate the presence of substantial matrix-filament bonding in this composite.

Even though the principal tensile stress direction in this transverse test was perpendicular to the axis of the boron filaments, a great amount of longitudinal filament breakup is seen in Fig. 4a. Close inspection of one filament in Fig. 4b shows substantial cracking of the filament perpendicular to the main fracture surface. These cracks extend undeflected through the tungsten boride core, indicating a high degree of continuity between the core and the surrounding boron. The cracks in the filament do not propagate continuously into the adjacent aluminum alloy matrix, however.

CROSS-PLY MATERIAL

Two different configurations of aluminum-boron cross-ply composite material have been examined in this study: ±30 deg with respect to loading axis, and 0/90 deg. In both cases 6061 aluminum alloy and four layers of filaments were fabricated symmetrically into 0.020-in.-thick panels.

The ±30-deg cross-ply composite shown in Fig. 5 had 50 volume percent of filaments, and failed at 86 ski in a room-temperature tension test. Longitudinal breakup of the filaments can be seen in Fig. 5a, particularly in one of the fibers nearly parallel to the fracture surface. Longitudinal layering of several fibers of the interior two rows can be observed also in both Figs. 5a and b.

Major problems at diffusion bond planes are apparent from the examination

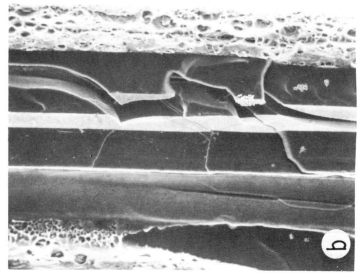

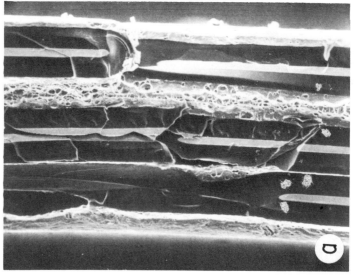

Fig. 4—Fracture surface of transversely loaded unidirectionally reinforced 50 volume percent composite.

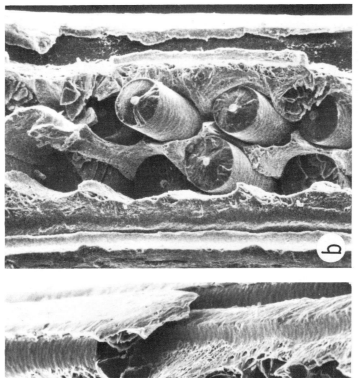

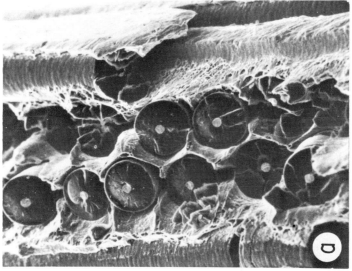

Fig. 5—Views of ±30-deg cross-ply aluminum-boron composite, showing imperfect initial diffusion bonding.

of fracture surfaces of the cross-ply material shown in Fig. 5. The amount and character of separation of diffusion bond planes between filaments in the inner two layers indicates that a diffusion bond between adjacent alunimum alloy foils never was formed. It appears that the matrix metal foils were not deformed enough to be brought into contact in the fabrication process. This observation is strongly reinforced by examination of the troughs left by the outer two layers of reinforcement in Fig. 5b. In both troughs the aluminum alloy matrix metal is seen to have never met to allow for diffusion bonding. In the more severe of these two cases, the aluminum has penetrated into the trough so little that the filament originally there must have been virtually in contact with the filament just below the trough.

The fact that the fibers nearly parallel to the fracture surface in Fig. 5 have pulled out of the matrix, rather than splitting as in Fig. 4, indicates little matrix-filament bonding in this ±30-deg cross-ply material. With little filament to matrix bond and little matrix-matrix contact or bonding at planes where diffusion bonding was to have occurred, this material scarcely seems held together.

The 0/90-deg. cross-ply composite shown in Figs. 6 and 7 had 45 volume percent of filaments. As can be seen by comparing the directions of reinforcement in these two figures, they represent tension tests conducted parallel to each reinforcement axis. From examination of Figs. 6a and b it appears that some diffusion bonding of the innermost matrix foils had occurred during fabrication and that debonding developed subsequently. On the same fracture surface, shown in more detail in Fig. 6c, can be seen areas where bonding never could have occurred because the aluminum foils had not been pressed into contact with each other. The two fibers perpendicular to the fracture surface have no aluminum between them. The two fibers parallel to the fracture surface in Fig. 6c can be seen to have had some matrix metal between them, but only pushed part way toward meeting to form a diffusion bonded joint.

Similar imperfect formation of diffusion bonded joints between original matrix foils is seen in Figs. 7a and b. Fibers perpendicular to the fracture surface are held simply in a crevice formed where the surrounding foils have not been pressed into contact with one another. The fibers parallel to the fracture surface illustrate the same phenomenon, with a long crevice indicating a not quite joined diffusion bond at one fiber location, and two fibers essentially in full contact in Fig. 7b with no aluminum between them.

The fibers in the two kinds of cross-ply material, shown in Figs. 5 through 7, are on average not as badly damaged as those in the well-bonded unidirectionally reinforced material shown in Fig. 2. There are exceptions, however, such as the set of fibers from 0/90-deg cross-ply material shown in Fig. 7c. Apparently these filaments were pressed together during fabrication with no matrix material between them.

MATRIX METAL

It has been hypothesized that a high degree of triaxiality may exist in the matrix metal of these composites, particularly between closely spaced fibers at high deformations [5]. Detailed examination of the aluminum alloy observable at the fracture surface in several specimens studied here revealed a cellular nature to the matrix metal in such constrained locations. A series of micrographs of such an area is shown in Fig. 8, where the central area of each photo is magnified successively in the next photo. This failure mode, involving microvoid formation and growth, is typical of ductile failure in metals. The observation that the matrix metal fails in a ductile mode does not indicate an absence of triaxial constraint in the matrix, but does indicate that any such constraint is not sufficient to lead to cleavage fracture.

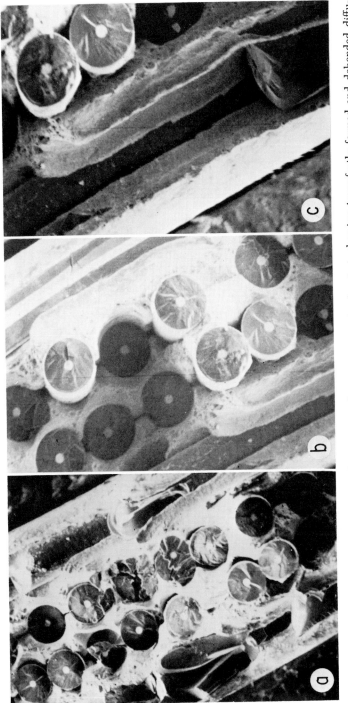

FIG. 6—Fracture surface of 0/90-deg cross-ply composite from transverse tension test, showing imperfectly formed and debonded diffusion bond planes.

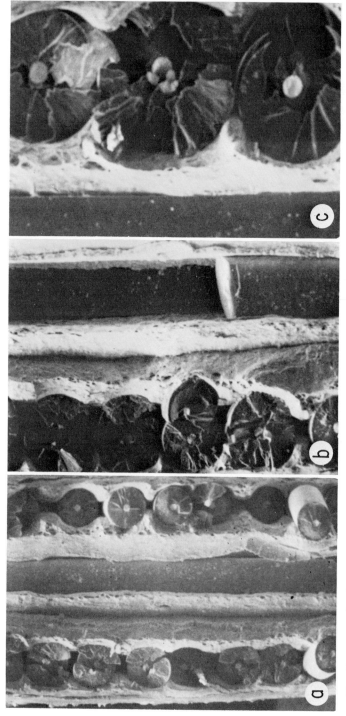

Fig. 7—Longitudinal tension test specimen of 0/90-deg cross-ply composite, with imperfectly bonded areas and some filament damage.

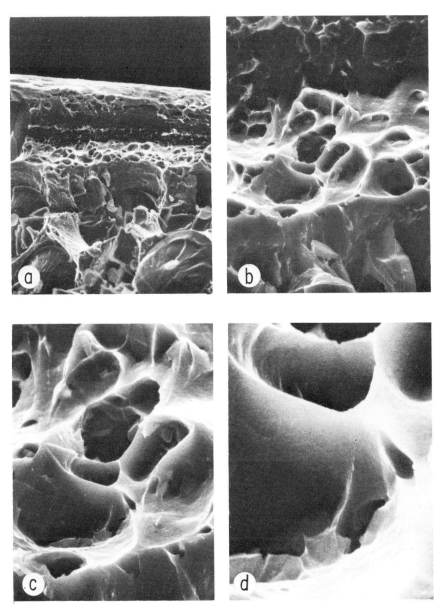

FIG. 8—Views of deformed matrix metal between filaments at fracture surface: (a) ×350, (b) ×1400, (c) ×2800, and (d) ×7000.

Matrix metal failure surfaces shown in Fig. 9 further illustrate the observed ductile mode of failure. The specimen examined here in some detail is the same 50 volume percent unidirectionally reinforced specimen previously shown in Fig. 3. At the between filament locations shown in Fig. 9, failure appears to occur largely by microvoid formation and growth, although some indications of shear sliding off can be found. The large outer shear lip shown in Fig. 3b is formed largely by shear sliding off.

Quantitative evaluation of the degree of triaxiality present in the matrix metal is not currently possible from such scanning electron microscopy examination. The qualitative difference between interior and exterior matrix material shown in Fig. 3b, however, indicates that some degree of constraint is present in matrix metal which is surrounded closely by stiff fibers.

FABRICATION PARAMETER TRADEOFF

Perhaps the simplest fabrication situation for diffusion bonded aluminum-boron composites is that of unidirectionally reinforced material. Volume fractions of reinforcement, alignment, and the bonding parameters should be controlled relatively readily. If temperature, pressure, atmosphere, and time of diffusion bonding are controlled properly, it should be possible to produce aluminum-aluminum joints which are scarcely discernible from originally continuous material.

Optimization of diffusion bonding parameters on the basis of producing the best possible aluminum-aluminum bond cannot be done without also considering the effect of these parameters on the boron filaments, however. If pressure is too great or is not uniform over the composite sheet, the relatively brittle filaments may be broken or shattered. If temperature is too high or too prolonged, chemical reaction at the filament-matrix interface may lead to degradation of filament properties. It is not clear that state-of-the-art diffusion bonding fabrication processes have succeeded in optimization of all of these factors, even in unidirectionally reinforced materials.

In cross-ply reinforced composites the tradeoff between optimization of diffusion bonding parameters and potential filament damage becomes even more critical. Filaments of differing orientation are pressed toward each other with only thin layers of matrix metal foil as cushions between them. It is understandable that fabricators of these cross-ply materials have reduced bonding pressures to lower the filament damage from stress concentration at points where fibers cross. The micrographs of cross-ply materials shown in this study indicate that alternative fabrication parameter balances to produce reasonable diffusion bond joints have not been developed as yet.

A comparison of the two unidirectionally reinforced composites studied, shown in Figs. 1 and 2, respectively, provides interesting insight into the macroscopic effects of bonding parameter tradeoffs. The two composites, fabricated by two different commercial producers of aluminum-boron composites, basically differ only in volume fraction of reinforcement, 25 versus 50 percent. Views of the fracture surfaces indicate that the 50 volume percent material is better bonded, but tension test values indicate that the 25 volume percent material approaches calculated rule-of-mixtures values much more closely. One explanation for the better bonding in the 50 volume percent case could be improved breakup of any oxide film on the original aluminum foils as they were deformed heavily during fabrication to accommodate the high volume percent of reinforcement. The lower relative strength of the 50 volume percent material apparently is due to substantial damage to the filaments during the diffusion bonding operation. Maximization of strength thus is seen to be a complex tradeoff of many factors, including volume fraction of reinforcement, control of placement of filaments during fabrication, and the bonding parameters of pressure, temperature, and time.

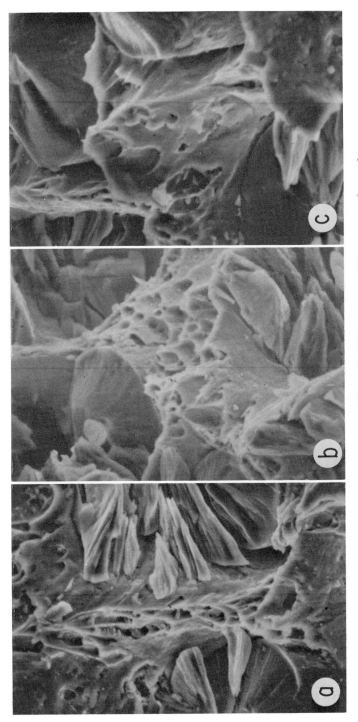

Fig. 9—Ductile failure of matrix metal between reinforcing fibers, largely by microvoid formation and growth.

FRACTURE INITIATION

In a simple strain-to-failure analysis of the components of the composite studied here, it would be predicted that the filaments would fracture first, followed by fracture of the more ductile aluminum alloy matrix. This interpretation can be made reasonably to some of the poorly bonded material observed in this study, such as the cross-ply material shown in Fig. 5b. It is not clear, however, that a similar sequence of events can be ascribed to the well-bonded composite shown in Figs. 2a and b. In this latter case the fracture surface appearance gives the impression that the fatal fracture crack propagated through the filaments and matrix at the same time in any given region. This would be expected in well-bonded material since the pullout length (load transfer length) is much smaller and the interface coupling of the propagating fracture is much better. The extreme overall raggedness of the fracture surface may indicate that fracture cracks nucleated at several different locations across the specimen, then joined together in final failure.

The most likely candidate for an initial fracture initiation site would be a defect or weak spot in a boron filament. Since substantial loss of material occurs due to shattering of the filaments near the fracture surface; however, even scanning electron microscopy does not allow identification of such sites. One interesting filament break is shown in Fig. 6a, though, where one of the filaments parallel to the fracture surface appears to have exploded. Since there are many filaments which statistically exhibit low fracture strengths, many filaments probably are broken prior to composite failure. It is thus doubtful that a single filament fracture would be a critical fracture initiation site. It is more likely that a critical fracture initiation site takes the form of coincidence of weak spots in several nearby filaments within some load transfer length.

Although the matrix material is considered generally to be more ductile than the reinforcement filaments, local triaxial constraint can reduce its ductility substantially. Voids due to imperfect bonding or cracks due to debonding can act as stress concentrators for the initiation of fracture cracks. In Fig. 6b, for example, a large void, debonding cracks, and gaps left in the matrix between filaments due to imperfectly formed diffusion bond joints are all observed at the fracture surface. Any of these could have provided a sufficient concentration of stress to start a fracture crack propagating through the triaxially constrained matrix.

It is not clear at this time whether debonding at diffusion bond planes is a cause contributing to premature failure of the composite or whether it is just an energy absorbing mechanism near the fracture surface as the final fracture crack propagates at ultimate failure. Studies are currently underway to determine at what point in the loading history of a tension specimen such debonding becomes prevalent.

SUMMARY AND CONCLUSIONS

Scanning electron microscopy provides a useful technique for study of the fracture surfaces of metal matrix composites such as the aluminum-boron system. By such factographic study, the micromechanisms of deformation and fracture of the components of the composite *in situ* can be determined. Interpretations of micrographs can lead to the formulation of realistic micromechanical models to explain the mechanical behavior of such composites.

Bond planes in diffusion bonded aluminum-boron composites have been identified as a problem area limiting the performance of such materials. In both unidirectionally and cross-ply reinforced materials, severe debonding of diffusion bond planes has been observed at fracture surfaces. In cross-ply materials, where bonding pressures have apparently been reduced to limit filament damage, incomplete joining and bonding of matrix metal between fibers has been observed on a large scale.

Observation of fracture surfaces containing large amounts of lack of bonding or debonding at diffusion bond planes leads to the conclusion that further optimization of fabrication parameters is needed to improve current state-of-the-art aluminum-boron materials. Particularly in cross-ply materials, the tradeoff between improvement of bonding parameters and potential filament damage must be shifted to provide better matrix-matrix and matrix-filament bonding.

Voids, debonding cracks, and filament defects are identified as potential fracture initiation sites. It is not clear at this time, however, whether debonding cracks are a contributory cause or an effect of the final fracture process.

ACKNOWLEDGMENTS

This study has been supported by the Massachusetts Institute of Technology Inter-American Program in Civil Engineering, which currently is funded largely by a grant from the Ford Foundation.

References

[1] Fairing, J. D., "Examination of Fracture Surfaces by Scanning Electron Microscopy," *Journal of Composite Materials*, Vol. 1, No. 2, 1967, pp. 208–210.

[2] Fairing, J. D., "Fractography of Composite Materials," *Scanning Electron Microscopy/1968*, IIT Research Institute, Chicago, 1968, pp. 121–129.

[3] Johari, O., "Scanning Electron Microscopy in Metallurgy," *Scanning Electron Microscopy/1968*, IIT Research Institute, Chicago, 1968, pp. 79–87.

[4] Johari, O., "Comparison of Transmission Electron Microscopy and Scanning Electron Microscopy of Fracture Surfaces," *Journal of Metals*, June 1968, pp. 26–32.

[5] Jones, R. C., "Deformation of Wire Reinforced Metal Matrix Composites," *Metal Matrix Composites, ASTM STP 438*, American Society for Testing and Materials, 1968, pp. 183–217.

[6] Jones, R. C., "Fractography of Metal Matrix Composites," *Proceedings*, 14th Refractory Composites Working Group Meeting, Air Force Materials Laboratory, 1968, pp. 295–308.

[7] Pruden, L. H., Korda, E. J., and Williams, J. P., "Characterization of Surface Topography with the Scanning Electron Microscope," *American Ceramic Society Bulletin*, 7 Aug. 1967, pp. 750–755.

Tensile Strength of Fiber-Reinforced Composites: Basic Concepts and Recent Developments

CARL ZWEBEN[1]

Reference:

Zweben, Carl, "Tensile Strength of Fiber-Reinforced Composites: Basic Concepts and Recent Developments," *Composite Materials: Testing and Design, ASTM STP 460,* American Society for Testing and Materials, 1969, pp. 528–539.

Abstract:

This paper presents a selective analysis of the subject of composite tensile strength, emphasizing basic principles including recent advances by the author. The importance of the statistical scatter in fiber strength and local fiber over- stress caused by fiber discontinuities are discussed. Recently developed failure criteria that give good correlation with experimental data are covered. The analysis of whisker-reinforced composites, in addition to giving good strength predictions, explains why predictions of previous theories are much higher than experimental values. The applicability of the "rule of mixtures" to composite strength is discussed, and the problem of notch sensitivity is mentioned briefly.

Key Words:

composite materials, fiber composites, whisker composites, mechanical properties, notch sensitivity, strength, fractures, materials, failure modes, statistics, evaluation, tests

One of the basic motivations for the increasing use of composite materials is the high tensile strength that can be achieved with them. The great tensile strength is, of course, a reflection of the properties of the fibers which carry most of the load. However, the way in which the properties of the constituents, the matrix, and especially the fibers, affect the strength of the composite is not widely understood and is also the subject of many misconceptions.

Much of the confusion centers about the applicability of the so-called "rule of mixtures" to composite strength. This relation is well presented in a paper by Kelly and Davies [1][2] as

$$\sigma_c = \sigma_f v_f + \sigma'_m(1 - v_f) \dots\dots\dots\dots\dots\dots\dots (1)$$

where:

σ_c = average stress in the composite at failure,
σ_f = ultimate strength of the fibers,
σ'_m = stress in the matrix just before the fibers fail, and
v_f = volume fraction of fibers.

The authors emphasize that this expression is based on the assumptions that all fibers have the same ultimate strength and that the strain is uniform throughout the composite. The requirement of uniform ultimate strength is not satisfied by

---

[1] Research engineer, Space Sciences Laboratory, General Electric Company, King of Prussia, Pa. 19101.
[2] The italic numbers in brackets refer to the list of references appended to this paper.

many of the fibers currently in use, such as boron, graphite, glass, alumina, etc. Therefore, there is no reason to expect the rule of mixtures to apply to composites made from them. In point of fact, the strength of composites made from these fibers generally does not follow the rule of mixtures using the mean strength of fibers whose gage length is the same as that of the composite specimens. The tungsten fibers used by Kelly and Davies exhibit very little scatter in strength, and composites using them do obey Eq 1. The dispersion in fiber strength greatly affects the strength of composites, and this topic is discussed extensively in the present paper.

The analysis of composite failure is one of great complexity involving many possible failure mechanisms, and the process is by no means completely understood. Therefore, the objective of this paper is to present what the author considers the basic principles involved in understanding the way the fiber and matrix properties interact to affect composite strength. The emphasis is on the influence of fiber properties and internal stress redistribution among the fibers because the fibers are the load-carrying elements in the composite.

In this paper we first consider basic fiber properties and their relation to the strength of bundles. Next, the effect of adding a matrix is considered, followed by an analysis of whisker-reinforced composites. Finally, the problem of notch sensitivity is discussed.

FIBER STRENGTH

Most high-strength fibers are brittle, and therefore they are sensitive to surface flaws and other defects. This fact accounts for the scatter in strength levels observed when fibers are tested at a constant gage length. The sensitivity to flaws is reflected also in the decrease in mean strength as the gage length is increased. These two phenomena are illustrated in Fig. 1, adapted from Ref 2. These data, which were obtained for boron fibers, are typical of the other high-strength fibers mentioned earlier. It easily can be seen that average fiber strength has no significance unless the gage length of the test specimens is reported.

In order to illustrate the significance of strength dispersion and length effect, their influences on the strength of bundles is considered before the complex problem of composite strength is discussed.

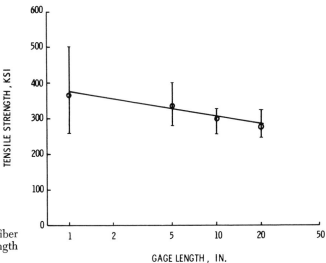

Fig. 1—Variation of fiber strength with gage length (after Herring).

GAGE LENGTH, IN.

BUNDLE STRENGTH

Consider the sequence of events associated with bundle failure. As the load is applied, fibers break randomly throughout the bundle. When they break, the fibers are completely ineffective in sustaining any of the load applied to the bundle. This is different from the situation in a composite where, because of the matrix, the effect of the fiber break is localized. Eventually, as the load on the bundle is increased, a point is reached at which no increase in load can be supported and failure occurs.

Coleman [3] studied the relation between the strength of a bundle of fibers and the scatter in strength of the individual fibers. His results are shown in Fig. 2 where the ratio of bundle strength σ_B to mean fiber strength $\bar{\sigma}_f$ is plotted against the fiber coefficient of variation $\bar{c}$, which is defined as the ratio of the standard deviation to the mean fiber strength. The results apply to fibers and bundles having the same gage length.

It can be seen that the normalized bundle strength decreases significantly as the coefficient of variation, which is a measure of scatter increases. Typically, $\bar{c}$ can be expected to be between 0.08 to 0.25. Therefore, even for a bundle of fibers, which is a much simpler configuration than a composite, one should expect to obtain strengths different from the mean strength of the fibers from which it is made.

The effect of comparing the strength of a bundle to the mean strength of fibers of a different length is shown in Fig. 3 adapted from Ref 4. L represents the gage length of the reference fibers and δ the length of the bundle. Since the strength of a bundle is related inversely to length, the normalized strength can be greater or less than unity, depending on the ratio δ/L.

STRENGTH OF COMPOSITES REINFORCED WITH CONTINUOUS FIBERS

Consider what happens to a bundle of fibers when a matrix is added. When load is applied, there will occur at some stress level the first fiber break. It seems reasonable to assume that one of three things may happen: (a) the fiber may debond over its entire length, (b) the fiber break may initiate a stress wave or a propagating crack in the matrix which causes additional fibers to break, resulting in overall composite failure, and (c) the effect of the break may be localized.

If complete debonding takes place, the composite is behaving essentially as a bundle which was considered earlier. However, at the current stage of development

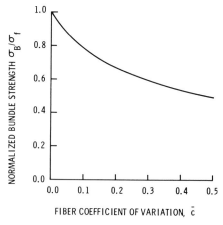

Fig. 2—Variation bundle strength with fiber coefficient of variation (after Coleman).

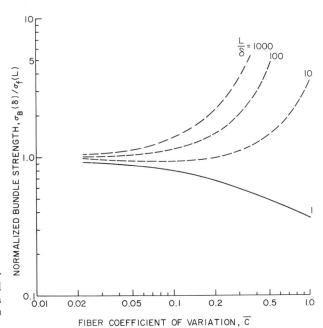

FIG. 3—Effect of fiber coefficient of variation and reference fiber length on bundle strength (taken from Ref 4).

of most composites, this type of behavior is unlikely. If a propagating crack occurs, the composite strength will be significantly weaker than a bundle of the same gage length. This type of phenomenon has been reported in Ref 5 for boron-epoxy and Ref 6 for graphite-epoxy composites. In Refs 6 and 7 it was observed that if the fiber-resin bond is weak, the fibers debonded locally, and catastrophic failure does not occur. Zweben [8], in a statistical analysis, demonstrated a correlation between the stress at which the first fiber break is expected and the observed failure stress in boron-epoxy composites. This provides further evidence for the existence of the catastrophic, or weakest link mode of failure. The stress at which the first fiber break is expected is a lower bound on the expected strength of the composite.

In order to achieve the full potential of the fiber strength, the catastrophic mode of failure initiated by one or a small number of isolated breaks must be prevented. This can be accomplished by weakening the fiber-matrix bond, but only at the penalty of decreased transverse strength properties. If weakest link failure is the result of a propagating crack, then an increase in the matrix fracture toughness may remedy the situation.

If catastrophic failure does not occur then the load on the composite can be increased, causing other fibers to break randomly throughout the composite before failure occurs. This type of behavior was demonstrated in a series of experiments by Rosen [9]. He loaded composites consisting of a single layer of 3.5 mil E-glass fibers in an epoxy matrix and observed them using transmitted polarized light. A typical sequence of photographs is shown in Fig. 4, taken at various percentages of the ultimate load. The fiber breaks show up as dark areas because the composite is unstressed or weakly stressed in this region. It can be seen that the reduction in stress is localized, and away from the area of the break the fractured fibers are carrying load. This is an important function of the matrix, to localize the effect of a fiber break and to permit the fiber to carry load over most of its length. This is in marked contrast to what happens in a bundle, where, when a fiber breaks, it is completely ineffective for its entire length.

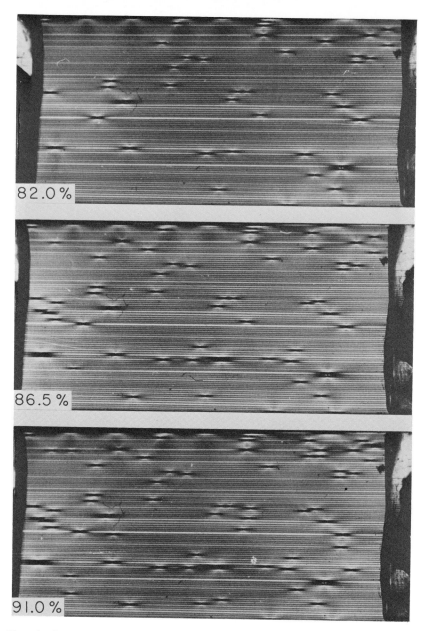

Fɪɢ. 4—Sequence of photographs illustrating the occurrence of scattered fiber breaks (taken from Ref 10).

Figure 5 shows a simplified version of the static pertubation in stress in the vicinity of a fiber break. The load in the broken fibers is transmitted to adjacent fibers in the cross section through the matrix, causing them to be overstressed locally. The interface shear stress, τ acts on the broken fiber and the axial stress, σ, increases from zero, approaching the average stress in the unbroken fibers. The ineffective, or transfer length is a measure of the distance over which the fiber

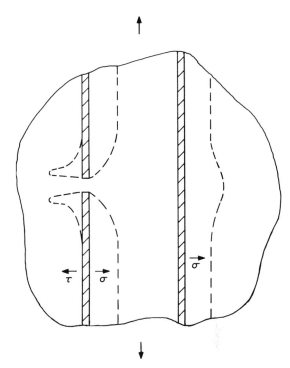

FIG. 5—Perturbation of
stresses in the vicinity of a
fiber break (after Rosen).

sustains a significant reduction in stress, and several expressions have been proposed
for its evaluation [8].

The increase in stress in fibers adjacent to a break appears in Fig. 4 as a local
brightening adjacent to the dark areas. The overstress that occurs in fibers adjacent
to broken ones has a significant effect on strength and will be discussed at some
length.

MODES OF FAILURE ASSOCIATED WITH SCATTERED FIBER BREAKS

One can think of two modes of failure that can occur because of the accumulation
of scattered fiber breaks: (a) a propagation of fiber breaks caused by the presence
of load concentrations and (b) failure caused by the inability of the weakest cross
section to sustain the applied load, that is, a bundle type of failure. These modes
of failure will now be considered. The geometrical model used in both cases is
the one suggested by Rosen [9]. As shown in Fig. 6, the composite is considered
to be made up of a chain of cross-sectional layers, each of which contains n-fiber
elements. The axial dimension of the layers is δ, the ineffective length. When
a fiber breaks within a layer it is assumed to be unloaded completely in that
layer but fully stressed in all other layers.

Fiber Break Propagation Mode

The significance of the overstress in fibers surrounding a broken one is that
it increases the probability that they will break. If an overstressed fiber does break,
the fibers adjacent to these two breaks are subjected to an even greater overstress,
resulting in the possibility of further breaks, and so on. Therefore, a possible mechanism of failure is the propagation of fiber breaks caused by load concentrations.
It should be emphasized that the matrix material between the broken fibers need
not be fractured at the initial stages of this phenomenon. This type of behavior
was observed by Cooper and Kelly [10] in a study of notch sensitivity of composites
containing brittle tungsten fibers.

The average increase in stress in fibers adjacent to i broken fibers in a two-dimensional composite (tape) was studied by Hedgepeth [11]. The first four concentration

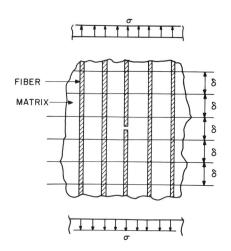

Fig. 6—Geometrical model for evaluation of composite strength.

factors were found to be $k_1 = 1.33$, $k_2 = 1.60$, $k_3 = 1.83$, $k_4 = 2.03$. This means, for example, that the stress in the two fibers adjacent to four broken fibers is more than twice the average applied stress. The validity of these concentration factors was experimentally verified by Zender and Deaton [12].

The effect of load concentrations on the strength of two-dimensional composites was studied by Zweben, in Ref 8. It was shown that there is a close correlation between composite failure and the probability that overstressed fibers will break. In Fig. 7 the expected number of groups containing at least i fractures, E_i, is plotted against fiber stress for $i = 1$ to 4 using the data of Ref 4. It can be seen that the expected number of multiple fractures rises sharply in the range of failure. The actual number of multiple-fracture groups was counted from photographs of these tests and is compared with the predicted number of Fig. 8. It can be seen that the observed behavior is consistent with the theory. Because of the correlation with experimental data, the stress at which the first multiple break is expected, σ_2, was proposed as a failure criterion for laboratory specimens exhibiting this mode of failure. Although agreement with small specimens is good,

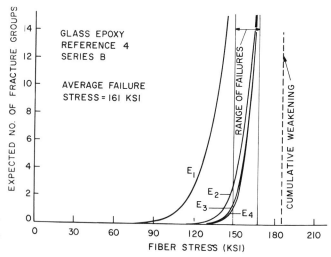

Fig. 7—Expected number of groups of fractures as a function of fiber stress.

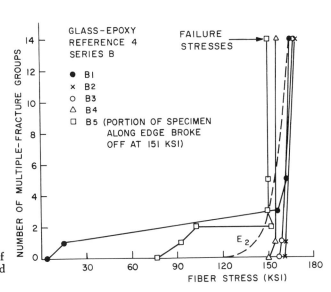

FIG. 8—Comparison of observed and predicted multiple fractures.

the criterion may only be a lower bound on the expected strength of large structures exhibiting an accumulation of fiber breaks before failure. More testing will be required to resolve this question.

This analysis has been extended recently to three-dimensional unidirectional composites [13]. Again, the first multiple break criterion provides good agreement with experimental data, as shown for boron-aluminum composites in Fig. 9. The curves for σ_1, which represents the expected first isolated break, and σ_2 were computed using data from recovered fibers as were the curves labeled "Rule of Mixtures" and "Cumulative Weakening." σ_1^* is the curve for the expected first break determined from virgin fiber data. Since it was observed [14] that a significant number of fiber breaks occurred before failure, it is evident that fiber degradation did occur in the fabrication process, which is consistent with data from recovered fibers. The potential for evaluating, in situ, fiber properties using this method should be obvious from this example.

It should be noted that the fiber break propagation failure criterion predicts a decrease in strength with composite size. For most materials, an order of magnitude increase in size is required to produce a significant drop in strength so that size effect may not be noticed in small laboratory specimens. On the other hand, the cumulative weakening criterion, which is discussed next, predicts little size dependence, in general.

Cumulative Weakening Failure Mode

The basic idea associated with this failure mode is that as the load on a composite is increased, random breaks occur in the various layers until a point is reached at which the weakest layer cannot sustain the applied load and failure occurs. This theory is due to Rosen [9]. In his analysis he assumed that when an element breaks the load on it is redistributed uniformly among all of the unbroken elements in the layer, neglecting the effects of localized fiber overstress. Neglecting load concentrations is a nonconservative approach, and the failure stresses predicted by this analysis are generally higher than those observed experimentally. This fact is demonstrated in Figs. 7 and 9. However, this analysis does provide a useful upper bound on expected composite strength.

Zweben on Fiber-Reinforced Composites   **535**

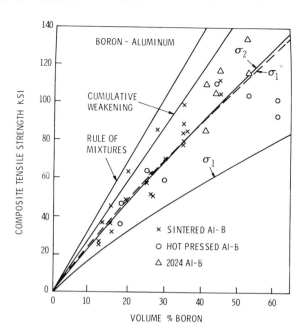

FIG. 9—Comparison of predicted and observed failure stress levels (data of Ref 15).

In Ref *13* the effects of load concentrations were introduced into the cumulative weakening theory using an approximate method. However, the predictions were not quite as good as the fracture propagation criterion and are inconclusive.

STRENGTH OF COMPOSITES REINFORCED WITH DISCONTINUOUS FIBERS

The rule of mixtures formula for the strength of whisker-reinforced composites is given usually as [*1*]

$$\sigma_c = \sigma_f(1 - l_c/2l) \dots\dots\dots\dots\dots\dots (2)$$

where l is the uniform length of the whiskers, and l_c is the "critical transfer length" which is the minimum length for which the interface shear stress can build up the stress in a fiber to its ultimate strength level which is assumed to be the same for all fibers. Again, the rule of mixtures formula does not apply to fibers that exhibit scatter in ultimate strength, and strength predictions for composites reinforced with such whiskers are generally many times higher than observed failure stresses.

The most important property of whisker-reinforced composites from the fracture standpoint is the tremendous number of fiber discontinuities. A typical specimen may contain millions of whiskers. As in the case of a broken continuous fiber within a composite, the load in each whisker is transferred to surrounding fibers at both ends. This causes local overstresses in adjacent fibers just as in the case of continuous fibers. As a result, there exists the possibility of a crack being initiated in the matrix which propagates, causing catastrophic failure before the limit of fiber strength is reached. If this does not occur, the two types of fracture associated with scattered fiber breaks described earlier must be considered.

FIBER BREAK PROPAGATION MODE

Because of the extermely large number of initial discontinuities, some overstressed elements can be expected to break at relatively low stress levels. However, these double breaks do not tend to propagate until a significantly higher stress level

536

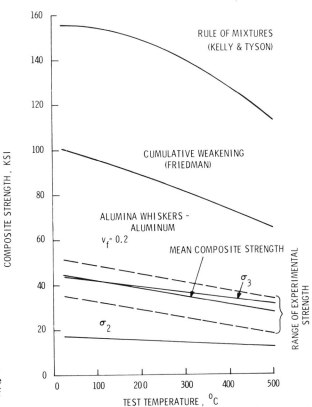

FIG. 10—Comparison of predicted and observed failure stress levels (data of Mehan).

is reached. But, when triple breaks are expected to occur, the probability of fiber breaks propagating becomes likely [13]. Because of this, the stress at which the first triple break occurs, σ_3, has been proposed as a failure criterion. Comparison of this prediction with experimental results for alumina whisker-aluminum composites at various temperatures is shown in Fig. 10. It can be seen that σ_3 gives good quantitative agreement with the data and successfully predicts the temperature dependence. In comparison, the rule of mixtures prediction is much higher and indicates a greater temperature dependence than is observed.

The stress at which the first double break occurs, σ_2, gives a lower bound, much as σ_1 does for continuous fiber composites.

CUMULATIVE WEAKENING FAILURE MODE

The cumulative weakening theory of Rosen was adapted to discontinuous fiber composites by Friedman [15]. His analysis includes the effects of variation in whisker length and diameter but not of load concentrations. Since no information is available about the statistical variation in size it was not possible to use his theory in its general form. However, the prediction including only the effects of discontinuities is shown in Fig. 10. Again, the cumulative weakening theory gives an upper bound.

NOTCH SENSITIVITY

The effect of fiber overstress is critical to the strength of notched filamentary structures. As we have stated earlier, if a notch that cuts i fibers is introduced into the center of an infinite plate, the stress in the fibers adjacent to the notch is $k_i\sigma$, where k_i is the load concentration factor corresponding to i broken fibers. If the fibers all have the same ultimate strength σ_u, the fibers adjacent to the notch will fail when the applied stress is σ_u/k_i. The failure of these fibers will precipitate overall failure of the composite in a fiber break propagation mode. Therefore, the critical factor that determines the failure stress associated with this mode is the number of broken fibers rather than the geometry of the crack. Crack propagation in the matrix is another possible failure mode, and this phenomenon will be affected by crack dimensions. However, whereas a matrix crack may be arrested in some manner, when fiber break propagation occurs the composite will fail since the fibers are the principal load-carrying elements. In studying crack propagation in composite materials it is essential to keep in mind that they are not isotropic or homogeneous, and, furthermore, one phase carries most of the load in uniaxial tension.

The problem of notch sensitivity in composites reinforced with fibers exhibiting scatter will be treated in a forthcoming paper.

SUMMARY

The following list summarizes the important points in this paper.

1. Most fibers of interest exhibit scatter in strength and a length effect.

2. The rule of mixtures does not apply to composites made from these fibers because of the dispersion in their strength.

3. There are at least three possible modes of composite failure: (a) crack propagation in the matrix associated with a few fiber breaks or other discontinuities, (b) the fiber break propagation mode, and (c) the cumulative weakening mode.

4. It is important to determine whether Mode a is occurring, because this indicates that a significant increase in composite strength may be possible by preventing this phenomenon.

5. The effects of local fiber overstress are extremely important for composites reinforced with continuous fibers or whiskers.

6. In studying notch sensitivity it is important to recognize that the fibers are carrying most of the load, and load concentrations are of critical importance.

CONCLUDING REMARKS

In this paper we have attempted to present an analysis of the most important factors affecting composite strength. Although much effort has been devoted to this subject it is far from a closed book. Despite the fact that there recently has been some success in predicting the strength of laboratory specimens using fracture propagation criteria [8,13], there remains a real question as to whether these criteria are applicable to large structures. That is, a large volume of material may be able to tolerate multiple breaks without failing. In this case the stress associated with the first multiple break may only be a lower bound on the actual failure stress. In any event, there will have to be much effort devoted to the study of tensile strength before the final page is written. This will require an interplay of careful experimentation in which the constituent materials are well characterized and the fracture behavior is observed closely together with a detailed theoretical analysis. Although the problem is a complex one, past success in predicting the behavior of laboratory specimens provides encouragement that the more general problem can be solved.

References

[1] Kelly, A. and Davies, G. J., "The Principles of the Fiber Reinforcement of Metals," *Metallurgical Reviews*, Vol. 10, 1965, pp. 1–77.

[2] Herring, H. W., "Selected Mechanical and Physical Properties of Boron Filaments," NASA TN D-3202, National Aeronautics and Space Administration.

[3] Coleman, B. D., "On the Strength of Classical Fibres and Fibre Bundles," *Journal of the Mechanics and Physics of Solids,* Vol. 7, 1958, pp. 60–70.

[4] Rosen, B. W., "Mechanics of Composite Strengthening," *Fiber Composite Materials,* American Society for Metals, Metals Park, Ohio, 1965.

[5] Mullin, J. et al, "Some Fundamental Fracture Mechanisms Applicable to Advanced Filament Reinforced Composites," *Journal of Composite Materials,* Vol. 2, No. 1 Jan. 1968, pp. 82–103.

[6] Wadsworth, N. J. and Spilling, I., "Load Transfer from Broken Fibers in Composite Materials," *British Journal of Applied Physics,* D. Vol. 1, 1968, pp. 1049–58.

[7] Gatti, A. et al, "Investigation of the Reinforcement of Ductile Metals with Strong High-Modulus Discontinuous Fibers," Final Report, Contract NASw-1543, National Aeronautics and Space Administration, 1968.

[8] Zweben, C., "Tensile Failure Analysis of Fibrous Composites," *Journal,* American Institute of Aeronautics and Astronautics, Vol. 6, No. 12, Dec. 1968, presented at the AIAA 6th Aerospace Sciences Meeting, Jan. 1968.

[9] Rosen, B. W., "Tensile Failure of Fibrous Composites," *Journal,* American Institute of Aeronautics and Astronautics, Vol. 2, No. 11, Nov. 1964, pp. 1985–1991.

[10] Cooper, G. A. and Kelly, A., "Tensile Properties of Fibre-Reinforced Metals: Fracture Mechanics," *Journal of the Mechanics and Physics of Solids,* Vol. 15, 1967, pp. 279–297.

[11] Hedgepeth, J. M., "Stress Concentrations in Filamentary Structures," NASA TN D-882, National Aeronautics and Space Administration, May 1961.

[12] Zender, G. and Deaton, J. W., "Strength of Filamentary Sheets with One or More Fibers Broken," NASA TN D-1609, National Aeronautics and Space Administration.

[13] Zweben, C. and Rosen, B. W., "A Statistical Theory of Material Strength with Application to Composite Materials," 7th Aerospace Sciences Meeting, American Institute of Aeronautics and Astronautics, New York, Jan. 1969.

[14] Kreider, K. G. and Leverant, G. R., "Boron Aluminum Composite Fabricated by Plasma Spraying," 10th National Symposium, Society of Aerospace Materials and Process Engineers, San Diego, Nov. 1966.

[15] Friedman, E., "A Tensile Failure Mechanism for Whisker Reinforced Composites," 22nd Annual Meeting of the Reinforced Plastics Division, Society of the Plastics Industry, Feb. 1967.

Fracture in Graphite Filament Reinforced Epoxy Loaded in Shear

R. C. NOVAK[1]

Reference:

Novak, R. C., "Fracture in Graphite Filament Reinforced Epoxy Loaded in Shear," *Composite Materials: Testing and Design, ASTM STP 460*, American Society for Testing and Materials, 1969, pp. 540–549.

Abstract:

The failure mode in graphite filament reinforced epoxy composites loaded in shear is dependent on the strength of the interfacial bond. Failure of composites made from untreated filaments occurs progressively by interfacial debonding and crack propagation through the resin. Shear strength of these composites is low and is inversely proportional to the filament volume fraction. Failure of composites made from treated filaments may occur at the interface, in the matrix, or in the filaments, and may be in shear or tension or a combination of both. Shear strength of these composites is high, but the factors leading to high shear strength also lead to brittle fracture of the material. Thus, the consequences of increasing shear strength on other properties such as impact strength must be studied carefully in light of the intended application.

Key Words:

fiber composites, fractures (materials), carbon fibers, graphite composites, epoxy laminates, mechanical properties, evaluation, tests

Structural graphite filaments are receiving widespread attention from many materials suppliers and users who recognize the potential of graphite filament reinforced composites to meet the demands of applications calling for strong and stiff yet lightweight materials. The large volume of experimental data which has been generated on the material has shown that the property frequently limiting its application is the shear strength parallel to the reinforcement direction. This problem has received a good deal of attention, and several investigators have been able to show improvements in the unidirectional composite shear strength. The purpose of this paper is to present the results of a study of fracture of these composites in order to further understand the problem of shear strength improvement. Baseline data are presented on the shear strength of composites reinforced with "as-received" graphite filaments. The results of these tests are analyzed from the standpoint of analytical and experimental micomechanics. A filament treatment is described which increases composite shear strength, and test data and failure of this composite are discussed. Finally, a general analysis is made of fracture in shear-loaded graphite filament composites, and conclusions based on the present work as well as recommendations for future work are presented.

EXPERIMENTAL RESULTS AND OBSERVATIONS

SPECIMEN PREPARATION

In order to employ an experimental approach to study the behavior of composite materials it is essential that the many variables due to processing and testing be

[1] Formerly, senior engineer, Aeronutronic Division of Philco-Ford, Newport Beach, Calif. 92663; now, research scientist, United Aircraft Research Laboratories, East Hartford, Conn. 06108.

540

controlled carefully or accounted for. Without this precaution the task of correctly attributing changes in material behavior to one cause or another becomes impossible. Consequently, the bulk of the experimental data reported in this paper were obtained from one composite system: Shell Chemical's Epon 828/1031/MNA/BDMA epoxy reinforced with Thornel 40 filaments produced by Union Carbide. Data for other composite systems are presented only to show the effect on composite behavior of filament or resin variables.

The solid rod torsion test developed by Adams et al [1][2] was used for the determination of shear behavior of the composites. The test specimen was a unidirectionally reinforced rod, ¼ in. in diameter by 4½ in. long, which was prepared in a cylindrical cavity mold. The impregnated graphite yarn was made by a wet winding technique in which carefully aligned impregnated yarn was wound on a rotating drum to form a unidirectional sheet. The sheet of impregnated yarn was b-staged in an air-circulating oven, then cut to the proper length, formed by hand into bundles of the correct weight, and placed in a mold where cure was carried out between the heated platens of a hydraulic press.

SHEAR TESTING OF AS-RECEIVED COMPOSITES

One of the most important composite variables in a test program is the volume fraction of filaments in the test specimens. To account for this factor, preliminary testing was carried out on specimens in which the filament volume fraction, v_f, was varied deliberately. Shear strength of the rods was determined by use of the following equation:

$$\tau = \frac{Tr}{J} \dots\dots\dots\dots\dots\dots\dots\dots\dots\dots\dots(1)$$

where:
τ = shear strength,
T = applied torque,
r = radius of rod, and
J = polar moment of inertia ($\pi d^4/32$; d = rod diameter).

The results of the preliminary tests are shown in Fig. 1 which represents the shear strength of unidirectional Thornel 40-Epon 828/1031 as a function of v_f. The decrease in shear strength with increasing filament volume may seem contrary to what one would expect since the filaments are stronger than the matrix. However,

[2] The italic numbers in brackets refer to the list of references appended to this paper.

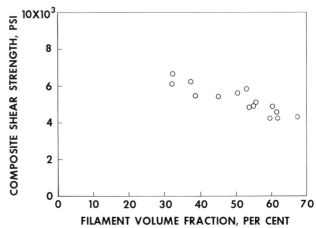

FIG. 1—Shear strength of as-received Thornel 40-Epon 828/1031 unidirectional composites.

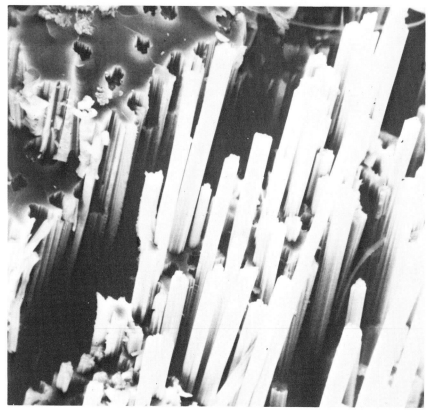

Fig. 2—Fracture surface of as-received Thornel 40-Epon 828/1031 torsion rod (×500).

a study of the failure of the specimens coupled with a micromechanics analysis provided a reasonable explanation of the behavior.

Through the courtesy of the Monsanto/Washington University ARPA Association, a scanning electron microscope was used to examine the specimen fracture surfaces. Figure 2 shows a large axial crack extending from lower left to upper right in the picture. There is abundant evidence of filaments being pulled out of the matrix; in fact, the far side of the large crack seems to consist of a series of interconnected holes caused by pulled-out filaments. The large number of bare filaments is good evidence of a weak interfacial strength.

Further evidence of a weak interface is provided by Figs. 3 and 4 which show the same specimen at higher magnifications. All the filaments which did not pull out were apparently debonded from the matrix. There is no evidence of the filaments failing in shear and very little shear failure in the resin. The resin failure which did occur was largely in the form of cracks proceeding from one debonded filament to another. Adams et al [1] conducted an analytical study of the stresses around filaments in composites subjected to external shear loads. It was found that local stresses in the immediate vicinity of the filaments were higher than the gross applied stress due to differences in the elastic properties of the filaments and the resin. The magnitude of the stress concentration factors depends on the ratio of filament shear modulus to the matrix shear modulus (G_f/G_m), the filament spacing which is controlled by filament volume fraction (v_f), and the filament shape. Assuming

Fig. 3—Filament pullout in failed torsion rod of as-received Thornel 40-Epon 828/1031 (×2000).

round filaments and using the values of $G_f = 2.1 \times 10^6$ psi and $G_m = 0.212 \times 10^6$ psi [2], the stress concentration factor for shear loading of Thornel 40-epoxy is 1.7 at $v_f = 30$ percent and 2.2 at $v_f = 60$ percent. This implies that cracks will tend to propagate through the resin in areas where the filaments are spaced very close since this is where the local stresses are the highest. An example of this behavior is evident in Fig. 3 in the area around the pulled-out filament in the center of the picture. The crack extending toward the upper left of the picture was arrested, possibly due to the low stress concentration in that area caused by the relatively large filament spacing. The crack extending toward the lower right propagated to the next filament due to the higher stresses caused by closer filament proximity.

These analyses and observations can be used to explain the variation of composite shear strength with filament volume fraction. The composite is considered to be made up of three phases: (1) the graphite filaments having a shear strength of approximately 20,000 psi[3], (2) the epoxy having a shear strength of approximately 10,000 psi, and (3) the filament-matrix interface having a shear strength much less than that of the resin since composite failure initiated at the interface. The strength of a composite which fails in shear will be governed generally by the behavior of the weakest phase. Increasing the filament volume increases the amount of interface in the composite and consequently the probability of failure occurring

[3] DeCrescente, M. A., United Aircraft Research Laboratories, personal communication.

FIG. 4—Interfacial debonding in failed torsion rod of as-received Thornel 40-Epon 828/1031 (×5000).

there. In addition the magnitude of the stress concentration factors around the filaments increases as filament volume increases. These factors combine to cause crack initiation at the interface and crack propagation through the resin to occur at lower applied stresses as filament volume is increased. It is interesting to note that the ratio of stress concentration factors at filament volume fractions of 60 and 30 percent is 1.3:1 and the ratio of shear strengths at 30 and 60 percent is 1.4:1. This indicates that the stress concentration factors account for the major portion of the change in shear strength with filament volume.

Composites utilizing other structural graphite filaments (HMG-25 from Hitco, HB-16 and 18 from the Royal Aircraft Establishment (RAE) of England) showed similar behavior to that of the Thornel 40 composites. It is noted that the Union Carbide and Hitco filaments are made from a rayon precursor while the RAE filaments are made from a polyacrylonitrile precursor. This had no apparent effect on the shear strength or the failure of the composites made from the as-received filaments.

SHEAR TESTING OF TREATED COMPOSITES

Due to the low shear strength of composites made from as-received filaments, an effort was made to devise a filament treatment which would increase the interfacial strength and thereby increase the shear strength. Various filament treatments were evaluated, and the one found most successful was oxidation of the filaments in air for 16 h at 475 C. After oxidation the filaments were incorporated into

Composite Materials

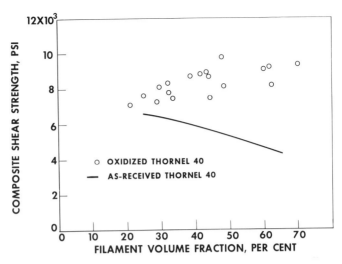

F<small>IG</small>. 5—Shear strength of Thornel 40-Epon 828/1031 unidirectional composites.

composites in the normal manner. Although the exact mechanism of interfacial strength improvement is unknown, the effectiveness of the treatment in increasing composite shear strength is unmistakable as shown in Fig. 5. The solid line represents the dependence of shear strength on filament volume for composites made from as-received filaments, while the open circles indicate data for the oxidized composites. There is an obvious difference in shear strength dependence on filament volume for the two composite systems. It was found, however, that the behavior of the oxidized composite could be explained using the same analysis employed to describe the behavior of the composites utilizing as-received filaments.

Figure 6 is a picture of the fracture surface of a failed torsion rod made from oxidized Thornel 40. There is no evidence of extensive filament pull out or bare filaments or interfacial debonding. In fact, resin can be seen adhered to the filaments in many places. Figure 7 shows the specimen at higher magnification, and the adhesion of the resin to the filaments is apparent. This picture also shows extensive failure in the matrix between the filaments; not the simple propagation of cracks from one debonded filament to another. Failure apparently occurred both at the interface and in the epoxy matrix, indicating that the strengths of the two phases were approximately the same.

As composite shear strength increased, an increasing amount of tensile failure was observed in the torsion rods. This type of failure can be observed in the upper right portion of Fig. 6. Tensile failure was accompanied generally by increasingly brittle behavior, especially at strength levels of approximately 8000 psi or greater. The as-received composites failed in a progressive manner, and failed specimens generally had the same gross appearance as untested specimens. The tendency toward brittle behavior as composite shear strength increased was noted also to a greater or lesser degree for composites made from treated filaments produced by Hitco and Morganite. Some experimentation was carried out to determine the effect of increasing the degree of oxidation of Thornel 40. This produced shear strengths in excess of 10,000 psi but the torsion specimens literally shattered at failure, and failure was almost totally in tension rather than shear. It should be noted that the oxidation treatment used to improve the shear strength did degrade the tensile strength of the filaments down to approximately 190,000 psi. The tensile strength of as-received filaments was 250,000 psi. The effect of this degradation of tensile properties on measured shear strength will be discussed later.

These observations led to a much more complicated model than the one for the as-received composites, since failure occurred in both shear and tension. Considering first the low volume fractions, failure tended to be a combination of shear and tension. The filaments were the strongest phase, and the strengths of the interface and the matrix were approximately equal. Increasing the filament volume increased the composite tensile strength with little change in the amount of the weaker phase since increases in interface were offset by decreases in matrix. The stress concentration factors again increased with filament volume, but as pointed out in Ref 1 the change is not linear. At low filament volumes (<50 percent) there is only mild dependence on filament volume. Thus, in this range one might expect the effect of increasing the amount of the stronger phase to be dominant, and indeed the composite shear strength increased with filament volume up to about 45 percent.

As filament volume was increased above 45 percent, tensile failure became increasingly prevalent, and it is quite possible that the limiting factor in the measured shear strength was actually the tensile strength of the composite. In addition, at high filament volumes the shear stress concentration factor increases rapidly (at $V_f = 75$ percent it is 3.7; at $V_f = 78$ percent it is 5.0) so one would expect local shear failures to initiate at lower applied loads. These effects would tend to make the measured shear strength level off, then decrease as filament volume increased above 45 percent. Unfortunately no composites were made having filament volume greater than 70 percent, and the decrease in shear strength was not observed.

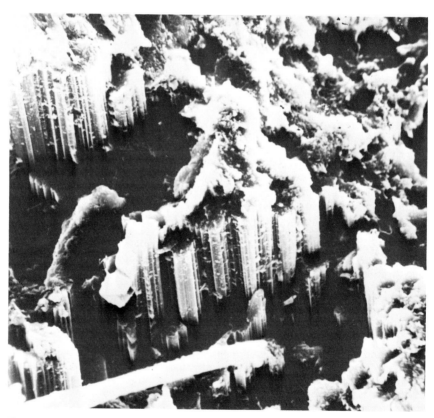

Fig. 6—Fracture surface of oxidized Thornel 40-Epon 828/1031 torsion rod ($\times 1000$).

FIG. 7—Fracture surface of oxidized Thornel 40-Epon 828/1031 torsion rod (×5200).

DISCUSSION

The types of failure which occurred in the graphite filament composites can be divided into three categories. The first type consisted of progressive interfacial debonding then crack propagation from one debonded filament to another. Failure was in shear, and this behavior was observed generally for specimens having shear strengths of 6000 psi or less. The important factor in this mode was the interfacial shear strength being much lower than the shear strength of the filaments or the matrix.

The second type of failure consisted of more or less coincident failure of the interface and the matrix, and occurred in treated composites having shear strengths of approximately 6000 to 8000 psi. Filaments were observed to have failed in tension, and failure tended to be brittle with the degree of this tendency dependent on filament type. This failure mode was largely due to the effects of the filament treatments which produced an interfacial strength approximately equal to that of the matrix without severely degrading the mechanical properties of the filaments.

The final type of observed behavior was a catastrophic failure in a combination of shear and tension. The stress level for the onset of this mode and the relative amounts of tensile and shear failure which occurred were dependent on the filament and the filament treatment. For the oxidized Thornel 40, this behavior was observed at approximately 8500 psi and above. Other investigators [3][4] have reported similar results at shear stress levels ranging from 7000 to 13,000 psi, depending on the

[4] Scola, D. A., United Aircraft Research Laboratories, personal communication.

system. This catastrophic failure was apparently a result of the filament treatment greatly strengthening the interface and degrading the mechanical properties of the filaments.

Further insight into this behavior can be gained by considering the factors which influence the toughness or resistance to cracking of composites. Kelly [4] has described two modes of crack propagation in a unidirectional composite subjected to longitudinal tensile loads. Since tensile failure can play an important part in the obtainable strength in a torsion test, his results are of interest to the present study. He experimentally observed that filaments whose ends were less than half the critical filament length (l_c) away from the advancing crack plane would pull out of the matrix rather than fracture. The toughness, G_c, for this type of failure is given by the following equation:

$$G_{c_1} = \frac{V_f \, d\sigma^2}{8\tau_y} \dots\dots\dots\dots\dots\dots\dots\dots\dots\dots\dots (2)$$

where:
  V_f = filament volume fraction,
  d = filament diameter,
  σ_f = filament tensile strength, and
  τ_y = yield strength of matrix.

Kelly was concerned with metal matrix composites and did not consider the effect of interfacial strength. Consequently, Eq 2 should be modified for graphite-epoxy by replacing τ_y with the smaller of the resin shear strength or the interfacial strength, since this governs the magnitude of the stresses which can be transferred to the filaments.

The second mechanism proposed by Kelly was that of the filaments breaking and the matrix plastically deforming then failing. Here the toughness would depend on the energy adsorbing characteristics of the matrix only, since fracture of the highly elastic filaments would absorb very little energy.

$$G_{c_2} = U_m \frac{d(1 - V_f)^2}{2V_f} \dots\dots\dots\dots\dots\dots\dots\dots\dots (3)$$

where U_m = energy to cause failure in the matrix.

A third mechanism by which composites can absorb energy was discussed by Cook and Gordon [5]. These authors considered a brittle material under longitudinal loading having planes of weakness (interfaces) perpendicular to the direction of an advancing crack. They concluded that with the correct cleavage strength of the weak planes, the advancing crack could be deflected along the planes and effectively blunted. The energy absorbed by this mechanism would be related directly to the energy required to create new surfaces at the filament-matrix interface, summed over the number of interfaces which fail.

$$G_{c_3} = \sum \Big|_0^i \gamma_i \dots\dots\dots\dots\dots\dots\dots\dots\dots\dots\dots\dots (4)$$

where γ = energy to cause failure at the interface.

The total energy absorbed by the composite during fracture is the sum of the contributions of the three mechanisms. Since conventional epoxies are fairly brittle the contribution of G_{c_2} is probably small. Equation 2 shows the G_{c_1} is inversely proportional to interfacial shear strength. From Eq 4 one can see that G_{c_3} would increase initially as interfacial strength increased since strength and surface energy should be related fairly closely. However, if the interfacial strength becomes too high, the mechanism described by Cook and Gordon will become inoperative and G_{c_3} will go to zero.

Thus, a high interfacial shear strength would tend to decrease G_{c_1} and eliminate G_{c_3} with the result that composite fracture would be very brittle. The experimental data definitely supported this theory. The composites made from as-received filaments had a low shear strength and failed in a progressive manner presumably due to filament pullout and crack deflection by the weak interface. Failure in this case occurred when filament debonding and crack propagation between filaments became so extensive that the specimen could no longer support loads.

As the interfacial strength was increased by filament treatments, the mechanisms for toughness became less effective, and failure became increasingly brittle. Fracture occurred at the interface and in the matrix in shear, and filaments failed in tension. At the highest values of composite shear strength, the material behaved much like a homogeneous brittle solid and failed in tension. The implication of this behavior is that the composites having high shear strength may be very notch sensitive and exhibit poor impact behavior. These properties are very important for many proposed applications of graphite composites, and, if the hypothesis is correct, the problem certainly requires a thorough study.

CONCLUSIONS AND RECOMMENDATIONS

Based on the results of this study, the mechanism of failure in graphite filament reinforced epoxy composites under shear loading is dependent on the interfacial strength between the filaments and the matrix. At low interfacial strength, failure is progressive, initiating at the interface and propagating from filament to filament. As interfacial strength is increased by filament treatment, failure occurs at the interface and in the matrix and tends to be increasingly brittle. At high interfacial strength, composite shear strength is high, but failure can be very brittle. This means that increasing the shear strength may be detrimental to other composite properties such as impact strength, and it may be necessary to make a trade-off between shear strength and toughness. Work along these lines is presently being conducted at United Aircraft Research Laboratories. Additional work is needed to determine the relation of variables such as resin toughness, filament strength, and interface strength to composite toughness. By this approach it may be possible to maintain toughness and increase shear strength. Finally, the general field of fracture in fibrous composites must be studied more thoroughly from both analytical and experimental viewpoints so that factors such as the dynamic effect of filament failure and the inelastic behavior of the matrix can be understood better.

ACKNOWLEDGMENT

The author expresses appreciation to the Naval Air Systems Command for support of the experimental work under Contract N00019-67-C-0354 to Aeronutronic Division of Philco-Ford Corporation. The contribution of D. Doner, Aeronutronic, to the analytical portion of the micromechanics study is greatly appreciated.

References

[1] Adams, D. F., Doner, D. R., and Thomas, R. L., "Mechanical Behavior of Fiber-Reinforced Composite Materials," Technical Report AFML-TR-67-96, Air Force Materials Laboratory, May 1967.

[2] Novak, R. C., "Investigation of Graphite Filament Reinforced Epoxies," Final Report for Naval Air Systems Command, Contract N00019-67-C-0354.

[3] Herrick, J. W., "Surface Treatments for Fibrous Carbon Reinforcements," Technical Report AFML-TR-66-178, Part II, Air Force Materials Laboratory, June 1967.

[4] Kelly, A., *Strong Solids*, Clarendon Press, Oxford, 1966.

[5] Cook, J. and Gordon, J., "A Mechanism for the Control of Crack Propagation in All-Brittle Systems," *Proceedings of the Royal Society*, Vol. 282A, 1965, p. 1391.

Electrical and Photographic Techniques for Measuring Microcracking in Some Fiber Reinforced Plastics

WILLIAM MAHIEU[1]

Reference:

Mahieu, William, "Electrical and Photographic Techniques for Measuring Microcracking in Some Fiber Reinforced Plastics," *Composite Materials: Testing and Design, ASTM STP 460,* American Society for Testing and Materials, 1969, pp. 550–561.

Abstract:

In an effort to find new techniques for detecting and measuring the microcracking process in fiber reinforced plastic composites, detectable electrical effects and, in some cases, luminescence have been observed. The nature of these effects at present is not fully known. However, knowledge of their existence makes possible the use of electrical and photographic techniques for investigating microcracking in fiber/plastic composites. This paper describes several experimental techniques that have been utilized and presents quantitative results for several composite materials of current interest. Determination of the potential usefulness of these techniques will require additional knowledge of the nature of the effects and detailed correlation with the composite microcracking process. Further work should be directed toward these ends.

Key Words:

composite materials, microcracking, fiber, reinforced plastics, matrix, interface, mechanical properties, electrical properties, luminescence, triboluminescence, photography, evaluation, tests

Composite materials have the potential for combining the characteristics of high strength and toughness. When these materials consist of brittle fibers in a brittle matrix, as is the case with most structural fiber reinforced plastics of current interest, toughness requires significant microcracking especially along the fiber-matrix interfaces or in the resin following fiber fracture [1].[2] It is desirable that most of this microcracking occur at high stress levels so that the initial properties can be utilized at a working stress that is a significant percentage of the composite's ultimate strength. The microcracking process is, therefore, extremely important in affecting the structural behavior of fiber reinforced plastics, with consequences that may be beneficial or undesirable.

Several techniques have been utilized to study this process in fiber/plastic composites. Probably the most precise but tedious method has been microscopic observation of microcracks in sections cut from prestressed composite specimens [2,3]. Crazing cracks have been observed in filament-wound vessels during pressurization that were of sufficient size to be directly visible [4]. Indirect methods of detecting microcracking that have been used include: monitoring the resulting acoustical

---

[1] Associate research engineer, Research Institute, University of Dayton, Dayton, Ohio. 45409.

[2] The italic numbers in brackets refer to the list of references appended to this paper.

vibrations [5,6], measuring the composite elastic properties and mechanical hysteresis [2], and such techniques as water absorption and dye penetration. However, there is a need for additional means of detecting microcracking in fiber reinforced plastics, especially methods of greater sensitivity that can be used during the loading process.

It was this need that lead to the discovery that at least some crack formation in these materials produces detectable electrical effects and, in some cases, luminescence [7]. Electrical effects have been observed in composites of epoxy resin reinforced with E-glass, S-glass, and graphite fibers. Similar effects were found when these fibers and epoxy resin castings were broken individually. The emission of visible light has occurred during cracking of polyester and epoxy resins reinforced with E-glass fibers.

Knowledge of these effects makes possible the use of electrical and photographic techniques to detect and record microcracks in fiber/plastic composites. Since both electrical and photographic instruments and methods are highly developed and readily available, their use has potential for significantly aiding investigations of microcracking in these materials.

This paper presents some of the electrical and photographic methods that have been used to detect and record microcracking in some fiber reinforced plastics. It has been divided into two main divisions. The first describes the electrical techniques that have been most useful for observing individual electrical signals produced during continuous tensile loading and for integrating them with time to obtain a measure of cumulative microcracking. The photographic techniques used to record the visible light emission during static and impact loadings are discussed in the second part of the body of this paper.

ELECTRICAL TECHNIQUES

NATURE OF ELECTRICAL EFFECT

The nature of the electrical effect observed during microcracking in fiber/plastic composites is not fully known at this time. It appears that the formation of at least some types of cracks produces electrostatic fields, suggesting a separation or polarization of electrical charge. It is also probable that the electrical effect occurs during crack formation as the larger electrical signals were accompanied by audible cracking sounds. All three types of microcracks (fiber, matrix, and interfacial) may contribute to the electrical effect. We have observed electrical signals during cohesive fracture of fibers and cast resin specimens when tested individually. While interfacial debonding has not been shown as yet to produce electrical effects, due to experimental difficulties, Trivisonno has measured charge separation during adhesive fracture of bonds between glass and an unsaturated polyester-styrene adhesive [8].

DETECTION OF ELECTRICAL EFFECT

The electrical effects of microcracking have been detected with a transducer consisting of two electrically conductive, but uncharged, plates located on (silver paint) or near (aluminum plates) the specimen gage section during tensile loading. A resistance of 10^6 to 10^{10} ohms was placed between the plates, one of which was grounded, or between both plates and ground. The voltages induced across this impedance were displayed on a cathode-ray oscilloscope screen. Still and movie cameras were used to record the displayed signals. This experimental technique is diagramed in Fig. 1. Conductive wire coils were tried as electrical input transducers but without success.

Electrical signals obtained with the above technique during tensile loading of bidirectional E-glass/epoxy (Scotchply Type 1002) composite specimens have been

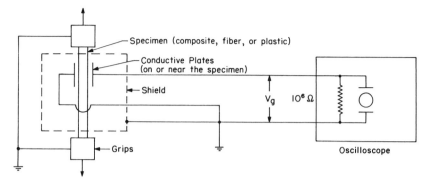

FIG. 1—Method used to observe individual electrical signals during continuous tension tests to fracture.

of two types. The first to occur, and the more recurrent at the lower stress levels, was a sudden change in either direction of the input voltage followed by an exponential decay in a time which appeared to be determined by the input impedance (Fig. 2). The same signals are shown in Fig. 3 at a faster scope sweep rate. In this case the scope was triggered to sweep only on a positive input signal, although some negative signals occurred before the earlier positive signals had decayed. The time required for these signals to reach maximum voltage was about 1 μs. This rise time is believed to have been determined by the cracking phenomenon, rather than limited by the instrumentation, since it responded in approximately 20 nanoseconds during the closure of a mercury switch. These fast signals are believed to have been due to filament fracture. At higher stress levels the signals were complex in form and persisted for longer periods of time (Fig. 4). They

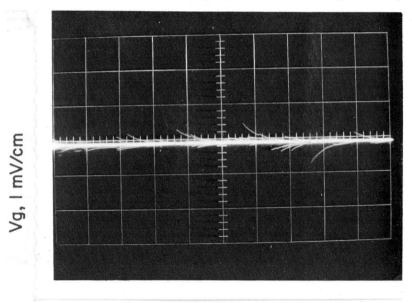

FIG. 2—Characteristic signals from bidirectional E-glass/epoxy composite tension specimen.

Composite Materials

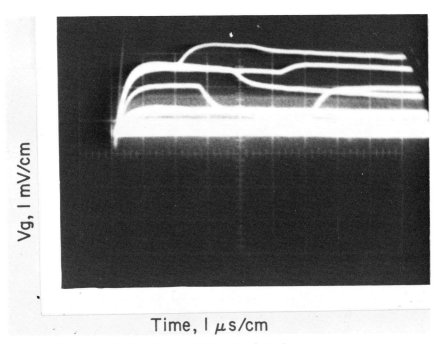

FIG. 3—Characteristic signals from bidirectional E-glass/epoxy tension specimen using triggered oscilloscope and fast sweep rate.

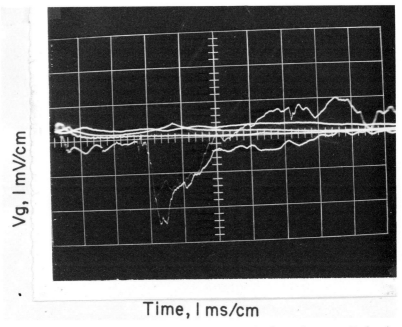

FIG. 4—Irregular signals that occurred at higher loads in the same E-glass/epoxy material.

Mahieu on Microcracking

were accompanied usually by cracking sounds audible to the unaided ear. They probably were produced by large cracks that were a combination of interfacial and matrix cracking.

The sequence of events relating oscilloscope signals to crack formation in the test specimen is believed to have been as follows: the occurrence of a small crack within the composite material produced an electrostatic field. A voltage was induced across the impedance to ground, producing a vertical deflection of the oscilloscope electron beam during its horizontal sweep. When formation of new crack surface had ceased, the induced voltage decreased toward zero in a time dependent on the input impedance. If the above speculation is valid, the number and duration of the displayed signals would have had some relationship to the number and size of material microcracks. However, the complexity of the process and the small amount of experimental investigation completed precludes a more definite statement of what that relationship was.

Time Integration of Electrical Effects

On the assumption that a relationship existed between the microcrack process and the induced electrical signals, however complex, these signals were electronically integrated with time to obtain a measure of cumulative microcracking throughout tension tests to fracture. The objective was to determine the area under the induced

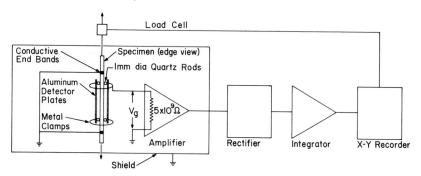

FIG. 5—Method used to obtain the integral with time of the absolute value of the electrical signals during tensile loading of bidirectional fiber/plastic composite specimens.

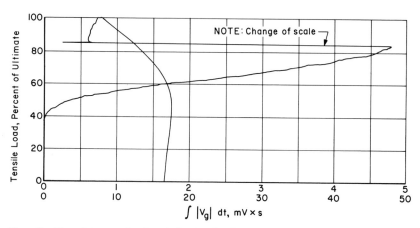

FIG. 6—Time-integrated electrical signals from a bidirectional E-glass/epoxy specimen during continuous tensile loading to fracture.

voltage (Vg)-time signals as it accumulated with increasing tensile load on the specimen. A block diagram of the apparatus designed for this purpose is shown in Fig. 5. In this case, both transducer plates were connected electrically. This did not change the shape of the induced signals. A much higher impedance to ground was used so that the signals would persist for longer periods of time, and the voltages were amplified; both done to increase the area under the voltage-time signals. The amplified voltages were rectified prior to time integration since it was desirable to add both positive and negative signals.'

Curves of time-integrated electrical effects versus load have been recorded during tension tests of the same E-glass/epoxy composite which was used to obtain the

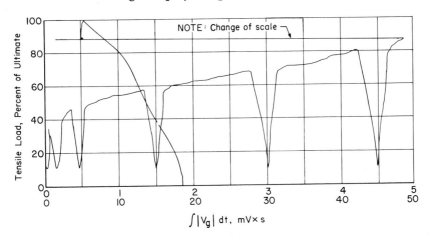

FIG. 7—Time-integrated electrical signals from a bidirectional E-glass/epoxy specimen during cyclic tensile loading to increasingly higher stresses until fracture.

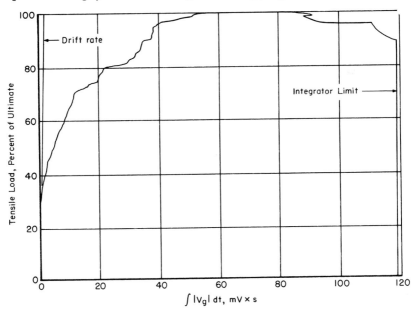

FIG. 8—Time-integrated electrical signals from a bidirectional S-glass/epoxy specimen during continuous tensile loading to fracture.

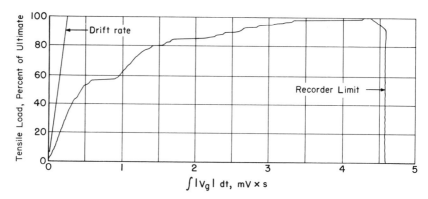

FIG. 9—Time-integrated electrical signals from a bidirectional graphite/epoxy specimen during continuous tensile loading to fracture.

induced voltage-time signals reported above. One such specimen was loaded continuously to fracture (Fig. 6), and another was loaded cyclically to increasingly higher load levels until fracture occurred (Fig. 7). Two observations on these results seem pertinent. Most of the electrical effects occurred at loads above 40 percent of the ultimate load. The electrical effects were much more significant during the first loading to any load level than on subsequent reloading to the same level. This was consistent with McGarry's findings that a disproportionate amount of the microcracking in glass fiber/epoxy composites occurred during the first loading cycle of tension fatigue tests [2]. It also is interesting to note that the electrical effects which occurred at fracture were greater than all that preceded in both of these tests.

The accumulated effects of microcracking have been recorded for a different epoxy resin system (ERL 2256: ZZLB 0820) reinforced with S-glass or graphite (Thornel 25) fibers. The results are shown in Figs. 8 and 9, respectively. The graphite composite exhibited an immediate and nearly linear increase in integrated signal, which was not found with either of the glass composites. This behavior was found on three different graphite specimens tested in this manner and apparently represents a characteristic difference in this material. The integrated effects of the S-glass composite were about twenty times greater than were either graphite in the same resin system or E-glass in a different plastic, both of which produced about the same total effect.

PHOTOGRAPHIC TECHNIQUE

NATURE OF THE LUMINESCENT EFFECT

The emission of visible light has been observed to occur as a result of some crack formation in composites of E-glass reinforced polyester and epoxy resins. While this phenomenon can be classified as an example of triboluminescence, I know nothing about the mechanism of light production in these materials. This light, which was directly observable in a dark room, has emanated from very localized volumes within the material as well as over large areas of delamination. This light has been intense enough to darken unexposed photographic film located near the specimen, allowing photographic recording with the method described below.

PHOTOGRAPHIC RECORDING OF TRIBOLUMINESCENCE

Light emission was recorded by locating unexposed photographic film near the composite specimen during crack production. A fast film (ASA 320) was used.

This technique intercepts much of the emitted light and requires no lens system or shutter mechanism. However, care must be taken not to mechanically damage the film emulsion as this will result in darkening when the film is developed. Also, extraneous light sources must be excluded. This requires sealing the specimen in a light-tight envelope or conducting the test in a dark room. Both methods have been used.

The luminescence appeared to originate from the crack surfaces and to have been fairly uniform in intensity, as evident in Fig. 10. The photo on the left of this figure was printed from a positive, making the luminescence appear dark. Both of these photographs were of planes of delamination in bidirectional unwoven E-glass/epoxy (Epon 828: D) composite specimens. The photographic film was located parallel with the delamination in one case (left photo) and perpendicular to it in the other.

A panel of this same material, when impacted with a hammer, produced very complex patterns of emitted light (Fig. 11). These photographs were obtained by placing the panel between two unexposed films and supporting the assembly horizontally on the end of a steel tube in a dark room. Some damage is evident from the impact on the upper surface film. The lines of luminescence (right photo) corresponded with cracks in the fiber direction and the areas of uniform light with planes of delamination. A similar panel impacted with less force produced less of both damage and luminescence.

One possible use of this technique is the study of glass/plastic composites subjected to impacting projectiles. Results are shown in Fig. 12 of one such impact. A woven E-glass roving/polyester panel (7.6 by 7.6 by 0.64 cm) was impacted with a .22 caliber lead bullet. The bullet passed through the front surface film and about a fourth of the panel thickness before disintegrating and spreading radially outward between delaminated plies. The shape of that delamination corresponded with the light pattern on the front surface photograph. The paths of some of the bullet fragments are even discernible. The light pattern on the side opposite the impacting projectile also is interesting. The intense pattern concentric with the dark spot was caused by shock wave damage of the film emulsion. Such damage has been eliminated by spacing the back-surface film 0.62 cm or more from the panel surface. The dark area in the center was directly opposite the point of impact, but the bullet did not penetrate the panel or the back-surface film. To avoid extraneous light sources, this panel and film was sealed in a black polyethylene bag and mounted in a light-tight cardboard box. The box was then shot, at the appropriate spot, on an indoor rifle range.

The use of the above technique for the study of cracking in fiber/plastic composites obviously will be limited to translucent composites in which the phenomenon occurs. It also appears most promising in those cases in which the cracking process is both rapid and intense. The lensless technique demonstrated above intercepts most of the emitted light provided film of sufficient size can be obtained to cover the area of interest. If lens systems must be resorted to, the light gathering capability of the system probably will be a critical consideration. Cameras have not been used as yet.

CONCLUSIONS

1. Electrical and photographic techniques have been presented which appear to be promising new tools for the study of the micrcracking process in some fiber reinforced plastic composites.

2. These techniques are based on recently observed electrical and luminescent effects that result from at least some crack formation in these materials.

3. Determination of the potential utility of these and similar methods will require additional knowledge of the underlying physical phenomena and detailed

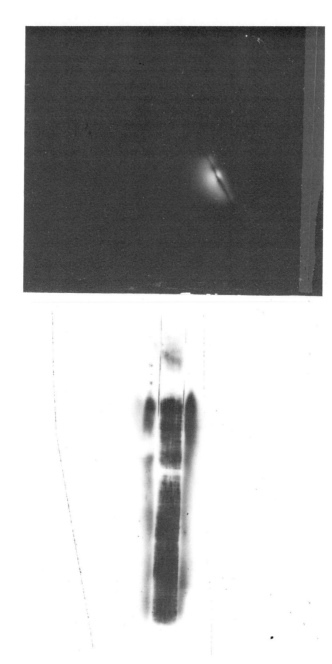

FIG. 10—Luminescence from flexural delaminations of bidirectional E-glass/epoxy specimens during static bending. Film was paralleled with a plane of delamination of left (negative print) and perpendicular to another on the right.

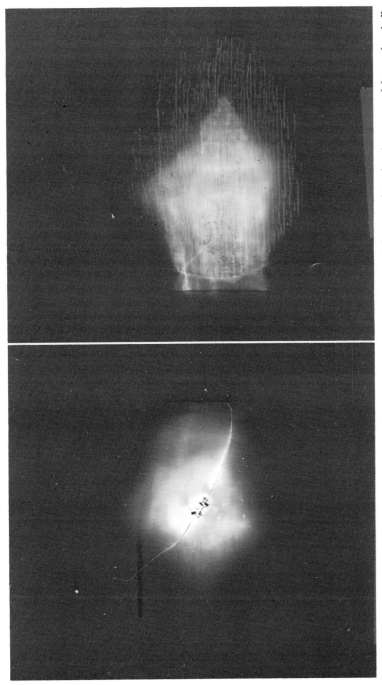

FIG. 11—Luminescence from impacted surface (left) and opposite surface of a bidirectional E-glass/epoxy panel hit with a ball-peen hammer.

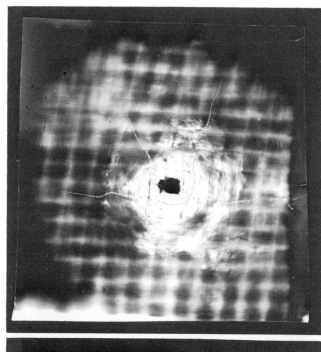

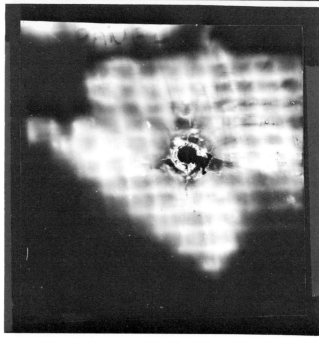

FIG. 12—Luminescence from impacted surface (left) and opposite surface of a woven E-glass roving/polyester panel impacted with a .22 caliber bullet.

correlation of the effects with the composite microcracking process. Future work should be directed toward those ends.

ACKNOWLEDGMENT

Several individuals on the Research Staff of the University of Dayton have made essential contributions to the experimental techniques and results presented in this paper. R. L. Fusek developed the method of recording the oscilloscope traces on movie film and conducted the tests utilizing it. C. E. Smith designed the electronic integrator. R. Larson constructed it. S. J. Hanchak conducted the time-integration tests and the photographic recording of luminescence during impacts.

This work was sponsored by the Nonmetallic Materials Division, Air Force Materials Laboratory, Wright-Patterson Air Force Base, Ohio. It was conducted under contract number AF 33(615)-3512.

References

[1] Kelly, A. and Davies, G. J., "The Principles of the Fiber Reinforcement of Metals," *Metallurgical Reviews*, Vol. 10, No. 37, 1965, pp. 45–51.

[2] McGarry, F. J., "Relationships Between Resin Fracture and Composite Properties," AFML-TR-66-288, Air Force Materials Laboratory, Wright-Patterson Air Force Base, Ohio, Sept. 1966.

[3] Broutman, L. J., "Failure Mechanisms for Filament Reinforced Plastics Subjected to Static Compression, Creep and Fatigue," *19th Proceedings*, Reinforced Plastics Division, The Society of the Plastics Industry, 1967, Section 9-C.

[4] Epstein, G. and Bandaruk, W., "The Crazing Phenomenon and its Effects in Filament-Wound Pressure Vessels," *19th Proceedings*, Reinforced Plastics Division, The Society of the Plastics Industry, 1967, Section 19-D.

[5] Steele, R. K. et al, "Use of Acoustical Techniques for Verification of Structural Integrity of Polaris Filament-Wound Chambers," *Proceedings of the 20th Conference*, Society of Plastics Engineers, Vol. X, 1964, Paper XXV-1.

[6] Noyes, J. V. and Jones, B. H., "Crazing and Yielding of Reinforced Composites," AFML-TR-68-51, Air Force Materials Laboratory, Wright-Patterson Air Force Base, Ohio, March 1968.

[7] Mahieu, W. "Electrical and Luminescent Effects of Fracture in Some Fiber Reinforced Plastics," AFML-TR-68-113, Air Force Materials Laboratory, Wright-Patterson Air Force Base, Ohio, May 1968.

[8] Trivisonno, N. M., "Electrostatic Effects in the Adhesion of Polyester Resin to Glass," Ph.D. thesis, Case Institute of Technology, Cleveland, Ohio, 1958, obtainable from University Microfilms, Inc., Ann Arbor, Mich.

Microstrain in Continuously Reinforced Tungsten-Copper Composites

E. J. HUGHES[1] AND
J. L. RUTHERFORD[1]

Reference:

Hughes, E. J. and Rutherford, J. L., "Microstrain in Continuously Reinforced Tungsten-Copper Composites," *Composite Materials: Testing and Design, ASTM STP 460*, American Society for Testing and Materials, 1969, pp. 562–572.

Abstract:

The microstrain characteristics of continuously reinforced tungsten-copper composites have been studied as a function of volume fraction of fibers and prestrain. In the unstrained condition, the dislocation friction stress of all of the specimens studied had a value of approximately 0.28 kgf/mm² (400 psi). The initial microyield stress, however, was dependent on the volume fraction of fibers, and a linear relationship was observed. Both the friction stress and microyield stress were prestrain dependent. The rate of increase of both of these quantities was dependent on the volume fraction of fibers. The results are discussed qualitatively in terms of high dislocation density regions in the copper matrix surrounding the fibers, which arise from stresses in the matrix as a result of the differential contraction between the tungsten and copper during fabrication.

Key Words:

strains, stresses, tungsten, copper, fiber composites, mechanical properties, fiber spacing, evaluation, tests

NOMENCLATURE

σ_E	Precision elastic limit
σ_{mys}	Microyield stress
ΔW_{irr}	Energy loss in generating a closed hysteresis loop
ϵ_d	Dislocation strain
σ_f	Lattice friction stress
V_f	Volume fraction of fibers
D	Fiber diameter
δ	Interfiber spacing

The properties of fiber reinforced composites consisting of both metallic and non-metallic matrices and fibers have been studied extensively over the past few years. These properties have been reviewed in detail by Kelly and Davies [1][2] and Cratchley [2]. A number of theoretical studies of the micromechanics of composites have been carried out recently. Those treatments usually assume an elastic and isotropic matrix, together with a simple geometric arrangement of fibers. The basic objective of these studies is to enable a prediction of the gross composite properties from a knowledge of the elastic constants of the components. This type of analysis has met with only limited success when applied to real systems due to the necessary

---

[1] Staff scientist and senior staff scientist, respectively, Materials Department, Aerospace Research Center, Singer-General Precision, Inc., Kearfott Group, Little Falls, N. J. 07424.

[2] The italic numbers in brackets refer to the list of references appended to this paper.

assumptions such as: the type of packing, elastic behavior of the constituents, and an isotropic system. These conditions, of course, never are fulfilled in practical systems. The experimental science of micromechanics of reinforced composite materials is less advanced than its theoretical counterpart, and to date the most success has been obtained using photoelastic techniques. In order to achieve useful results, the theories of micromechanics must be verified by experiment. Furthermore, if the nonideal behavior of composites can be determined experimentally, it should be possible to include the results of such measurements in the theoretical analysis and so lead to a more meaningful model for the behavior of composites under the influence of an applied stress.

The objective of the present paper was to determine the mechanical behavior of unidirectionally reinforced composites under the influence of a tensile stress by using a microstrain technique which is capable of measuring very small changes in specimen dimensions. The tungsten wire/copper matrix system was chosen for study because: (1) the properties of the constituents are well known and reproducible, (2) copper wets tungsten, and (3) there are no detrimental interfacial reactions between fiber and matrix.

EXPERIMENTAL PROCEDURE

SPECIMEN PREPARATION

Specimens were prepared by a liquid infiltration technique. The 0.13-mm (0.005-in.)-diameter tungsten (General Electric Type 218 CS) were cut into 7.5 cm (3 in.) lengths and placed in a graphite crucible having a 9.53 mm (0.375 in.) inside diameter. The crucible fitted into a graphite collar which served as a reservoir for the molten copper. The reservoir was filled with rods of OFHC copper of 99.999 percent purity, and the whole assembly was placed in a quartz tube which was evacuated to a pressure of 10^{-3}-mm Hg. The copper and crucible were heated slowly to 1100 C using an rf induction coil. The molten copper infiltrated the wires in the crucible, and finally the whole assembly was lowered slowly through the induction coil to reduce any porosity.

MECHANICAL TESTING

Tension specimens were prepared from the as-cast ingots by electrodischarge machining. They had a gage length of 12.7 mm (0.5 in.) and a gage diameter of 3.18 mm (0.125 in.). The specimens were tested in an Instron tensile machine at a strain rate of 1.7×10^{-4}/s. Elongations were measured with a capacitance type extensometer [3] which has a measuring sensitivity of 2.54×10^{-5} mm (10^{-6} in.). The extensometer consists of a pair of circular copper plates that are mounted concentrically on the specimen. These capacitor plates form one leg of a capacitance bridge. At the start of the test the bridge is balanced; as the specimen is loaded, the distance between the plates changes, unbalancing the bridge. The out-of-balance voltage thus generated is fed to the X-axis of an X-Y recorder. One of the capacitor plates is mounted on a precision micrometer thread to provide for calibration. The load signal from the Instron load cell is fed into the Y-axis of the recorder. Thus load-extension curves may be recorded at very high sensitivities. To minimize specimen misalignment, the load is applied through ball-and-socket grips. A small bias stress is maintained to preserve axiality. In studying the microstrain behavior, a series of load-unload cycles are drawn at increasing stress amplitudes (Fig. 1). At low stresses the deformation is elastic; at higher stresses the precision elastic limit (σ_E) is exceeded. Deviations from elasticity may be in the form of hysteresis loops with no permanent set or by plastic strain. The stress (σ_{mysx}) required for permanent strain is generally larger than σ_E. The load-unload tests then are repeated until an open loop is generated. Predetermined amounts of permanent strain are

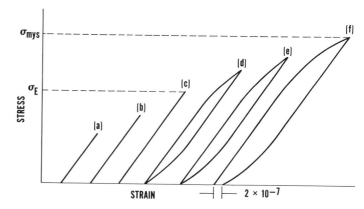

FIG. 1—Schematic of microstrain behavior in load-unload tests.

put into the specimen, and the σ_{mys} is remeasured by carrying out load-unload tests. This procedure is repeated up to the fracture stress of the specimen. Following fracture, the specimens were sectioned transversely, away from the fracture and outside the necked region. The sections were mounted and polished, and the volume fraction was determined from a wire count.

Typical microstrain behavior is shown schematically in Fig. 1. Up to σ_E, the load-unload cycles are elastic. The first hysteresis loops are observed at a stress slightly greater than σ_E. At increasingly higher stresses, the strain amplitudes increase until an open loop is observed, indicating permanent set. In the anelastic region, the behavior is reproducible, and identical loops can be obtained by repeated load-unload cycles drawn to the same stress. During anelasticity, the area of the hysteresis loop is a measure of the energy loss which is attributed to bowing of dislocations. The lattice resistance to dislocation motion may be considered as a friction stress which dissipates energy. The energy lost (ΔW_{irr}) in generating a closed loop (determined by the loop area) is related to the dislocation strain (ϵ_d) by:

$$\Delta W_{irr} = 2\sigma_f \epsilon_d \dots\dots\dots\dots\dots\dots\dots\dots(1)$$

In Eq 1, σ_f is the friction stress and can be obtained by taking half the slope of the curve obtained by plotting ΔW_{irr} versus ϵ_d. The dislocation strain is the deviation of the loop from the initial elastic slope. The friction stress also can be determined by a method arising from a theory proposed by Cottrell [4]. From this theory it follows that the friction stress (σ_f) is one half of the elastic limit (σ_E), which is the minimum stress required to produce a hysteresis loop. Thus, a plot of loop area against stress amplitude should give a value of $2\sigma_f$ on extrapolation to zero loop area. Both methods of determining σ_f were used in the present work and gave comparable results. Examples of both types of calculations are shown in Figs. 2 and 3.

RESULTS

A series of composite specimens were prepared using 0.13-mm (0.005-in.)-diameter tungsten wires with volume fraction of fibers ranging from 0.11 to 0.61. The friction stress and microyield stress were measured at room temperature as a function of prestrain.

FRICTION STRESS (σ_f)

The variation of σ_f with prestrain was examined for each of the composites (Fig. 4). In each case, the friction stress increased with increasing prestrain. How-

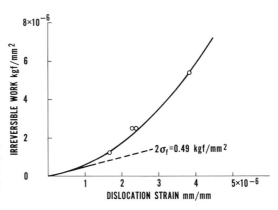

FIG. 2—Plot of irreversible work against forward plastic strain for specimen containing 21.8 percent tungsten fibers.

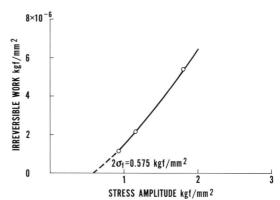

FIG. 3—Plot of irreversible work against stress amplitude for specimen containing 21.8 percent tungsten fibers.

ever, the slope of the σ_f versus prestrain curve was found to increase with increasing volume fraction of fibers. No friction stress values could be obtained for the specimen containing 61.4 volume percent, since this specimen showed no hysteresis prior to yielding. Only after considerable prestrain (2.24×10^{-4}) did this specimen show hysteresis loops. It is interesting to note that although all of the composites showed different rates of increase of σ_f with prestrain, they all had approximately the same friction stress in the unstrained state, as shown in Table 1.

Table 1—*Friction stress of composites prior to straining.*

Fibers, volume %	$\sigma_f (kgf/mm^2)$
61.4	—
30.0	0.281 (400 psi)
23.8	0.281 (400 psi)
21.8	0.287 (438 psi)
16.1	0.263 (375 psi)
11.2	0.239 (340 psi)

MICROYIELD STRESS

The microyield stress of each composite was determined in the unstrained state and also as a function of prestrain. In unstrained specimens the microyield stress increased with increasing volume fraction in a linear manner as illlustrated in Fig.

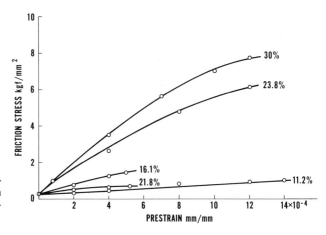

FIG. 4—Variation of friction stress with prestrain in tungsten-copper composites.

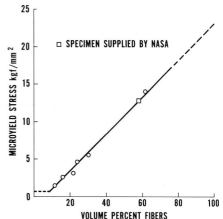

FIG. 5—Microyield stress of copper reinforced with 0.127 mm (0.005 in.) diameter tungsten fibers.

5. Also included in Fig. 5 is the microyield stress of a composite specimen supplied by NASA/Lewis,[3] and tested in this laboratory. The microyield stress of this specimen is in good agreement with those fabricated and tested in the present program. The microyield stress of pure copper has been determined independently by a number of workers [5–7] who obtained values ranging from 0.492 to 0.703 kgf/mm² (700 to 1000 psi). The σ_{mys} versus volume percent fibers relationship does not extrapolate to this value, indicating there is a critical volume fraction of fibers (approximately 0.10) which must be exceeded if reinforcement is to take place. This is in agreement with the findings of other workers [8,9] who found a similar effect in the macroyield stress versus volume percent fibers relationship. When extrapolated to 100 volume percent fibers, a microyield stress of 21.97 kgf/mm² (31,250 psi) was obtained for the tungsten fibers. For prestrains up to 2 × 10⁻⁴ the microyield stress increased in a manner similar to that of the friction stress. This is illustrated in Fig. 6, which also shows that the rate of increase gets larger as the volume fraction of fibers is increased. At higher amounts of prestrain (Fig. 7), the microyield stress-prestrain relationship showed a sharp discontinuity, which occurred at prestrains that decreased with increasing volume fraction and at stresses which increased with increasing volume fraction. Table 2 lists values for the stress on the fibers (calculated from elasticity theory) and the prestrain corresponding to the discontinuities of Fig. 7.

[3] Signorelli, R. W., private communication.

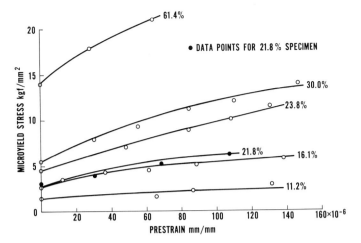

FIG. 6—Variation of microyield stress with prestrain after small amounts of prestrain.

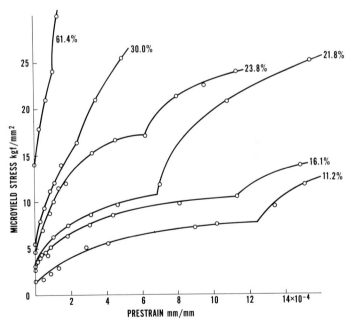

FIG. 7—Variation of microyield stress with prestrain after large prestrains.

Fibers, volume %	Stress on Fibers at Discontinuity, kgf/mm²	Prestrain
11.2.............	28.71(40 840 psi)	12.4 × 10⁻⁴
16.1.............	35.15(50 000 psi)	11.1
21.8.............	31.79(45 220 psi)	6.9
23.8.............	47.76(67 930 psi)	6.2
30.0.............	33.71(47 940 psi)	2.4
61.4.............	36.35(51 700 psi)	1.0

DISCUSSION

FRICTION STRESS

The friction stress (σ_f) measured in the unstrained composites was essentially the same regardless of the volume fraction of fibers. These results indicate that the presence of tungsten fibers has little or no influence on the initial friction stress of the copper matrix. The value of approximately 0.28 kgf/mm² (400 psi) obtained in the present work in the unstrained composites is comparable with the friction stress of 0.422 kgf/mm² (600 psi) obtained by Galyon [6] on polycrystalline OFHC copper. In all the composites tested, the friction stress was dependent on the prestrain, but, in the lower volume fraction materials, the effect was diminished. In the higher volume fraction composites (0.238 and 0.300) a much greater dependence was observed. After a prestrain of 5 × 10⁻⁴, σ_f had values of 3.23 kgf/mm² (4600 psi) and 4.22 kgf/mm² (6000 psi), respectively, in the above two composites compared with a maximum value of 1.41 kgf/mm² (2000 psi) observed in the other three lower volume fraction specimens. Galyon's work with copper showed increases in σ_f to a maximum of 0.703 kgf/mm² (1000 psi) at a strain of 50 × 10⁻⁴.

The mechanism of energy dissipation giving rise to the friction stress may result from one or more of the following dislocation mechanisms:

1. Lattice resistance such as the Peierls-Nabarro force.
2. Pinning of dislocations by point defects or impurities.
3. Internal stress fields.

When a dislocation passes from one equilibrium position to the next, it must pass through a high energy configuration. The Peierls-Nabarro energy is the difference between the initial position and the high energy position; the stress to overcome this energy barrier is the Peierls-Nabarro stress. The Peierls-Nabarro stress is short range and, in effect, represents the resistance of the lattice to dislocation motion. If the friction stress measured in the present work arose from the Peierls-Nabarro force, it should be temperature dependent and prestrain independent. Although all the experiments in the present work were carried out at room temperature, the fact that a prestrain dependence was observed indicates that the Peierls-Nabarro force is not a major contributor to the friction stress in the tungsten-copper composites.

The second mechanism for energy dissipation is due to pinning of dislocations by point defects. When a point defect is located at a dislocation, it reduces the strain energy of the dislocation in its immediate vicinity. If the dislocation is to move, its strain energy must be increased at the place where it will leave the point defect behind. Thus, point defect pinning will necessitate a higher stress to move the dislocations through the lattice. Once again, however, this mechanism should be independent of prestrain because the impurity content does not change as the material is deformed. Since there is a prestrain dependence, the dislocation pinning mechanism cannot be operative. Furthermore, since high-purity copper was used in the experiments, this mechanism is not likely.

Bonfield and Li [10] recently suggested that the friction stress in beryllium was dependent on the elastic interaction of the dislocation stress fields. Since the latter are influenced by the dislocation density, then σ_f should be prestrain dependent. In their study of the friction stress in both annealed and hot pressed beryllium, σ_f was independent of prestrain in the hot pressed beryllium as opposed to the annealed material which was prestrain dependent up to approximately 10^{-4} prestrain. These observations were combined with an electron microscopy study which indicated that σ_f varied with the dislocation density, which in turn caused an increase in the local internal stress fields due to the elastic interaction of parallel dislocations.

From Fig. 4, in specimens containing less than 23.8 percent fibers, σ_f behaved in a similar manner to pure copper and showed only a small dependence on prestrain. Assuming the initial value of 0.281 kgf/mm² (400 psi) was due to dislocation bowing in the copper matrix alone, then the small prestrain dependence of σ_f in these specimens indicates that for the prestrains measured all of the dislocations which were influenced by the applied stress were probably in the matrix material. The influence of volume fraction on the prestrain dependence of the friction stress can be understood qualitatively if it is assumed that each filament is surrounded by a high stress region of the matrix with a dislocation density higher than the other matrix areas. These regions are similar to those observed by Hancock and Grosskreutz [11] in the steel-aluminum system that arise from the differential contractions of the filaments and matrix. These workers recently conducted an electron microscopy study on a 2024 aluminum alloy, reinforced with axially aligned stainless steel wires, solution treated and quenched prior to examination. It was seen that the matrix was in a strain hardened condition, prior to deformation, due to the differential contraction of the filaments and the matrix. Furthermore, by observing the annihilation of helical dislocations, it was established that plastic flow in the matrix began at stresses well below the macroyield stress of the composite. Hancock and Grosskreutz observed high dislocation densities at the interface indicating that local stresses exist there. The dislocation density decayed rapidly with distance from the interface and became approximately constant beyond about 19.3×10^{-3} mm (8×10^{-4} in.).

The situation envisaged in the present work is shown schematically in Fig. 8a, which illustrates tungsten fibers in the copper matrix. On the application of a

(a) VOLUME FRACTION < 0.23   (b) VOLUME FRACTION ~ 0.23

FIG. 8—Schematic of variation of fiber spacing with volume fraction of composites, showing gradual impingement of high dislocation density stress fields.

(c) VOLUME FRACTION ~ 0.47

stress the dislocations first activated would be those well away from the filament-matrix interface in the strain free areas of the copper matrix which have a comparatively low dislocation density. Dislocations in these regions will be able to bow with little or no intersection with other dislocations, giving rise to a behavior similar to that observed in pure copper. As the volume fraction is increased, the interfiber distance decreases, until a stage is reached where the fiber stress fields touch leaving very small areas of the strain free, low dislocation density matrix. This situation is shown in Fig. 8b. The applications of a stress in this case will give rise first to dislocation bowing in the copper matrix where low dislocation density regions still exist. Thus, the measured friction stress will still be that of pure copper. Prestraining will generate more dislocations in these small, low dislocation density regions causing a rapid increase in the dislocation density. This will allow dislocation intersections and interactions on the applications of a stress. Following the arguments of Bonfield and Li, the friction stress will be expected to increase appreciably with increasing prestrain when such a situation exists. This behavior was observed in the specimen containing 23.8 percent fibers. Between approximately 22 and 24 percent fibers, the friction stress behavior in the present work changed from being almost prestrain independent to prestrain dependent. It is believed, therefore, that in this region the fiber stress fields have approached each other until they are touching or just overlapping. Assuming that touching occurs between these two volume fractions, that is, about 23 percent, then the size of the stress field surrounding the fibers can be calculated.

The relationship between fiber volume fraction and interfiber spacing has been derived by Snide [12] for cubic, hexagonal, and rectangular packing. Although the packing in the present work was not ideally uniform, a cubic arrangement was assumed. Snide derived the following expression for cubic packing.

$$V_f = \frac{\pi}{4} \frac{D^2}{(\delta + D)^2}. \quad\dots\dots\dots\dots\dots\dots\dots\dots\dots(2)$$

where D is the fiber diameter and δ the interfiber spacing.

Using Eq 2 gives an interfiber spacing of 17.8×10^{-3} mm (7×10^{-4} in.) for a 61.4 volume percent composite reinforced with 0.127-mm (0.005-in.)-diameter fibers. This increases to 20.3×10^{-2} mm (8×10^{-3} in.) for a 11.21 volume percent composite.

Since the diameter of the fibers and the volume fraction at which the stress fields touch (approximately 0.23) are known, then using Eq 2 the size of the stress field can be determined by a straight forward calculation of the interfiber spacing.

Using the above quantities, δ is calculated to be 0.107 mm (0.0042 in.). At 23 volume percent the stress fields surrounding the fibers just touch so that the interfiber spacing, δ, is twice the radius of the stress fields, which therefore have a radius of 0.053 mm (0.0021 in.). This compares with about 0.020 mm (0.0008 in.) observed for the steel-aluminum system which has a differential contraction ratio of 2 compared with 3.6 in the tungsten-copper system. Higher fiber volume fractions will result in increased overlapping of the stress fields leaving a decreasing volume of low dislocation density copper. Composites in this range, following the above arguments will be characterized by a σ_f equal to that of pure copper, together with an increasing prestrain dependence of σ_f which was observed. Eventually a volume fraction will be reached when the stress fields completely overlap, as shown in Fig. 8c. By simple geometry this corresponds to an interfiber distance of 0.038 mm (0.0015 in.). Substituting this value in Eq 2 gives $V_f = 0.465$. Thus at approximately 46.5 percent fibers the matrix material should contain a uniform, high-density dislocation network.

From the above discussion it appears that the tungsten fibers are surrounded by a region containing a high density of dislocations. When the specimen is loaded, the initial dislocation movement takes place outside these regions where they can move more freely. As the volume fraction of fibers increases, the high-density regions approach each other and touch at an interfiber spacing of about 0.107 mm (0.004 in.). Further increases in the volume fraction of fibers cause overlapping of the stress fields. However, at volume fractions less than about 0.465 there are still areas of the original matrix in which the dislocations can easily bow, Without intersecting other dislocations. According to this hypothesis, as the volume fraction of fibers is increased above about 0.465 there should be a stage at which a sudden increase in σ_f will be observed, indicating the complete interaction of the high dislocation density regions throughout the matrix. This type of behavior was observed in the 61 volume percent specimen.

MICROYIELD STRESS

The microyield stress increased linearly with volume fraction of fibers, as did the ultimate tensile strength, indicating that these properties are affected by the reinforcing fibers in a similar manner to the macroyield stress. It is not possible at this stage to predict whether or not the "rule of mixtures" is obeyed fully since the microyield stress of the tungsten fibers is not known. The variation of the microyield stress with prestrain is in keeping with observations on a number of materials including aluminum [13,14] nickel [15], brass [5] beryllium [10], and copper [6,7,14]. The microyield stress is the first indication of long range dislocation motion and can therefore be taken as the stress at which dislocation sources are activated. Prestraining a composite specimen increases the dislocation density in the copper matrix.

The microstrain mechanisms leading to this type of behavior have been described and discussed in detail in other publications [10,14,15]. It will suffice, therefore, to state that the increased dislocation density arising from prestraining results in a microwork hardening process similar to that observed in the macroregion. The rate of change of microyield stress with prestrain increased with fiber volume fraction, as is shown in Fig. 6. This can be attributed to the decreasing volume of strain-free matrix material with an increase in fiber volume fraction. Consequently, for a given amount of prestrain the dislocation density of the matrix material should increase with increasing volume of fibers (provided that the fibers deform elastically). This will lead to increasing dislocation interactions at higher volume fractions, so giving rise to an increase in the microyield stress of the composite.

An interesting feature observed in the present work is the transition observed in the microyield stress versus prestrain curve (Fig. 7). The transition occurs at increasingly lower strains and higher stresses as the volume fraction of fibers increases. The reasons for such behavior are obscure, but it could signify the onset of plastic flow (microyield stress) in the tungsten fibers. The stress on the fibers at the discontinuity was calculated, and the values shown in Table 2 were obtained. For this calculation, a tungsten modulus of 35.15×10^3 kgf/mm^2 (50×10^6 psi) was used. Since the copper was in a plastic state, a modulus value of 5.48×10^3 kgf/mm^2 (7.8×10^6 psi) was taken from a stress-strain curve for copper similar to that used in the matrix. With one exception, the stress on the fibers at the discontinuity was in the range 28.83 to 35.86 kgf/mm^2 (41,000 to 51,000 psi). If the discontinuity was due to the onset of plastic flow in the tungsten fibers, then the stress on the fibers in each composite at the discontinuity should be the same. It is not surprising that the tungsten fiber microyield stress obtained by this method differs from 21.97 kgf/mm^2 (31,250 psi) obtained by extrapolating the values of Fig. 5. This method gives a value for the microyield stress of tungsten embedded in copper, whereas that of Fig. 5 is the microyield stress of tungsten by itself. It has been shown by a number of workers that the mechanical properties of materials are affected strongly by their environment.

ACKNOWLEDGMENTS

The authors want to thank Charles Maccia for fabrication and metallographic examination of the composite specimens and Kenneth Zwoboda for machining them.

References

[1] Kelly, A., and Davies, G. J. "The Principles of the Fibre Reinforcement of Metals," *Metallurgical Reviews*, Vol. 10, 1965, p. 79.

[2] Cratchley, O., "Experimental Aspects of Fibre-Reinforced Metals," *Metallurgical Reviews*, Vol. 10, 1965, p. 1.

[3] Roberts, J. and Brown, N., "Microstrain in Zinc Single Crystals," *Transactions*, American Institute of Mining, Metallurgical, and Petroleum Engineers, Vol. 218, 1960, p. 454.

[4] Cottrell, A. H., *Dislocations and Plastic Flow in Crystals*, Oxford University Press, London, 1953, p. 112.

[5] Rutherford, J. L., "Observations of Dislocation Behavior in Copper and Alpha Brass, in the Temperature Range 298° to 5°K, using a Microstrain Technique," Ph.D. thesis, University of Pennsylvania, Philadelphia, Pa., 1963.

[6] Galyon, G. T., "Microstrain in Copper," M. S. thesis, University of Pennsylvania, Philadelphia, Pa., 1963.

[7] Thomas, D. A. and Averbach, B. L., "The Early Stages of Plastic Deformation in Copper," *Acta Metallurgica*, Vol. 7, 1959, p. 69.

[8] McDanels, D. L., Jech, R. W., and Weeton, J. W., "Stress-Strain Behavior of Tungsten-Fiber-Reinforced Copper Composites," NASA Technical Note D 1881, National Aeronautics and Space Administration, 1963.

[9] Kelly, A. and Tyson, W. R., *Proceedings of the 2nd International Materials Symposium*, Wiley, New York and London, 1964.

[10] Bonfield, W. and Li, C. H., "The Friction Stress and Initial Micro-Yielding of Beryllium," *Acta Metallurgica*, Vol. 13, 1965, p. 317.

[11] Hancock, J. R. and Grosskreutz, J. C., "Plastic Yielding and Strain Distribution in Filament-Reinforced Metals," *Metal Matrix Composites, ASTM STP 438*, American Society for Testing and Materials, 1968, pp. 134–149.

[12] Snide, J. A., "Relation of Volume Fraction to Reinforcement Spacing and Sheet Thickness of Diffusion Bonded Metal Matrix Composites," Technical Report AFML-TR-67-122, Air Force Materials Laboratory, 1967.

[13] Ku, R. C., "Micro-Deformation in Aluminum," M. S. thesis, University of Pennsylvania, Philadelphia, Pa.

[14] Rosenfield, A. R. and Averbach, B. L., "Initial Stages of Plastic Deformation in Copper and Aluminum," *Acta Metallurgica*, Vol. 8, 1960, p. 624.

[15] Carnahan R. D. and White, J. E., "The Microplastic Behavior of Polycrystalline Nickel," *Philosophical Magazine*, Vol. 10, 1964, p. 513.

The Role of Bond Strength in the Fracture of Advanced Filament Reinforced Composites

A. GATTI,[1] **J. V. MULLIN,**[1]
AND J. M. BERRY[1]

Reference:

Gatti, A., Mullin, J. V., and Berry, J. M., "The Role of Bond Strength in the Fracture of Advanced Filament Reinforced Composites," *Composite Materials: Testing and Design, ASTM STP 460*, American Society for Testing and Materials, 1969, pp. 573–582.

Abstract:

A great number of studies have been made to establish the longitudinal strength characteristics of composites, with success being measured in many instances by the degree to which "rule of mixtures" predictions have been attained. Several investigators report good agreement for modulus and tensile strength in metallic filament reinforced composites. Agreement for strength predictions is less frequent in systems using continuous and discontinuous nonmetallic filament arrays. This lack of ability to predict strength of uni-axially reinforced specimens points up the need for further research on the factors contributing to premature and sometimes catastrophic failure of filament reinforced specimens.

As a result of using single and multiple filament specimens of boron in epoxy novolac, methods for localizing the damage resulting from individual filament fractures were developed.

Optimization of the bonding mechanism is one area of research which has yielded some very interesting results. Contrary to the concept that bond strength should be as high as possible, these studies show that, although a good bond is essential to the reinforcing mechanism, too high a bond strength may render the composite severely matrix limited. Further, it has been demonstrated in this work that varying the nature of the bond surface can be most effective in isolating the sudden energy release at a filament fracture.

Key Words:

fiber composites, boron, epoxy laminates, fracture properties, bonding strength, tests, evaluation

The potential afforded by composite strengthening of metals and plastics with advanced filamentary materials presently available or being developed has led to extensive effort to produce, utilize, and understand such materials. A good approximation of the mechanical properties to be expected is afforded by the "rule of mixtures" prediction which depends on each constituent of a composite to contribute its properties on a volume fraction basis. Thus, modulus measurements which require very small stress during determination adhere to the rule almost universally. However, the results begin to vacillate when larger strains typical of design stress levels are reached. Descriptive terms such as "synergism," catastrophic, damage, efficiency, and the like, begin to appear. The problem introduces a complexity which the rule of mixtures cannot handle and indeed, was never intended to explain. The obvious answer is, of course, the role which the interaction of the composite components play during service.

[1] Metallurgist, material scientist, and mechanical metallurgist, respectively, Space Sciences Laboratory, Missile and Space Division, General Electric Co., Philadelphia, Pa. 19101.

Briefly then, a composite is not a mixture of linearly behaving constituents but a combination of filamentary variable and matrix variables which combined in some complex fashion to produce yet another group of composite variables. The enumeration of these variables is in itself complex; 15 or 20 can be listed without difficulty. This combined behavior or "mechanical compatibility" was the subject of a recent paper presented in Toronto.[2]

Some of the more pertinent parameters were enumerated, and their relationship to the expected tensile behavior of a composite was indicated. Bond strength and the products of bond strength and crack sensitivity were noted to be important considerations. This and other work[3] had delineated the failure modes associated with filament failures in epoxy specimens. It was shown that the least damaging failure mode was unbonding when compared to either a high-speed disk shaped tensile crack or a slow-growing resolved shear crack.

The purpose of the work described in this paper was to test the inference that bond strength should be optimized, not necessarily maximized, for effective utilization of filament strength.

EXPERIMENTAL PROCEDURES

Single and multiple filaments of boron on tungsten (B/W) imbedded in an epoxy novolac (DEN 438)[2] were used to study bonding parameters. As it turned out, the multiple filament specimens and the single filament specimens behaved identically so that only the single filament results will be reported. A prerequisite to this study was to devise a means by which the bond strength between the filament at the matrix could be varied without affecting mechanical, physical, and chemical properties of the matrix. The problem of varying bond strength independent of these variables was solved by treating the surface of the filaments with graphite or Teflon so that intermittent bonds would form. These materials are inert in the novolac formulation, and it is to be noted that many other organic or inorganic lubricants or sizings (such as MoS_2, WS_2, etc.) could be equally effective. The

[2] Berry, J. M., Gatti, A. and Mullin, J. V., "The Role of Mechanical Compatibility in Advanced Filament Composites," *International Symposium on Macromolecular Chemistry*, 3–5 Sept. 1968, Toronto, Ont., Canada.
[3] Mullin, J. V., Berry, J. M., and Gatti, A., "Some Fundamental Fracture Mechanisms Applicable to Advanced Filament Reinforced Composites," *Journal of Composite Materials*, Vol. II, No. 1, Jan. 1968.

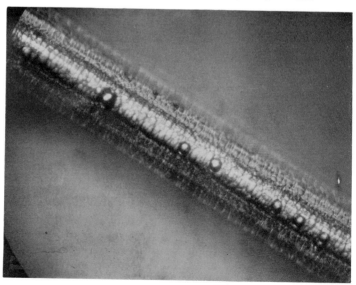

FIG. 1—Photograph of "corn-cob" surface (×202).

treated filaments were then cast into epoxy novolac, cured, and several small "dog-bone" tension specimens were then sawed from cast epoxy. Specimen dimensions were approximately 2 in. long with a gage length of 1 in. and a cross section of 0.060 by 0.125 in. with a 2-in. filament centered within its cross section. All tension testing was performed on an Instron testing machine at strain rates of 0.02 and 2 in./in./min.

THE INTERMITTENT BOND CONCEPT

Control of the bond strength between filament and matrix by the mechanical application of lubricants such as graphite to the surface of boron filament has provided some interesting results. The "corn-cob"-like surface of the filament causes nonuniform application of the graphite on the filament with a resulting pattern of intermittent coated and uncoated regions. This condition provides bonding in the low spots, while the high points are shielded from the matrix by the graphite coating.

To better appreciate the experimental results to be presented it is important to analyze the way in which the reinforcing mechanism is influenced by this intermittent coating of the filament.

Figure 1 shows the surface of a boron filament at high magnification, and the knobby appearance is quite representative of all points along the length of the filament.

The usual model used to describe the shear stress distribution at the interface in a uniformly bonded filament is shown in Fig. 2. Here the distribution of the shear stress is assumed to be continuous along the interface between filament and matrix, reaching its highest values near the ends.

Actually there are probably local variations in the shear stress distribution which result from surface imperfections and foreign materials such as minute dirt particles which are present on the surface. These can be considered of secondary importance, however, compared to the case of the graphite coated filament where relatively large parts of the surface are prevented from bonding. In the simplest model of an intermittently bonded interface we can construct the condition shown in Fig. 3. Here the shear loads are considered as a series of concentrated forces rather

SHEAR STRESS AT INTERFACE

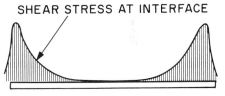

FIG. 2—Distribution of shear stresses in a uniformly bonded filament.

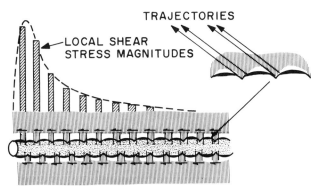

TRAJECTORIES

LOCAL SHEAR STRESS MAGNITUDES

FIG. 3—Model of intermittently bonded filament.

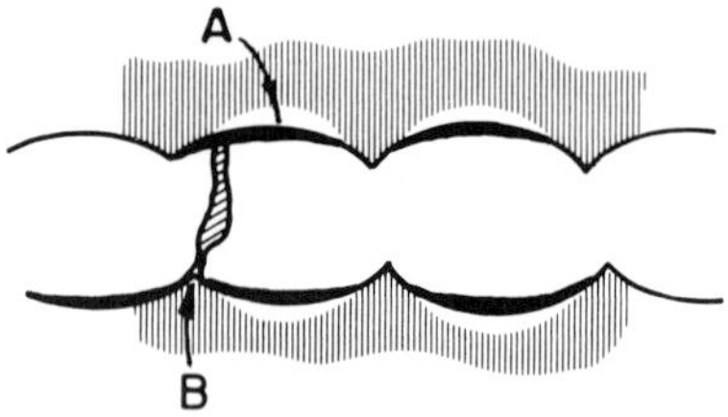

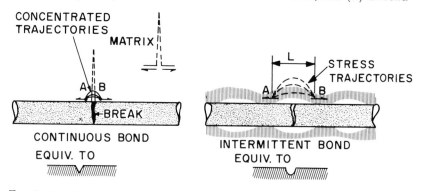

Fig. 4—Details of intermittent bonding and its effect on matrix crack sensitivity (schematic).

than the continuously distributed forces assumed previously. This has the effect of concentrating the stress trajectories in the regions of bonding and increasing the shear stress at these sites. As each bond area fails its effect on the adjacent areas is damped to some extent by the unbonded region between them. This condition tends to inhibit a continuous and sudden crack growth mechanism both at the interface and in the matrix.

Now consider what happens when the filament fractures. The fracture may occur at any point in the loaded region, and the matrix, being only partially bonded at the filament fracture site, is not influenced as suddenly by the fracture as in the continuously bonded condition. Consider Fig. 4. At Point A there is no bond to initiate the sudden crack, and at Point B the matrix has no re-entrant corner to make it crack sensitive.

Further, one can apply a stress concentration argument to the incidence of cracking in the matrix adjacent to a filament break. First consider what happens in the continuously bonded filament shown at the left of Fig. 5. The immediate redistribution of load requires very high shears on each side of the break at Points A and B. These are very close to the crack and quite high, causing the matrix to experience very high stress concentrations at the point of the break. In the intermittent bond condition, shown at the right of Fig. 5, the regions immediately adjacent to the fracture site are not bonded, and therefore the new shears which are applied to the matrix at Points A and B are further away. Distribution of the new local matrix stress over a greater length L in the matrix causes less stress concentration at the filament fracture site and therefore minimizes the possibility of a matrix crack.

BOND STRENGTH—A COMPROMISE

The experimental work to be described has been designed to demonstrate the critical nature of the filament-matrix bond, which is important for two reasons: (1) an understanding of the critical nature of the bond is necessary for a general understanding of composite performance, as will be discussed later, and (2) unbond-

Fig. 5—Distribution of stress trajectories as a function of bonding (schematic).

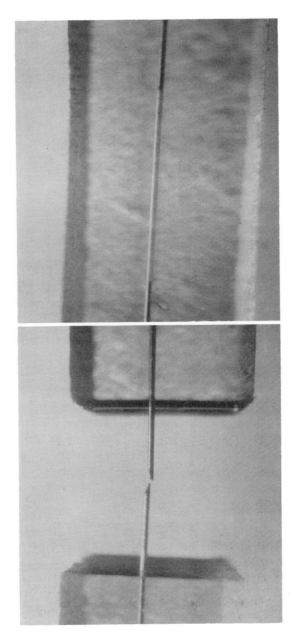

To filament failures with extensive unbonding and filament pullout and no filament-associated matrix cracking. Specimen failure originated in matrix at edge of specimen.

Fig. 6—Profile of failures in Teflon coated single B/W filament-epoxy specimen tested in tension at $\epsilon = 2$ in./in./min (Specimen 715, ×17).

ing is a rather common practical problem in metal matrix composites and has led to considerable efforts toward improved bonding.

Since unbonding had not been a major experimental difficulty in filament-epoxy specimens, special attempts were made to reduce bond strength in order to test the hypothesis that the maximization of bond strength may not be always necessary or even desirable. Figure 6 shows failure in a single filament-epoxy specimen in which the filament had been Teflon coated to drastically reduce the bond strength. Neither of the two filament cracks propagated into the matrix; long unbonded regions were evident, the extent of pullout being evident from the fact that two ends project from the fracture surface. The final fracture started in the matrix at the edge of the specimen. As will be discussed later, the mechanical behavior of this specimen was similar to that of plain epoxy.

Figure 7 shows the fracture surfaces and profiles of single discontinuous filament-

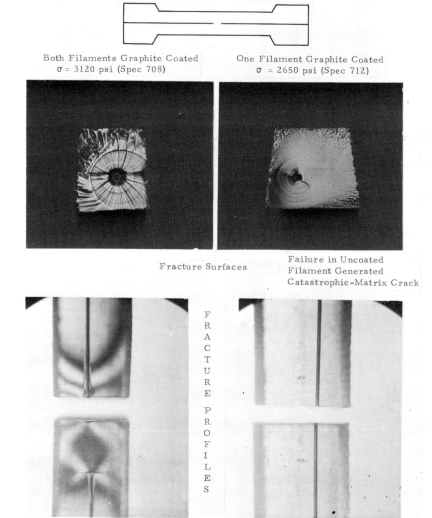

Both Filaments Graphite Coated
σ = 3120 psi (Spec 708)

One Filament Graphite Coated
σ = 2650 psi (Spec 712)

Fracture Surfaces

Failure in Uncoated
Filament Generated
Catastrophic-Matrix Crack

F R A C T U R E   P R O F I L E S

FIG. 7—Discontinuous single B/W filaments tested at $\dot{\epsilon} = 0.02$ in./in./min (×17). Reduced one third for reproduction.

epoxy specimens. When both segments of the filament were graphite coated (Specimen 708, Fig. 7) failure was by the slow propagation of a matrix tensile crack from an end of one of the filament segments. The other cracks in the filament segments resulted in local unbonding or slowly propagating (noncatastrophic) matrix tensile cracks or both; when closely coupled, the alternation (of unbonding and cracking) may be characterized as a "slip-stick" fracture mode. This specimen was approximately as strong as plain epoxy (also true in tests at $\dot{\epsilon} = 2$ in./in./min) and considerably stronger than the other discontinuous specimen (No. 712) having only one of the segments graphite coated. The latter failed in a brittle manner by the rapid propagation of a matrix tensile crack from a crack in the uncoated filament (as is evident in Fig. 7); that is, this specimen was not significantly different from a continuous filament specimen similarly tested. In regard to the mechanical behavior and fracture mode, the fact that these specimens had discontinuous rather than continuous filaments was of less importance than the bond strength.

Duplicate specimens (each containing a single continuous graphite coated B/W filament in epoxy) were tested in tension at 2 in./in./min, a strain rate which always results in weak brittle specimens if the filaments are uncoated (that is, well bonded). The fracture surfaces and profiles are shown in Fig. 8. In addition, each specimen had five noncatastrophic filament failures which exhibited the "slip-stick" fracture mode, an example of which is shown in Fig. 9. It is evident (Fig. 8) that considerable unbonding was associated with the catastrophic failure in Specimen 701; eventually it stuck enough to generate the fatal, slowly propagating, tensile crack (perhaps in a region of incomplete graphite coating). In the case of Specimen 702, two matrix cracks (associated with the slip-stick region, evident in the profile in Fig. 8, were in the process of growth. However, before either of these could propagate sufficiently to separate the specimen, a matrix crack originating at the specimen edge (see fracture surface, Fig. 9) intervened catastrophically. The mechanical behavior of these specimens was similar and superior to

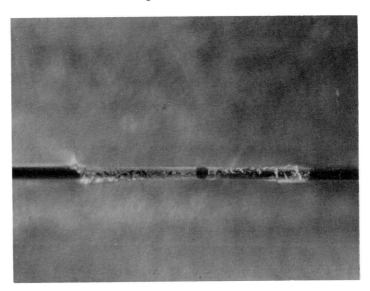

Note combination of unbonding and small matrix cracks.

Fig. 8—A typical fracture profile of one five noncatastrophic cracks in graphite coated B/W filament-epoxy specimen tested at $\dot{\epsilon} = 2$ in./in./min (Specimen 701, ×35).

Gatti et al on Filament Reinforced Composites

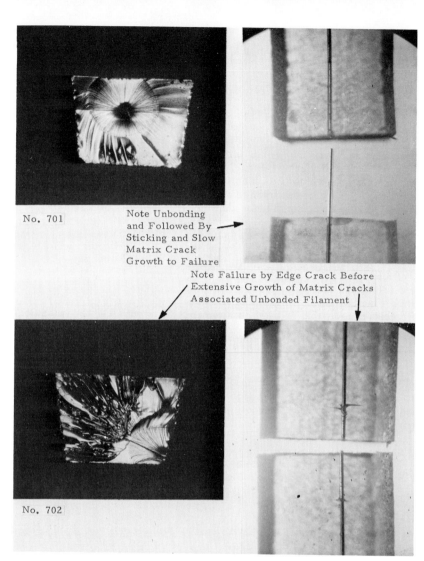

No. 701

Note Unbonding and Followed By Sticking and Slow Matrix Crack Growth to Failure →

Note Failure by Edge Crack Before Extensive Growth of Matrix Cracks Associated Unbonded Filament

No. 702

FIG. 9—Graphite coated, continuous single B/W filament tested at $\dot{\epsilon} = 2$ in./in./min ($\times 17$).

specimens exhibiting unlimited unbonding (for example, Fig. 6) or very good bonding (for example, Fig. 7, Specimen 712).

The relationship between tensile behavior and bond strength for single filament-epoxy specimens is illustrated by the stress-elongation curves in Fig. 10. The tension tests were made at a strain rate of 2 in./in./min, a rate known to result in weak brittle behavior for well-bonded specimens.[4] As expected, the well-bonded specimen (uncoated filament) failed at the first filament failure and was both brittle and weak relative to the plain epoxy. This specimen is an example of a "matrix limited" composite. Specimen 715, having the Teflon coated filament, exhibited two filament

[4] Gatti, A. et al., "Investigation of the Reinforcement of Ductile Metals with Strong, High Modulus Discontinuous, Brittle Fibers," Final Report, NASA Contract No. NASw-1543, National Aeronautics and Space Administration, Dec. 1968.

Composite Materials

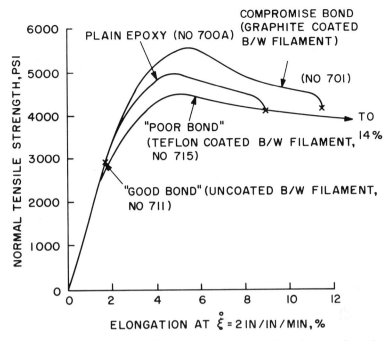

FIG. 10—The effect of bond strength on the tensile behavior of single B/W filament-epoxy Novolac specimens.

cracks and unlimited unbonding (Fig. 6). It was slightly weaker than the plain epoxy (but not brittle). Specimen 701 exhibited cumulative damage (many filament cracks prior to failure) and was somewhat stronger than the plain epoxy.

It is evident from these results that the slip-stick mode of failure associated with early filament failure in graphite coated filament is preferable to either unlimited unbonding or to catastrophically "good" bonding. For this particular set of circumstances, the graphite coating is the best compromise bond. First, the bond is weak enough so that filament failure does not generate a rapidly propagating catastrophic matrix tensile crack. Therefore, the load can continue to rise as the filament breaks up, and this ensures some utilization of the stronger portions of the filament. This point is, of course, crucial if "real" composites are to reflect the average rather than the lowest filament strength. Second, the bond is strong enough to "stick" and prevent the unlimited unbonding which also precludes load redistribution.

CONCLUDING REMARKS

In summary, the "slipping" prevents catastrophy, and the "sticking" is necessary for effective load redistribution. While the foregoing results are definitively applicable only to these specimen materials and configurations, they serve to emphasize the point that bond strength should be an appropriate, not necessarily the maximum attainable, value. The bond strength required will depend upon the matrix crack sensitivity, and it will be noted that the product of the two has been identified as an important parameter in previous work.[2]

There are ambiguities in the nature of materials that may frustrate attempts to make statements that are both rigorous and general. For example, in the case of epoxy novolac, there is an ambiguity in the term "matrix crack sensitivity." This epoxy novolac exhibits two kinds of tensile cracks. One propagates very rapidly, creating smooth surfaces, and may propagate to specimen failure while the load-elongation curve remains linear; therefore, it is logical to refer to this kind of crack as "brittle." The other kind of matrix crack propagates less rapidly or even quite slowly, creating linearly marked surfaces, and can do so during nonlinear load-deflection behavior; this kind of crack propagation has, from time to time, been characterized as "ductile." When, by a change in formulation, the epoxy is made more flexible, it appears to become more resistant to the formation or propagation of "brittle" cracks or both but less resistant to the growth of "ductile" cracks. These divergent characteristics lead to some ambiguity in some early attempts to modify bond strength by changing epoxy formulation. The expedient of coating the filaments with graphite or Teflon worked, because the bond strength could be changed independently of either of the two matrix crack sensitivities. The situation might be considerably more difficult to evaluate and control in metal matrix composites if, for example, bond-enhancing coatings diffuse into the matrix and adversely affect its capacity for plastic deformation or its crack sensitivity.

ACKNOWLEDGMENTS

We acknowledge the many discussions and review of the manuscript by W. H. Sutton.

Special acknowlegement is extended to NASA Headquarters and J. J. Gangler, program monitor, without whose encouragement and support this fundamental segment of NASA Contract NASw-1543 could not have been accomplished.